Pneumatic Conveying

Peter Hilgraf

Pneumatic Conveying

Basics, Design and Operation of Plants

Springer

Peter Hilgraf
Hamburg, Germany

ISBN 978-3-662-67222-8 ISBN 978-3-662-67223-5 (eBook)
https://doi.org/10.1007/978-3-662-67223-5

This Springer imprint is published by the registered company Springer-Verlag GmbH, DE, part of Springer Nature.
The registered company address is: Heidelberger Platz 3, 14197 Berlin, Germany

Das Papier dieses Produkts ist recyclebar.

Preface

The author of the present text has spent approximately 40 years of his professional life dealing with the handling of various bulk materials. This includes, among other things, their comminution, agglomeration, fluidization, mixing, calcination, as well as storage and conveying. Problems to be solved were approached both from a practical side, i.e., through extensive systematic experiments in the pilot plant, commissioning and optimization of operating facilities, *trouble shooting*, as well as theoretically by creating calculation models and design programs based on them for various devices and processes.

A partial task that occurs in almost all of these systems is the transport of the solid material to be processed from one process step to the next or through them. In the individual treatment stages, the properties of the bulk material often change. The pneumatic conveying, which is frequently used for transport and discussed in depth in this book, generally reacts relatively sensitively to such influences and therefore requires a bulk material-specific consideration. Compared to competing mechanical conveying methods, this is an obvious disadvantage, as it significantly complicates the calculation of pneumatic transport. The characteristic values required for its design, conveying gas velocity and conveying line pressure loss, as well as the flow pattern of the two-phase mixture gas/solid that occurs in the conveying pipe, are determined, among other things, by particle size, their size distribution, solid density, and grain shape. A suitable classification system for the bulk material classification in this regard is presented in the text.

Another disadvantage of pneumatic conveyors compared to mechanical transport systems for comparable tasks is their generally significantly higher drive power requirement. By developing and using so-called dense phase conveying methods, which are offered in various forms, the energy consumption of pneumatic conveying systems has been drastically reduced in some cases. Dense phase conveyors are characterized by low conveying gas velocities, which are close to the respective conveying limit, and high loadings of the transport gas with solid material. The function and application areas of such systems, the determination of the minimum conveying gas velocity not to be undercut in individual cases, and the flow conditions that occur in the conveying pipe are discussed in detail.

The main advantages of pneumatic conveying systems are the simple system structure with few moving parts and low space requirements, the closed transport route, which minimizes influences on the bulk material and/or the environment, and an extremely flexible route layout, which allows horizontal and vertical conveying sections as well as branches to various target points without special transfer devices and dedusting. The transport of flammable or explosive bulk materials can be realized relatively easily and safely by conveying under protective/inert gas. Automation and integration into higher-level processes can be implemented without any problems. All of this means that, for the same task, the investment costs for a pneumatic conveying system are generally significantly lower than for a mechanical system. When faced with a specific transport task, it is recommended to select the most suitable method using a neutrally designed decision matrix. This and the determination of the ongoing operating costs are discussed in detail.

Essential for the dimensioning of a pneumatic conveying system is the design of the actual conveying route. For this purpose, two variables must be known or determined: the smallest conveying gas velocity at which safe conveying operation is just barely possible, and the pressure difference of the conveying gas necessary for solid transport. Both variables are influenced by the specific properties of the respective bulk material as well as by the chosen operating conditions. The calculation approaches and models required for the design are presented and discussed. Furthermore, the selection and dimensioning of the solid locks to be used is of importance. Their areas of application and designs are described in detail.

For a deeper understanding of the operating behavior of pneumatic conveying systems, the present text is supported by a large number of detailed calculation and practical examples as well as a complete conveying system design. Also, techniques related to or influencing the field of pneumatic transport, such as gas flow through stationary and moving bulk materials, the behavior of various design variants of fluidized beds, wear in conveying systems, are integrated into the presentation as detailed as necessary for understanding. An introduction to the conservation laws of multiphase systems and the scale-up of plants complete the explanations. Details can be found in the table of contents.

The author has endeavored to present his experiences, the diverse influences, and the approach to the obviously complex design and dimensioning of pneumatic conveying systems in a comprehensible and clear manner. Hints on inadequacies and improvements or additions are welcome. Unclear and improvable explanations are solely the responsibility of the author.

I would like to thank my wife Jelica Hilgraf for her constant drive, my friend Prof. Dr.-Ing. Theodor Hesse for the first proofreading and the never-ending willingness to discuss, as well as the employees of Springer-Verlag for the pleasant and uncomplicated cooperation.

Hamburg, Germany
September 2018

Peter Hilgraf

List of Symbols

a	[m]	Impression radius
$a_{i,j}$	[–]	Exponents, auxiliary variables
A	[–]	Constant
A, A_R	[m^2]	Area, pipe cross-sectional area
Ar	[1]	Archimedes number
B	[–]	Constant
B_S	[m]	Strand width
c	[m]	Half crack length
c_P	[J/(kg K)]	Specific isobaric heat capacity
c_W	[1]	Drag coefficent, particle resistance coefficient
C	[1]	Solid/gas velocity ratio
C, C_x	[–]	Constants, auxiliary quantities
d	[1]	Number of basic dimensions of a problem
d_S	[m]	Particle diameter
$d_{S,50}$	[m]	Particle diameter at sieve residue $R = 0{,}50$
d'_S	[m]	Particle diameter at sieve residue $R = 0{,}368$
D	[1]	Sieve passage
D_R	[m]	Inner pipe diameter
e^f	[J/kg]	Mass-specific energy, formed with enthalpy
e^v	[J/kg]	Mass-related specific energy, formed with internal energy
E_i	[J]	Energy content of the phase i
$\dot{E}_i$	[J/s]	Energy flow of the phase i
$E_P, E_W, E_\otimes$	[N/m^2]	Elastic modulus of solid, of wall, reduced E-modulus
Eu	[1]	Euler number
f	[1]	Particle shape factor
f_C	[N/m^2]	Uniaxial compressive strength

f_Q	[1]	Pressure vessel surcharge factor
f_{SF}	[1]	Safety factor
f_1, f_2	[N]	Auxiliary variables
F	[N]	Force, general
F_A	[N]	Buoyancy force
F_G	[N]	Gravitational force
F_H	[N]	Interparticle adhesive force
F_R	[N]	Frictional force
F_{Strand}	[N]	Shear force on strand surface
F_T	[N]	Inertial force
F_W	[N]	Resistance force
$F_{\Delta p}$	[N]	Pressure difference force
F_σ	[N]	Force due to normal stress
F_τ	[N]	Force due to shear stress
Fr	[1]	Froude number, various definitions possible
Fri	[1]	Friction number = extended Froude number, strand conveying
FF_C	[1]	Jenike's flow function
g	[m/s^2]	Gravitational acceleration
h	[Nm/kg]	Specific enthalpy
H	[m]	Height
H_P, H_W	[N/m^2]	(Vickers-) hardness of solid particles, wall
H_{crit}	[m]	Critical vertical conveying height
H_R	[m]	Vertical conveying height
H_{SS}	[m]	Bulk material height
i	[Nm/kg]	Mass-related internal energy
$\mathfrak{I}_i$	[kgm/s]	Momentum of phase/component *i*
$\dot{\mathfrak{I}}_i$	[kgm/s^2]	Momentum flow of phase/component *i*
k	[1]	Coordination/contact number, exponent
K	[1]	Stress transmission coefficient
K_0	[1]	Resting pressure coefficient
K_R	[–]	Velocity-dependent resistance function, cf. (5.90) ff.
K_{St}	[bar m/s]	Explosion characteristic
K_x	[–]	Various constants, auxiliary variables
K_I	[Pa m$^{1/2}$]	Stress intensity factor
l	[m]	Exponent
L, L_h, L_V	[m]	Length, horizontal length, vertical length

$L_{äq}$	[m]	Equivalent length
L_R	[m]	Total length of conveyor line
L_P	[kg]	Plug length
m, m_F, m_S	[kg]	Mass, gas mass, solid mass
m_D	[kg]	Steam mass
m_P	[kg]	Particle mass
$\dot{m}, \dot{m}_F, \dot{m}_S$	[kg/s]	Mass flow, gas mass flow, solid mass flow
$\dot{m}_{S,n}$	[kg/s]	Nominal design mass flow in pressure vessel conveyance
n	[1]	Exponent, number of process variables of a problem, slope of the RRSB line =scattering parameter
n_W, n_{ZR}	[1/s]	Speed of screw feeder/rotary valve lock
N	[1]	Number of parallel conveyance lines
N_{Ch}	[1/h]	Hourly batch number of a pressure vessel
O	[m^2]	Surface
O_V	[m^2]	Surface of a volume equivalent sphere
OEG	[N/m^2]	Upper explosion limit
$p, \Delta p$	[N/m^2]	Pressure, pressure difference
p_B	[N/m^2]	Pressure in the pressure vessel
$p_D(T_F)$	[N/m^2]	Partial steam pressure at the gas temperature T_F
$p_D^*(T_F)$	[N/m^2]	Saturation pressure at the gas temperature T_F
p_{ex}	[N/m^2]	Explosion overpressure
$p_{F,\mathrm{dry}}$	[N/m^2]	Partial pressure of the dry gas
p_V	[N/m^2]	Pressure at compressor/pressure generator
Δp_R	[N/m^2]	Total pressure loss of the conveying line
Δp_S	[N/m^2]	Pressure loss caused by solid
Δp_F	[N/m^2]	Pressure loss caused by conveying gas
P, P_{Pneu}	[W]	Power, total power requirement of a pneumatic conveyance
P_{Belt}	[W]	Power requirement of a conveyor belt
P_{Mech}	[W]	Power requirement of mechanical transport systems
P_{Est}	[W]	Power requirement of the dust extraction system
P_K	[W]	Coupling power of the pressure generator/compressor
P_{Sch}	[W]	Power requirement of the lock system
P_{sp}	[kW h/ (t· 100 m)]	Specific total power requirement based on 100 m conveying distance
$P_{th,\mathrm{spez}}$	[kW h/ (t · 100 m)]	Theoretical specific power requirement, $\eta_V = 1$

P_V	[W]	Power requirement of the pressure generation system
$\dot{q}$	[m³/(m²s)]	Specific aeration gas flow
$\dot{q}_e$	[W/kg]	Mass-related thermal power
q_r	[1/m]	Density distribution curve of a grain size distribution
$Q, \dot{Q}$	[–]	Conservation quantity, flow of conservation quantity
$\dot{Q}_e$	[W]	Supplied/removed thermal power
Q_r	[1]	Sum curve of a grain size distribution
r	[–]	Marking of the type of quantity
r_{O_2}	[1]	Oxygen volume concentration in the dust/gas mixture
$\dot{r}_Q$	[–]	Production rate of the conservation quantity Q per unit of time and volume
R	[1]	Sieve residue
R	[m]	Radius of a pipe bend
R	[J/(mol K)]	Universal gas constant
R_C	[1]	Carr index
R_x	[J/(kg K)]	Specific gas constant of the gas/vapor x
R_H	[1]	Hausner index
$\dot{R}_Q$	[–]	Production rate of the conservation quantity Q per unit of time
Re	[1]	Reynolds number, various definitions possible
s	[m]	Distance, gap
S_m	[m²/kg]	Mass-related specific surface area
S_V	[m²/m³]	Volume-related specific surface area
t	[°C]	Temperature
T	[K]	Temperature
T_F	[K]	Gas temperature
T_M	[K]	Mixed gas/solid temperature
T_S	[K]	Solid temperature
T_T	[K]	Dew point temperature
T_U	[l]	Turbulence level
u	[m/s]	True speed
u_F	[m/s]	True gas velocity
u_S	[m/s]	True solids velocity
U, U_R	[m]	Circumference, pipe circumference
UEG	[g/m³]	Lower explosion limit
v	[m/s]	Empty pipe velocity
v_F	[m/s]	Empty pipe gas velocity

$v_{F,L}$	[m/s]	Minimum fluidization velocity
$v_{F,\mathrm{min}}$	[m/s]	Minimum conveying gas speed, conveying limit
$\Delta v_{F,\mathrm{min}}$	[m/s]	Distance of the operating speed to the conveying limit
$v_{F,\mathrm{treib}}$	[m/s]	Driving nozzle gas speed of a jet conveyor
$V_{B,\mathrm{Brutto}}$	[m^3]	Gross volume pressure vessel
$V_{B,\mathrm{netto}}$	[m^3]	Net volume pressure vessel = filling volume
V_F	[m^3]	Gas volume
V_P	[m^3]	Single particle volume
V_S	[m^3]	Solid volume
V_{SS}	[m^3]	Bulk material volume
V_{ZR}	[m^3]	Total chamber volume of a rotary valve
$\dot{V}_F$	[m^3/s	Gas volume flow
$\dot{V}_{F,\mathrm{Leck}}$	[m^3/s	Leakage gas volume flow
$\dot{V}_{SS}$	[m^3/s]	Bulk solid volume flow
$\dot{V}_V$	[m^3/s]	Volume flow of pressure generator
w_T, $w_{T,\varepsilon}$	[m/s]	Terminal velocity single particle, particle cloud
$\dot{w}_e$	[W/kg]	Mass-related mechanical power
W_x	[–]	Wear characteristics
$\dot{W}_e$	[W]	Supplied/removed mechanical power
x	[kg/kg]	Moisture loading of the dry substance
x_{pl}	[m]	Size of the plastic deformation at crack tips
x_i, y_i	[–]	Auxiliary variables
x, y, z	[m]	Coordinates
α_i	[°]	Angle
α_{RRSB}	[°]	Inclination angle of the RRSB straight line
$\alpha_{S,W}$	[°]	Impact angle
α_U	[°]	Deflection angle of a pipe redirection
α_{WS}	[°]	Inclination angle of a fluidzed channel against the horizontal
β_R	[1]	Resistance/friction coefficient
Γ_{strand}	[1]	Extension factor for strand resolution
ε	[m^3/m^3]	Volume fraction
ε_F	[m^3/m^3]	Volume fraction of gas in a bulk material = relative void volume
η_F	[Pa s]	Dynamic viscosity of gas
η_V	[1]	Compressor/pressure generator efficiency

η_V	[1]	Volumetric transport efficiency of a screw feeder
κ	[1]	Adiabatic exponent, wedging force factor according to (4.9)
λ_F	[1]	Friction coefficient gas flow
λ_S, λ_S^*	[1]	Resistance coefficients solid flow
μ	[(kg/s)/(kg/s)]	Loading = solid mass flow/conveying gas flow
μ_{strand}	[1]	Loading of a strand
μ_W	[1]	Wall friction coefficient
ξ_{Ch}	[1]	Utilization of the available batch time, pressure vessel
ξ_i	[1]	Resistance coefficients gas flow
ξ_Z	[1]	Utilization degree screw lock
ξ_{ZR}	[1]	Filling degree rotary valve
Π_i	[1]	Dimensionless number i
ϱ	[kg/m^3]	Density
ϱ_b	[kg/m^3]	Actual bulk density
ϱ_F	[kg/m^3]	Gas density
ϱ_P	[kg/m^3]	Particle density
ϱ_S	[kg/m^3]	Solid density
ϱ_{SS}	[kg/m^3]	Loose bulk density = bulk density
ϱ_{SR}	[kg/m^3]	Vibrated bulk density
σ	[N/m^2]	Normal stress
σ_S	[N/m^2]	Normal stress due to bulk material/solid
σ_W	[N/m^2	Wall normal stress
$\sigma_{1,\ldots,3}$	[N/m^2]	Principal stresses
σ_1	[N/m^2]	Greatest principal stress
σ_3	[N/m^2]	Lowest principel stress
τ	[N/m^2]	Shear stress
τ_C	[N/m^2]	Cohesion
τ_W	[N/m^2]	Wall shear stress
τ	[s]	Time
τ_E	[s]	Deaeration time of a fluidized bed
$\Delta\tau$	[s]	Time difference
$\Delta\tau_{Ch}$	[s]	Batch time of a pressure vessel
$\Delta\tau_{\text{för}}$	[s]	Conveying time during a pressure vessel batch
$\Delta\tau_{\text{i}}$	[s]	Individual times during a pressure vessel batch
$\Delta\tau_{\text{tot}}$	[s]	Dead time during a pressure vessel batch

φ	[1]	Relative humidity
φ_e	[°]	Effective friction angle according to Jenike
φ_i	[°]	Internal friction angle at the onset of flow
φ_{SF}	[°]	Internal friction angle at steady flow
φ_W	[°]	Friction angle bulk material ↔ surrounding wall
Φ	[–]	Variable
Φ	[m^2/m^2]	Cross-sectional/area ratio
ψ	[1]	Sphericity
ψ	[1]	Bulk material filling share of the conveyor line, (5.87)

Indices

ab	shearing off
an	shearing on
ax	axial
A	Beginning of the conveyor line
B	Acceleration, Container, Bubbles, Operating condition
ch	Condition: *choking*
crit	Critical condition, Limit
C	Condition at system exit
dp	Condition: *dense phase* Fluidized bed
drag	Flow resistance
D	Distributor
E	End of the conveyor line
f	Friction, rubbing, fine-grained
fluid	fluidized
F	Gas
g	Coarse-grained
G	State at the outlet of the line
h	horizontal
H	Lift
$i, j, 1, 2, \ldots$	Counter, running number
in	into the calculation section
L	Loosening state
max	maximum
min	minimum
mp	State: Pressure minimum
n	nominal
out	out of the calculation section
O	Surface
pf	State: *fast fluidization*
pu	State: *pickup*

P	Particle, Plug
r	Quantity type
rad	radial
rel	relative
R	Pipe
sa	Condition: *saltation*
st	Condition: blockage, blockage limit
S	Solid matter
SD	Sauter diameter
tr	Transport, Transition
tot	total
T	Nozzle
U	Deflection
v	Vertical
V	Volume
W	Wall
WS	Fluidized bed
x	Variable
0	Reference value, e.g. ambient or standard condition
I	Characteristics of a strand
II	Characteristics of the free space above the strand
#, *	(superscript) special condition, e.g. partial load
–	(overlined) mean value

Self-explanatory indices and symbols are not listed.

Abbreviations

MF	mechanical conveyance
PF	pneumatic conveyance
DPF	dense phase pneumatic conveyance,
LPF	lean phase pneumatic conveyance

Contents

Introduction

1

1.1 Task Description

In a variety of industries, bulk materials are processed, refined, or handled in some form. These solids must be fed to individual process steps, introduced into them, removed after treatment, and transported to the next processing step. Along such a process chain, the properties of the bulk material, e.g., its grain size or grain size distribution, and thus also its handling behavior, can change significantly. Transports between the various process steps generally take place at ambient pressure. However, deviating pressure levels are also possible in individual process stages. Example: Feeding fine coal against pressures of approx. 6 bar(g) into the wind box of a blast furnace. The temperatures of the bulk materials can reach up to approx. 500 °C, depending on the process. Entry and removal from a process step generally occur in a controlled manner. Usual in-plant conveying distances range from a few meters to the kilometer range. In this context, solid mass flows of a few kg/h up to several 100 t/h must be managed. The particle sizes of the goods to be transported can range from a few micrometers to several centimeters, with a tendency towards increasingly finer-grained products being discernible.

The transport, entry, and discharge systems to be used in the current process must be planned and selected according to the indicated requirements.

1.2 Conveying Methods

There are three basic ways to realize bulk material transport. The bulk material drive can

- mechanically, i.e., by conveyor belts, scraper conveyors, screws, vibrating troughs, etc.,

P. Hilgraf, *Pneumatic Conveying*, https://doi.org/10.1007/978-3-662-67223-5_1

- pneumatically, i.e., by flowing around the bulk material particles in a channel with a compressible gas, usually air,
- hydraulically, i.e., by flowing around the bulk material particles in a channel with an – for practical considerations – incompressible liquid, usually water,

take place.

Hydraulic conveying is generally only used where the bulk material is already present in a liquid phase and is being processed. It will not be further considered here, except for the following note: A calculation of hydraulic transport is possible with the later presented approaches of pneumatic conveying if the liquid density, which is about three orders of magnitude larger than the gas density, is appropriately taken into account (→ effects on the conveying behavior, among other things, due to the larger particle buoyancy). Both are two-phase fluid/solid flows.

When selecting the most suitable transport system for a given application, the advantages and disadvantages of mechanical (MF) and pneumatic (PF) conveying methods must be weighed against each other. The following comparison provides initial selection criteria:

- **Energy demand**: This is, even with an optimally designed PF, many times greater than with an MF.
- **Conveying path**: Almost any route consisting of horizontal and vertical elements with integrated deflections in the PF. Inclined upward paths should be avoided (→ see Sect. 4.10.2). The PF is excellently adaptable to structural conditions. By installing switches, several feed/target points can be approached. In the case of MF, generally only straight paths can be executed. Deflections must be realized with two conveyors and an intermediate transfer station. Example: Horizontal belt conveying transfers to vertical bucket elevator.
- **Bulk material influence**: A PF is designed for a specific range of allowable bulk material properties in operation, e.g., particle size and size distribution, particle density, moisture content, etc. Deviations from these properties lead to reduced throughput, unstable conveying behavior, and in extreme cases, pipe blockages. MF systems are less sensitive or even unaffected by such changes.
- **Particle size**: In PF, limited to approx. 10 mm, in special cases up to approx. 20 mm. In principle, unlimited for MF.
- **Product stress**: High in PF with high conveying gas velocity (lean phase conveying), low in slow velocity conveying. However, the latter is not feasible with all bulk materials. In most MF processes, the stress is low.
- **System wear**: Strongly depends on the properties of the solid material being transported, especially its particle size, shape, and hardness. In PF systems, wear is generally low. Particularly vulnerable areas are flow deflections/bends and mechanically sealing locks/solid feeders. Some MF variants show significantly higher wear

compared to PF when handling hard particles, e.g., where clamping and sliding effects occur or at transfer stations.

- **Explosion protection**: In PF, it can be ensured by conveying with an inert gas. A gas recirculation system can be easily implemented to reduce gas consumption. When conveying with air, the ratio of solid mass flow to gas mass flow, the so-called loading, can often be adjusted so that a value above the upper explosion limit of the respective product is obtained along the transport route. The explosion-prone areas of the system are then reduced to the zones where the bulk material enters and exits the system, e.g., the feeding lock and the upper space of the receiving silo. Ensuring explosion protection in MF systems is difficult and extremely complex to implement and control.
- **Back pressure**: Conveying against pressures significantly above/below ambient pressure is easier to achieve with PF than with MF. In many MF processes, such operation is not possible at all.
- **Product change**: Product mixing during a change of material can be avoided in PF, e.g., by sufficient purging, but requires special cleaning devices in most MF systems.
- **High bulk material temperature**: Critical system elements of PF are the bulk material inlet systems (→ for pressure conveying) or outlet systems (→ for suction conveying). Not all locks are suitable. Specifically, systems that seal the conveying pressure through mechanical gaps, e.g., rotary valve locks, must be operated close to the design temperature. They react to strongly varying product temperatures with impairments of the conveying process, due to changes in gap dimensions and thus leakage gas losses, and possibly also with mechanical seizing/jamming. MF allows high-temperature transport only with some of the available methods, e.g., scraper belt or trough chain conveyors.
- **Environmental aspects**:
  a) Dust generation: The actual transport using PF is dust-free. The two critical areas of bulk material input and output can be safely dedusted using exhaust filters. In MF, often all transfer stations must be equipped with dedusting, possibly aspiration systems. Not all methods are dust-tight to the environment. The issue of dust generation during the loading and unloading of bulk material also exists here. Overall, the efforts for dust-free conveying in MF are significantly higher than in PF.
  b) Noise emission: The pressure generators required for PF – blowers, compressors – must always be installed with sound insulation, i.e., encapsulated and/or in specially prepared rooms. With fine-grained bulk materials, the flow noises in the conveying line are low and do not require special noise protection measures. Hard, coarse-grained bulk material particles can cause significant noise emissions in the pipeline. MF methods, where the bulk material is pushed over a surface, e.g., scraper belt conveyors, generate significantly higher noise emissions than systems where the product is carried, e.g., belt conveyors.

c) Product contamination: PF systems are closed and can be operated without deposits and dead zones, avoiding contact with the environment. Transport under protective gas is easily possible. With many MF methods, this is not guaranteed or possible.

- **Investment costs**: In the case of PF, the expenses for pressure generators and locks are high, while those for the actual conveying route (→ pipe with support) are low. In the case of MF, the ratios are exactly the opposite: lower expenses for the drive (→ gear motor), high for the conveying route/conveying means. For the same task, the investment costs for a PF are generally lower than those for an MF.
- **Maintenance costs**: Are generally significantly lower for a PF than for an MF.
- **Additional benefits**: The PF allows physical or chemical processes to be carried out during transport. Examples: drying fine-grained solids in flow dryers, calcination of gypsum in *flash calciners*, cooling or heating of bulk material. Such is generally not possible with MF.

It is recommended, when faced with a specific transport task, to create a decision matrix based on and weighted by the above and possibly other influencing factors for evaluating the applicable systems [1]. Such a matrix allows a comprehensible and largely neutral selection of the most suitable method.

1.3 Pneumatic Conveying

A variety of natural phenomena show that flowing gases, under certain conditions, are capable of carrying specifically heavier solids and transporting them over greater distances. This ability is specifically exploited for the pneumatic conveying of dust-fine to coarse-grained bulk materials through pipelines/channels. Implemented systems realize conveying distances from a few meters up to currently about 3500 m. In an industrial test facility, conveying tests over approximately 5000 m have already been carried out [2]. Solid throughput rates from a few kg/h up to approx. 500 t/h are possible. Particles with sizes from a few micrometers to several centimeters can be conveyed.

The main advantages of a pneumatic conveying system are:

- simple system structure, few moving parts, low space requirement,
- closed transport route, i.e., influences on the bulk material = product and/or the environment are minimal,
- extremely flexible route design, e.g., branches to different target points are possible,
- easily achievable conveying under protective/inert gas, i.e., safe transport of flammable or explosive bulk materials,
- simple automation and integration into higher-level processes.

Main disadvantages of a pneumatic conveying system are:

- high power demand,
- pipe wear,
- particle breakage and/or abrasion,
- risk of conveying line blockage due to incorrect operation and/or incorrect design.

In order to minimize the inherent disadvantages, current developments in the field of pneumatic conveying are particularly concerned with reducing the drive power requirements and improving the design and operational safety. The former is possible, among other things, by using so-called dense phase conveying methods, which are offered in various forms, and the latter by more accurate calculation and scale-up methods, with which the test results determined on a test track can be safely transferred to planned operating plants. Dense phase conveying is characterized by very low conveying gas velocities and, as a result, high loadings μ = solid mass flow $\dot{m}_S$/gas mass flow $\dot{m}_F$. Both are discussed in detail in Sect. 4.3.

The conveying process consists of the following steps:

- Introduction of the solid into the conveying gas stream,
- Transport through the conveying line,
- Separation of solid and conveying gas ($\rightarrow$ is not always necessary; example: feeding a reactor or burner)

This results in the basic structure of a pneumatic conveying system shown in Fig. 1.1. The driving energy is supplied to the system by the conveying gas itself. The gas flows around the bulk material particles, which can move as relatively isolated individual particles or as denser particle accumulations, e.g., as plugs or strands, and drives them through the drag forces generated. Thus, the solid velocity u_S is always smaller than the gas velocity u_F, i.e., it holds: $0 < C = {}^{u_S}/_{u_F} < 1$. The gas pressure decreases in the conveying direction, while the gas velocity increases due to gas expansion.

According to the above-mentioned partial steps, a pneumatic conveying system consists of the components

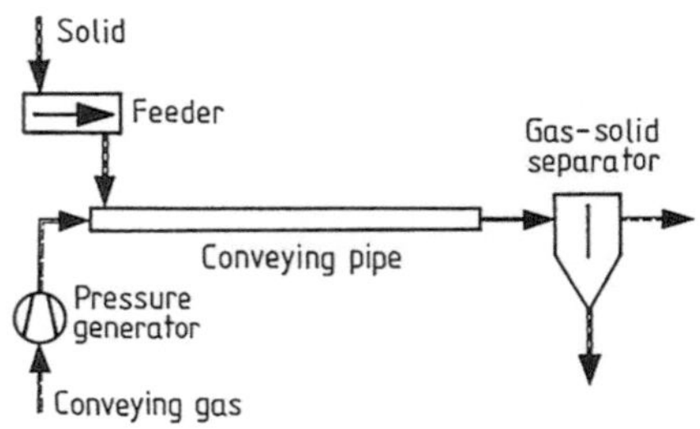

Fig. 1.1 Structure of a pneumatic pressure conveying system

- Solid lock (*feeder*): pressure vessel, rotary valve, screw lock, etc.,
- Conveying gas supplier (*pressure generator*): blower, compressor, compressed air network/accumulator system, etc.,
- Conveying section (*conveying pipe*): generally circular conveying pipe,
- Solid separator (*gas-solid separator*): cyclone, fabric filter, gravity separator, e.g., free upper space in the receiving silo, etc.

There are two basic types of systems/circuit variants possible, which can also be combined, see Fig. 1.2:

Pressure conveying systems The pressure generator is arranged in the conveying direction in front of the solid lock, see Fig. 1.2a. Advantages: any conveying pressures can be realized, simple discharge of the solid. Disadvantages: solid introduction against overpressure, sealing of the overpressure by lock required. Such systems are advantageous when several consumers are to be supplied from a central supplier.

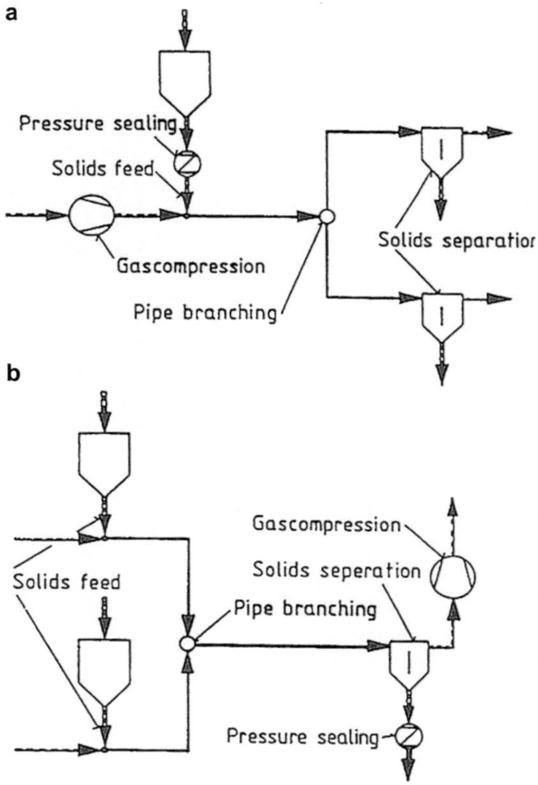

Fig. 1.2 Circuit variants of pneumatic conveying systems. **a** Pressure conveying system, **b** Suction conveying system

Suction conveying systems The pressure generator is arranged in the conveying direction behind the solid separator, see Fig. 1.2b. Advantage: simple solid material feeding at ambient pressure. Disadvantages: solid material discharge from negative pressure to ambient pressure, sealing of negative pressure by sluice system required, pressure difference at the conveying line limited to approx. 0.50 bar. Such systems are advantageous to use when a central consumer is to be supplied from several suppliers.

Basic for the dimensioning of a pneumatic conveying system is the design of the actual conveying route. For this purpose, two quantities must be known or determined:

- the smallest conveying gas velocity at which safe conveying operation is just still possible,
- the pressure difference of the conveying gas necessary for solid material transport.

Both quantities are influenced by the specific properties of the respective bulk material as well as by the chosen operating conditions.

A task, i.e., the transport of the solid mass flow $\dot{m}_S$ over the distance L_R at a given gas velocity level, can generally be realized with various combinations of pressure difference Δp_R at the conveying pipe and conveying pipe diameter D_R. The individual $(\Delta p_R, D_R)$ -working points lead to different power demand values P_{pneu}. By determining the minimum of $P_{\text{pneu}} = P_{\text{pneu}}(\Delta p_R, D_R)$ the respective conveying system can be energetically optimized. This also applies to dense phase systems.

Depending on various factors, such as the type of solid material to be conveyed and the chosen conveying gas velocity, different flow forms – suspension, strand, dune, conglobation, plug flow conveying – occur in the conveying section. Their behavior covers the range of steady, quasi-steady to non-steady conveyance and is also influenced by the pipeline routing. The transitions are fluid.

With knowledge of the data of the conveying section (→ solid material throughput, conveying gas velocity, conveying pipeline pressure loss, conveying pipe diameter) and taking into account the expected flow conditions in the conveying pipe as well as the specific requirements of the bulk material, e.g. with regard to wear, the suitable type and size of the solid material lock, the pressure generator, and the separator can be determined.

References

[1] Dikty, M., Schwei, P.: Entscheidungsmatrix für den Schüttguttransport. ZKG Int. **60**(7), 56–66 (2007)

[2] Göcke, V.: Long-distance conveyance using a pneumatic system. ZKG Int. **66**(12), 56–65 (2013)

General Principles

2

The following explanations are intended to provide the reader/user with some necessary basics and at the same time show the integration of pneumatic conveying into related/ adjacent techniques. Only the facts needed in the further course of the text are presented.

First, some definitions:

- Unless otherwise noted, the pipe flows considered here are treated as continuous one-dimensional flows with suitably averaged characteristic values over the current pipe cross-section.
- Differences $\Delta\Phi$ and differentials $d\Phi$ of a variable Φ are, as usual in mathematics, defined as differences of the quantity flowing out of the calculation section Φ_{out} minus the inflowing quantity Φ_{in}, i.e. $(\Delta\Phi,\ \mathrm{d}\Phi) = (\Phi_{\text{out}} - \Phi_{\text{in}})$. For the pressure drop in a conveying pipe section, this results in due to $p_{\text{out}} < p_{\text{in}}$ a negative pressure difference Δp, while the pressure increase by, for example, a compressor provides a positive Δp.
- With the symbol "*u*", true velocities are denoted, with "*v*" so-called empty pipe velocities are designated. For the gas phase in the pneumatic conveying pipe, the following applies:

$$v_F = \frac{\dot{V}_F}{A_R} = \varepsilon \cdot u_F \tag{2.1}$$

with:

$\dot{V}_F$ current operating gas volume flow,

A_R pipe cross-sectional area,

ε relative gap volume in operation = volume fraction of the gas in the considered conveying pipe volume element $(= \Delta V_F / (A_R \cdot \Delta L_R)) \equiv$ area fraction of the conveying pipe cross-section flowed through by the gas. The fraction $(1 - \varepsilon)$ is covered by the solid.

P. Hilgraf, *Pneumatic Conveying*, https://doi.org/10.1007/978-3-662-67223-5_2

Since all pneumatic conveyors and a large part of dense phase conveyors have relative void volumes of $\varepsilon \geq 0.95$, in these cases, with justifiable accuracy, $v_F \cong u_F$ can be set. In practice, the easier-to-determine gas empty pipe velocity v_F is used. The determination of the ε that occurs during operation will be discussed later.

- Pressures "p" are given as absolute pressures; exceptions will be pointed out. With "T", absolute temperatures [K] are denoted, and with "t", Celsius temperatures °C are indicated.

2.1 Conveying Gas

Carrier gases are preferably and predominantly air, in special cases nitrogen N_2, carbon dioxide CO_2 or oxygen O_2 and very rarely also available exhaust gases are used. All these gases behave like Newtonian fluids. The description of the gas behavior primarily requires its density, viscosity, and the moisture content contained in the gas.

2.1.1 Density

In the application range of pneumatic conveyance, the carrier gases can be treated with sufficient accuracy as ideal gases. From the ideal gas law

$$p_F \cdot V_F = m_F \cdot R_F \cdot T_F \tag{2.2}$$

follows with $\varrho_F = m_F / V_F$ for the gas density:

$$\varrho_F = \frac{p_F}{R_F \cdot T_F} \tag{2.3}$$

with:

V_F, m_F	Gas volume, gas mass,
p_F	Gas pressure,
T_F	Gas temperature,
R_F	specific gas constant (= $(\mathcal{R}/M_F)$, $\mathcal{R} = 8.3145\,\text{J}/(\text{mol K})$ = universal gas constant, M_F = molar mass of the gas).

By referring to a reference state "0", the result from (2.3):

$$\varrho_F = \varrho_{F,0} \cdot \frac{p_F}{p_{F,0}} \cdot \frac{T_{F,0}}{T_F} \tag{2.4}$$

2.1.2 Viscosity

In terms of viscosity, a distinction must be made between kinematic viscosity ν_F and dynamic viscosity η_F The following applies:

$$\nu_F = \frac{\eta_F}{\varrho_F} \tag{2.5}$$

Since the kinematic viscosity ν_F is dependent on both pressure and temperature, while η_F depends on the temperature, but up to about $p_F \cong 10\,\text{bar}$ it does not depend on the pressure, so it is recommended to replace ν_F with (η_F/ϱ_F). The temperature dependence of the dynamic viscosity of gases can be described in the temperature range of interest by the Sutherland equation [1].

$$\eta_F = \eta_{F,0} \cdot \frac{T_{F,0} + C}{T_F + C} \cdot \left(\frac{T_F}{T_{F,0}}\right)^{3/2} \tag{2.6}$$

with:

$T_{F,0}$ Reference temperature,
$\eta_{F,0}$ dynamic viscosity of the considered gas at $T_{F,0}$ and $p_{F,0} = 1.0\,\text{bar}$,
C Sutherland constant.

Examples: Gas = Air, $T_{F,0} = 293\,\text{K}$, $\eta_{F,0} = 18.21 \cdot 10^{-6}\,\text{Pa s}$, from this: $C = 122.34\,\text{K}$. Gas = Nitrogen, $T_{F,0} = 293\,\text{K}$, $\eta_{F,0} = 17.60 \cdot 10^{-6}\,\text{Pa s}$, from this: $C = 114.77\,\text{K}$. The viscosity of a gas increases with rising temperature T_F greater. This increases its ability to carry solid matter.

2.1.3 Moisture

When operating a pneumatic conveying system with conveying air drawn from the environment, this air contains varying proportions of water vapor, depending on the location of the system and weather conditions. In a pressure conveying system, for example, this air-water vapor mixture is first compressed to the required pressure, then possibly cooled to a temperature suitable for the bulk material and/or various system components, e.g., valves, to be mixed with the colder or warmer bulk material and led through the conveying line under pressure reduction. If the dew point temperature T_T of the vapor is exceeded, condensate will precipitate. This, in turn, requires suitably positioned liquid separators before the actual conveying section and can, in case of failure in the conveying line, lead to conveying problems and/or impairments of product quality. In such cases, special treatment = drying of the conveying gas is required. The indicated problem applies to all gases containing a readily condensable vapor.

The moist conveying gas = gas-vapor mixture "F" consists of the dry gas "F,dry" and the vapor "D". Both components can be treated as ideal gases and thus follow Dalton's law. Here it applies:

$$p_F = p_{F,\text{dry}} + p_D = \frac{m_{F,\text{dry}} \cdot R_{F,\text{dry}} \cdot T_F}{V_F} + \frac{m_D \cdot R_D \cdot T_F}{V_F} \tag{2.7}$$

To describe the moisture content, the relative gas humidity

$$\varphi = \frac{p_D(T_F)}{p_D^*(T_F)} \quad \text{with } 0 \le \varphi \le 1 \tag{2.8}$$

with:

$p_D(T_F)$ Partial pressure of the vapor at the gas temperature T_F,
$p_D^*(T_F)$ Saturation pressure of the vapor at the gas temperature T_F (→ the pressure dependence of the saturation pressure can generally be neglected)

is used. A moist gas with $\varphi < 1$ is unsaturated. If it is kept at constant total pressure p_F and thus also constant p_D cooled, when reaching the dew point temperature T_T, i.e., at $p_D^*(T_T) = p_D$, the first liquid or, depending on operating conditions, also solid condensate (→ for water vapor at $t_D^* < 0\,°\text{C}$ e.g., in the form of frost or ice fog). The relative gas humidity can thus also be represented by

$$\varphi = \frac{p_D^*(T_T)}{p_D^*(T_F)} \tag{2.9}$$

Figure 2.1 shows the described relationships.

For practical calculations, the moisture load

$$x = \frac{m_D}{m_{F,\text{dry}}} = \frac{R_{F,\text{dry}}}{R_D} \cdot \frac{p_D}{p_F - p_D} = \frac{R_{F,\text{dry}}}{R_D} \cdot \frac{p_D^*(T_F)}{\frac{p_F}{\varphi} - p_D^*(T_F)} \tag{2.10}$$

is generally better suited, since when the state of the moist gas changes, its reference size $m_{F,\text{dry}}$ does not change. The value of x is, as (2.10) shows, not only influenced by

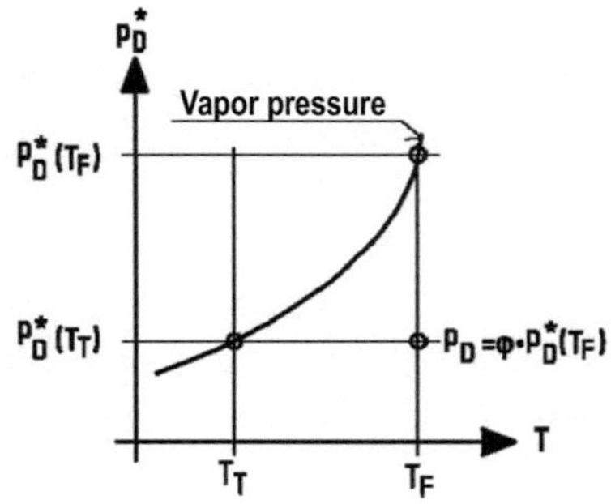

Fig. 2.1 Relationship between vapor partial pressure p_D, relative humidity φ and the saturation pressures $p_D^*(T_F)$ and $p_D^*(T_T)$

the temperature T_F, but also from the respective total pressure p_F determined. The maximum loading x^* of a dry gas with moisture in the vapor state = saturated gas results from (2.10) for $\varphi = 1$. If the loading x exceeds the value x^*, then the moisture mass $x \cdot m_{F,\text{dry}}$ is present as saturated vapor and the one $(x - x^*) \cdot m_{F,\text{dry}}$ as condensate. For more in-depth considerations, please refer to the technical literature, e.g., [2].

Required for the evaluation of the above relationships is the knowledge of the vapor pressure curve of the present vapor, i.e., the dependency $p_D^*(T_F)$. This can be taken from steam tables or represented by equations for limited temperature ranges. The Antoine equation

$$\log(p_D^*) = A - \frac{B}{C + T_F} \tag{2.11}$$

is frequently used, with its three constants A, B, C tabulated for many vapors [3]. It should be noted that both the decimal and the natural logarithm as well as different units for pressure and temperature are used. For the "ubiquitous" water vapor, the following applies:

$$\log_{10}\left(\frac{p_D^*}{\text{mbar}}\right) = 8.19621 - \frac{1730.630}{233.426 + \frac{t_F}{°\text{C}}}\text{, range of validity: } (1 \ldots 100)\,°\text{C} \tag{2.12}$$

(2.11) can for a given p_D explicitly solved for the temperature corresponding to the boiling temperature.

If condensate forms in the conveying line, the effects depend on the amount of liquid falling out in relation to the solid matter being absorbed, the type of bulk material, and the flow pattern that develops. Each bulk material is capable of absorbing a limited amount of liquid before its handling behavior changes. The level of this critical moisture loading $X_{S,\text{crit}}$ can be simplified as the bulk moisture required to fill particle capillaries, pores, etc. Only the liquid amount exceeding this level reaches the particle contact points. Below $X_{S,\text{crit}}$, Van der Waals forces between the individual particles of the bulk material are essentially effective, above this limit they are increasingly overlaid and reinforced by the binding forces of an increasing number of liquid bridges in the particle contacts. The same applies to particle-pipe wall contacts. $X_{S,\text{crit}}$ approaches zero for smooth spherical bulk material particles with a closed surface and becomes larger the more porous and fissured the particles are [4]. It is evident that the respective flow pattern and the level of conveying speed also have an impact on the resulting behavior.

2.1.4 Mixture Temperature

At the entrance to the conveying section, bulk material and conveying gas are combined and mixed. Both can have different temperatures. With which gas temperature in the pipeline must in the 2.1.3 described humidity analyses be carried out? From a heat balance around the supplied mass flows of gas $\dot{m}_F$ and solid $\dot{m}_S$ with the specific isobaric heat capacities $(c_{p,F}, c_{p,S})$ and the temperatures (T_F, T_S) their joint mixing temperature can be determined.

$$T_M = \frac{\dot{m}_F \cdot c_{p,F} \cdot T_F + \dot{m}_S \cdot c_{p,S} \cdot T_S}{\dot{m}_F \cdot c_{p,F} + \dot{m}_S \cdot c_{p,S}} \tag{2.13}$$

can be calculated. In (2.13), simplifying temperature-independent specific heat capacities were used. This is generally permissible with sufficient accuracy and avoids iterations. Furthermore, an unsaturated moist gas with the moisture load x is assumed. $c_{p,F}$ is thus composed proportionally of the heat capacities of the dry gas and the vapor. Details can be found in [2]. By introducing the solid/gas loading $\mu = \dot{m}_S/\dot{m}_F$ can be transformed from (2.13) into

$$T_M = \frac{T_F + \mu \cdot \frac{c_{p,S}}{c_{p,F}} \cdot T_S}{1 + \mu \cdot \frac{c_{p,S}}{c_{p,F}}} \tag{2.14}$$

Since in pneumatic conveying systems, the solid/gas loading is usually significantly above $\mu = 10$ and the ratio of specific heat capacities is on the order of $c_{p,S}/c_{p,F} \cong 1$, it follows from (2.14) that the mixing temperature T_M adjusts close to the solid temperature T_S. Often, $T_M = T_S$ can be set. For fine-grained bulk materials, average particle diameter $d_{S,50} < 100\,\mu\text{m}$ (median value), the temperature equalization occurs almost instantaneously at the beginning of the pipeline. With increasing particle size, the pipe length over which the temperature equalization takes place is extended. Since the pressure p_F of the gas decreases in the conveying direction and thus, according to (2.10), its vapor absorption capacity increases, setting the mixing temperature to the state at the beginning of the pipeline "*R*", i.e. $T_M = T_R, p_F = p_R$, the maximum possible condensate dropout. With the values for (T_R, p_R) the corresponding saturation loading x^* can be determined from (2.10). If this is smaller than the supplied x, then the proportion $(x - x^*)$ drops out as condensate (→ depending on the gas temperature, liquid or solid). The released phase transition enthalpy (condensation, sublimation heat) leads to a, in general, negligibly small temperature increase of the gas-solid mixture.

The generally low possible mixture cooling by heat dissipation via the pipe surface to the environment is discussed in Chap. 9. In case of greater cooling, the condition of the moist gas should also be checked at the end of the pipeline, "*G*", i.e., at (T_G, p_G), to obtain information on the processes in the conveying line.

2.1.5 Calculation Example 1: Condensation Along a Conveying Line

$\dot{m}_S = 50\,\text{t/h}$ of a dry fine-grained solid ($d_{S,50} \cong 12\,\mu\text{m}$) with $\dot{m}_F = 1020\,\text{kg/h}$ of air over $L_R = 125\,\text{m}$. The loading is therefore $\mu = 49.0\,\text{kg S/kg F}$.

The conveying air is drawn from the surroundings at $t_0 = 30\,°\mathrm{C}$ and $\varphi_0 = 0.60$ at $p_0 = 1.00\,\mathrm{bar}$ suction. It is compressed oil-free to the required operating pressure in a single-stage pressure generator and then, for safety reasons and to maintain product quality, it is $t_1 = 80\,°\mathrm{C}$ cooled down. The associated pressure behind the cooler is $p_1 = 2.50\,\mathrm{bar}$ and is identical to the pressure p_R at the beginning of the conveying line. The back pressure at the end of the line has the value $p_G = p_0 = 1.00\,\mathrm{bar}$. The inlet temperature of the bulk material corresponds to the ambient temperature, i.e. $t_S = t_0 = 30\,°\mathrm{C}$. The moisture distribution along the conveying gas path should be checked.

Suction state "0":

- Saturation pressure at $t_0 = 30\,°\mathrm{C}$, (2.12):

$$\log_{10}\left(\frac{p^*_{D,0}}{\mathrm{mbar}}\right) = 8.19621 - \frac{1730.630}{233.426 + \frac{30\,°\mathrm{C}}{°\mathrm{C}}} = 1.62651 \rightarrow p^*_{D,0} = 42.316\,\mathrm{mbar}$$

- Current vapor partial pressure, (2.8):

$$p_{D,0} = \varphi_0 \cdot p^*_{D,0} = 0.60 \cdot 42.316\,\mathrm{mbar} = 25.390\,\mathrm{mbar}$$

- Current moisture load, (2.10):
  With the specific gas constants of dry air $R_{F,\mathrm{dry}} = 287.06\,\mathrm{J/(kg\,K)}$, of water vapor $R_D = 461.52\,\mathrm{J/(kg\,K)}$ and the total pressure $p_0 = 1000\,\mathrm{mbar}$ follows:

$$x_0 = 0.622 \cdot \frac{p_{D,0}}{p_0 - p_{D,0}} = 0.622 \cdot \frac{25.390\,\mathrm{mbar}}{1000\,\mathrm{mbar} - 25.390\,\mathrm{mbar}}$$
$$= 16.204 \cdot 10^{-3}\,\frac{\mathrm{kg}\,D}{\mathrm{kg}\,F,\mathrm{dry}}$$
$$x_0 = 16.204\,\frac{\mathrm{g}\,D}{\mathrm{kg}\,F,\mathrm{dry}}$$

- Saturation loading at $t_0 = 30\,°\mathrm{C}$, (2.10):

$$x_0^* = 0.622 \cdot \frac{p^*_{D,0}}{p_0 - p^*_{D,0}} = 0.622 \cdot \frac{42.316\,\mathrm{mbar}}{1000\,\mathrm{mbar} - 42.316\,\mathrm{mbar}}$$
$$= 27.484 \cdot 10^{-3}\,\frac{\mathrm{kg}\,D}{\mathrm{kg}\,F,\mathrm{dry}}$$
$$x_0^* = 27.484\,\frac{\mathrm{g}\,D}{\mathrm{kg}\,F,\mathrm{dry}}$$

- Dew point temperature $t_{T,0}$, (rearranged 2.12):

$$\frac{t_{T,0}}{°C} = \frac{1730.630}{8.19621 - \log_{10}(\frac{p_{D,0}}{\text{mbar}})} - 233.426$$
$$= \frac{1730.630}{8.19621 - \log_{10}(\frac{25.390\,\text{mbar}}{\text{mbar}})} - 233.426 = 21.40$$
$$t_{T,0} = 21.40\,°C$$

- Mass flow rates of dry air $\dot{m}_{F,\text{dry}}$ and of water vapor $\dot{m}_D$, (2.10):

$$\dot{m}_F = \dot{m}_{F,\text{dry}} + \dot{m}_D = \dot{m}_{F,\text{dry}} \cdot (1 + x_0) \to \dot{m}_{F,\text{dry}} = \frac{\dot{m}_F}{1 + x_0}$$
$$= \frac{1020\,\frac{\text{kg}}{\text{h}}}{1 + 16.204 \cdot 10^{-3}\,\frac{\text{kg}}{\text{kg}}}$$
$$\dot{m}_{F,\text{dry}} = 1003{,}74\frac{\text{kg}}{\text{h}}$$

$$\dot{m}_D = \dot{m}_F - \dot{m}_{F,\text{dry}} = (1020 - 1003.74)\frac{\text{kg}}{\text{h}} = 16.26\,\frac{\text{kg}}{\text{h}}$$

State "1" after cooler:

- Saturation pressure at $t_1 = 80\,°C$, (2.12):

$$\log_{10}\left(\frac{p^*_{D,1}}{\text{mbar}}\right) = 8.19621 - \frac{1730.630}{233.426 + \frac{80\,°C}{°C}} = 2.67456 \to p^*_{D,1} = 472.668\,\text{mbar}$$

- Saturation load at $t_1 = 80\,°C$ and $p_1 = 2500\,\text{mbar}$, (2.10):

$$x^*_1 = 0.622 \cdot \frac{p^*_{D,1}}{p_1 - p^*_{D,1}} = 0.622 \cdot \frac{472.668\,\text{mbar}}{2500\,\text{mbar} - 472.668\,\text{mbar}}$$
$$= 145.018 \cdot 10^{-3}\,\frac{\text{kg}\,D}{\text{kg}\,F,\text{dry}}$$
$$x^*_1 = 145.018\,\frac{\text{g}\,D}{\text{kg}\,F,\text{dry}}$$

- Current moisture load:

$$x_1 = x_0 = 16.204 \cdot 10^{-3}\,\frac{\text{kg}\,D}{\text{kg}\,F,\text{dry}}$$
$$x_1 = 16.204\,\frac{\text{g}\,D}{\text{kg}\,F,\text{dry}}$$

- Current relative humidity φ_1, (2.10) rearranged:

$$\varphi_1 = \frac{x_1}{0.622 + x_1} \cdot \frac{p_1}{p^*_{D,1}} = \frac{16.204 \cdot 10^{-3}}{0.622 + 16.204 \cdot 10^{-3}} \cdot \frac{2500\,\text{mbar}}{472.668\,\text{mbar}} = 0.1343$$

- Current vapor partial pressure, (2.8):

$$p_{D,1} = \varphi_1 \cdot p^*_{D,1} = 0.1343 \cdot 472.668\,\text{mbar} = 63.475\,\text{mbar}$$

- Dew point temperature $t_{T,0}$, (rearranged 2.12):

$$\frac{t_{T,1}}{°\text{C}} = \frac{1730.630}{8.19621 - \log_{10}(\frac{p_{D,1}}{\text{mbar}})} - 233.426$$

$$= \frac{1730.630}{8.19621 - \log_{10}(\frac{63.475\,\text{mbar}}{\text{mbar}})} - 233.426 = 37.26$$

$$t_{T,1} = 37.26\,°\text{C}$$

No condensate dropout at position "1" behind the cooler.

State "*R*" after the entry of the conveying line:

- Incoming moisture load: $x_1 = 16.204 \cdot 10^{-3}\,\text{kg}\,D/\text{kg}\,F, \text{dry}$
- Mixed temperature of bulk material and conveying gas, (2.13): With the boundary conditions of Sect. 2.1.4 and the specific heat capacities [2] of dry air, $c_{p,F,\text{dry}} = 1.004\,\text{kJ}/(\text{kg K})$, and of superheated water vapor, $c_{p,D} = 1.86\,\text{kJ}/(\text{kg K})$, the specific heat capacity of moist air can be calculated as follows:

$$c_{p,F} = \frac{c_{p,F,\text{dry}} + x_1 \cdot c_{p,D}}{1 + x_1} = \frac{1.004\,\frac{\text{kJ}}{\text{kg K}} + 16.204 \cdot 10^{-3} \cdot 1.86\,\frac{\text{kJ}}{\text{kg K}}}{1 + 16.204 \cdot 10^{-3}} = 1.018\,\frac{\text{kJ}}{\text{kg K}}$$

The specific heat capacity of the solid is $c_{p,S} = 0.84\,\text{kJ}/(\text{kg K})$, the solid/gas loading $\mu = 49.0\,\text{kg}\,S/\text{kg}\,F$. With this data, the following results from (2.14):

$$t_R = t_M = \frac{t_1 + \mu \cdot \frac{c_{p,S}}{c_{p,F}} \cdot t_S}{1 + \mu \cdot \frac{c_{p,S}}{c_{p,F}}} = \frac{80\,°\text{C} + 49.0 \cdot \frac{0.84\,\text{kJ}/(\text{kg K})}{1.018\,\text{kJ}/(\text{kg K})} \cdot 30\,°\text{C}}{1 + 49.0 \cdot \frac{0.84\,\text{kJ}/(\text{kg K})}{1.018\,\text{kJ}/(\text{kg K})}} = 31.21\,°\text{C}$$

- Saturation pressure at $t_R = 31.21\,°\text{C}$, (2.12):

$$\log_{10}\left(\frac{p^*_{D,R}}{\text{mbar}}\right) = 8.19621 - \frac{1730.630}{233.426 + \frac{31.21\,°\text{C}}{°\text{C}}} = 1.65647 \rightarrow p^*_{D,R} = 45.339\,\text{mbar}$$

- Saturation load at $t_R = 31.21\,°\text{C}$ and $p_R = 2500\,\text{mbar}$, (2.10):

$$x^*_R = 0.622 \cdot \frac{p^*_{D,R}}{p_R - p^*_{D,R}} = 0.622 \cdot \frac{45.339\,\text{mbar}}{2500\,\text{mbar} - 45.339\,\text{mbar}}$$

$$= 11.489 \cdot 10^{-3}\,\frac{\text{kg}\,D}{\text{kg}\,F, \text{dry}}$$

$$x^*_R = 11.489\,\frac{\text{g}\,D}{\text{kg}\,F, \text{dry}}$$

Since $x^*_R < x_1$ is, condensed liquid water is formed. This is absorbed by the solid.

- Precipitating condensate/water quantity "K":

$$x_{K,R} = (x_1 - x_R^*) = (16.204 \cdot 10^{-3} - 11.489 \cdot 10^{-3}) \frac{\text{kg}\,D}{\text{kg}\,F,\text{dry}}$$

$$= 4.715 \cdot 10^{-3} \frac{\text{kg}\,K}{\text{kg}\,F,\text{dry}}$$

$$\dot{m}_{K,R} = x_{K,R} \cdot \dot{m}_{F,\text{dry}} = 4.715 \cdot 10^{-3} \frac{\text{kg}\,K}{\text{kg}\,F,\text{dry}} \cdot 1003.74 \frac{\text{kg}\,F,\text{dry}}{\text{h}} = 4.733 \frac{\text{kg}\,K}{\text{h}}$$

- Moistening of the solid:

$$X_{S,R} = \frac{\dot{m}_{K,R}}{\dot{m}_S} = \frac{4.733 \frac{\text{kg}\,K}{\text{h}}}{50.000 \frac{\text{kg}\,S}{\text{h}}} = 0.095 \cdot 10^{-3} \frac{\text{kg}\,K}{\text{kg}\,S}$$

$$X_{S,R} \cong 0.01\,\text{M.-\%}$$

In the case of the solid at hand, this moistening has no influence on the conveying behavior.

- Released condensation heat:
The released condensation enthalpy $r_D = 2500\,\text{kJ/kg}\,K$ of water vapor heats the gas/solid mixture up slightly. From a heat balance around the gas and solid streams without considering heat transfer to the body of the conveying pipe, the following heating can be derived:

$$\Delta t_R = \frac{\dot{m}_{K,R} \cdot r_D}{\dot{m}_F \cdot c_{P,F} + \dot{m}_S \cdot c_{p,S}} = \frac{4.733 \frac{\text{kg}\,K}{h} \cdot 2500 \frac{\text{kJ}}{\text{kg}\,K}}{1020 \frac{\text{kg}}{h} \cdot 1.018 \frac{\text{kJ}}{\text{kg}\,K} + 50{,}000 \frac{\text{kg}}{h} \cdot 0.84 \frac{\text{kJ}}{\text{kg}\,K}}$$

$$= 0.27\,\text{K}$$

The above calculations are sufficiently accurate, and no correction is necessary. Note: For larger values of Δt_R must (2.13) be considered in view of an expected condensation/phase change, i.e. $x_R > x_R^*$, must be re-established. Since x_R^* is still unknown, this leads to an iterative solution.

- State of the gas phase after mixing with the solid:
Relative humidity: $\varphi_R = 1$; Vapor partial pressure: $p_{D,R} = p_{D,R}^* = 45.339\,\text{mbar}$, Moisture load: $x_R = x_R^* = 11.489 \cdot 10^{-3}\,\text{kg}\,D/\text{kg}\,F,\text{dry}$, Dew point temperature: $t_{T,R} = t_R = 31.21\,°\text{C}$.

State "G" at the end of the conveying line:

- Cooling of the solid/gas mixture along the conveying line:
With the given conveying pipe inlet temperature of $t_R = 31.21\,°\text{C}$, cooling to ambient temperature $t_G = t_0 = 30\,°\text{C}$ is expected.

- Current moisture load:

$$x_G = x_R = 11.489 \cdot 10^{-3} \, \frac{\text{kg}\,D}{\text{kg}\,F, \text{dry}}$$

- Saturation pressure at $t_G = 30\,°\text{C}$:

$$p^*_{D,G} = p^*_{D,0} = 42.316\,\text{mbar}$$

- Saturation loading at $t_G = 30\,°\text{C}$ and $p_G = 1000\,\text{mbar}$:

$$x^*_G = x^*_0 = 27.484 \cdot 10^{-3} \, \frac{\text{kg}\,D}{\text{kg}\,F, \text{dry}}$$

- Current relative humidity φ_G, (rearranged from 2.10):

$$\varphi_G = \frac{x_G}{0.622 + x_G} \cdot \frac{p_G}{p^*_{D,G}} = \frac{11.489 \cdot 10^{-3}}{0.622 + 11.489 \cdot 10^{-3}} \cdot \frac{1000\,\text{mbar}}{42.316\,\text{mbar}} = 0.4286$$

- Current vapor partial pressure, (2.8):

$$p_{D,G} = \varphi_G \cdot p^*_{D,G} = 0.4286 \cdot 42.316\,\text{mbar} = 18.136\,\text{mbar}$$

- Dew point temperature $t_{T,G}$, rearranged from (2.12):

$$\frac{t_{T,G}}{°\text{C}} = \frac{1730.630}{8.19621 - \log_{10}(\frac{p_{D,G}}{\text{mbar}})} - 233.426$$

$$= \frac{1730.630}{8.19621 - \log_{10}(\frac{18.136\,\text{mbar}}{\text{mbar}})} - 233.426 = 16.03$$

$$t_{T,G} = 16.03\,°\text{C}$$

- Escaping gas flow:

$$\dot{m}_{F,G} = \dot{m}_F - \dot{m}_{K,R} = (1020 - 4.733)\,\frac{\text{kg}}{\text{h}} \cong 1015.3\,\frac{\text{kg}}{\text{h}}$$

  No further condensate dropout in the conveying line.

Following the above scheme, the operating conditions occurring in a conveying system can be systematically analyzed.

2.2 Bulk Material

In the following, an overview of the methods of bulk material characterization relevant to the present topic is given. Since pneumatic conveying systems consist not only of a conveying pipe but also of a lock and a feeding and receiving system, the bulk material behavior in these system components must also be taken into account. Their design must be adapted to the given or locally expected bulk material properties. A detailed discussion of the bulk material influences, specifically with regard to pneumatic conveying behavior, follows in Sect. 4.5.

The handling behavior of a bulk material is determined by its dispersion state, i.e., by particle size, shape, strength, relative void volume, surface properties, etc. It can either be treated as a system of individual particles interacting with each other and the surrounding walls or alternatively as a continuum with derived parameters, e.g., internal friction angle, wall friction angle, compressive strength. Available calculation models can be divided in the same way. Since pneumatic flight conveying transports relatively isolated individual particles, strand conveying transports loosened flowing bulk material strands with a free surface, and plug conveying transports compact solid plugs completely filling the conveying pipe cross-section, both approaches are necessary. The individual particles of flight conveying and their wall impacts can be described with the corresponding motion and impact equations, while the behavior of a plug can be described with a continuity approach capturing the stress transfer within the plug and to the surrounding pipe wall.

2.2.1 Resting Pressure Coefficient

The force transmission in bulk materials cannot be described with the laws applicable to liquids or solids. To characterize the "aggregate state" of a substance, the so-called at-rest pressure coefficient K_0 from soil mechanics can be used [5]. This results, as shown in Fig. 2.2, as the ratio of the perpendicular pressure stresses σ_x and σ_z, which act on a volume element at depth z of an undisturbed infinite half-space made of isotropic

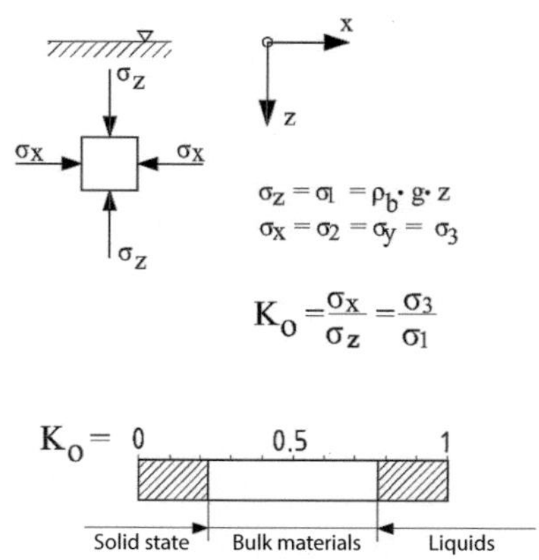

Fig. 2.2 At-rest pressure coefficient K_0

material. The mentioned stresses are simultaneously principal stresses (→ Indices 1, 2, 3), i.e., the cutting planes are free of shear stress. Thus, it applies:

$$K_0 = \frac{\sigma_x}{\sigma_z} = \frac{\sigma_3}{\sigma_1} \tag{2.15}$$

The greater the internal cohesion, i.e. the attraction forces between the particles, of a material, the smaller is σ_x. This implies:

ideal (Newtonian) fluid:	$\sigma_x = \sigma_z \Rightarrow K_0 = 1$,
ideal rigid solid:	$\sigma_x = 0 \Rightarrow K_0 = 0$,
bulk material:	$0 < \sigma_x < \sigma_z \Rightarrow 0 < K_0 < 1$,
	common values: $K_0 \cong 0.4$–$0{,}6$.

Limit states of bulk materials are:

fluidized bulk material:	$\sigma_x \rightarrow \sigma_z \Rightarrow K_0 \rightarrow 1$,
briquette, tablet, strongly compressed:	$\sigma_x \rightarrow 0 \Rightarrow K_0 \rightarrow 0$.

Bulk materials thus transmit stresses or forces perpendicular to their load direction. K_0 is a function of the internal friction angle φ_i and becomes smaller as it increases. According to [6], the following applies:

$$K_0 \cong 1 - \sin \varphi_i \tag{2.16}$$

A bulk material surrounded by walls parallel to its load direction thus transfers normal stresses to these walls, which in turn leads to wall shear stresses. These are directed against the load direction. This results, for example, in vertical and horizontal stress profiles in a cylindrical silo, which, starting from the bulk material surface, follow the course of an exponential function and approach a limit value with increasing silo depth [7]. In a silo filled with water, $K_0 = 1$, the stress profiles would be linear. In pneumatic plug conveying, the axial compressive stresses imprinted on the compact individual plug in the conveying direction are also partially redirected to the conveying pipe wall. They generate a wall shear stress or frictional force directed against the solid movement, which is superimposed on the frictional force caused by the weight of the material (in horizontal pipes) and must be additionally overcome by the driving conveying gas.

To estimate the required stress transmission coefficient $K = \sigma_W/\sigma_{\text{axial}}$ can be used (2.16). As the internal bulk friction angle φ_i the internal friction angle φ_{SF} during steady state flow, approximately the effective friction angle φ_e, is to be inserted, see also Sect. 2.2.4. Alternatively, K can be measured in suitable apparatus [7].

When assessing a bulk material, it is essential to consider that its current behavior is significantly influenced by its history/pre-treatment. This must be included in an evaluation.

2.2.2 Individual Particles

Basic building block of a bulk material. A distinction must be made between the physical/chemical properties, which determine the interactions with the conveying gas and the surrounding pipe walls, and the geometric properties of the particles. The latter influence, among other things, the structure of the bulk material formed from the individual particles. Essential particle characteristics are:

- **Particle density** ϱ_P:
  This is defined as

$$\varrho_P = \frac{m_P}{V_P} \tag{2.17}$$

  with:

  m_P Particle mass,
  V_P Particle volume, including internal pores → volume enclosed by the geometric surface.

  As particle density, the solid density ϱ_S can often be used. This is no longer possible for particles with larger internal porosity. Example: Sandy alumina, Al_2O_3, $d_{S,50} \cong 80\,\mu m$, has a solid density of $\varrho_S \cong 4350\,kg/m^3$. Depending on the manufacturing process, particle densities down to $\varrho_P \cong 2200\,kg/m^3$ are determined.
- **Particle size** $d_{S,x}$:
  In general, irregularly shaped particles are present. Their detailed description is complex and difficult, but in many cases not necessary. Therefore, so-called equivalent diameters are used. These are spherical diameters calculated from measured features, e.g., projection area, surface area, volume, settling velocity, with the same physical property as determined on the original particle, i.e., spheres of equal projection area $d_{S,P}$, equal surface area $d_{S,O}$, equal volume $d_{S,V}$, equal settling velocity $d_{S,T}$ etc. The characteristic feature to be used must be chosen according to the conditions/influencing factors of the current handling situation and represent the application case in a representative manner. In pneumatic conveying, the volume-equivalent sphere diameter, which corresponds to the mass-equivalent diameter, is generally used.
- **Particle shape**:
  This is often only characterized qualitatively, as exact determination methods are very complex and generally not necessary for technical processes. Figure 2.3 shows an example of the grain shape description according to FEM 2582 [8].
  For a quantitative description using so-called shape factors, the ratios of various equivalent diameters $d_{S,x}$ are suitable. Frequently used here is the sphericity ψ according to [9], which is defined as follows:

Symbol	Classification	
I	Sharp edges with oddly equal dimensions in the three spatial directions Example: CUBE	
II	Sharp edges, one dimension is significantly longer than those in the other two spatial directions Example: LONG PRISMA	
III	Sharp edges, one dimension is significantly smaller than those in the other two directions. Example: PLATE, DISH	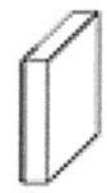
IV	Rounded edges with approximately equal dimensions in the three spatial directions. Example: BALL	
V	Rounded edges, with one dimension significantly longer than those in the other two spatial directions Example: CYLINDER, STAKES	
VI	Fibrous, threadlike, curly intertwined. Example: FIBER, THREAD	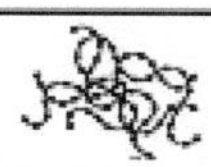

Fig. 2.3 Grain shape description according to FEM 2582

$$\psi = \frac{\text{Surface of the volume-equivalent sphere } O_V}{\text{Surface of the real particle } O} \equiv \left(\frac{d_{S,V}}{d_{S,O}}\right)^2 \tag{2.18}$$

ψ corresponds to the square of the ratio of the diameters of the volume- and surface-equivalent spheres of the real particle. In Table 2.1, measured and calculated ψ values compiled [10]. It applies: $\psi \leq 1$, with $\psi_{\text{Sphere}} = 1$.
Example for determining sphericity: A cylindrical particle with diameter d and length $l = 2 \cdot d$ is considered:
Volume of the cylinder: $V = \frac{\pi}{4} \cdot d^2 \cdot 2 \cdot d = \frac{\pi}{2} \cdot d^3 = \frac{\pi}{6} \cdot d_{S,V}^3$,
Diameter of the volume-equivalent sphere: $d_{S,V} = 3^{1/3} \cdot d$,
Surface area of the volume-equivalent sphere: $O_V = \pi \cdot d_{S,V}^2 = 3^{2/3} \cdot \pi \cdot d^2$,
Surface of the cylinder: $O = 2 \cdot \frac{\pi}{4} \cdot d^2 + d \cdot \pi \cdot 2 \cdot d = 2.5 \cdot \pi \cdot d^2 = \pi \cdot d_{S,O}^2$,
Diameter of the sphere with equal surface area: $d_{S,O} = 2.5^{1/2} \cdot d$,
thus: $\psi = \frac{O_V}{O} = \frac{3^{2/3} \cdot \pi \cdot d^2}{2.5 \cdot \pi \cdot d^2} = 0.832$, alternatively: $\psi = (\frac{d_{S,V}}{d_{S,O}})^2 = (\frac{3^{1/3} \cdot d}{2.5^{1/2} \cdot d})^2 = 0.832$.

The sphericity ψ is used, among other things, to convert the resistance coefficients $c_W(Re_P)$ determined by a "standard curve" for the flow around spherical particles to deviating particle shapes: With the same inflow conditions, i.e., constant Re_P-number, the c_W-value increases with decreasing sphericity [11]. An overview of further possible form factor definitions can be found in [12].

Table 2.1 Calculated and measured values of sphericity ψ

Particle type	ψ		ψ
Calculated values:		Measured values:	
Ball	1.000	Tungsten powder	0.85
Cylinder h = d	0.874	Sugar	0.85
regular octahedron	0.846	Fly ash (roundish)	0.82
Cylinder	0.832	Potash salt	0.70
Cube	0.806	Sand (roundish)	0.70
Tetrahedron	0.670	Cocoa powder	0.61
		Coal dust	0.61
		Cement	0.57
		Glass dust (angular)	0.53
		Cork particles	0.51
		Mica dust	0.11

- **Other influencing factors:**
  These include, among others:
  - Particle hardness,
  - Breaking strength,
  - Abrasion resistance,
  - Surface texture/roughness,
  - Coefficient of restitution

  etc.

Examples of the effects of various particle characteristics: Particle density, size, and shape determine the dynamic properties of the particles, i.e., their movement behavior and the gas velocities required for transport. The coefficient of restitution describes the interaction of the particles with the conveying pipe wall. Grain shape and hardness, in addition to conveying speed, determine the extent of wear in a pneumatic conveying system: Sharp-edged and hard particles increase both pipe and the solids feeder wear. With low breaking and/or abrasion resistance, the conveyed material can be altered in its granulometry or quality by the conveying process, e.g., because an undesired fine fraction is generated.

2.2.3 Bulk Solid

Accumulation of interacting individual particles, generally of different sizes and shapes. A description of dry bulk solids is possible through:

- **Packing structure**:
  To determine this, the relative void volume (also: void fraction, porosity) is used:

$$\varepsilon_F = \frac{\text{Gas-/Void volume } V_F}{\text{Bulk-/Total volume } V_{SS}} = \frac{V_F}{V_F + V_S} = 1 - \frac{V_S}{V_{SS}} \tag{2.19}$$

with:
$V_S \quad (= \sum V_P)$, solid volume, volume of all particles in the measurement volume,

and the coordination number k = number of contact points of a particle with the surrounding neighboring particles. The following distinctions are made:

- Regular packing structures: particle arrangements with periodic repetition of an elementary scheme. Regular packings of equal-sized spheres, e.g. cubic primitive, hexagonal, cubic face-centered structures, can have a minimum of $k = 6$ and a maximum of $k = 12$ immediately adjacent particles, i.e. $6 \leq k \leq 12$. The associated relative gap volumes lie between $\varepsilon_F \cong 0.48$ at $k = 6$ and $\varepsilon_F \cong 0.26$ at $k = 12$.
- Uniform/complete random packings: no preferential direction of the particles; in each cross-sectional plane through the packing, the area porosity determined there ε_A is equal to the total porosity ε_F. This also assumes, among other things, that no segregation occurs according to particle size, since $\varepsilon_F = \varepsilon_F(d_S)$ is. Complete random packings of equal-sized spheres have a minimum of $k = 3$ and a maximum of $k = 8$ immediate neighboring particles, thus: $3 \leq k \leq 8$ [10].
- Irregular/random particle arrangements: No predictions can be made here.

All three types of packing can coexist in partial areas of a bulk material. The description of practically occurring packing structures is generally based on the assumption of a complete random packing. Coordination numbers can be estimated here by the approximation equation

$$k \cdot \varepsilon_F \cong \pi \tag{2.20}$$

[13]. The relative void volume is determined experimentally and depends, among other things, on the particle size. It increases from $\varepsilon_F \cong (0.40-0.45)$ for particle diameters $d_S \gtrsim 200\,\mu\text{m}$ monotonically with decreasing particle size up to $\varepsilon_F \cong 0.80$. Figure 2.4 shows, as an example, the dependence $\varepsilon_F(d_{S,50})$ for a limestone in different grinding. The particle diameter $d_{S,50}$ is the median value of the respective particle size distribution, i.e., 50 M.-% of the distribution is finer, 50 M.-% is coarser than $d_{S,50}$. Other influencing factors on ε_F include, among others: the width of the particle size distribution, the grain shape, the size of the plant part in which the bulk material is located, e.g., the measuring apparatus or in the case of plug conveying the conveying pipe diameter. The reason for the size influence: The porosity depends on the distance to the surrounding walls. Directly at a wall, a local relative void volume of $\varepsilon_F \to 1$ is established for spherical rigid particles and decreases with increasing wall distance to a constant value. This tendency is also measured for other particle shapes. For a ratio of the wall distance y_W to the mean particle diameter $d_{S,50}$ greater than $y_W/d_{S,50} \gtrsim 10$ can the influence on the container cross-section averaged ε_F be neglected. Nevertheless, in flow processes, an edge walk-

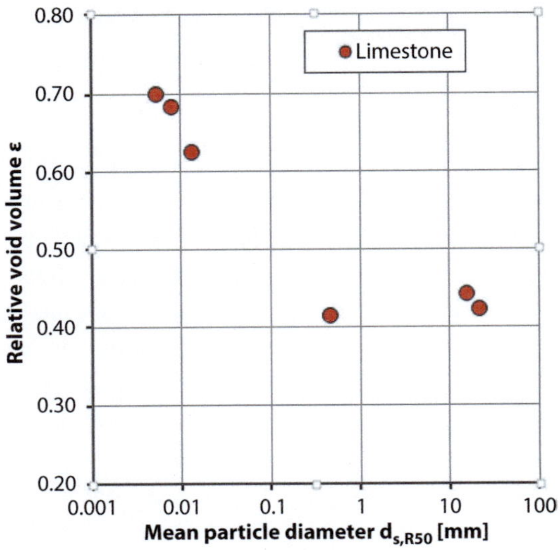

Fig. 2.4 Dependency of the relative void volume on particle size; limestone, loosely poured

ability of the flowing medium should be taken into account. With a widening grain size distribution, the relative void volume, compared to that of a narrow distribution, becomes smaller, as the finer particles partially fill the void between the coarser grains [14].

Between bulk density ϱ_{SS} and the relative void volume ε_F there is the relationship:

$$\varrho_{SS} = (1 - \varepsilon_F) \cdot \varrho_P + \varepsilon_F \cdot \varrho_F \cong (1 - \varepsilon_F) \cdot \varrho_P \cong (1 - \varepsilon_F) \cdot \varrho_S \quad \text{with: } \varrho_F \ll \varrho_P \text{ and } \varrho_S \cong \varrho_P \tag{2.21}$$

By introducing vibrations, stamping, or applying a load to a stationary bulk material, it becomes denser, i.e., the bulk density increases and the relative void volume decreases. In this case, the changes caused by vibrations and stamping are greater than those caused by static loads and greater for fine-grained products than for coarse-grained products. The limestone shown in Fig. 2.4 was loosely poured into the measuring apparatus [15]. The generally highest achievable bulk density is the so-called vibration density ϱ_{SR}. This is determined by defined shaking of a bulk material mass located in a measuring cup until its height becomes constant. Examples: Cement, $d_{S,R50} = 13\,\mu\text{m}$, bulk density $\varrho_{SS} = 1170\,\text{kg/m}^3$ (loosely filled), vibrated density $\varrho_{SR} = 1700\,\text{kg/m}^3$ (compacted by tamping), ratio $\varrho_{SR}/\varrho_{SS} = 1.453$, limestone chippings, $d_{S,R50} = 460\,\mu\text{m}$, $\varrho_{SS} = 1600\,\text{kg/m}^3$, $\varrho_{SR} = 1700\,\text{kg/m}^3$, $\varrho_{SR}/\varrho_{SS} = 1.063$.

The compressibility of a bulk material provides information on its handling behavior. As an example, the classification according to Carr [16] is presented. Based on the characteristic for compressibility

$$R_C = \frac{\varrho_{SR} - \varrho_{SS}}{\varrho_{SR}} \tag{2.22}$$

the following classification is proposed:

$R_C < 0.2$	Bulk material shows good flowability, is free-flowing,
$0.2 \leq R_C \leq 0.3$	Bulk material shows poor flowability, is cohesive,
$R_C > 0.3$	Bulk material shows extremely poor flowability, is very cohesive, mechanical discharge/flow aids should be considered.

The classification according to Carr is consistent with the corresponding converted ordering scheme according to Hausner ($\rightarrow$ Hausner ratio $R_H = \varrho_{SR}/\varrho_{SS}$) [17]. R_C allows, among other things, statements about storage and compaction behavior, starting up after longer downtimes, or time solidification [4].

Since the bulk density can assume various values depending on the stress situation, it is generally denoted in the following text with ϱ_b. ϱ_{SS} and ϱ_{SR} are thus only special manifestations of ϱ_b. (2.21) can be used accordingly.

- **Particle size distribution**:
  Usually, polydisperse bulk materials are present. The distribution of particle sizes can be determined by a variety of different grain size analysis methods. A prerequisite for meaningful results in all cases is a representative sampling for the entire bulk material. The proportions assigned to the various particle sizes (= equivalent diameter) or size classes are determined. These can be represented in two ways [18], namely as

Distribution cumulative curve $Q_r(d_S)$ Compare with Fig. 2.5a. $Q_r(d_S)$ represents the quantity fraction of all particles with equivalent diameters smaller than d_S relative to the total amount, i.e., normalized. It holds:

$$Q_r(d_S) = \frac{\text{Partial amount } (d_{S,\min} \text{ bis } d_S)}{\text{Total amount } (d_{S,\min} \text{ bis } d_{S,\max})} \tag{2.23}$$

with: $Q_r(d_S \leq d_{S,\min}) = 0$ and $Q_r(d_S \geq d_{S,\max}) = 1$.

Distribution density curve $q_r(d_S)$ Compare to Fig. 2.5b. $q_r(d_S)$ indicates the proportion of a particle size class ($d_{S,1}$ to $d_{S,2}$) related to the respective class width Δd_S i.e.:

$$\begin{aligned} q_r(d_{S,1}, d_{S,2}) &= \frac{\text{Proportion } (d_{S,1} \text{ to } d_{S,2})}{\text{Class width } (d_{S,2} - d_{S,1})} = \frac{\Delta Q_r(d_{S,1}, d_{S,2})}{\Delta d_S} \\ &= \frac{Q_r(d_{S,2}) - Q_r(d_{S,1})}{d_{S,2} - d_{S,1}} \end{aligned} \tag{2.24}$$

Dimension of $q_r(d_S)$: [1/Length]. The total area under the $q_r(d_S)$-curve sums up to the value 1. Since, in general, measurement values of proportions are only available in discrete size classes, the course of $q_r(d_S)$ is approximated by a histogram. The same applies to $Q_r(d_S)$.

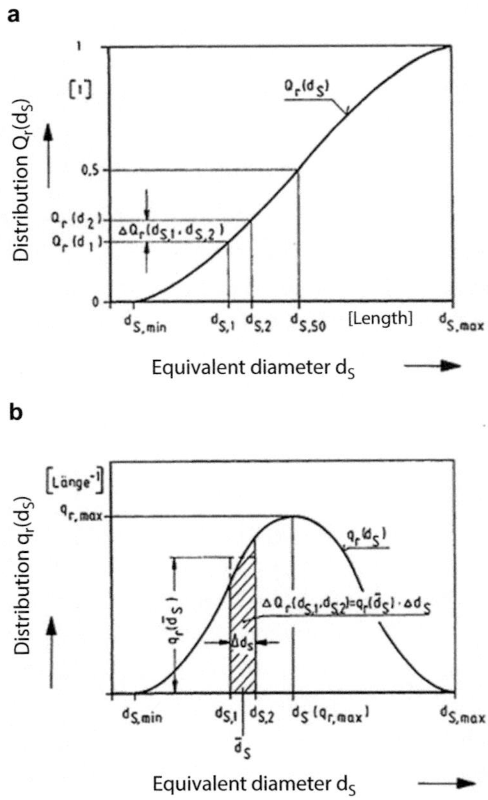

Fig. 2.5 Representation of particle size distributions. **a** Distribution cumulative curve $Q_r(d_S)$, **b** Distribution density curve $q_r(d_S)$

If $q_r(d_S)$ is a continuously differentiable function, then the following applies:

$$q_r = \frac{dQ_r}{d(d_S)} \tag{2.25a}$$

$$Q_r = \int_{d_{S,\min}}^{d_S} q_r \cdot d(d_S) \tag{2.25b}$$

Depending on the grain size analysis method used, different types of quantities are measured. The designation of the type of quantity is indicated by the index "r" of $Q_r(d_S)$ or $q_r(d_S)$ and is necessary for calculations with or conversions of distributions. The following are used (with: L = Länge):

$r = 0 \rightarrow$ Number distribution $(= \mathrm{L}^0)$,
$r = 1 \rightarrow$ Length distribution $(= \mathrm{L}^1)$,
$r = 2 \rightarrow$ Area distribution $(= \mathrm{L}^2)$,
$r = 3 \rightarrow$ volume distribution $(= \mathrm{L}^3)$, mass distribution $(= \varrho_S \cdot \mathrm{L}^3)$;
both are identical in the case of $\varrho_S = $ constant.

With the frequently used sieve analysis, for example, mass fractions are measured, i.e. $r = 3$, thus $Q_3(d_S)$ and $q_3(d_S)$. For $Q_3(d_S)$ the term passage $D(d_S)$ = mass fraction below the sieve with the mesh size d_S has become established. The fraction above the sieve is called residue $R(d_S)$. It holds: $D(d_S) + R(d_S) = 1$.

To capture the particle size distribution of a bulk material with only a few parameters, a description using mathematical functions is necessary. Such approximation functions can be represented by straight lines in special grid papers [18]. Among others, power distribution [19], logarithmic normal distribution [20], and RRSB distribution according to Rosin et al. [21] are used. The appropriate grid must be chosen according to the task or the given particle size distribution.

In Fig. 2.6, the RRSB grid, which is frequently used in industry and well-suited for particle size distributions resulting from comminution processes, is shown as an example. The associated distribution function is:

$$R(d_S) = 1 - D(d_S) = 1 - Q_3(d_S) = \exp\left[-\left(\frac{d_S}{d'_S}\right)^n\right] \tag{2.26}$$

with:

d'_S Location parameter: Particle diameter at residue $R = 0.368$ or passage $D = 0.632$,
n Dispersion parameter: Measure of the distribution width = slope of the RRSB line.

Grain size distributions that can be approximated by (2.26) result in the RRSB grid (abscissa division: $\lg(d_S)$, ordinate division: $\lg[\lg(1/R(d_S))]$), Fig. 2.6, as lines with the slope n. At the edge scales, this slope n and the volume-related specific surface S_V of the present particle collective can be read. The shape factor f takes into account the deviation of the particles from the spherical shape. It applies $f \geq 1$, with $f_{\text{Sphere}} = 1$. For the determination of n and S_V the line of the measured grain size distribution must be shifted parallel to the pole [21].

To describe the dependencies of the properties of a polydisperse bulk material on its grain size distribution and to simplify the characterization of the granulometric state of a particle collective, it is generally necessary to reduce/compress the information content of measured grain size distributions to only one representative numerical value. A characteristic particle size or a specific surface is usually used. Depending on the requirement, the following are used, for example:

- **Median value** $d_S(Q_r = 0.50) = d_{S,50,r}$: Central value, 50% of the mass fraction is larger or smaller than $d_{S,50,r}$. The value of $d_{S,50,r}$ can be read directly from the distribution cumulative curve, see Fig. 2.5a. Example: Sieve analysis measurement, $d_{S,50.3} = d_S(D = R = 0.5)$. $d_{S,50.3}$ will be referred to as $d_{S,50}$ from now on.
- **Mode** $d_S(q_{r,\max})$: Particle size at which $q_r(d_S)$ has a maximum. $d_S(q_{r,\max})$ is the most common particle size and can be read directly from the distribution density curve, see Fig. 2.5b.

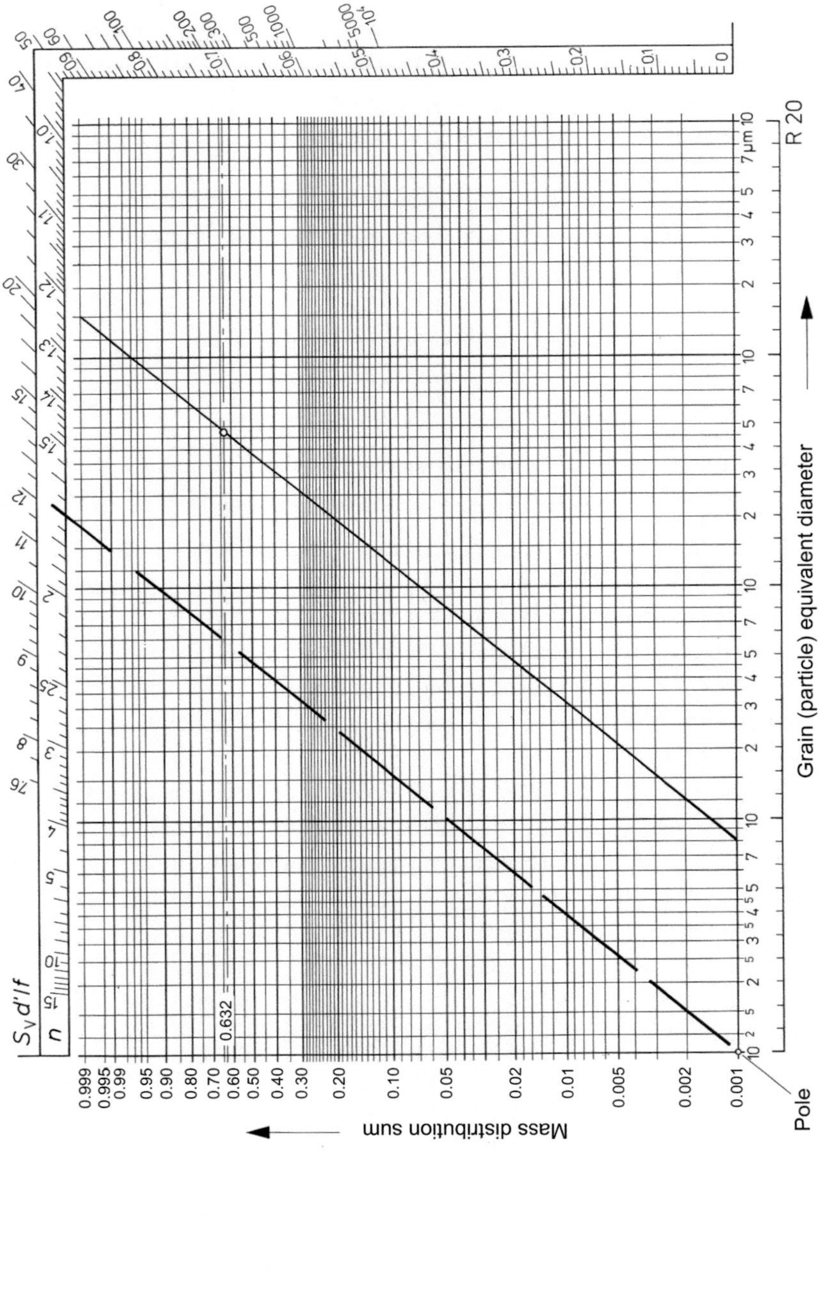

Fig. 2.6 RRSB grain size distribution network

- **Sauter diameter** $d_{S,SD}$: Sphere diameter of a monodisperse bulk that has the same total solid volume V_S and the same total solid surface area O_V as the real polydisperse material system idealized by volume-equivalent spheres:

$$d_{S,SD} = \frac{6 \cdot V_S}{O_V} \tag{2.27}$$

  $d_{S,SD}$ is always smaller than $d_{S,50}$. In determining $d_{S,SD}$, it is assumed that the bulk material system to be characterized and the comparison system consist of spherical particles. The adaptation to real systems can be done by correcting with a shape factor. Multiplication with the sphericity $\psi = O_V/O$ yields the modified Sauter diameter:

$$d^*_{S,SD} = \psi \cdot d_{S,SD} = \frac{6 \cdot V_S}{O} \tag{2.28}$$

- **Specific Surface Area**: Surface areas related to the volume or mass of the particle collective, S_V or S_m, are used. It holds:

$$S_V = \frac{O}{V_S}, S_m = \frac{O}{m_S} = \frac{6 \cdot \varphi}{\varrho_P \cdot d_S} \quad \text{with: } \varphi = \text{shape factor} \tag{2.29}$$

Reference values for the shape factor φ is included in Table 2.2 [22].

Often, directly measured surfaces, e.g., the Blaine- or the BET surface, are used for S. The measured values obtained in this way generally do not agree with those calculated from the particle size distributions, as different surfaces are detected by the different measurement methods. The Blaine surface S_{Blaine} [23], often used in the minerals industry, provides a flow-relevant surface, with which $d^*_{S,SD}$ can be estimated.

The mentioned and other characteristic sizes can be calculated from the so-called moments of the individual distributions. Details can be found, among others, in [18, 22, 24, 25].

Table 2.2 Shape factors φ for (2.29)

Bulk material	Form factor φ	Density ϱ_S [kg/m³]
Fly ash	2.26	2280
Glass	1.90	2570
Mica	9.27	2800
Coal dust	2.12	1300
Cork	1.98	300
Sand	1.43	2640
Tungsten powder	1.18	17300

2.2.4 Consideration as a Continuum

In classical thermodynamics, the properties of a gas in volume V are described by its state variables pressure p, temperature T, internal energy U etc., i.e., the gas is treated as a continuum. In the context of statistical thermodynamics, these quantities are then traced back to the behavior of the gas molecules constituting the system. A similar approach can be taken when describing the behavior of bulk materials. However, calculating the interactions between individual particles of a technical system is currently only possible to a limited extent due to the large number of particles ($\rightarrow$ 1 m^3 bulk volume contains approx. 10^{12} particles with a diameter of 100 μm), their possibly different size, shape, varying material composition, etc. Reasons for this are, in addition to the extremely complex task, the currently available computing power, which only allows simulations with relatively low particle numbers.

One considers the bulk material as a continuum. This is possible when the dimensions of the examined volume in the three spatial directions are larger than approximately 25 times the average particle diameter [26]. Within the scope of measurements, forces are applied to the boundary surfaces of the volume, and the resulting system responses, e.g., deformations, required stresses, etc., are measured and analyzed. The forces between the individual particles are thus recorded integrally, i.e., by means of averages.

The following characteristic values are required to describe the mechanical properties as well as the compression and flow behavior of bulk materials. They represent the input variables in the bulk material mechanics theory of Jenike [27] and its further developments [26, 28] and are used, among other things, for the flow-optimized design of the storage containers and locks of a pneumatic conveying system, for calculating the drive power of mechanical lock systems, for determining the friction behavior on the pipe bottom of sliding solid strands, or for the processes at a bulk material plug.

The required measurements are usually carried out with shear devices. In industrial practice, the Jenike translation shear device or so-called ring shear devices are frequently used, the later are easier to operate and automate. In Fig. 2.7, the shear cell of the Jenike device is shown. It consists of a lid and two concentric rings, the lower one of which is closed on the bottom side. The bulk material sample in the shear cell is loaded with a normal force F_σ and sheared by moving the upper ring. The required shear force F_τ is measured. Division of the forces F_σ and F_τ by the cross-sectional area A of the shear cell results in pressure stresses σ and shear stresses τ, which are used for further work. Sign

Fig. 2.7 Jenike shear tester

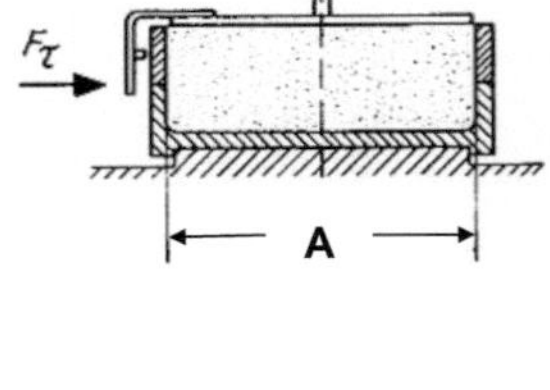

convention: Pressure stresses and contractions are defined as positive, tensile stresses and expansions as negative.

Since the behavior of a bulk material is significantly influenced by its stress history (→ bulk materials have a "memory"), it must be subjected to a defined reference pre-treatment before the actual measurement in order to obtain comparable results. The state of "stationary flow" = flow of the bulk material under constant stress and volume is chosen as the history. This corresponds, for example, to the bulk material state during free outflow from a container. The term flow refers to the irreversible plastic deformation of the stressed bulk material. To achieve this, the dissolution of particle contacts, i.e., the overcoming of adhesive forces, is required.

The stress history is set for each individual measurement by a defined pre-compaction and subsequent so-called shearing on. The actual shearing off of the prepared sample takes place under a lower load than the shearing on- = compaction load and provides measurement points "beginning of flow" = flow under volume increase. For one shearing on load σ_{an} several shearing off points with different shearing loads σ_{ab} are determined. The connecting line of the (σ, τ) value pairs of such a measurement series is called flow locus and is exemplarily shown in Fig. 2.8.

From a flow locus, the following bulk material properties can be determined, see Fig. 2.8:

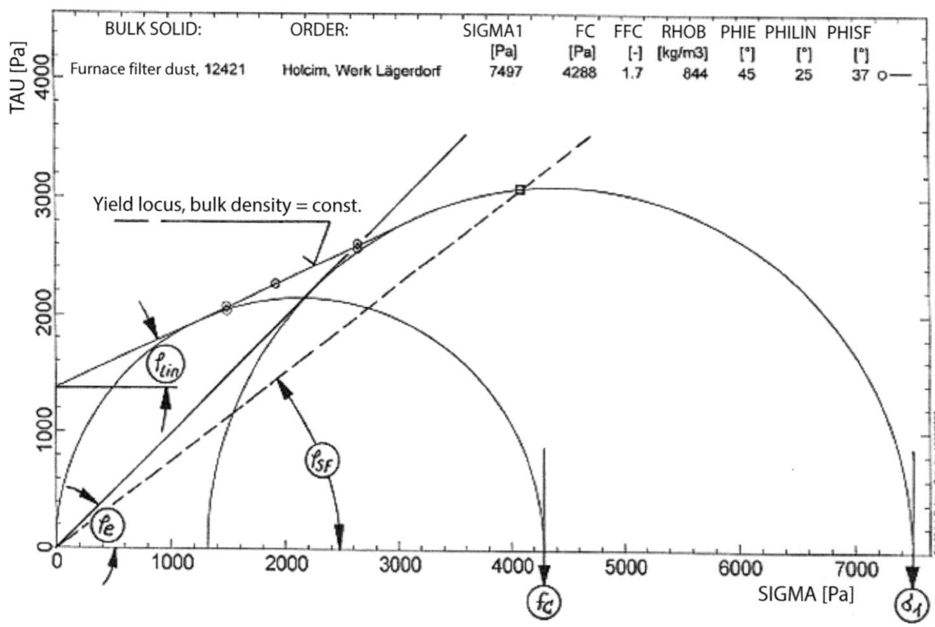

Fig. 2.8 Characteristics of a flow locus

- **Maximum Consolidation Stress during shearing on** σ_1: The compression state set during shearing on of the sample is the "cause" of all characteristic values shown below. It can be described by a Mohr's stress circle that touches the flow locus and runs through its endpoint = shearing on point. The Mohr circle is fully determined by the size of its two principal stresses (σ_1, σ_3) ($\rightarrow$ these act in shear stress-free planes and correspond to the abscissa intersections in Fig. 2.8). To characterize the consolidation state, the largest principal stress σ_1 of the Mohr circle is used [7].
- **Bulk Material Compressive Strength** f_C: This refers to the uniaxial compressive strength of the bulk material compressed by σ_1. It is determined as the largest principal stress f_C of that Mohr circle which touches the yield locus and passes through the coordinate origin ($\rightarrow$ for the smaller principal stress, thus $\sigma_3 = 0$, this corresponds to a free surface). Figure 2.9 illustrates the relationship using a thought experiment: compression of the bulk material sample supported by a frictionless wall with σ_1, removal of the wall, gradual increase of the load on the unsupported sample, failure at f_C. The compressive strength f_C determines, among other things, the necessary dimensions of the pre-container outlets or lock inlets to prevent bridging or funnel formation.
- **Bulk density** ϱ_b: Density to which the bulk material was compressed by the consolidation stress σ_1 before shearing ($\rightarrow$ filling mass/shear cell volume). Since the same pre-consolidation is always set along a flow locus, ϱ_b uniquely characterizes the flow locus.
- **Internal Friction Angle** φ_i **at the onset of flow**: Describes the transition of the bulk material from the resting state to the moving state and is defined as the local inclination angle of the flow locus against the σ-axis. If a curved flow location with $\varphi_i(\sigma)$ is present, the inclination angle φ_{lin} of the linearized flow location can be used as a simplification. This results as the tangent to the two Mohr circles with the largest principal stresses σ_1 and f_C. Note: φ_i is not an actual friction angle, but a quantity to characterize the inclination/shape of the flow location.
- **Internal Friction Angle** φ_{SF} **at steady flow**: This (true) friction angle is the angle that the straight line from the coordinate origin through the shearing on point $(\sigma_{\text{an}}, \tau_{\text{an}})$= endpoint of the flow locus with the σ-axis. It characterizes the internal friction in the shear plane during steady flow and is used, among other things, to calculate the forces for driving sluice and discharge devices. Associated friction coefficient: $\mu_{SF} = \tan \varphi_{SF}$.

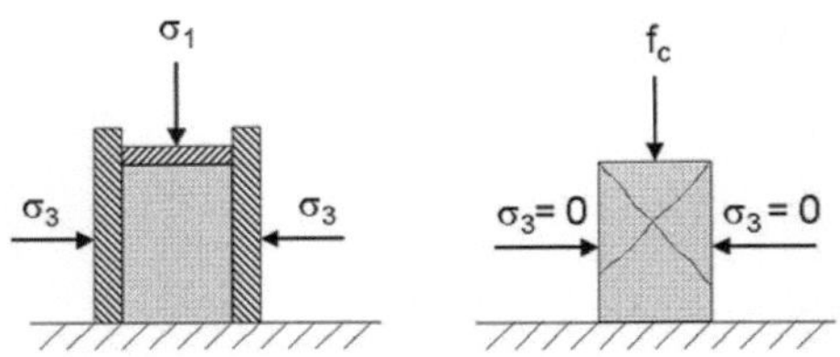

Fig. 2.9 Idealized uniaxial compression test

- **Effective Friction Angle** φ_e: The tangent from the coordinate origin to the Mohr's stress circle with the largest principal stress σ_1 is referred to as the effective yield locus according to Jenike [27]. It forms the the effective friction angle φ_e and represents a mathematically simple measure for the internal friction during steady state flow. The effective yield locus can be described by the principal stresses (σ_1, σ_3) of the Mohr's circle of shear on stress. It holds:

$$\sin \varphi_e = \frac{\sigma_1 - \sigma_3}{\sigma_1 + \sigma_3} \tag{2.30}$$

  φ_e is required, among other things, for silo dimensioning. The effective yield locus characterizes the stationary flow with constant $(\varrho_b, \sigma, \tau)$-values, while the individual flow loci characterize the beginning of flow, which is always associated with an increase in volume. Beginning flow ends after passing through a transition phase in the state of steady flow.
- **Cohesion** τ_C: The shear stress τ_C at the intersection of the flow locus with the τ-ordinate is called cohesion and is a measure of the interparticle adhesive forces of the bulk material precompressed with σ_1, but without load, i.e., at $\sigma = 0$, sheared bulk material. If $\tau_C = 0$, the flow point runs through the coordinate origin and the bulk material is cohesionless/free-flowing and does not build its own compressive strength f_C. The corresponding f_C-stress circle degenerates to a point with $f_C = 0$. The comparison with the effective flow locus clarifies that stationary flow is also assumed to be cohesionless. This is corrected in more advanced theories [26, 28].

The extension of the flow locus shown in Fig. 2.8 into the (negative) tensile stress range shows that bulk materials can also transmit tensile forces to a small extent. Stress states of a bulk material characterized by Mohr circles that lie below the flow locus only lead to reversible elastic deformations. For Mohr circles that are tangent to the flow locus, flow occurs by definition. Stress circles that intersect the flow locus are not possible, as flow sets in beforehand.

To determine the characteristic values described above for the expected stress range in a system, flow properties with different consolidation stresses σ_{an} are set and measured by systematically varying the shear stress σ_1 and bulk densities ϱ_b, see Fig. 2.10. The resulting characteristic values can then be represented as a function of the consolidation state σ_1.

Since the bulk material also interacts with the surrounding walls, it is necessary to measure the resulting behavior as well. This is done by determining a so-called wall flow locus, which represents the required wall shear stress τ_W as a function of the applied wall normal stress σ_W. Figure 2.11 shows an example of a wall yield locus against steel St 1.203 with a surface roughness of (2–3) µm. The shear force $F_{\tau,W}(\tau_W)$ is measured at different normal loads $F_{\sigma,W}(\sigma_W)$. The wall yield locus can be determined, for example, using the Jenike shear tester, if its lower shear cell ring, see Fig. 2.7, is replaced by a plate made of the wall material to be investigated. Since there is generally no

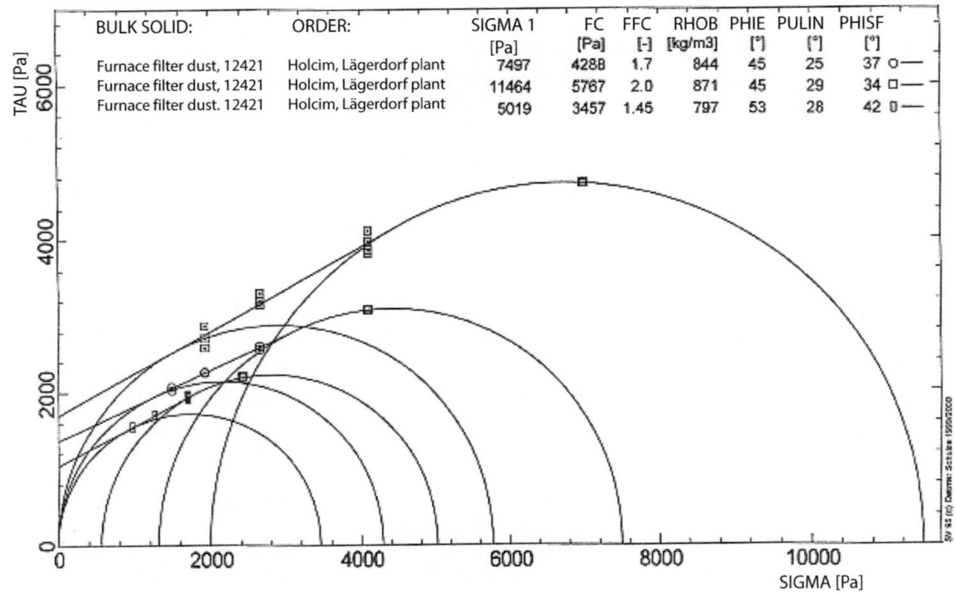

Fig. 2.10 Flow properties with different shear on loads

dependence of the wall yield locus on the bulk density ϱ_b, the test procedure is straightforward: no pre-compaction, no shearing on, change of normal load $F_{\sigma,W}$ during the experiment, etc. For calculations/designs, the wall friction angle φ_W is required:

- **Wall friction angle** φ_W: This can be calculated using the equation

$$\varphi_{W,i} = \tan^{-1}\left(\frac{\tau_{W,i}}{\sigma_{W,i}}\right) = \varphi_W(\sigma_W) \quad \text{with: } i = i\text{-th measurement point} \qquad (2.31)$$

  from the wall flow location. φ_W is only identical to the inclination angle of the wall flow locus and a constant value independent of σ_W when the flow locus represents a straight line through the coordinate origin. This also applies to the friction coefficient $\mu_{W,i} = \tan \varphi_{W,i}$. Metallic wall materials often provide a constant friction angle φ_W, see Fig. 2.11.

The measurement of the flow/shear characteristics described above can be carried out in such a way that essential operating factors influencing the later plant operation, such as varying storage times, changing moisture contents or temperatures of the bulk material, etc. can be simulated/set and thus quantitatively described.

To qualitatively describe the flowability of bulk materials, Jenike introduced the flow function FF_C [27]. This is defined as the ratio of the generating compaction stress σ_1 to the generated uniaxial compressive strength f_C, i.e., as:

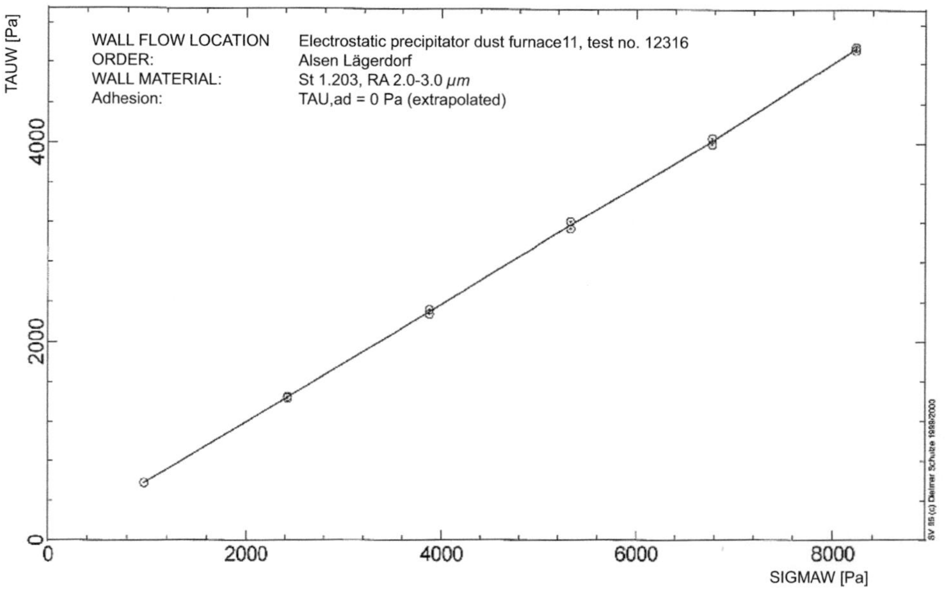

Fig. 2.11 Wall flow locus against steel St 1.203

$$FF_C = \frac{\sigma_1}{f_C} \tag{2.32}$$

Compare also Fig. 2.9. Large FF_C values describe good flowability, while small values indicate poor flowability. The classification proposed by Jenike in an extended form is shown in bold in Box 2.1.

Box 2.1: Bulk material classification with d'_S in µm

$10 \leq FF_C < \infty$	$\rightarrow$	$20 \leq (n \cdot \sqrt{d'_S})$	**free flowing**
$4 \leq FF_C < 10$	$\rightarrow$	$10 \leq (n \cdot \sqrt{d'_S}) < 20$	**lightly flowing**
$2 \leq FF_C < 4$	$\rightarrow$	$5 \leq (n \cdot \sqrt{d'_S}) < 10$	**cohesive**
$1 \leq FF_C < 2$	$\rightarrow$	$2.5 \leq (n \cdot \sqrt{d'_S}) < 5$	**very cohesive**
$FF_C < 1$	$\rightarrow$	$(n \cdot \sqrt{d'_S}) < 2.5$	**non-flowing, hardening**

The FF_C value is generally not a constant, but a function of the compression stress σ_1. As a reference consolidation stress, the value $\sigma_1 = 10\,\text{kPa}$ is used.

The essential problem of the FF_C value and the classification based on it is that extensive shear tests must be carried out to determine σ_1 and f_C. If the results of these tests are available, they can be applied directly to the respective problem. Classification is then no longer necessary. It would be more sensible to be able to estimate the possible behavior of a bulk material in a planned process before the plant design. For this purpose, the FF_C value must be determined from other, more easily accessible bulk material characteristics. In [29, 30, 31] FF_C is correlated with the parameters of an RRSB grain size distribution [21]:

$$FF_C = FF_C(n \cdot \sqrt{d'_S}) \tag{2.33}$$

The approach takes into account both a "mean" grain size ($\rightarrow d'_S$ [μm]) as well as the width of the grain size distribution ($\rightarrow n$). Both parameters significantly influence the bulk material behavior. The assignment shown in black in Box 2.1 is suggested for the FF_C-values [31]. Figure 2.12 illustrates that a bulk material assessment using this method provides a conservative estimate, i.e., the respective product is rather unfavorably assessed in its flow behavior. The large scattering of the measurement points in the range $FF_C > 10$ is not a concern, as the bulk material is then free-flowing anyway. The particle size distribution parameters n and d'_S can be determined without any problems, see Sect. 2.2.3. Further investigations show that the integration of additional parameters, e.g., the bulk density ϱ_b or the number of contact points k, in (2.33) does not significantly improve their accuracy.

Further details and applications will be discussed at the appropriate point.

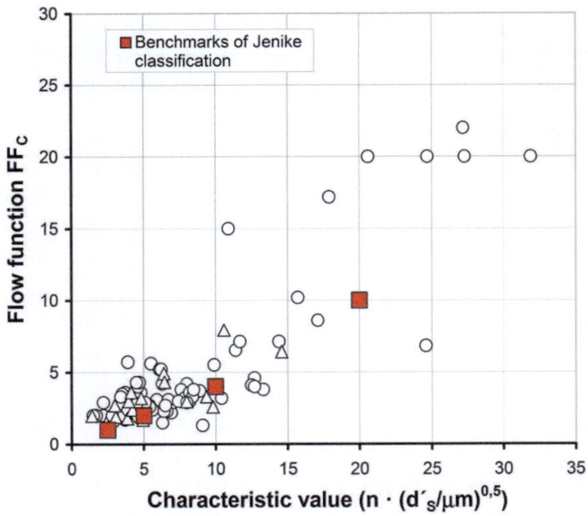

Fig. 2.12 Measured correlation $FF_C = FF_C(n \cdot \sqrt{d'_S})$

2.2.5 Adhesive Forces

The behavior of particularly fine-grained bulk materials is significantly determined by the adhesive/interaction forces between the particles. External, compressive forces acting on the bulk material assembly cause neighboring particles to approach each other and be deformed in the contact area, i.e., the contact surfaces increase. Both lead to an adhesive force amplification compared to the unloaded contact. The interaction forces generally result in a sticking together or a more or less firm bond between the particles. While this behavior is specifically used in agglomeration or granulation technology to build free-flowing, non-dusting, well-wetting, etc. (coarse-grained) agglomerates from non-flowing, dusting, poorly wetting, etc. (fine-grained) dusts, the effects on the flow and handling behavior are generally negative. If the adhesive forces transferred in the particle contacts are sufficiently large in relation to the forces stressing the bond during handling, the bulk material becomes cohesive, i.e., it is capable of transmitting tensile stress to a limited extent, and it exhibits a non-zero compressive strength f_C. This enables it to form stable free surfaces even under load and to initiate typical flow and silo problems. Equally important are the adhesive/adhesion forces between bulk material particles and the surrounding/adjacent walls.

Figure 2.13 illustrates that the cohesiveness of a bulk material must be defined as the ratio of the adhesive forces acting between the particles to the external forces trying to dissolve (or strengthen) the bond. The bulk material particle in Fig. 2.13a behaves adhesively, i.e., it adheres to the wall, if the adhesive force F_H between it and the wall is greater than its own weight force. F_G. If a gas stream blows parallel to the wall, see Fig. 2.13b, then the particle is additionally subjected to the flow force F_W exerted by the gas when it flows around the particle. As long as $F_H >$ effect $(F_G + F_W)$ remains, the particle continues to adhere to the wall. However, it is obvious that in case “b” it will be detached from the wall more easily than in case “a”. Again, different conditions arise when, for example, the particle is initially pressed against the wall for a short time with a force F_σ. This increases the effective adhesive force to $F_{H,\kappa} = \kappa \cdot F_H$, $\kappa > 1$, with the corresponding consequences. From this example, it follows that the same bulk material behaves differently in terms of cohesiveness and adhesiveness in different handling situ-

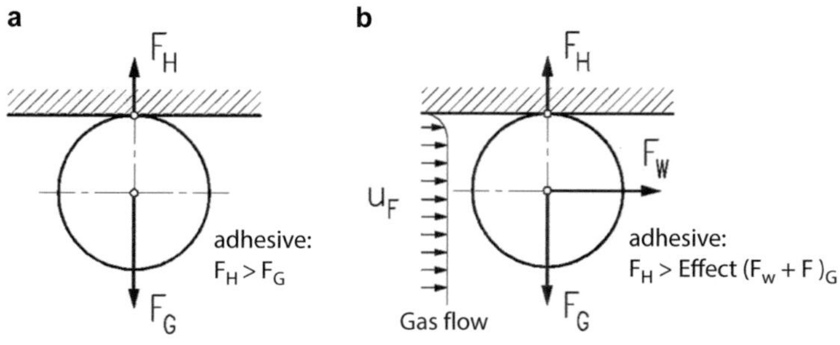

Fig. 2.13 Forces on a bulk material particle [4]

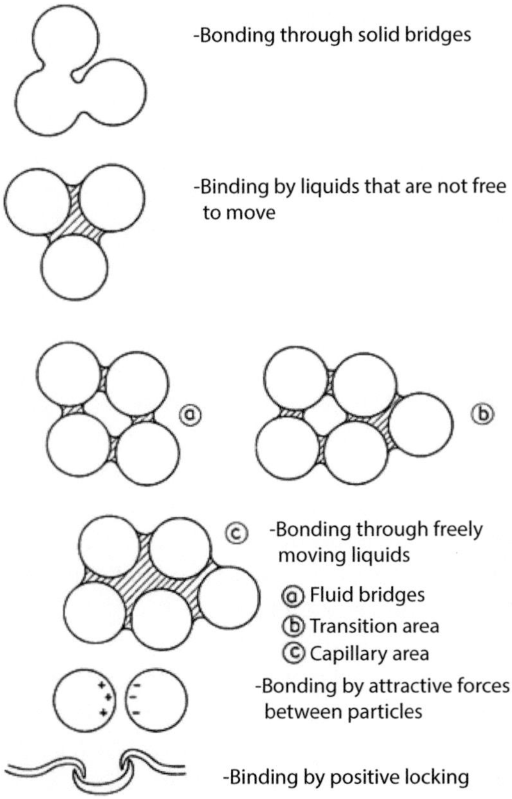

Fig. 2.14 Schematic representation of possible bonding mechanisms

ations depending on the external forces currently acting on it. Cohesiveness must therefore always be described with respect to the specific application. Cohesiveness increases with decreasing particle size.

The overview of possible bonding mechanisms shown in Box. 2.2 and Fig. 2.14 is based on the classification proposed in [32–34]. Bulk materials in a gas atmosphere are considered.

Box 2.2: Adhesive force bonding mechanisms

Solid bridges

- **Sintering**: Sufficiently high temperatures required; for initial estimates: ≥ approx. 60% of the absolute melting temperature.
- **Chemical reaction**: Interfacial oxidation; bonding processes during cement storage due to water absorption from pore and ambient air.
- **Structural change**: Recrystallization of amorphous surface layers, e.g., of powdered sugar, due to low water vapor absorption from the environment; leads to agglomeration.

- **Melting adhesion, cold welding**: Melting of roughness peaks and contact points due to friction and/or pressure; materials with low melting points are at risk.
- **Hardening binders**: Lime-containing products bind in a humid atmosphere.
- **Crystallization of dissolved substances during drying**: For example, salts, sugar, etc.

Bonding by non-freely moving liquids
- **Highly viscous binders, adhesives**: Essentially adhesion forces between the bonding partners.
- **Adsorption layers**: For example, water vapor from the surrounding atmosphere; a few molecular layers thick, micro-roughness is covered, both adhesion force enhancement and reduction possible.

Bonding by freely moving (wetting) liquids
- **Liquid bridges between particles**: There is only a small amount of liquid present, which accumulates at the particle contact and proximity points due to capillary forces. At these points, capillary condensation occurs preferentially in originally dry bulk material, given a sufficiently high moisture content of the surrounding atmosphere, due to the vapor pressure reduction in capillaries, narrow pores, gaps, etc. Capillary condensation of water vapor can already occur at relative humidities of $\varphi = (55-65)\%$.
- **Liquid bridges and liquid-filled pores**: Transition area; increased amount of liquid in the bulk material compared to the bridge area, saturation degree of the liquid $S \geq (0.2-0.4)$.
- **Pore volume filled with liquid**: Saturation degree $S \geq 0.8$; capillary area, no liquid bridges; only the force holding the entire agglomerate together can be determined/calculated (also applies to the transition area).

Binding by attractive forces between particles
- **Van der Waals forces**: Result of permanent or induced electrical dipole moments of the interacting atoms/molecules; always present.
- **Electrostatic forces**: Result of opposite charges of the two adhesion partners. Causes are a) electron transfer at particle contact due to different work functions of the adhesion partners (contact potential), especially for electrical conductors, b) excess charge, e.g., as a result of a previous treatment stage (friction, comminution, etc.), especially for electrical non-conductors.

Binding by form fit
- **Hooking/entanglement**: Mechanical form fit of fibrous, platelet-shaped, or other irregularly shaped particles.

Of the binding mechanisms mentioned, those by Van der Waals forces, by electrostatic forces and those, in general, due to local capillary condensation caused adhesive forces by liquid bridges are of primary importance for the problem at hand. The liquid bridge force consists of two components: the capillary force in the bridge due to the capillary negative pressure (attractive) or positive pressure (repulsive) and the always attractive force due to the surface tension of the liquid [35]. Considerable problems can also be caused by form-fitting bonds, the problem of which, in general, is already visually recognizable. The remaining adhesion mechanisms shown in Box 2.2 usually only occur under special operating conditions. However, it is appropriate to systematically analyze the relevance of the individual binding mechanisms for a given task within the framework of the search for a solution in order to prevent surprises.

Van der Waals, electrostatic, and liquid bridge forces can be theoretically calculated for idealized simple systems, consisting, for example, of ideally smooth spherical particles, smooth flat walls, geometrically clearly defined roughness elevations, etc., see, among others, [10, 35–37]. Figure 2.15a shows adhesive forces $F = F_H$ as a function of particle diameter d_S for the model system sphere/plate with the constant contact distance $a_0 = 0.4\,\text{nm}$. Of dominant importance here are the liquid bridge and Van der Waals forces, while electrostatic and particle weight forces are orders of magnitude lower. The order and size relations of the forces to each other are confirmed by practice. If in Fig. 2.15a instead of the absolute adhesive force F_H the ratio of adhesive force F_H/particle weight F_G is plotted, the curves decrease with increasing particle diameter, i.e., the bulk material becomes, as expected, less critical.

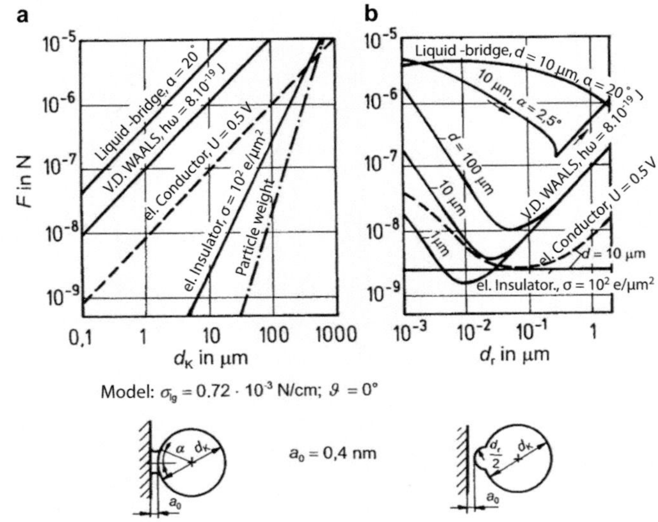

Fig. 2.15 Comparison of adhesive forces of various bonding mechanisms [38]. **a** ideal sphere, **b** rough sphere

In Fig. 2.15b, the influence of real, i.e., rough, contact surfaces is simulated by introducing a hemispherical roughness elevation with variable diameter d_r. Figure 2.15b illustrates how sensitive the sphere/plate system reacts to such influences. More practical calculations can be found in, among others, [37]. Overall, the size relations between the individual forces are roughly preserved.

In pneumatic conveying systems, the indicated cohesion and adhesion effects can cause, among other things, the following problems:

- Design of the conveying system for a bulk material that forms solid agglomerates due to its cohesiveness on feeding units, e.g. flow or vibration channels, i.e., a coarser product with a generally broader particle size distribution is present than was assumed for the dimensioning. This can lead to reduced throughput, conveying problems, etc.
- Entry of lumps, falling wall deposits, etc. from upstream plant areas into the conveying system. Consequence: lock system and conveying problems,
- Formation of wall deposits in the conveying system, especially in deflections/bends, due to bulk material impact or deposits resulting from adhering, non-flowing strand areas on the pipe bottom. Consequence: increased pressure loss or reduced solid throughput due to reduction of the conveying pipe cross-section; extreme pressure peaks when deposits fall off and/or deposits start moving again due to increased gas velocities.

Changes in product quality must also be expected. Figure 2.16 shows, as an example, the behavior of a bulk material sample that was subjected to a normal force in a wall shear test after the removal of the shear cell.

In general, it is recommended to evaluate a representative bulk material sample not only based on the measured values presented above but also to assess the product manually, i.e., touch, compress, flow/pour, etc.it. This develops a sense of the handling issues.

Fig. 2.16 Compacted bulk material sample after a wall shear test

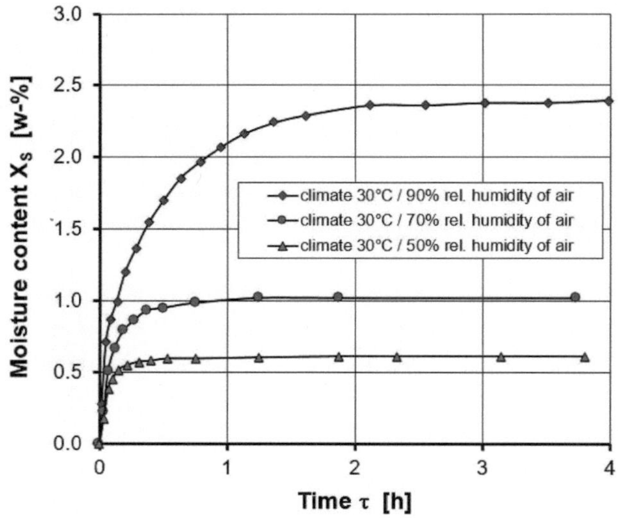

Fig. 2.17 Adsorption behavior of granulated kaolin, $d_{S,50} = 190\,\mu m$

2.2.6 Sorption Behavior

If a small amount of a dry bulk material is applied as a single-particle layer on a substrate, deposited in a climate chamber adjusted to the temperature T and the relative humidity φ, and the temporal weight gain of the bulk material is measured, this leads to dependencies as exemplified in Fig. 2.17. The loading X_S of the dry solid with moisture is plotted against the contact time τ at various relative humidities φ. The weight increase of the bulk material is here solely due to the adsorption of H_2O molecules from the surrounding gas on the accessible solid surface. Due to their ability to form hydrogen bonds, H_2O adsorption dominates over that of other air (gas) components. Mineral substances are initially always covered by an H_2O adsorption layer. Hydrophilic behavior is favored by a high polarity of the adsorptive and a low content of organic components in the adsorbent.

Adsorption layers influence the contact geometry and the distance behavior of adjacent particles: for example, it is no longer just particle roughness that interacts, but also adsorption layers that cover this roughness; consequence: increase in contact area due to overlapping adsorption layers. In the case of aqueous adsorption layers, this generally leads to an increase in adhesive forces compared to those without adsorption. The influence of adsorption layers on the size of the resulting adhesive forces cannot be described universally.

The endpoints of the $X_S(\varphi)$-curves in Fig. 2.17 represent equilibrium points on a sorption isotherm, see Fig. 2.18 [39]. It is obvious that in the area of the already mentioned capillary condensation, $\varphi \gtrsim 0.60$, the strongest influences on adhesive forces can be expected.

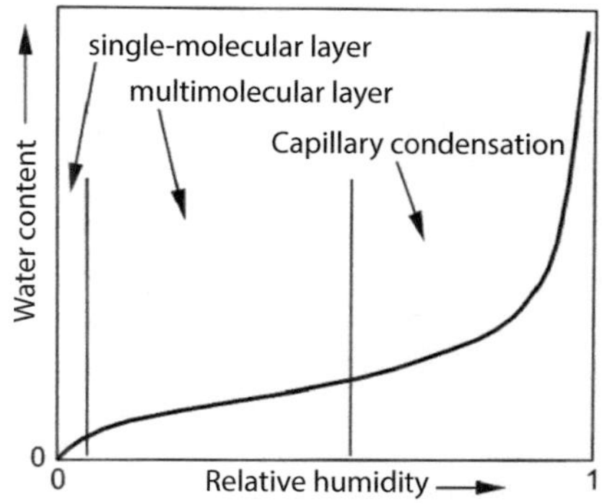

Fig. 2.18 Possible course of an H_2O-adsorption isotherm, T = const.

For a significant moistening of the bulk material shown in Fig. 2.17, contact times on the order of several minutes are required. The solid residence time in a pneumatic conveying line, e.g. $\tau_{\mathrm{pneu}} = L_R/u_S = 100\,\mathrm{m}/10\,\mathrm{m/s} = 10\,\mathrm{s}$, L_R = conveying distance, u_S = solid velocity, is not sufficient for this purpose. The same applies to other bulk materials: even with hygroscopic products, drying the conveying gas to avoid moistening in the pipeline is not absolutely necessary. However, this is only true as long as the bulk material is fed directly to a downstream treatment stage. If the process stages are decoupled, e.g., by conveying into an intermediate silo, there is sufficient long contact between the surface of the stored bulk material and the headspace gas = conveying gas for moisture absorption. In this case, either conditioning of the conveying gas or the silo headspace is required.

The climate cabinet experiment described above provides direct indications of possible effects of moistening, e.g., by adhesions to the carrier, agglomerations, crusts, etc. If a very moist bulk material is fed into the climate cabinet, it dries (desorbs) to the equilibrium value corresponding to the set relative humidity φ (deviations due to hysteresis effects!).

2.2.7 Combustion and Explosion Characteristics

Bulk materials can be combustible and thus also explosive. Combustion and explosion are exothermic oxidation reactions of combustible dusts with an oxidizing agent, usually, and exclusively considered in the following, the oxygen contained in the surrounding air. They differ essentially in the reaction rate and the pressure increase associated with the explosion due to the rapid heating and volume expansion of the reaction gases. Deposited combustible dusts can be ignited both by self-ignition—minimum temperature required—and by local supply of external energy. If a sufficiently strong ignition source is present in a swirled up dust/air mixture becomes effective, a dust explosion with a progressing pressure and flame front, similar to a gas explosion, can occur. In Fig. 2.19, the basic parameters are compiled, with which the combustion and explosion behavior

Definitions/ parameters

Pressure rise, maximum time

Highest value occurring during the explosion of a dust for the pressure increase over time in a closed container at optimum concentration (see also "Cubic law").

Ignition temperature of a dust layer, minimum

Lowest temperature of an exposed hot surface at which ignition of a dust layer of defined thickness occurs on this surface (at 5 mm layer thickness identical with glow temperature).

Explosion

Fast combustion with recognizable pressure increase.

Explosion pressure, maximum

The highest pressure value occurring during the explosion of a dust/air mixture of optimum concentration in a closed container.

Explosion limits

Lower explosion limit (LEL) and upper explosion limit (UEL) are the limit concentrations of a dust in air between which the dust/air mixture can be caused to explode by external ignition.

Glow temperature

Lowest temperature of an exposed hot surface at which a layer of dust 5 mm thick on it will ignite within two hours.

Hybrid mixture

Mixture of combustible dust with combustible gas and/or vapor in air

K_{St} value

Dust and test method-specific parameter calculated from the cubic law. It is numerically equal to the value for the maximum time pressure increase in the 1 m^3 container under the test conditions specified in the VDI 3673 guideline and in ISO 6184/1 (8; 9). The KSt value depends in particular on the particle size distribution and the surface structure of the dust

Cubic law

Volume dependence of the maximum temporal pressure increase

$$(dp/dt)_{max} \cdot V^{1/3} = konst = K_{St}$$

Because of the relationship between volume V and maximum pressure rise over time (dp/dt)max , data for maximum pressure rise over time are not sufficient without simultaneous volume data.

Median value

Value for the median particle size: 50% by weight of the dust is coarser and 50% by weight is finer than the median value

Minimum ignition energy

Lowest value of electrical energy stored in a capacitor which, under variations of the discharge circuit parameters, is just sufficient to ignite the most ignitable mixture of dust and air at atmospheric pressure and room temperature during discharge (10).

Limiting oxygen concentration

Experiments# determined oxygen concentration in a dust/air/inert gas mixture at which just no more explosion is possible, regardless of the dust concentration.

Spontaneous combustion

Process in which a dust deposit ignites (volume-dependent) after self-heating under the influence of heat from all sides and in the presence of air.

Smoldering point

Lowest temperature at which a dust releases combustible vaporous or gaseous products ("smoldering gases") in such quantities that they can be ignited by a small flame in the air space above the fill.

Dust

Finely divided solid material of any shape, structure and density below a grain size of approx. 500 m.µ

Dust explosion classes

Classes into which dusts are classified on the basis of their K_{St} values [8].

Dust explosion class	K_{St} in bar m s^{-1}
St 1	> 0 to 200
St 2	> 200 to 300
St 3	> 300

Dust/air mixture

Dust whirled up in the air. The characteristic variable is the dust concentration.

Decomposition, exothermic

A reaction that also takes place without atmospheric oxygen, which can lead to self-heating and, in the case of gas release in closed apparatus, to an increase in pressure volume-dependent).

Ignition temperature

Lowest temperature of a hot surface at which the most ignitable mixture of the dust in question with air will just ignite.

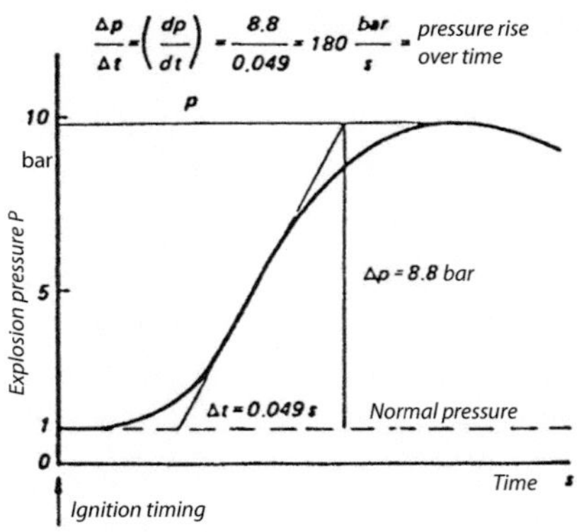

Determination of the pressure rise over time of a dust explosion at any concentration

Fig. 2.19 Combustion and explosion characteristicsn of dusts, definitions [40]

of a dust is described. Corresponding characteristic data for more than 1000 products are systematically listed in [40]. Characteristics of unknown/limiting products must be determined anew by experiment. The required measurement methods are also described in [40]. All characteristic data apply to a triggering/operating condition in the plant of: Pressure $p_0 =$ (0.9–1.1) bar, O_2-content $r_{O_2,0} \cong 21$ Vol.-%, temperature $T_0 =$ (0–30) °C. They are not applicable if a so-called hybrid mixture of combustible dust and combustible gas or vapor is present. Maximum explosion overpressure $p_{ex,\max}$, maximum rate of pressure increase $(dp/d\tau)_{\max}$ and thus also the K_{St}-value are the maximum values achievable during the explosion of a dust/air mixture with optimally adjusted dust concentration and under optimal ignition conditions in a closed container. Practical values may be lower.

Influences of bulk material and pneumatic conveying on explosion behavior:

- Explosions are only possible with (approximately homogeneous) dust concentrations distributed in air above the lower explosion limit (UEG) and below the upper explosion limit (OEG). The UEG is in the range of (15–60) g/m.^{3}. At lower values, the energy released during the reaction of a particle is not sufficient to ignite neighboring particles due to the large distances and the associated heating of the surrounding air. The OEG is in the range of (6000–10,000) g/m^3, sometimes higher. At concentrations above the OEG, the energy released during the reaction of a particle is not sufficient to bring the mass of surrounding particles to ignition temperature. The dust concentration of the OEG can be converted into a loading μ_{OEG} characterizing pneumatic conveying using the equation

$$\mu_{\text{OEG}} \cong \frac{\text{OEG} \cdot C}{\varrho_F}, \quad \text{with: } C \cong \frac{u_S}{v_F} \tag{2.34}$$

  With OEG = 10 kg/m^3, $C \cong 0.70$ and $\varrho_F = 1.20$ kg/m^3 follows from (2.34): $\mu_{\text{OEG}} \cong 6.0\,\text{kg}\,S/\text{kg}\,F$. In practice, pneumatic conveying systems are usually designed for significantly higher loadings μ, i.e. $\mu > \mu_{\text{OEG}}$. Thus, an explosion cannot occur in the actual conveying section, i.e. the conveying pipe. However, it should be noted that, for example, when entering and especially when discharging from the conveying line, the explosive concentration range between UEG and OEG can be passed through, so there is a risk of explosion. Example: When conveying into the free upper space of a storage silo, the gas/solid jet leaving the conveying pipe expands/dissolves with increasing distance from the entry point, i.e., the local dust concentration decreases.
- Combustible dusts above a limit diameter of $d_S \gtrsim (400-500)\,\mu\text{m}$ cannot be ignited even with a strong ignition initial. Finer dusts of the same bulk material generally react more violently than coarser ones, with this having a stronger effect on the pressure increase rate $(dp/d\tau)$ = explosion violence than on the explosion pressure p_{ex} affects. Already admixtures of (5÷10) M.-% of an explosible fine dust mixed with a non-explosible coarse dust of the same material is sufficient to form an ignitable mixture. In this case, almost the same maximum explosion pressure $p_{ex,\max}$ is achieved as

with the fine dust alone. Only the explosiveness $(dp/d\tau)$ is influenced by the coarse/fine mixture ratio: decrease with increasing coarse content. Explosion hazard is generally present whenever the proportion of explosible fine dust contained in the non-explosible coarse dust exceeds the UEG [40]. It is essential to note that in almost all processes involving the handling of bulk materials, grain breakage and/or fine-grained abrasion is generated. This also applies to pneumatic conveying, where in particular the filter dusts produced are critical.

- Below a bulk material-specific and inert gas type-dependent limiting oxygen concentration, dust explosions are no longer possible: Examples of inerting with nitrogen N_2: limiting oxygen concentration for hard coal: $r_{O_2,\mathrm{grenz}} = 14$ vol.-%, limiting oxygen concentration for lignite: $r_{O_2,\mathrm{grenz}} = 12$ vol.-%, limiting oxygen concentration for aluminum dust: $r_{O_2,\mathrm{grenz}} = 6$ vol.-%. Explosive bulk materials can also be converted into non-explosive mixtures by adding inert dusts, e.g., rock salt. However, inert dust additions of more than 50 mass-% are required.
- If the above-mentioned "standard" conditions (p_0, r_0, T_0) are not maintained during the triggering of an explosion in an operating system, the "standard" explosion parameters must be converted to the current operating conditions or measured at these conditions. For conversions, physically justified approaches can be used [41]. In pneumatic pressure conveying systems, for example, triggering pressures $p > p_0$ occur. The associated explosion characteristics can be calculated from the values determined under standard conditions as follows:

$$p_{ex,\max}(p);\quad \left(\frac{dp}{d\tau}\right)_{\max,p};\quad K_{St}(p) = \left(p_{ex,\max}(p_0);\ \left(\frac{dp}{d\tau}\right)_{\max,p_0};\ K_{St}(p_0),\right)\cdot\frac{p}{p_0} \tag{2.35}$$

with: $r_{O_2,0} \cong 21$ Vol.-%, $T_0 = (0\text{–}30)$ °C.

Absolute pressures must be used. Comparable approaches exist for temperature, volume, fill level, etc.

Example: A lignite dust with the characteristic data $p_{ex,\max}(p_0) = 11$ bar, $(dp/d\tau)_{\max,p_0} = 150$ bar/s, $K_{St}(p_0) = 150$ bar m/s, $p_0 = 1.0$ bar is mixed with air and $p = 3.0$ bar conveyed, temperature and oxygen content correspond to the values at standard conditions. With the pressure ratio $p/p_0 = 3.0\,\mathrm{bar}/1.0\,\mathrm{bar} = 3$ result with (2.35) the following operating parameters: $p_{ex,\max}(p) = 33$ bar, $(dp/d\tau)_{\max,p} = 450$ bar/s, $K_{St}(p) = 450$ bar m/s. The lignite dust thus rises from the dust explosion class 1 to the more dangerous/difficult to control class 3.

From the above explanations, it follows that the following conditions must be met simultaneously, i.e., at the same time and place, to trigger an explosion:

- Presence of a combustible bulk material in dust-fine form, i.e., $d_S \lesssim 400\,\mu$m,
- Presence of a sufficient amount/concentration of atmospheric oxygen,

- Presence of an explosive, i.e., a swirled up dust/oxygen mixture within the explosion limits UEG and OEG,
- Presence of a sufficiently strong ignition source.

The possibilities for explosion protection measures that can also be derived from the above considerations are not discussed here, see, for example, [42].

2.3 Conservation Laws of Multiphase Systems

In general form, the conservation law for a conserved quantity Q (here: mass, momentum, energy) can be represented as follows [43]:

$$\frac{\text{Accumulation of } Q \text{ in the system}}{\text{Time unit}} = \frac{\text{amount of } Q_{\text{in}} \text{ flowing into the system}}{\text{Time unit}} - \frac{\text{amount of } Q_{\text{out}} \text{ flowing out of the system}}{\text{Time unit}} + \frac{\text{Production } R_Q \text{ of } Q \text{ in the system}}{\text{Time unit}} \tag{2.36}$$

A system can, for example, be an entire factory, an apparatus, part of an apparatus, a pipeline, or an infinitesimally small length element/volume of this line, see Fig. 2.20.

From the verbal description, (2.36), it follows:

$$\frac{dQ}{d\tau} = \dot{Q}_{\text{in}} - \dot{Q}_{\text{out}} + \dot{R}_Q \tag{2.37}$$

With

ΔV_C size of the control volume,
q^v volumetric mean value of the volume-related (specific) conservation quantity,
$q^f_{\text{in/out}}$ Flow mean of the volume-related (specific) conservation quantity entering ΔV_C and exiting ΔV_C,

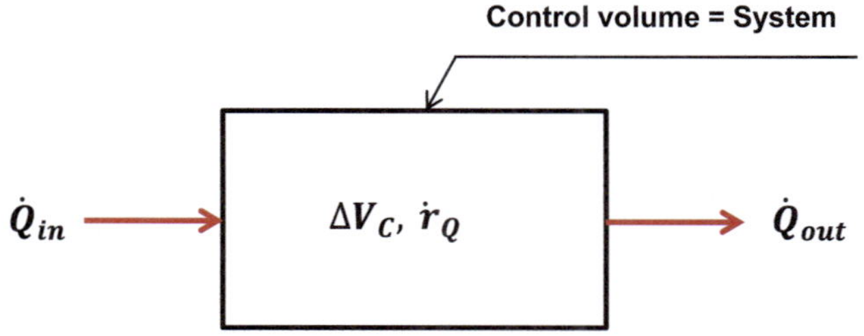

Fig. 2.20 System/Control volume of a macro balance

$\dot{V}_{in/out}$ Volume flows entering and exiting ΔV_C,

$\dot{r}_Q$ Production rate of the conservation quantity per unit of time and volume, can (2.37) also in the form

$$\Delta V_C \cdot \frac{dq^v}{d\tau} = \dot{V}_{in} \cdot q^f_{in} - \dot{V}_{out} \cdot q^f_{out} + \Delta V_C \cdot \dot{r}_Q \tag{2.38}$$

be written. If there are several phases "i" present, the respective conservation law must be applied to each of these phases. The following considerations are limited to one-dimensional pipe flows with averaged characteristic values over the current pipe cross-section.

2.3.1 Mass Conservation

With $Q = m_i$, $\dot{R}_Q = \dot{R}_{m,i}$ follows for phase, i:

$$\frac{dm_i}{d\tau} = \dot{m}_{i,in} - \dot{m}_{i,out} + \dot{R}_{m,i} \tag{2.39}$$

as well as:

$$\Delta V_C \cdot \frac{d(\varepsilon_i \cdot \varrho_i)}{d\tau} = (\varepsilon_i \cdot A_R \cdot \varrho_i \cdot u_i)_{in} - (\varepsilon_i \cdot A_R \cdot \varrho_i \cdot u_i)_{out} + \Delta V_C \cdot \dot{r}_{m,i} \tag{2.40}$$

ε_i describes the local volume concentration, $(\varepsilon_i \cdot A_R) = A_i$ represents the flow cross-section, ϱ_i the density and u_i the velocity of phase i. Other representations are possible. "Production" of mass in a multiphase system occurs when mass is transferred from one phase to another, e.g., when in a gas/liquid flow, a part of the liquid evaporates or steam condenses. Another example is the drying of bulk material in a pneumatic conveyor dryer [44]. For the sum of all individual production rates $\dot{r}_{m,i}$ follows according to Lavoisier:

$$\sum_i (r_{m,i}) = 0 \tag{2.41}$$

What is supplied to one phase (+), is taken away from the other C (−). For the solid (S)/gas (F) pipe flows considered here, in the steady state (= *steady-state flow*) with $d\dot{m}_i/d\tau = 0$ and usually $\dot{r}_{m,i} = 0$:

$$\dot{m}_{i,in} = (\varepsilon_i \cdot A_R \cdot \varrho_i \cdot u_i)_{in} = \dot{m}_{i,out} = (\varepsilon_i \cdot A_R \cdot \varrho_i \cdot u_i)_{out} = \text{const.} \quad \text{with: } i = S, F \tag{2.42}$$

Thus, the total mass flow also applies:

$$\dot{m}_{tot} = \sum_i (\dot{m}_i) = \text{const.} \tag{2.43}$$

Furthermore, from $\sum_i(\varepsilon_i) = 1$ follows the relationship $\varepsilon_S = (1 - \varepsilon_F)$.

2.3.2 Momentum Conservation

The momentum is a vector and is characterized by both its direction and magnitude, i.e., it is described by its three components in x-, y- and z-direction. Since momentum conservation must be fulfilled in all three coordinate directions, this leads to three separate equations. In the following, only the approaches in x-direction = conveying direction are considered.

With $Q = \Im_i = (m_i \cdot u_i)$, $\dot{Q} = \dot{\Im}_i = (\dot{m}_i \cdot u_i)$ and $\dot{R}_Q = \sum_k (F_{i,k})=$ sum of all forces acting on phase i F_k results for phase i in x-direction:

$$\frac{d\Im_i}{d\tau} = \dot{\Im}_{i,\text{in}} - \dot{\Im}_{i,\text{out}} + \sum_k (F_{i,k}) \tag{2.44}$$

or:

$$\frac{d(m_i \cdot u_i)}{d\tau} = (\dot{m}_i \cdot u_i)_\text{in} - (\dot{m}_i \cdot u_i)_\text{out} + \sum_k (F_{i,k}) \tag{2.45}$$

as well as:

$$\Delta V_C \cdot \frac{d(\varepsilon_i \cdot \varrho_i \cdot u_i)}{d\tau} = (\varepsilon_i \cdot A_R \cdot \varrho_i \cdot u_i^2)_\text{in} - (\varepsilon_i \cdot A_R \cdot \varrho_i \cdot u_i^2)_\text{out} + \sum_k (F_{i,k}) \tag{2.46}$$

Other representations are possible. The production term here is identical to the sum of all forces acting on the considered phase. For gas/solid pipe flows in the steady state, i.e. at $d\Im/d\tau = 0$:

$$(\dot{m}_i \cdot u_i)_\text{in} - (\dot{m}_i \cdot u_i)_\text{out} + \sum_k (F_{i,k}) = 0 \quad \text{with: } \dot{m}_{i,\text{in}} = \dot{m}_{i,\text{out}} = \dot{m}_i \tag{2.47}$$

The forces acting on phase i include, among others, pressure, weight, friction, and forces exerted by other phases on phase i. Applying (2.47) separately to the solid and conveying gas in a horizontal conveying pipe element with the cross-sectional area A_R and the length dL, see Fig. 2.21, results in:

Balance on the solid:

$$(\dot{\Im}_{S,\text{in}} - \dot{\Im}_{S,\text{out}}) + (F_{S,p,\text{in}} - F_{S,p,\text{out}}) + F_{S,\text{drag}} - F_{S,f} = 0 \tag{2.48}$$

with:

$F_{S,p,\text{in/out}}$	Pressure forces,
$F_{S,\text{drag}}$	Force due to conveying gas flow around the solid particles = flow resistance, drag force,
$F_{S,f}$	Forces due to solid wall friction and particle impact.

Fig. 2.21 Forces on the solid (S) and on the conveying gas (F) in a conveying pipe element ($dL \cdot A_R$)

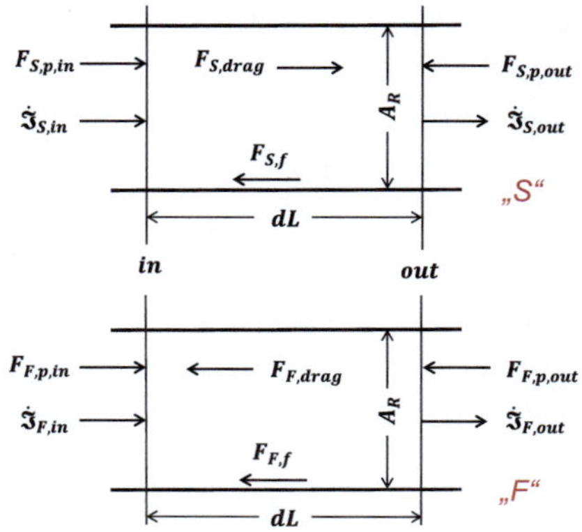

Furthermore, it applies:

$$(\dot{\mathfrak{I}}_{S,\text{in}} - \dot{\mathfrak{I}}_{S,\text{out}}) = \dot{m}_S \cdot (u_{S,\text{in}} - u_{S,\text{out}}) = -\dot{m}_S \cdot du_S \tag{2.49}$$

$$(F_{S,p,\text{in}} - F_{S,p,\text{out}}) = (1 - \varepsilon_F) \cdot A_R \cdot (p_{\text{in}} - p_{\text{out}}) = -(1 - \varepsilon_F) \cdot A_R \cdot dp \tag{2.50}$$

On calculation approaches for $F_{S,\text{drag}}$ and $F_{S,f}$ will be discussed in Chap. 3. From (2.48–2.50) it follows:

$$-\dot{m}_S \cdot du_S = (1 - \varepsilon_F) \cdot A_R \cdot dp - F_{S,\text{drag}} + F_{S,f} \tag{2.51}$$

Balance on the conveying gas:

$$(\dot{\mathfrak{I}}_{F,\text{in}} - \dot{\mathfrak{I}}_{F,\text{out}}) + (F_{F,p,\text{in}} - F_{F,p,\text{out}}) - F_{F,\text{drag}} - F_{F,f} = 0 \tag{2.52}$$

With

$$(\dot{\mathfrak{I}}_{F,\text{in}} - \dot{\mathfrak{I}}_{F,\text{out}}) = \dot{m}_F \cdot (u_{F,\text{in}} - u_{F,\text{out}}) = -\dot{m}_F \cdot du_F \tag{2.53}$$

$$(F_{F,p,\text{in}} - F_{F,p,\text{out}}) = \varepsilon_F \cdot A_R \cdot (p_{\text{in}} - p_{\text{out}}) = -\varepsilon_F \cdot A_R \cdot dp \tag{2.54}$$

follows:

$$-\dot{m}_F \cdot du_F = \varepsilon_F \cdot A_R \cdot dp + F_{F,\text{drag}} + F_{F,f} \tag{2.55}$$

The two equations (2.51) and (2.55) allow for the determination of the design parameters of a pneumatic conveying line, i.e., the necessary pressure difference Δp_R and the resulting solid velocity u_S, or the velocity ratio $C = u_S/u_F$.

Balance at the pipe element:

An analogous consideration for the entire pipe element leads to:

$$-(\dot{m}_S \cdot du_S + \dot{m}_F \cdot du_F) = A_R \cdot dp + F_{S,f} + F_{F,f} \tag{2.56}$$

Since the forces acting between the solid and gas phases $F_{S/F,\text{drag}}$ are equal in magnitude and opposite in direction ($F_{S,\text{drag}} = -F_{F,\text{drag}}$), they do not appear in (2.56): They are virtually invisible from the outside, see the image parts (S) and (F) in Fig. 2.21. This applies to all interphase forces. (2.56) is also obtained by adding (2.51) and (2.55) and is generally used to determine the pressure loss Δp_R.

The impulse current changes ($\dot{m}_{S/F} \cdot du_{S/F}$) in the preceding equations describe the instantaneous acceleration of the considered phase/mass. Neglecting these terms assumes an incompressible flow and the balances may only be solved over short lengths dL, in which the velocity changes are correspondingly small. Pneumatic conveyances are compressible flows: the incompressible solid is coupled to the compressible conveying gas. Neglecting the impulse current changes leads to simple force balances. In vertical or inclined conveying pipe elements, the weight force acting on the considered phase must be taken into account.

2.3.3 Energy Conservation

With $Q = E_i = m_i \cdot (i + \frac{1}{2} \cdot u^2 + g \cdot z)_i = m_i \cdot e_i^v$, i = mass-specific internal energy, z = elevation, $\dot{Q} = \dot{E}_i = \dot{m}_i \cdot (i + \frac{1}{2} \cdot u^2 + g \cdot z + \frac{p}{\varrho})_i = \dot{m}_i \cdot (h + \frac{1}{2} \cdot u^2 + g \cdot z)_i = \dot{m}_i \cdot e_i^f$, h = mass-related specific enthalpy, $\dot{R}_Q = (\dot{W}_e + \dot{Q}_e)_i$ = sum of all mechanical and thermal energies per time unit supplied to (+) or removed from (−) phase i in the form of mechanical ($\dot{W}_e$) and thermal ($\dot{Q}_e$) energies per time unit:

$$\frac{dE_i}{d\tau} = \dot{E}_{i,\text{in}} - \dot{E}_{i,\text{out}} + \dot{W}_{e,i} + \dot{Q}_{e,i} \tag{2.57}$$

$$\frac{d(m_i \cdot e_i^v)}{d\tau} = (\dot{m}_i \cdot e_i^f)_{\text{in}} - (\dot{m}_i \cdot e_i^f)_{\text{out}} + \dot{W}_{e,i} + \dot{Q}_{e,i} \tag{2.58}$$

$$\Delta V_C \cdot \frac{d(\varepsilon_i \cdot \varrho_i \cdot e_i^v)}{d\tau} = (\varepsilon_i \cdot A_R \cdot \varrho_i \cdot u_i \cdot e_i^f)_{\text{in}} - (\varepsilon_i \cdot A_R \cdot \varrho_i \cdot u_i \cdot e_i^f)_{\text{out}}$$

$$+\Delta V_C \cdot \varepsilon_i \cdot \varrho_i \cdot (\dot{w}_e + \dot{q}_e)_i \tag{2.59}$$

$(\dot{w}_e + \dot{q}_e)_i$ are the specific values of i related to the mass unit of $(\dot{W}_e + \dot{Q}_e)_i$. Note: In $\dot{W}_{e,i}$ and $\dot{Q}_{e,i}$ all those energies are included that act on the individual phases or are exchanged between them. In particular, $\dot{W}_{e,i}$ includes all those portions that are dissipated as heat, e.g., frictional forces multiplied by their effective distance or velocity. This takes into account non-isentropic behavior. An example is presented below.

For solid/gas pipe flows in steady state, i.e., at $dE_i/d\tau = 0$:

$$(\dot{m}_i \cdot e_i^f)_{\text{in}} - (\dot{m}_i \cdot e_i^f)_{\text{out}} + \dot{W}_{e,i} + \dot{Q}_{e,i} = 0 \quad \text{with: } \dot{m}_{i,\text{in}} = \dot{m}_{i,\text{out}} = \dot{m}_i \tag{2.60}$$

(2.60) applied to the solid in a horizontal conveying pipe element, i.e., at $z_{\text{in}} = z_{\text{out}}$, with the cross-sectional area A_R and the length dL, yields for an isothermal flow with $t_S = t_F = t_M$, see Sect. 2.1.4, and $\dot{Q}_{e,i} = 0$ the approach:

$$\dot{m}_S \cdot \left(h_{S,\text{in}} + \frac{1}{2} \cdot u_{s,\text{in}}^2 \right) - \dot{m}_S \cdot \left(h_{S,\text{out}} + \frac{1}{2} \cdot u_{s,\text{out}}^2 \right) + \dot{W}_{e,S} = 0 \tag{2.61}$$

With

$$h_{S,\text{in}} = h_{S,\text{out}} = c_{p,S} \cdot t_M,$$

$$\left(\frac{u_{s,\text{in}}^2}{2} - \frac{u_{s,\text{out}}^2}{2} \right) = -d\left(\frac{u_s^2}{2} \right) = -u_S \cdot du_S$$

as well as

$$\begin{aligned} \dot{W}_{e,S} &= F_{S,\text{drag}} \cdot u_S + (F_{S,p,\text{in}} - F_{S,p,\text{out}}) \cdot u_S - F_{S,f} \cdot u_S \\ &= (F_{S,\text{drag}} - (1 - \varepsilon_F) \cdot A_R \cdot dp - F_{S,f}) \cdot u_S \end{aligned}$$

inserted into (2.61) results in:

$$-\dot{m}_S \cdot du_S = (1 - \varepsilon_F) \cdot A_R \cdot dp - F_{S,\text{drag}} + F_{S,f} \tag{2.62}$$

(2.62) is identical to the one derived from a momentum balance (2.51). An analogous consideration yields for the conveying gas (2.55).

Note: The balance calculations shown above can be extended, depending on the flow pattern established in the conveying pipe, to gain further insights into the respective problem. This is indicated below using the example of pneumatic strand conveying through a horizontal pipe. Two flow areas form here: a dense strand with a high solid and low gas content above the pipe bottom, and a pneumatic conveying with a low solid but high gas content above it. Both areas can be balanced separately. Thus, four phases must be considered: solid and gas in the strand as well as solid and gas in the upper pneumatic conveying cross-section. Both areas are coupled via the forces transferred at their common boundary/separating surface. Suitable force approaches must be provided or developed for this purpose. In order to be able to solve the resulting system of equations for a conveying section, realistic boundary or initial conditions are also required, e.g., the proportion of the conveying pipe cross-section covered by the strand at the beginning of the line.

2.3.4 Calculation Example 2: Pressure Loss of a Compressible Gas Flow

The pressure loss of the mass flow $\dot{m}_F$ of an ideal gas through a horizontal, straight, hydraulically smooth pipe with a constant cross-sectional area $A_R = (\pi/4) \cdot D_R^2$ and the length L_R must be calculated. The flow is assumed to be isothermal with the temperature T_F and the pressure at the end of the line is given by p_{out}.

- Momentum balance on the pipe element dL, (2.55) with $F_{F,\text{drag}} = 0$ and $\varepsilon_F = 1$:

$$-\dot{m}_F \cdot du_F = A_R \cdot dp + F_{Ff}$$

- Approach for pipe friction force, (2.76), Sect. 2.5.1:

$$F_{Ff} = A_R \cdot \lambda_F \cdot \frac{1}{D_R} \cdot \frac{\varrho_F}{2} \cdot u_F^2 \cdot dL$$

with:
$\lambda_F = \lambda_F(Re)$ Friction coefficient = function of the Reynolds number Re

- Combination of the equations results in:

$$dp = -\left(\frac{\dot{m}_F}{A_R} \cdot du_F + \lambda_F \cdot \frac{1}{D_R} \cdot \frac{\varrho_F}{2} \cdot u_F^2 \cdot dL\right)$$

- From the continuity equation, $\dot{m}_F = (A_R \cdot \varrho_F \cdot u_F)$, and the ideal gas law, (2.3), it follows:

$$\varrho_F \cdot u_F^2 = (\varrho_F \cdot u_F)^2 \cdot \frac{1}{\varrho_F} = \left(\frac{\dot{m}_F}{A_R}\right)^2 \cdot \frac{R_F \cdot T_F}{p}$$

as well as:

$$u_F = u_{F,\text{out}} \cdot \frac{p_{\text{out}}}{p} = \left(\frac{\dot{m}_F}{A_R} \cdot R_F \cdot T_F\right) \cdot \frac{1}{p}$$

$$du_F = -\left(\frac{\dot{m}_F}{A_R} \cdot R_F \cdot T_F\right) \cdot \frac{dp}{p^2}$$

- Inserting the above equations into the basic equation results in:

$$dp = \left(\frac{\dot{m}_F}{A_R}\right)^2 \cdot R_F \cdot T_F \cdot \frac{dp}{p^2} - \frac{1}{2} \cdot \frac{\lambda_F}{D_R} \cdot \left(\frac{\dot{m}_F}{A_R}\right)^2 \cdot \frac{R_F \cdot T_F}{p} \cdot dL$$

- Before integrating the equation, the change in the pipe friction coefficient $\lambda_F(Re)$ along the conveying path must be checked and, if necessary, incorporated. As can be seen from the Reynolds number

$$Re = \frac{u_F \cdot D_R \cdot \varrho_F}{\eta_F} = \frac{\dot{m}_F}{A_R} \cdot \frac{D_R}{\eta_F} \quad \text{with: } \eta_F \neq \eta_F(p), \text{ if } p \lesssim 10\,\text{bar}$$

this does not change along the conveying path for the present isothermal flow. Thus, it holds: $\lambda_F = \text{const.}$

- By variable separation, the above differential equation can be integrated. The result, transformed into a dimensionless representation, is:

$$\frac{p_{\text{in}}^2 - p_{\text{out}}^2}{(\frac{\dot{m}_F}{A_R})^2 \cdot R_F \cdot T_F} - 2 \cdot \ln\left(\frac{p_{\text{in}}}{p_{\text{out}}}\right) = \lambda_F \cdot \frac{L_R}{D_R}$$

  The pressure difference $\Delta p_R = (p_{\text{in}} - p_{\text{out}})$ in the pipeline or the pressure p_{in} at its beginning ($\rightarrow$ given the end pressure p_{out}) can thus only be determined iteratively.

- In order to determine the pressure curve along the conveying line, instead of the fixed values (p_{in}, L_R) the corresponding variables (p_x, L_x) must be inserted into the above determination equation. The same applies to the velocity curve. Here, additionally, p_x must be replaced by

$$p_x = \frac{\dot{m}_F}{A_R} \cdot R_F \cdot T_F \cdot \frac{1}{u_{F,x}}$$

The pressure loss $\Delta p_R = (p_{\text{in}} - p_{\text{out}})$ consists of two parts in this example: one part to overcome the pure friction loss and a second part for the permanent acceleration of the gas due to its expansion in the conveying direction. Under the operating conditions of pneumatic conveying, the contribution to gas acceleration is generally negligibly small. However, the acceleration loss of the solid must always be taken into account due to its greater mass.

2.4 Scale-up, Dimensional Analysis

This section provides a condensed introduction to the theory of model transfer and the necessary dimensional analysis. For a deeper understanding of the relationships, reference is made, among others, to [45–50].

2.4.1 Model Transfer

A common task in engineering practice is to design a planned large-scale plant based on results from laboratory or pilot plant investigations that are yet to be determined or already available. Alternatively, the case may arise where an existing operating plant is not running satisfactorily and is to be optimized through suitable model-scale experiments. The transfer of existing experimental results to the plant to be dimensioned or the design of representative model experiments within the framework of plant optimization is based on similarity theory/model theory. This states that

> Model and large-scale execution are only similar in their behavior if, under similar boundary conditions, the relevant dimensionless characteristic numbers for the model and large-scale execution have the same numerical values.

Both geometric and mechanical, chemical, and thermal similarity are required here. The determination of the corresponding characteristic numbers is carried out using dimensional analysis, which, from the functional relationship of the process variables

$$y_1 = f(y_2, \ldots, y_n) \quad \text{with: } y_1 = \text{Target} \tag{2.63}$$

a functional relationship between dimensionless characteristic variables

$$\Pi_1 = F(\Pi_2, \ldots, \Pi_m) \quad \text{with: } \Pi_1 = \text{dimensionless target variable} \tag{2.64}$$

can be generated. In this case,

$$m < n \tag{2.65}$$

i.e., the number of dimensionless characteristic numbers is less than that of the dimensioned process variables. This results in a significantly reduced experimental effort, as only the "m" dimensionless characteristic variables ($\Pi_1, \ldots, \Pi_m$) need to be systematically measured. Moreover, it is not necessary, especially in complex problems, for an arithmetic combination of the relevant y_i- or Π_i-variables to be present. It is sufficient for the Π_i-values in the area of interest to determine and transfer with constant values to the target system. Process engineering relationships can be captured by measuring dependencies of the Π_i-values, generally in the form of power products. The flexibility of the experimental design is increased by the fact that the dimensionless numbers represent products of several process variables, which can be set separately to achieve a given value range. Example: The Reynolds number $Re = (u \cdot D_R \cdot \varrho)/\eta$ can be achieved by a suitable choice of geometry ($\rightarrow D_R$), operating conditions ($\rightarrow u$) and the type of flowing medium ($\rightarrow \varrho, \eta$). Dimensionless numbers can also be interpreted as ratios of forces, heat flows, etc. This leads to a deeper understanding of the underlying relationships. Overall, the dimensionless representation provides a compressed description of the problem at hand.

Not always are all determined dimensionless numbers to be maintained simultaneously for the model and operating system. If only a part of the common numbers has the same numerical values, this is called partial similarity. Which numbers are then most relevant/important for the considered process and must be kept constant can only be decided based on the knowledge of the physical relationships.

Often problematic is also the dimensional-analytical treatment of variable material properties, e.g., as a result of a temperature field in the system. For this, the reader is referred to the literature.

2.4.2 Dimensional Analysis

Every physical quantity consists of a numerical value multiplied by a unit of measurement, e.g., diameter: $D_R = 0.2\,\text{m}$, mass flow: $\dot{m} = 3\,\text{kg/s}$, energy: $E = 10\,\text{J} = 10\,\text{N m} = 10\,(\text{kg m}^2)/\text{s}^2$. Various measurement systems/units are possible.

Table 2.3 Base units and base dimensions of the SI system

Base quantity	Base dimension	Base unit	
Length	L	m	Meter
Mass	M	kg	Kilogram
Time	T	s	Second
Thermodynamic temperature	Θ	K	Kelvin
Amount of substance	N	mol	Mole
Electric current	I	A	Ampere
Luminous intensity	Iv	Cd	Candela

Table 2.4 Derived units and their dimensions

Secondary size	Dimension	Unit of measurement	Abbreviation
Force	$\mathrm{M\,L\,T^{-2}}$	$\mathrm{kg\,m\,s^{-2} = N}$	Newton
Pressure	$\mathrm{M\,L^{-1}T^{-2}}$	$\mathrm{kg\,m^{-1}\,s^{-2} = Pa}$	Pascal
Energy	$\mathrm{M\,L^2\,T^{-2}}$	$\mathrm{kg\,m^2\,s^{-2} = J}$	Joule
Power	$\mathrm{M\,L^2\,T^{-3}}$	$\mathrm{kg\,m^2\,s^{-3} = W}$	Watt

A length can be given in meters, millimeters, inches, miles, light years, etc. There are defined basic units of measurement and derived units of measurement. The basic quantities used here are those of the SI system (→ Système International d'Unités), see Table 2.3. Derived units are exemplified in Table 2.4. They result from power products of the basic units.

To qualitatively describe the physical properties of a situation, the "dimension" is introduced. It is independent of the chosen measurement system and its specific units and indicates which types of quantities describe the respective problem at hand. Table 2.3 contains the basic dimensions, Table 2.4 shows examples of dimensions for derived units of measurement. Notation: The pipe diameter D_R has the dimension of a length, written: $[D_R] = \mathrm{L}$, for the control volume it applies: $[\Delta V_R] = \mathrm{L}^3$, for the conveying gas velocity: $[u_F] = \mathrm{LT}^{-1}$. Relevant for the topic "pneumatic conveying" considered here are, in general, only the basic types of quantities length L, mass M, and time T. The thermodynamic temperature Θ is not required if a constant mixture temperature and thus temperature-independent material values are assumed in the conveying pipe.

From the relationship represented by the dimensionally homogeneous size equation (2.63), the dimensionless characteristic numbers of (2.64) follow as power products of the process variables.y_i with the exponents x_i:

$$\Pi_i = y_1^{x_1} \cdot y_2^{x_2} \cdot \dots \cdot y_n^{x_n} \tag{2.66}$$

The exponents x_i must be chosen in such a way that

$$[\Pi_i] = 1 \tag{2.67}$$

results. The following is based on the basic dimensions L, M, T, and Θ. The exponents of the base units in the units of measurement are denoted with a_{ik}. This results in:

$$\begin{aligned} y_1 &= \{y_1\} \cdot [\mathrm{L}^{a_{11}} \cdot \mathrm{M}^{a_{12}} \cdot \mathrm{T}^{a_{13}} \cdot \Theta^{a_{14}}] \\ y_2 &= \{y_2\} \cdot [\mathrm{L}^{a_{21}} \cdot \mathrm{M}^{a_{22}} \cdot \mathrm{T}^{a_{23}} \cdot \Theta^{a_{24}}] \\ &\dots \\ y_n &= \{y_n\} \cdot [\mathrm{L}^{a_{n1}} \cdot \mathrm{M}^{a_{n2}} \cdot \mathrm{T}^{a_{n3}} \cdot \Theta^{a_{n4}}] \end{aligned} \tag{2.68}$$

with:

$\{y_i\}$ Numerical value of y_i.

(2.68) introduced in (2.66) and the exponents belonging to the same basic dimensions combined yields:

$$\begin{aligned} y_1^{x_1} \cdot y_2^{x_2} \cdot \dots \cdot y_n^{x_n} = &\{y_1\} \cdot \{y_2\} \cdot \dots \cdot \{y_n\} \\ &\cdot [\mathrm{L}^{a_{11} \cdot x_1 + a_{21} \cdot x_2 + \dots + a_{n1} \cdot x_n} \cdot \mathrm{M}^{a_{12} \cdot x_1 + a_{22} \cdot x_2 + \dots + a_{n2} \cdot x_n} \\ &\cdot \mathrm{T}^{a_{13} \cdot x_1 + a_{23} \cdot x_2 + \dots + a_{n3} \cdot x_n} \cdot \Theta^{a_{14} \cdot x_1 + a_{24} \cdot x_2 + \dots + a_{n4} \cdot x_n}] \end{aligned} \tag{2.69}$$

(2.69) will only become dimensionless if the individual sums of the exponents of all basic units/dimensions become zero. This requires the determination of linearly independent solutions of the linear homogeneous system of equations:

$$\begin{aligned} a_{11} \cdot x_1 + a_{21} \cdot x_2 + \dots + a_{n1} \cdot x_n &= 0 \\ a_{12} \cdot x_1 + a_{22} \cdot x_2 + \dots + a_{n2} \cdot x_n &= 0 \\ a_{13} \cdot x_1 + a_{23} \cdot x_2 + \dots + a_{n3} \cdot x_n &= 0 \\ a_{14} \cdot x_1 + a_{24} \cdot x_2 + \dots + a_{n4} \cdot x_n &= 0 \end{aligned} \tag{2.70}$$

Instructions for handling such systems of equations can be found in the relevant literature, e.g., [51]. It is evident that the number of equations of (2.70) is identical to the number of basic dimensions determining the problem. The system of equations (2.70) provides $(n - r)$ independent solutions, with: r = rank of the coefficient matrix a_{ik}, see Table 2.5, and thus determines the number of dimensionless quantities with which the respective problem is completely described. Usually, the rank r of the coefficient matrix corresponds to the number d of basic dimensions of the problem. The number of dimen-

Table 2.5 Dimension-/Coefficient matrix

	Core matrix			Residual matrix		
	y_1	y_2	y_3	y_4	y_5	y_6
L	a_{11}	a_{12}	a_{13}	a_{14}	a_{15}	a_{16}
M	a_{21}	a_{22}	a_{23}	a_{24}	a_{25}	a_{26}
T	a_{31}	a_{32}	a_{33}	a_{34}	a_{35}	a_{36}

sionless quantities/groups m is then obtained as the difference between the n process variables and the d basic dimensions:

$$m = n - d \tag{2.71}$$

Exceptions to this rule are rare, but possible ($r < d$), and should be checked [51]. To determine the solutions of the system of equations (2.70), for r (d) equations with n unknowns, $m = (n - r)$ unknown x_i values must be specified. Since the target variable y_1 generally should only appear in one of the dimensionless numbers, the associated exponent x_1 for one of the solutions to $x_1 = 1$, for the remaining ones to $x_1 = 0$ can be set. By choosing the free exponents x_i any number of dimensionless groups can be formed, all of which equivalently describe the present situation. Products of powers of the determined m independent characteristic numbers and thus new characteristic numbers can be formed. Example: $\Pi_i^* = \Pi_i \cdot \Pi_k^2$. The new characteristic number Π_i^* replaces Π_i, Π_k remains in the set of key figures, as its scope m is fixed. In this way, for example, known key figures or key figures whose process variables can be measured more accurately can be generated.

The determined Π_{n-r} dimensionless numbers are functionally linked according to (2.64). Dimensional analysis does not provide any information about the type/form of this function. Often, the relationships are represented as power products

$$\Pi_1 = K \cdot \Pi_2^a \cdot \Pi_3^b \cdot \ldots \cdot \Pi_m^l \quad \text{with: } K = \text{Constant} \tag{2.72}$$

The above-described procedure for determining the Π_i-values represents the classical method. In general, a faster result can be achieved with the more formalized procedure according to [45], which will be briefly introduced below.

For example, a problem with $n = 6$ process variables and $d = r = 3$ basic dimensions is considered. This leads to $m = (6 - 3) = 3$ dimensionless numbers. From the dimensions of the process variables y_i the dimension matrix, which corresponds to the coefficient matrix of (2.70), is formed, see Table 2.5. In its header row, the individual influencing variables y_i, in the first column the occurring base dimensions, here L, M, T. The intersections of rows and columns contain the exponents a_{ik}, with which the base dimensions/units appear in the dimension/unit of measurement of the respective variable. Example: y_1 = gas velocity u_F, $[u_F] = \mathrm{L}^1\,\mathrm{M}^0\,\mathrm{T}^{-1} \rightarrow a_{11} = 1, a_{21} = 0, a_{31} = -1$.

This dimension matrix, as shown in Table 2.5, is divided into a square core matrix and a residual matrix. The process variables of the residual matrix appear in the numerator of only one of the dimensionless characteristic numbers to be determined, while the elements of the core matrix serve as "filling variables" and appear in the denominators of all characteristic numbers. The residual matrix should therefore be filled with the essential process variables, in particular the target variable and the parameters whose influence is still unclear. If one of the latter proves to be unimportant for the present problem, e.g., showing no or only negligibly small influence in model tests, the characteristic number

Table 2.6 Core matrix as unit matrix and modified residual matrix

	Core matrix			Residual matrix		
	y_1	y_2	y_3	y_4	y_5	y_6
L	1	0	0	a_{14}^*	a_{15}^*	a_{16}^*
M	0	1	0	a_{24}^*	a_{25}^*	a_{26}^*
T	0	0	1	a_{34}^*	a_{35}^*	a_{36}^*

in which this process variable (in the numerator) appears can be omitted without replacement [47].

In a next step, the core matrix must be transformed into a unit matrix using suitable linear transformations, which has ones on its main diagonal and otherwise only zero values, see Table 2.6. In this process, the exponents a_{ik} of the residual matrix change to a_{ik}^*. The application of the Gaussian algorithm [51] additionally provides the rank r of the dimension matrix (zero-free main diagonal, only zeros below).

Once this representation is achieved, the dimensionless numbers Π_i can be formed according to the following scheme: The process variables of the residual matrix are successively divided by the process variables of the unit matrix raised to the corresponding exponents a_{ik}^*. The exponents assigned to the process variables of the unit matrix a_{ik}^* are located in the column below the considered process variable of the residual matrix and are taken for the process variables of the unit matrix along the unit diagonal. Below are the three dimensionless characteristic values resulting from Table 2.6, which illustrate the procedure:

$$\Pi_1 = \frac{y_4}{y_1^{a_{14}^*} \cdot y_2^{a_{24}^*} \cdot y_3^{a_{34}^*}}, \quad \Pi_2 = \frac{y_5}{y_1^{a_{15}^*} \cdot y_2^{a_{25}^*} \cdot y_3^{a_{35}^*}}, \quad \Pi_3 = \frac{y_6}{y_1^{a_{16}^*} \cdot y_2^{a_{26}^*} \cdot y_3^{a_{36}^*}} \tag{2.73}$$

These key figures can, as already demonstrated, be combined with each other. Calculation example 3, Sect. 2.4.3, shows the practical application.

Essential for the success of a dimensional analysis is the knowledge of the process variables influencing the respective problem, which are compiled in a so-called relevance list. It is recommended to work through this list with regard to the following aspects/criteria:

- Target variable, e.g. pressure loss,
- geometric influencing variables, e.g. conveying pipe diameter,
- material influencing variables, e.g. gas and particle density, particle diameter,
- process-related influencing variables, e.g. bulk material throughput, loading,
- other influencing variables/dimensioned constants, e.g. gravitational acceleration.

Not considering influencing variables leads to incorrect results, superfluous influencing variables lead to increased experimental effort to identify and eliminate them. The development of the relevance list requires physical understanding, experience, and great care.

2.4.3 Calculation Example 3: Model Transfer of Pneumatic Conveying Experiments

In a reactor pneumatically loaded with bulk material, a process engineering process takes place, which is to be modified/optimized. For this purpose, the previous used conveying gas flow must be reduced as much as possible. Investigations in this regard on the operating plant (P) are not possible and must therefore be carried out on a test plant (M). For this purpose, it must be clarified,

a) which dimensionless numbers are relevant for the present problem,
b) with which process parameters the tests are to be carried out specifically,
c) how the test results from (M) are to be transferred to (P).

The operating conveying system is a positive pressure conveying system with a horizontal, unstepped pipeline, which has a 30 m long straight start-up section behind the bulk material lock. The smallest, i.e. critical, conveying gas velocity is the one at the beginning of the pipeline. If the gas velocity is successively reduced, then below a limit/minimum velocity $v_{F,\text{in,min}}$ of the conveying gas, no solid transport is possible anymore: The mass flow $\dot{m}_S$ decreases with decreasing gas velocity towards zero. $v_{F,\text{in,min}}$ is the absolute conveying limit to be determined here. Details on this conveying limit are discussed in Sect. 4.6.2.

Relevant data of the operating plant:

- Conveying pipe diameter: $D_R^{(P)} = 300\,\text{mm}$,
- Conveying pressure at the beginning of the pipeline: $p_{R,\text{in}}^{(P)} = 3.0\,\text{bar}$,
- Temperature at the beginning of the pipeline $\cong$ mixture temperature, (2.13): $T_{\text{in}}^{(P)} \cong T_M^{(P)} = 293\,\text{K}$,
- Conveying gas$^{(P)}$: air,
- Bulk material data:
  average particle diameter: $d_{S,50}^{(P)} = 60\,\mu\text{m}$,
  Particle- = solid density: $\varrho_P^{(P)} = 2700\,\text{kg/m}^3$,
  Dispersion parameter of the particle size distribution, (2.26): $n^{(P)} = 1.2$.

For a) Dimensional analysis:

- Relevance list:
  Practical experience shows that the pipe diameter D_R, the particle size $d_{S,50}$ as well as the gas state at the beginning of the pipeline, in particular the gas density $\varrho_{F,\text{in}}$, have a significant influence on $v_{F,\text{in,min}}$. Using the ideal gas law, (2.3), $\varrho_{F,\text{in}} = \varrho_{F,\text{in}}(p_{R,\text{in}}, T_{\text{in}})$, the process variables are determined by $\varrho_{F,\text{in}}\,(p_{R,\text{in}}, T_{\text{in}})$. Thus, the following applies:

- Target variable: $v_{F,\mathrm{in,min}}$,
- geometric influencing variables: D_R,
- material influencing factors: Gas: $\varrho_{F,\mathrm{in}}$, $\eta_{F,\mathrm{in}}$, Solid: ϱ_P, $d_{S,50}$, n, ψ, ξ_S,
- process-related influencing variables: $p_{R,\mathrm{in}}$, $T_{\mathrm{in}} \rightarrow$ not relevant, as covered by $\varrho_{F,\mathrm{in}}$,
- other influencing variables/dimensioned constants: $g = 9.81\ \mathrm{m/s^2}$.

With the influencing variable ξ_S the resistance behavior of the bulk material, with the sphericity ψ its grain shape should be described. Based on experiences with similar problems, it is known that the sizes (n, ψ, ξ_S) have such a small influence on the result that they can be neglected. Therefore, consideration is not considered necessary, but should be checked if necessary.

Note: The quantities (n, ψ, ξ_S) are already dimensionless numbers. Therefore, they do not need to be included in the dimension matrix.

- Dimension matrix:
  The resulting dimension matrix is shown in the upper part of Table 2.7.
  In the parts following the dimension matrix in Table 2.7, the core matrix is transformed into a unit matrix. Here, $ZX =$ row X.
- Dimensionless numbers:
  The core matrix shows that the rank of the dimension matrix $r = 3$ is equal to the number of basic dimensions d. The $n = 7$ influencing process variables can thus be reduced to $m = (n - r) = (7 - 3) = 4$ dimensionless characteristics to be investigated. From the lower part of Table 2.7, rows Z7–Z9, it follows:

Table 2.7 Dimension matrix (the upper three rows) and transformations of the core matrix into a unit matrix

	$d_{S,50}$	ϱ_P	g	$v_{F,\mathrm{in,min}}$	D_R	$\varrho_{F,\mathrm{in}}$	$\eta_{F,\mathrm{in}}$	
L	1	−3	1	1	1	−3	−1	
M	0	1	0	0	0	1	1	
T	0	0	−2	−1	0	0	−1	
Z1	1	0	1	1	1	0	2	$= \mathrm{L} + 3 \cdot \mathrm{M}$
Z2	0	1	0	0	0	1	1	$= \mathrm{M}$
Z3	−1	0	1	0	−1	0	−1	$= -\mathrm{T} - \mathrm{Z1}$
Z4	1	0	0	1/2	1	0	3/2	$= 1/2 \cdot (\mathrm{Z1} - \mathrm{Z3})$
Z5	0	1	0	0	0	1	1	$= \mathrm{Z2}$
Z6	−1	0	1	0	−1	0	−1	$= \mathrm{Z3}$
Z7	1	0	0	1/2	1	0	3/2	$= \mathrm{Z4}$
Z8	0	1	0	0	0	1	1	$= \mathrm{Z5}$
Z9	0	0	1	1/2	0	0	1/2	$= \mathrm{Z4} + \mathrm{Z6}$

$$\Pi_1 = \frac{v_{F,\text{in, min}}}{d_{S,50}^{1/2} \cdot \varrho_P^0 \cdot g^{1/2}} = \frac{v_{F,\text{in, min}}}{\sqrt{g \cdot d_{S,50}}}$$

$$\Pi_2 = \frac{D_R}{d_{S,50}^1 \cdot \varrho_P^0 \cdot g^0} = \frac{D_R}{d_{S,50}}$$

$$\Pi_3 = \frac{\varrho_F}{d_{S,50}^0 \cdot \varrho_P^1 \cdot g^0} = \frac{\varrho_F}{\rho_P}$$

$$\Pi_4 = \frac{\eta_F}{d_{S,50}^{3/2} \cdot \varrho_P^1 \cdot g^{1/2}} = \sqrt{\frac{\eta_F^2}{d_{s,50}^3 \cdot \varrho_P^2 \cdot g}}$$

To work with key figures that are common in bulk material technology and also conform to existing experience, the following transformations are made

$$\Pi_1^* = \Pi_1^2 \cdot \Pi_2^{-1} = \frac{v_{F,\text{in, min}}^2}{g \cdot d_{S,50}} \cdot \frac{d_{S,50}}{D_R} = \frac{v_{F,\text{in, min}}^2}{g \cdot D_R} = Fr_{R,\text{min}} \rightarrow \text{Froudezahl}$$

$$\Pi_2^* = \Pi_2 = \frac{D_R}{d_{S,50}} \rightarrow \text{Durchmesserverhaltnis}$$

$$\Pi_3^* = \Pi_3 = \frac{\varrho_{F,\text{in}}}{\varrho_P} \rightarrow \text{Dichteverhaltnis}$$

$$\Pi_4^* = \Pi_4^{-2} \cdot \Pi_3 = \frac{d_{S,50}^3 \cdot \varrho_P^2 \cdot g}{\eta_F^2} \cdot \frac{\varrho_{F,in}}{\varrho_P}$$

$$= \frac{d_{S,50}^3 \cdot g}{\eta_F^2} \cdot \varrho_{F,in} \cdot \varrho_P = Ar \rightarrow \text{Archimedeszahl}$$

In the Archimedes number Ar is $\varrho_{F,\text{in}} \cdot (\varrho_P - \varrho_{F,\text{in}})$ replaced by $\varrho_{F,\text{in}} \cdot \varrho_P$ since $\varrho_P \gg \varrho_{F,\text{in}}$ is. Thus, the following applies:

$$Fr_{R,\text{min}} = F\left(\frac{D_R}{d_{S,50}}, \frac{\varrho_{F,\text{in}}}{\varrho_P}, Ar\right)$$

For b) Experiment execution:

- Operating parameters of the plant:
  Conveying gas$^{(P)}$: Air,
  Gas density, (2.4), reference state: $\varrho_{F,0} = 1.20\,\text{kg/m}^3$, $p_{F,0} = 1.0\,\text{bar}$, $T_{F,0} = 293\,\text{K}$:

$$\varrho_{F,\text{in}}^{(P)} = \varrho_{F,0} \cdot \frac{p_{R,\text{in}}^{(P)}}{p_{F,0}} \cdot \frac{T_{F,0}}{T_{\text{in}}^{(P)}} = 1.20\,\frac{\text{kg}}{\text{m}^3} \cdot \frac{3.0\,\text{bar}}{1.0\,\text{bar}} \cdot \frac{293\,\text{K}}{293\,\text{K}} = 3.60\,\frac{\text{kg}}{\text{m}^3}$$

dynamic viscosity: $\eta_{F,\text{in}}^{(p)} = 18.26 \cdot 10^{-6}\,\text{Pa s}$.

With these values, the following results are obtained for the independent parameters:

$$\left(\frac{D_R}{d_{S,50}}\right)^{(P)} = \frac{300\,\text{mm}}{60\cdot 10^{-3}\,\text{mm}} = 5000$$

$$\left(\frac{\varrho_{F,\text{in}}}{\varrho_P}\right)^{(P)} = \frac{3.60\,\frac{\text{kg}}{\text{m}^3}}{2700\,\frac{\text{kg}}{\text{m}^3}} = \frac{1}{750.0}$$

$$Ar^{(P)} = \frac{d_{S,50}^3\cdot g}{\eta_{F,\text{in}}^2}\cdot \varrho_{F,\text{in}}\cdot\varrho_P$$

$$= \frac{(60\cdot 10^{-6}\,\text{m})^3\cdot 9.81\,\frac{\text{m}}{\text{s}^2}}{(18.26\cdot 10^{-6}\,\text{Pa s})^2}\cdot 3.60\,\frac{\text{kg}}{\text{m}^3}\cdot 2700\,\frac{\text{kg}}{\text{m}^3} = 61.77$$

- Test facility data:
  Pipe with sufficiently long horizontal straight starting section,
  Conveying pipe diameter: $D_R^{(M)} = 100\,\text{mm}$,
  Conveying gas$^{(M)}$: Air, alternatively nitrogen available,
  Mixture temperature: $T_\text{in}^{(M)} = T_\text{in}^{(P)} = 293\,\text{K}$.
- Possible test settings:
  Maintaining the dimensionless parameters of the large-scale plant (*P*) also in the test (*M*) is difficult, as, for example, the particle diameter $d_{S,50}$ occurs both in the first and third power and this influence cannot be easily compensated by another conveying gas or bulk material, as shown below. This leads to partial similarities. Therefore, the following cases are examined:
- Test execution 1: The same operating settings and bulk material as in the operating system are used, i.e., conveying gas = air, $d_{S,50}^{(M)} = d_{S,50}^{(P)}$, $\varrho_P^{(M)} = \varrho_P^{(P)}$, $p_{R,\text{in}}^{(M)} = p_{R,\text{in}}^{(P)}$, $T_\text{in}^{(M)} = T_\text{in}^{(P)}$, $\varrho_{F,\text{in}}^{(M)} = \varrho_{F,\text{in}}^{(P)}$, $\eta_{F,\text{in}}^{(M)} = \eta_{F,\text{in}}^{(P)}$. From this, the following dimensionless parameters result:

$$\left(\frac{D_R}{d_{S,50}}\right)^{(M)} = \frac{100\,\text{mm}}{60\cdot 10^{-3}\,\text{mm}} = 1666.7 = \frac{1}{3}\cdot\left(\frac{D_R}{d_{S,50}}\right)^{(P)} \neq \left(\frac{D_R}{d_{S,50}}\right)^{(P)}$$

$$\left(\frac{\varrho_{F,\text{in}}}{\varrho_P}\right)^{(M)} = \left(\frac{\varrho_{F,\text{in}}}{\varrho_P}\right)^{(P)}$$

$$Ar^{(M)} = Ar^{(P)}$$

- Experiment 2: The bulk material used in the operating system is ground to $d_{S,50}^{(M)} = 20\,\mu\text{m}$ in size. The remaining process variables are set as in Experiment 1. This results in:

$$\left(\frac{D_R}{d_{S,50}}\right)^{(M)} = \left(\frac{D_R}{d_{S,50}}\right)^{(P)}$$
$$\left(\frac{\varrho_{F,\text{in}}}{\varrho_P}\right)^{(M)} = \left(\frac{\varrho_{F,\text{in}}}{\varrho_P}\right)^{(P)}$$
$$Ar^{(M)} = \frac{d_{S,50}^3 \cdot g}{\eta_{F,\text{in}}^2} \cdot \varrho_{F,\text{in}} \cdot \varrho_P = \frac{(20 \cdot 10^{-6})^3\,\text{m}^3 \cdot 9.81\,\frac{\text{m}}{\text{s}^2}}{(18.26 \cdot 10^{-6}\,\text{Pa s})^2} \cdot 3.60\,\frac{\text{kg}}{\text{m}^3} \cdot 2700\,\frac{\text{kg}}{\text{m}^3}$$
$$= 2.29 = \frac{1}{27} \cdot Ar^{(P)} \neq Ar^{(P)}$$

- Experiment 3: Experiments 1 and 2 only show partial similarities. To set all three independent dimensionless numbers identically in the experimental and production plants, the following procedure is required.
  With $d_{S,50}^{(M)} = 20\,\mu\text{m}$ results in:

$$\Pi_2^* = \left(\frac{D_R}{d_{S,50}}\right)^{(M)} = \left(\frac{D_R}{d_{S,50}}\right)^{(P)} = 5000$$

Consequence for $\Pi_4^* = Ar$:

$$\frac{Ar^{(M)}}{Ar^{(P)}} = 1 = \left(\frac{1}{3}\right)^3 \cdot \left(\frac{\eta_F^{(P)}}{\eta_F^{(M)}}\right)^2 \cdot \left(\frac{\varrho_{F,\text{in}}^{(M)}}{\varrho_{F,\text{in}}^{(P)}}\right) \cdot \left(\frac{\varrho_P^{(M)}}{\varrho_P^{(P)}}\right)$$
$$= \frac{1}{27} \cdot \left(\frac{\eta_F^{(P)}}{\eta_F^{(M)}}\right)^2 \cdot \left(\frac{\varrho_{F,\text{in}}^{(M)}}{\varrho_{F,\text{in}}^{(P)}}\right) \cdot \left(\frac{\varrho_P^{(M)}}{\varrho_P^{(P)}}\right)$$

that is, by changing the solid and/or gas medium or state, the factor $1/27$ must be compensated. Assumption: There is a ground solid with $d_{S,50}^{(M)} = 20\,\mu\text{m}$ and $\varrho_P^{(M)} = 2 \cdot \varrho_P^{(P)}$ available. This implies:

$$1 = \frac{1}{13.5} \cdot \left(\frac{\eta_F^{(P)}}{\eta_F^{(M)}}\right)^2 \cdot \left(\frac{\varrho_{F,\text{in}}^{(M)}}{\varrho_{F,\text{in}}^{(P)}}\right) \tag{2.74}$$

Compliance with the performance indicator $\Pi_3^* = \varrho_{F,\text{in}}/\varrho_P$ requires:

$$\frac{(\frac{\varrho_{F,\text{in}}}{\varrho_P})^{(M)}}{(\frac{\varrho_{F,\text{in}}}{\varrho_P})^{(P)}} = 1 = \left(\frac{\varrho_{F,\text{in}}^{(M)}}{\varrho_{F,\text{in}}^{(P)}}\right) \cdot \left(\frac{\varrho_P^{(P)}}{\varrho_P^{(M)}}\right)$$

This leads to $\varrho_P^{(M)} = 2 \cdot \varrho_P^{(P)}$ on

$$\frac{\varrho_{F,\text{in}}^{(M)}}{\varrho_{F,\text{in}}^{(P)}} = \left(\frac{\varrho_P^{(P)}}{\varrho_P^{(P)}}\right) = 2 \tag{2.75}$$

In the (2.74) for the ratio of Ar-numbers inserted, this results in:

$$1 = \frac{1}{6.75} \cdot \left(\frac{\eta_F^{(P)}}{\eta_F^{(M)}} \right)^2$$

Thus, to achieve the condition $Ar = \text{idem}$ for the experimental setup, a carrier gas with a dynamic viscosity of

$$\eta_F^{(M)} = \frac{\eta_F^{(P)}}{\sqrt{6.75}} = 0.385 \cdot \eta_F^{(P)}$$

must be provided. So here: $\eta_F^{(M)} = 0.385 \cdot 18.26 \cdot 10^{-6} \text{Pa s} = 7.03 \cdot 10^{-6}\,\text{Pa s}$. From the dimensionless number Π_3^*, or (2.75), it also follows that the operating condition of $\varrho_{F,\text{in}}^{(M)}$ during the experiments accordingly

$$\frac{\varrho_{F,\text{in}}^{(M)}}{\varrho_{F,\text{in}}^{(P)}} = 2 = \left(\frac{p_{R,\text{in}}^{(M)}}{p_{R,\text{in}}^{(P)}} \right) \cdot \left(\frac{T_{\text{in}}^{(P)}}{T_{\text{in}}^{(M)}} \right) \cdot \left(\frac{R_F^{(P)}}{R_F^{(M)}} \right)$$

must be adjusted. With the operating data of the production plant $\varrho_{F,\text{in}}^{(P)}$, i.e. $p_{R,\text{in}}^{(P)}$, $T_{\text{in}}^{(P)}$, $R_F^{(P)}$, as well as the chosen carrier gas $R_F^{(M)}$ and the temperature $T_{\text{in}}^{(M)}$, the delivery pressure $p_{R,\text{in}}^{(M)}$ can be calculated for the experiments.

- The implementation of experimental procedure 3-complete similarity-is practically hardly feasible for technical and cost reasons. To adjust the viscosity $\eta_F^{(M)}$ an exotic conveying gas, e.g. hydrogen or ammonia, with the corresponding handling problems would be required. This would, among other things, place special safety requirements on the test facility, and in the case of a ground-up different bulk material, possibly further dimensionless parameters, e.g. the sphericity ψ, could become relevant. The model tests are therefore carried out simplifyingly under partial similarity.
- In experimental procedure 1, the diameter ratio $(D_R/d_{S,50})^{(M)} = 1/3 \cdot (D_R/d_{S,50})^{(P)}$ does not fulfill the similarity conditions, while in experimental procedure 2, the Ar-number $Ar^{(M)} = 1/27 \cdot Ar^{(P)}$ does not meet the similarity conditions. The Ar-number can be interpreted as a product of the relevant inertial, weight, and viscous forces and thus ultimately describes the considered gas/solid system more completely than the diameter ratio. $(D_R/d_{S,50})$. This is with $(D_R/d_{S,50})^{(M)} = 1666.7$ also in a range that does not expect significant influence on the test result. The model tests are therefore carried out according to test procedure 1.
- Since a conveying limit is determined with $v_{F,\text{in,min}}$, the flow patterns and their throughput fluctuations influencing the reactor operation should also be determined during the stepwise reduction of the gas velocity. An indicator for this is the easily measurable pressure pulsations in the delivery system. The irregularities in the bulk material flow determine the operating speed that can be practically realized in the production plant.

For c) Result transfer:

The conveying gas limit velocity measured in the test according to test procedure 1 is $v_{F,\text{in},\text{min}} = 3.75\,\text{m/s}$. From this, the $Fr^{(M)}_{R,\text{min}}$-number of the experiment to:

$$Fr^{(M)}_{R,\,\text{min}} = \left(\frac{v^2_{F,\,\text{in},\,\text{min}}}{g \cdot D_R}\right)^{(M)} = \left(\frac{\left(3.75\,\frac{\text{m}}{\text{s}}\right)^2}{9.81\,\frac{\text{m}}{\text{s}^2} \cdot 0.100\,\text{m}}\right)^{(M)} = 14.335$$

From the equation $Fr^{(P)}_{R,\text{min}} = Fr^{(M)}_{R,\text{min}} = 14.335$ then follows for the conveying limit in the delivery pipe $D^{(P)}_R$ of the production facility:

$$v^{(P)}_{F,\,\text{in},\,\text{min}} = \sqrt{Fr^{(P)}_{R,\,\text{min}} \cdot g \cdot D^{(P)}_R} = \sqrt{14.335 \cdot 9.81\,\frac{\text{m}}{\text{s}^2} \cdot 0.300\,\text{m}}$$

$$v^{(P)}_{F,\,\text{in},\,\text{min}} = 6.50\,\frac{\text{m}}{\text{s}}$$

2.5 Pressure Losses of Single-phase Fluids

Only the well-known and subsequently required calculation approaches are compiled below. Details can be found in the relevant literature, e.g. [52–56].

2.5.1 Horizontal Pipe Flow

The pressure loss of an *incompressible* fluid in a steady state through a straight pipe is calculated as follows:

$$-\Delta p = (p_\text{in} - p_\text{out}) = \lambda_F \cdot \frac{L_R}{D_R} \cdot \frac{\varrho_F}{2} \cdot u_F^2 = \lambda_F \cdot \frac{L_R}{D_R} \cdot \frac{1}{2} \cdot \left(\frac{\dot{m}_F}{A_R}\right)^2 \cdot \frac{1}{\varrho_F} \tag{2.76}$$

As a result of a dimensional analysis, (2.76) can be written as follows:

$$\left(\frac{-\Delta p}{\varrho_F \cdot u_F^2}\right) \cdot \left(\frac{D_R}{L_R}\right) = \frac{1}{2} \cdot \lambda_F = \lambda_F\left(Re_R = \frac{u_F \cdot D_R \cdot \varrho_F}{\eta_F}, \frac{k_R}{D_R}\right) \tag{2.77}$$

with:

$Eu = -\Delta p/(\varrho_F \cdot u_F^2)$	Euler number = dimensionless pressure loss,
λ_F	Friction coefficient of fluid flow, see Fig. 2.22,
k_R	Pipe roughness, average height of roughness elevations.

Since the pressure drop Δp is proportional to the length L_R under the considered boundary conditions, the geometry ratio (D_R/L_R) can be treated as a multiplicative factor. λ_F

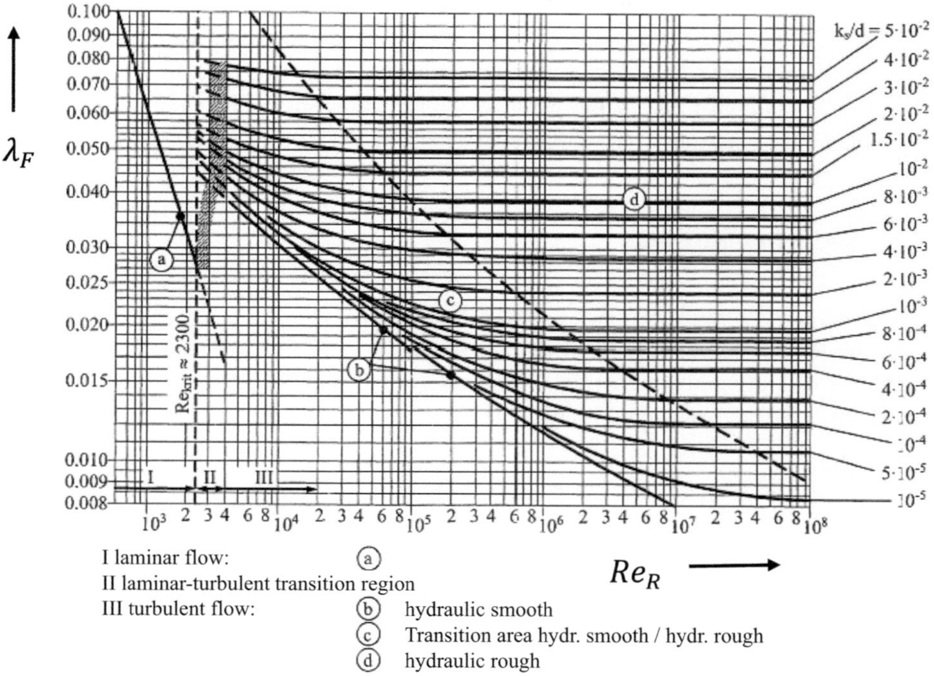

Fig. 2.22 Resistance coefficient $\lambda_F(Re_R = \frac{u_F \cdot D_R \cdot \varrho_F}{\eta_F}, \frac{k_R}{D_R})$ for technically rough straight pipes [52]

values for different flow forms, or Reynolds numbers Re_R, and relative pipe roughness (k_R/D_R) can be found in Fig. 2.22. In pneumatic conveying lines, turbulent gas flow and, after an transitional phase, hydraulically smooth pipes can always be assumed.

As shown in Calculation Example 2, Sect. 2.3.4, the corresponding pressure loss of an isothermal *compressible* fluid = gas is calculated by

$$\frac{p_{\text{in}}^2 - p_{\text{out}}^2}{(\frac{\dot{m}_F}{A_R})^2 \cdot R_F \cdot T_F} - 2 \cdot \ln\left(\frac{p_{\text{in}}}{p_{\text{out}}}\right) = \lambda_F \cdot \frac{L_R}{D_R} \tag{2.78}$$

Under the operating conditions of pneumatic conveyance, the second term on the left side of (2.78) is generally negligible, i.e., the flow is then treated as quasi-steady state. With this simplification and given p_{out}, (2.78) can be transformed as follows:

$$-\Delta p = (p_{\text{in}} - p_{\text{out}}) = p_{\text{out}} \cdot \left(\sqrt{1 + \lambda_F \cdot \frac{L_R}{D_R} \cdot \left(\frac{\dot{m}_F}{A_R}\right)^2 \cdot \frac{R_F \cdot T_F}{p_{\text{out}}^2}} - 1\right) \tag{2.79}$$

An analogous calculation equation results for a given pressure p_{in}.

The continuous course of the gas state changes along the pipeline. It can be approximated by a staircase if the pressure loss is calculated incompressibly in sections over shorter pipe sections with (2.76), starting from known/desired operating conditions, e.g., at the beginning or end of the pipeline. The calculation accuracy can be increased through recursive iteration.

2.5.2 Vertical Pipe Flow

In addition to the pressure loss due to friction, Δp_f, a pressure loss Δp_G must be applied to support the fluid weight $F_{F,G} = \Delta m_F \cdot g = H_R \cdot A_R \cdot \varrho_F \cdot g$ that is currently in the straight vertical pipe of height H_R. With $\Delta p_G = F_{F,G}/A_R$, for *incompressible* fluids, from an impulse/force balance follows:

$$-\Delta p = (p_{\text{in}} - p_{\text{out}}) = (\Delta p_f + \Delta p_G) = \varrho_F \cdot H_R \cdot \left(\lambda_F \cdot \frac{u_F^2}{2 \cdot D_R} + g\right) \tag{2.80}$$

The pressure loss of a *compressible* fluid = gas in steady state operation can also be calculated with (2.80) if for ϱ_F and u_F the average fluid density and the average gas velocity in the vertical pipe are used. Both must be determined by iterative calculation. A sectional calculation is also possible.

2.5.3 Deflections, Other Pressure Losses

These are generally caused by a dynamic gas pressure $(\varrho_F \cdot u_F^2)/2$ proportional resistance coefficient ξ described. With the local gas density ϱ_F follows:

$$-\Delta p = \xi \cdot \frac{\varrho_F}{2} \cdot u_F^2 = \xi \cdot \frac{1}{2} \cdot \left(\frac{\dot{m}_F}{A_R}\right)^2 \cdot \frac{1}{\varrho_F} \tag{2.81}$$

For example, the size of the ξ-value in turbulent fluid flow through a pipe bend is significantly influenced by the ratio of the bend radius R_U/pipe diameter D_R, the deflection angle α_U and their geometric design, e.g., execution as an arc, elbow, segment, is determined. In the case of cross-sectional changes of the conveying pipe, ξ is a function of the respective area ratio (A_1/A_2) and can be calculated using the conservation laws of single-phase flows, etc. A clear compilation of ξ values for various pipe elements/flow situations is contained in [56].

Note:
In general, in pneumatic conveying lines, the proportion of pressure loss caused by the conveying gas alone is small compared to that of the solid material and can often be neglected. This is particularly true for dense phase conveying.

References

1. Hirschfelder, J., Curtiss, C.F., Bird, R.B.: Molecular theory of gases and liquids. Wiley, New York (1954)
2. Baehr, H.D.: Thermodynamik: Eine Einführung in die Grundlagen und ihre technische Anwendung. Springer, Berlin (1992)
3. Lüdecke, D., Lüdecke, C.: Thermodynamik: Physikalisch-chemische Grundlagen der thermischen Verfahrenstechnik. Springer, Berlin (2000)
4. Hilgraf, P.: Einfluss von Schüttgütern im Hinblick auf ihre Handhabung. VDI-Berichte 1918., pp. 63–93 (2006)
5. Kezdi, A.: Grundlagen einer allgemeinen Bodenphysik. Z. Ver. Dt. Ing. **108**(5), 161–166 (1966)
6. Kezdi, A.: Erddrucktheorien. Springer, Berlin (1962)
7. Schulze, D.: Pulver und Schüttgüter: Fließeigenschaften und Handhabung. Springer, Berlin (2006)
8. FEM-Dokument 2582: Allgemeine Schüttguteigenschaften hinsichtlich der Klassifizierung und der Symbolisierung. Ausgabe D, VDMA-Fachgemeinschaft Fördertechnik, Frankfurt 1984
9. Waddell, H.: Volume, shape and roundness of quartz particles. J. Geol. **43**, 250–280 (1935)
10. Schubert, H., Heidenreich, E., Liepe, F., Neeße, T.: Mechanische Verfahrenstechnik I. VEB Deutscher Verlag für Grundstoffindustrie, Leipzig (1977)
11. Haider, A., Levenspiel, O.: Drag coefficient and terminal velocity of spherical and non-spherical particles. Powder Technol. **58**, 63–70 (1989)
12. Pahl, M.H., Schädel, G., Rumpf, H.: Zusammenstellung von Teilchenbeschreibungsmethoden, Teil 1. Aufbereitungs-Technik **14**(5), 257–264 (1973). Teil 2: Nr. 10, S. 672–683, Teil 3: Nr. 11, S. 759–764
13. Smith, W.O., Foote, P.D., Busang, P.F.: Packing of homogeneous spheres. Phys. Rev. **34**, 1271–1274 (1929)
14. Hilgraf, P.: Handling of bulk material mixtures. ZKG Int. **61**(11), 46–59 (2008)
15. DIN EN 459-2: Baukalk – Teil 2: Prüfverfahren. Beuth Verlag GmbH, Dezember 2010
16. Carr, R.L.: Evaluating flow properties of solids. Chem. Eng. **18**, 163–168 (1965)
17. Rastogi, S., Dhodapkar, S.V., Cabrejos, F., Baker, J., Weintraub, M., Klinzing, G.E.: Survey of characterization of dry ultrafine coals and their relationships to transport, handling and storage. Powder Technol. **74**, 47–59 (1993)
18. DIN ISO 9276-1: Darstellung der Ergebnisse von Partikelgrößenanalysen – Teil 1: Grafische Darstellung. Beuth Verlag GmbH, September 2004
19. DIN 66143: Darstellung von Korn-(Teilchen-)größenverteilungen; Potenznetz. Beuth Verlag GmbH, März 1974
20. DIN 66144: Darstellung von Korn-(Teilchen-)größenverteilungen; Logarithmisches Normalverteilungsnetz. Beuth Verlag GmbH, März 1974
21. DIN 66145: Darstellung von Korn-(Teilchen-)größenverteilungen; RRSB-Netz. Beuth Verlag GmbH, April 1976
22. Pahl, M.H.: Lagern, Fördern und Dosieren von Schüttgütern. TÜV Rheinland, Köln (1989)
23. DIN 66126, Teil 2: Bestimmung der spezifischen Oberfläche disperser Feststoffe mit Durchströmungsverfahren – Verfahren und Gerät nach Blaine. Beuth Verlag, Februar 1989
24. Löffler, F., Raasch, J.: Grundlagen der Mechanischen Verfahrenstechnik. Vieweg, Braunschweig, Wiesbaden (1992)
25. Bohnet, M. (Eds.): Mechanische Verfahrenstechnik. Wiley-VCH, Weinheim (2004)

26. Tomas, J.: Modellierung des Fließverhaltens von Schüttgütern auf der Grundlage der Wechselwirkungskräfte zwischen den Partikeln und Anwendung bei der Auslegung von Bunkeranlagen. TU Bergakademie, Freiberg (1991). Dissertation B
27. Jenike, A.W.: Storage and flow of solids. Bull. No. 123, Utah Engng. Exp. Station. Univ. of Utah, Salt Lake City (1964)
28. Molerus, O.: Schüttgut-Mechanik. Springer, Berlin (1985)
29. Höhne, D., Schünemann, U.: Untersuchung zum Einfluss der Korngrößenverteilungs-parameter auf die Fließfähigkeit feinkörniger Schüttgüter. European Symposium „Particle Technology 1986", Amsterdam, pp. 968–980. Preprints, Volume B, 1986
30. Hilgraf, P., Kleibs, R.: Schüttguteigenschaften und Lagerzeit als Einflussgrößen bei der Auswahl des Silierungskonzeptes von Flugaschen. VGB Kraftwerkstech. **6**, 479–486 (1990)
31. Hilgraf, P.: Kostenminimierung bei der Projektierung von Schüttgut-Anlagen. Technische Akademie Wuppertal, Hamburg (1998). Lehrgangshandbuch zum gleichnamigen Seminar
32. Rumpf, H.: Grundlagen und Methoden des Granulierens. Chem.-Ing.-Tech. **30**(3), 144–158 (1958). Nr. 5, S. 329–336
33. Rumpf, H., Herrmann, W.: Eigenschaften, Bindungsmechanismen und Festigkeit von Agglomeraten. Aufbereitungs Tech. **11**(3), 117–127 (1970)
34. Schubert, H.: Grundlagen des Agglomerierens. Chem.-Ing.-Tech. **51**(4), 266–277 (1979)
35. Schubert, H.: Kapillarität in porösen Feststoffsystemen. Springer, Berlin (1982)
36. Israelachvilli, J.: Intermolecular and surface forces. Academic Press, London (1991)
37. Pahl, M.H.: Haftkräfte in kompaktierten Schüttgütern. Aufbereitungs Tech. **29**(7), 371–380 (1988)
38. Rumpf, H.: Die Wissenschaft des Agglomerierens. Chem.-Ing.-Tech. **46**(1), 1–11 (1974)
39. Schubert, H.: Kapillarkräfte – Modellierung und Anwendung in der Partikeltechnologie. Aufbereitungs Tech. **25**(1), 39–49 (1984)
40. Forschungsbericht Staubexplosionen: Brenn- und Explosions-Kenngrößen von Stäuben. Schriftenreihe des Hauptverbandes der gewerblichen Berufsgenossenschaften e. V., Bonn 1980
41. Wiemann, W.: Influence of temperature and pressure on the explosion characteristics of dust/air and dust/air/inert gas mixtures. Industrial dust explosions. ASTM STP 958. American Society for Testing and Materials, Philadelphia, pp. 33–44 (1987)
42. Bartknecht, W.: Explosionen – Ablauf und Schutzmaßnahmen. Springer, Berlin (1980)
43. Beek, W.J., Muttzall, K.M.K., van Heuven, J.W.: Transport phenomena, 2nd edn. Wiley, Chichester (1999)
44. Hilgraf, P.: Berechnung der Trocknung von Feststoffen im Stromtrockner. Aufbereitungs Tech. **17**(6), 288–294 (1976)
45. Pawlowski, J.: Die Ähnlichkeitstheorie in der physikalisch-technischen Forschung – Grundlagen und Anwendungen. Springer, Berlin (1971)
46. Görtler, H.: Dimensionsanalyse – Theorie der physikalischen Dimensionen mit Anwendungen. Springer, Berlin (1975)
47. Zlokarnik, M.: Scale-up – Modellübertragung in der Verfahrenstechnik. Wiley-VCH, Weinheim (2000)
48. Palmer, C.A.: Dimensional analysis and intelligent experimentation. World Scientific Publishing, Singapore (2008)
49. Gibbings, J.C.: Dimensional analysis. Springer, London (2011)
50. Zierep, J.: Ähnlichkeitsgesetze und Modellregeln der Strömungslehre. G. Braun Buchverlag, Karlsruhe (1991)
51. Papula, L.: Mathematik für Ingenieure und Naturwissenschaftler. Vieweg, Wiesbaden (2001)

52. Merker, G.P., Baumgarten, C.: Fluid- und Wärmetransport: Strömungslehre. Teubner, Stuttgart (2000)
53. Rennels, D.C., Hudson, H.M.: Pipe flow—A practical and comprehensive guide. Wiley, Hoboken (2012)
54. Truckenbrodt, E.: Lehrbuch der angewandten Fluidmechanik. Springer, Berlin (1988)
55. Crowe, C.T., Elger, D.F., Roberson, J.A.: Engineering fluid mechanics. Wiley, Hoboken (2005)
56. VDI-Gesellschaft Verfahrenstechnik und Chemieingenieurwesen (Eds.): VDI-Wärmeatlas, 10. edn. Springer, Berlin (2006). Abschn. Laa1-Lac9

Gas/Solid Systems

3

The following provides a condensed overview of those gas/solid systems that border the field of pneumatic conveying and are relevant for the description/calculation of specific manifestations of the operating or flow conditions that occur. Among other things, the flow around isolated individual particles, that of solid particles in bulk clouds, the flow of gas through bulk materials, the operating behavior of various fluidized bed systems, etc., are considered.

3.1 Single Particles in the Gas Stream

Initially, ideally spherical, rigid, non-rotating solid particles are considered. An incompressible fluid with Newtonian behavior is assumed as the gas. The flow around the particle is laminar, uniform, and unaffected by solid (pipe) walls, free surfaces, or adjacent particles.

3.1.1 Particle Flow Around

If such a sphere is moved in the resting gas or is flowed around by it in a stationary manner, the resistance force $\vec{F}_W$ must be overcome. This is composed of varying proportions of friction and pressure resistance. From a dimensional analysis, it follows that, under the stated conditions, $\vec{F}_W$ depends solely on the relative velocity = difference velocity gas/particle $\vec{u}_{\text{rel}} = (\vec{u}_F - \vec{u}_S)$, the gas density ϱ_F, the gas viscosity η_F and the particle diameter d_S. From the five influencing factors of this relevance list and the associated three basic dimensions, two dimensionless characteristic numbers result:

$$c_W = \frac{F_W}{\frac{\pi}{4} \cdot d_S^2 \cdot \frac{\varrho_F}{2} \cdot u_{\text{rel}}^2}, \tag{3.1}$$

P. Hilgraf, *Pneumatic Conveying*, https://doi.org/10.1007/978-3-662-67223-5_3

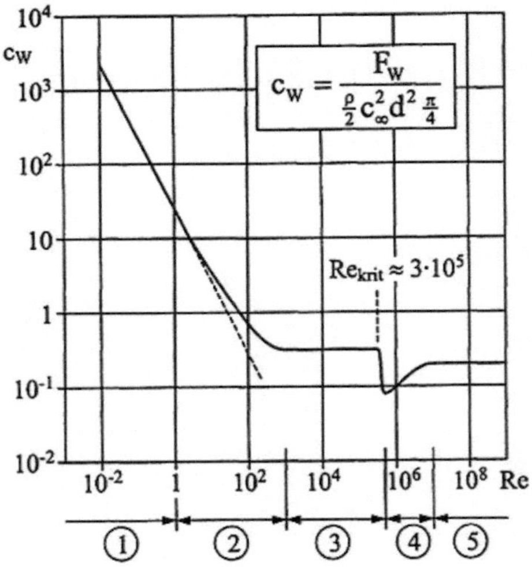

Fig. 3.1 Course of $c_W(Re_{S,rel})$ for spherical particles [1]

$$Re_{S,rel} = \frac{u_{rel} \cdot d_S \cdot \varrho_F}{\eta_F} \tag{3.2}$$

The Reynolds number $Re_{S,rel}$ is formed with a characteristic particle dimension, here the sphere diameter d_S, and as a reference/approach area A_P the projection area of the body perpendicular to the flow direction is used, here: $A_P = \pi/4 \cdot d_S^2$. In the case of one-dimensional parallel or antiparallel flow, the vector representation can be omitted, and the resistance force F_W is calculated as:

$$F_W = c_W(Re_{S,rel}) \cdot \frac{\pi}{4} \cdot d_S^2 \cdot \frac{\varrho_F}{2} \cdot |u_F - u_S| \cdot (u_F - u_S) \tag{3.3}$$

Fig. 3.1 shows the measured curve of $c_W(Re_S)$ for spherical single particles.
Depending on $Re_{S,rel}$, various flow forms around a particle and resulting drag coefficients c_W occur. For the following considerations, refer to Figs. 3.1 and 3.2.

Region (1), $Re_{S,rel} \leq 1$: The streamlines flow symmetrically around the sphere; the boundary layer is laminar and does not separate. In this "creeping" flow, the friction forces are much greater than the inertial forces. The analytical solution of Stokes neglecting the inertial forces leads to the resistance law:

$$c_W(Re_{S,rel}) = \frac{24}{Re_{S,rel}} \tag{3.4}$$

(3.4) is well confirmed by measurements up to $Re_{S,rel} \cong 0.25$; it can be applied with acceptable error up to $Re_{S,rel} \cong 1$, possibly even higher, if necessary (→ Stokes region).

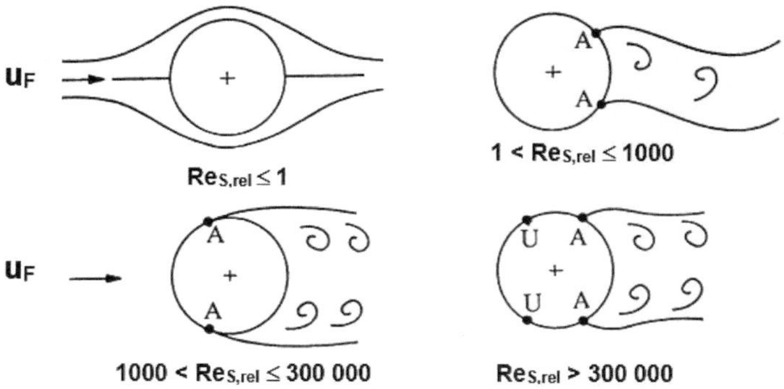

Fig. 3.2 Flow forms around a sphere; A = separation point, U = transition point

Region (2), $1 < Re_{S,\text{rel}} \leq 1000$**:** In this transitional range, the laminar boundary layer detaches from the sphere surface, points A in Fig. 3.2. Up to $Re_{S,\text{rel}} \cong 200$, a stationary detachment area forms behind the sphere. For larger $Re_{S,\text{rel}}$ numbers, the detachment of unsteady vortices begins, which are carried away by the laminar flow. The influence of inertial forces increases significantly.

Region (3), $10^3 < Re_{S,\text{rel}} \leq 3 \cdot 10^5$**:** The separation point A of the still laminar boundary layer moves up to just before the sphere's equator, i.e., onto the inflow side of the particle. As a result, the size of the separation area hardly changes, but at the same time, the wake flow becomes turbulent. The pressure in the separation area is thus lower than the static pressure in the undisturbed flow. This results in a force on the sphere proportional to the dynamic pressure ($\varrho_F \cdot u_{\text{rel}}^2/2$) and in an approximately constant drag coefficient (→ Newton region). This has the value:

$$c_W(Re_{S,\text{rel}}) \cong 0.44 \tag{3.5}$$

Region (4), $3 \cdot 10^5 < Re_{S,\text{rel}} \leq 10^7$**:** In the range of the value referred to as critical $Re_{S,\text{crit}} \cong 3 \cdot 10^5$ of the Reynolds number, a transition of the boundary layer from laminar to turbulent occurs before the separation point A, see points U in Fig. 3.2. This results in a more intense momentum exchange with the external flow = energy supply to the boundary layer and shifts the separation point A downstream, i.e., the separation area is reduced under pressure recovery. The resistance coefficient decreases from $c_W(Re_{S,\text{rel}}) \cong 0.44$ to $c_W(Re_{S,\text{rel}}) \cong 0.09$. Further increase of the $Re_{S,\text{rel}}$-number drives both the separation point A and the transition point U again upstream, causing $c_W(Re_{S,\text{rel}})$ to rise again.

Region (5), $Re_{S,\text{rel}} > 10^7$**:** The laminar/turbulent boundary layer transition U occurs very early on the upstream side of the sphere, and the position of the separation point A no longer changes. As a result, the boundary layer is essentially turbulent over the entire sphere surface, and constant flow conditions with the constant drag coefficient $c_W(Re_{S,\text{rel}}) \cong 0.20$ are established, regardless of $Re_{S,\text{rel}}$.

For the pneumatic conveying systems considered here, only areas (1) to (3) are relevant, i.e., $Re_{S,\mathrm{rel}} < Re_{S,\mathrm{crit}}$. In addition to the area equations (3.4) and (3.5), for the entire Reynolds number range $Re_{S,\mathrm{rel}} < Re_{S,\mathrm{crit}}$ a series of empirical $c_W(Re_{S,\mathrm{rel}})$-approaches are available, with which the present measurement results can be represented more or less accurately. Known are, among others, the approaches according to [2]:

$$c_W(Re_{S,\mathrm{rel}}) = \frac{24}{Re_{S,\mathrm{rel}}} + \frac{4}{\sqrt{Re_{S,\mathrm{rel}}}} + 0.4 \tag{3.6}$$

or [3]

$$c_W(Re_{S,\mathrm{rel}}) = \frac{1}{3} \cdot \left(\sqrt{\frac{72}{Re_{S,\mathrm{rel}}}} + 1\right)^2 \tag{3.7}$$

In the transition region (→ region (2)) adapted approaches of the form

$$c_W(Re_{S,\mathrm{rel}}) = \frac{A}{{Re_{S,\mathrm{rel}}}^n} \tag{3.8}$$

for different $Re_{S,\mathrm{rel}}$-ranges are often used.

3.1.2 Influences on c_W

The assumptions underlying the "standard" resistance function discussed in Sect. 3.1.1 $c_W(Re_{S,\mathrm{rel}})$—including, among others, frictionless external flow (→ boundary layer theory according to Prandtl), ideally spherical particles, no limiting walls, etc.—are generally not achievable in practice. This changes the resulting c_W values. Examples of this are:

- **Turbulence in external flow**: Turbulent fluctuating movements in the external flow provide additional energy to the boundary layer. This leads to a reduction of the critical Reynolds number $Re_{S,\mathrm{crit}}$ with strongly varying c_W values in the Newton range and a generally significant c_W increase in the transition region [4]. The determining factor for the behavior is the turbulence level of the external flow $Tu = \sqrt{u_F'^2}/u_{\mathrm{rel}}$, with: $u_F'^2$—Effective value of the fluid's fluctuating movement. The indicated deviations become larger with increasing Tu.
- **Particle shape**: Its apparent influence is usually described by the sphericity ψ, (2.18). Decreasing sphericity ψ (→ $\psi_{\mathrm{Sphere}} = 1$) leads to larger c_W-values. Figure 3.3 shows a summary of measurement results from [5]. Further results/calculation equations can be found in [4, 6].
- **Boundary Walls**: Particles near the wall, e.g., in a conveying pipe, are asymmetrically approached and flowed around by the gas due to the adhesion of the fluid/gas to the wall (→ $u_{F,\mathrm{Wall}} = 0$) , see Fig. 3.4. This leads to a higher $c_W$value than in undisturbed flow. At the same time, with a constant total pressure $p = (p_{\mathrm{dyn}} + p_{\mathrm{stat}})$, the

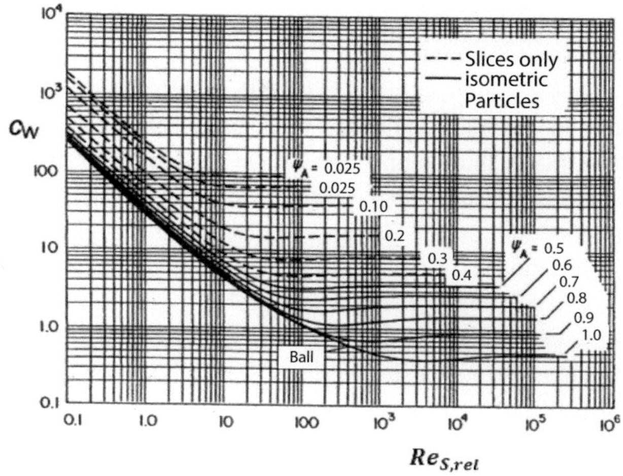

Fig. 3.3 Dependency $c_W(Re_{S,\text{rel}}, \psi)$ for isometric particles [4]

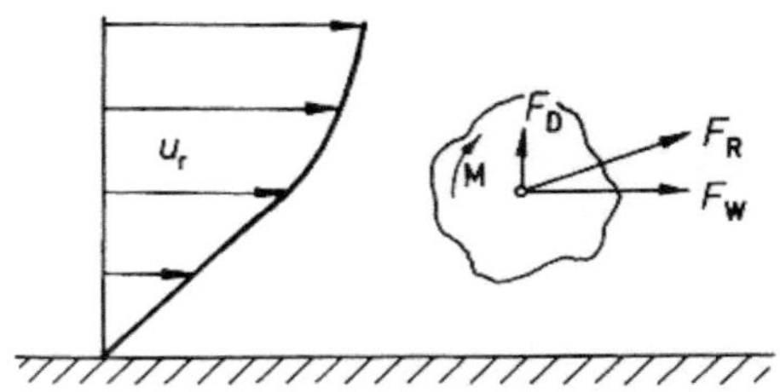

Fig. 3.4 Flow around a particle near a wall [4]

effect of the static pressure p_{stat} is always directed away from the wall and into the core flow, and is capable of driving wall-close particles in the direction towards the center of pipe when sufficiently large. Reason: across the flow cross-section, a higher static pressure is established on the wall-side of the particles due to the lower gas velocity there than on the wall-opposite side (→ Bernoulli). The resulting dynamic buoyancy force F_D is perpendicular to the resistance force F_W.

- **Particle rotation**: Such rotation can be caused, among other things, by an asymmetric gas flow around the particle, see Fig. 3.4, particle/wall and particle/particle collisions, as well as the specific particle feed into the gas stream. A dynamic lift force F_D (→ Magnus force) is exerted on a particle rotating around an axis perpendicular to the flow direction of the fluid, even in the case of symmetric flow. This lateral force acts in the direction in which the particle rotation and fluid flow directions are aligned. Additionally, the drag coefficient c_W increases compared to a non-rotating particle [7].
- **Other factors:** In case of pneumatic conveying the surface roughness of the particles, the incompressibility of the transport gas (→ Mach number $Ma = u_{\text{rel}}/a_{\text{Sound}} < 0.1$, with: a_{Sound}—speed of sound) and its possibly time-varying flow-around-the particle velocity (→ unsteady resistance component ≪ steady component) can be neglected. The continuum character of the conveying gas is generally always ensured (→ Knudsen number $Kn = \lambda/d_S \ll 0.1$, with: λ—mean free path of gas molecules). Details can be found in, among others, [4, 6, 8, 9].

The influence of neighboring particles, or the behavior of particles in a particle cloud, is discussed separately in Sect. 3.4.

For a significant part of the effects described above on c_W or the resistance F_W there are no generally valid and sufficiently reliable measurement results that would allow integration into engineering design models for pneumatic conveying processes without additional measurements and/or adjustment variables. Example: Pneumatic conveying systems are designed without considering the—naturally existing—particle rotation. It is currently not possible, in particular, to define suitable initial/boundary conditions for their triggering and size. These problems also exist in computer programs for numerical flow calculation (CFD). In practice, overall resistance coefficients are used, which are calculated from experiments and correlated accordingly.

3.1.3 Settling Velocity

For the reasons indicated in the previous section, a complete representation of the unsteady motion equation of a single particle is omitted here. Calculation example 4, Sect. 3.1.4, shows its analytical solution path using a simple task. Since the stationary settling velocity w_T of a characteristic single particle in an undisturbed resting fluid = gas without influence from nearby walls is of particular interest for modeling a pneumatic conveying process, it is examined more closely here. It follows from a force balance on the particle, see Fig. 3.5. Since the individual forces act parallel to the coordinate z, it can be written as a scalar equation. It holds:

$$F_W - (F_G - F_A) = 0 \tag{3.9}$$

with:

F_W Resistance force according to (3.3),

$(F_G - F_A)$ particle weight reduced by the static buoyancy.

Thus, it holds:

$$c_W(Re_{S,\text{rel}}) \cdot \frac{\pi}{4} \cdot d_S^2 \cdot \frac{\varrho_F}{2} \cdot |u_{\text{rel}}| \cdot (u_{\text{rel}}) - \frac{\pi}{6} \cdot d_S^3 \cdot (\varrho_P - \varrho_F) \cdot g = 0 \tag{3.10}$$

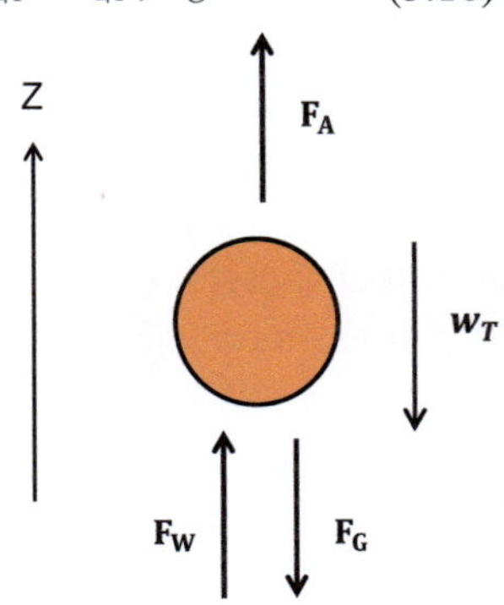

Fig. 3.5 Forces on a single particle falling freely in a stationary gas

For $u_F = 0$ follows: $u_{\text{rel}} = (u_F - u_S) = -u_S = |w_T|$. The settling velocity w_T is equal in magnitude to the relative velocity u_{rel} but always acts in the direction of gravitational acceleration. Inserted into (3.10) this results in:

$$w_T = \sqrt{\frac{4}{3} \cdot g \cdot \frac{d_S}{c_W(Re_{S,\text{rel}})} \cdot \frac{\varrho_P - \varrho_F}{\varrho_F}} \tag{3.11}$$

The Reynolds number $Re_{S,\text{rel}}$ is formed here with the settling velocity w_T as the characteristic velocity, i.e. $Re_{S,\text{rel}} = Re_{S,T} = w_T \cdot d_S \cdot \varrho_F / \eta_F$. By appropriate extensions, (3.11) can be written dimensionless:

$$Re_{S,T} = \sqrt{\frac{4}{3} \cdot \frac{Ar}{c_W(Re_{S,T})}} \tag{3.12}$$

with: Archimedes number $Ar = d_S^3 \cdot g \cdot \varrho_F \cdot (\varrho_P - \varrho_F)/\eta_F^2 \rightarrow$ Material constant of the gas/solid system.

In the Stokes range, with (3.4):

$$Re_{S,T} = \frac{1}{18} \cdot Ar \quad \text{or} \tag{3.13}$$

$$w_T = \frac{1}{18} \cdot g \cdot d_S^2 \cdot \frac{\varrho_P - \varrho_F}{\eta_F} \tag{3.14}$$

and with (3.5) in the Newton range:

$$Re_{S,T} \cong \sqrt{3 \cdot Ar} \quad \text{or} \tag{3.15}$$

$$w_T \cong \sqrt{3 \cdot g \cdot d_S \cdot \frac{\varrho_P - \varrho_F}{\varrho_F}} \tag{3.16}$$

As expected, in the Stokes range, the gas viscosity η_F and in the Newton range, the gas density ϱ_F is dominant. Over the entire Re range $Re_{S,T} < 3 \cdot 10^5$ only the equation based on (3.7) allows a closed solution: (3.17).

$$Re_{S,T} = 18 \cdot \left(\sqrt{1 + \frac{1}{9} \cdot \sqrt{Ar}} - 1\right)^2 \tag{3.17}$$

With (3.6) and other approaches suggested in the literature, only iterative calculations of w_T are possible, since the sought-after velocity w_T is also required simultaneously for the determination of $c_W(Re_{S,\text{rel}})$.

Figure 3.6 shows the movement of a single particle in the vertical gas flow at different speeds and flow directions of the gas. Here, u_F and u_S are absolute velocities with respect to the stationary z-axis, while w_T describes the particle velocity relative to the flowing gas.

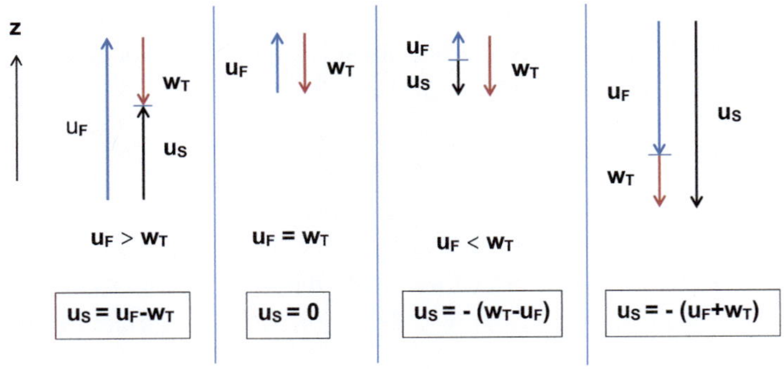

Fig. 3.6 Movement of a single particle in the vertical gas flow

When simultaneously feeding two particles with different settling velocities ($w_{T,1}, w_{T,2}$) into the gas flow of Fig. 3.6, the gas velocity u_F can be adjusted so that the particle with the lower settling velocity is discharged upwards, and the other one downwards. This effect, used for the separation of bulk materials, must also be taken into account in pneumatic conveying.

3.1.4 Calculation Example 4: Acceleration Distance of a 10 µm Particle

A spherical bulk material particle is introduced into a vertically upward flowing gas and accelerated from its resting position by it. After what distance is the acceleration completed?

Relevant Data:

- Gas data:
  Gas type: Air,
  Gas velocity: $u_F = 10.0\,\text{m/s} = \text{const.}$,
  Gas pressure: $p_F = 1.0\,\text{bar}$,
  Gas temperature: $T_F = 293\,\text{K}$,
  from this:
  Gas density, (2.3): $\varrho_F = 1.20\,\text{kg/m}^3$,
  dynamic viscosity, (2.6): $\eta_F = 18.26 \cdot 10^{-6}\,\text{Pa s}$.
- Particle data:
  Particle diameter: $d_S = 10\,\mu\text{m}$,
  Particle- = solid density: $\varrho_P = 2700\,\text{kg/m}^3$,
  initial vertical velocity of the particles: $u_{S,\text{in}} = 0\,\text{m/s}$.

General approach:
(2.48) applied to the motion behavior of a single particle, taking into account (3.9), leads to the force balance:

$$F_T - F_W + (F_G - F_A) = 0 \tag{3.18}$$

with:

F_T Inertial force.
F_T is calculated with the particle volume V_P as follows:

$$F_T = V_P \cdot (\varrho_P + j \cdot \varrho_F) \cdot \frac{du_S}{d\tau}, \quad j \cong 0.5 \tag{3.19}$$

During the acceleration process, a part of the gas shell moving with the particle is also accelerated. This is taken into account in the approach for F_T by the gas mass fraction $(V_P \cdot j \cdot \varrho_F)$.

(3.19) and (3.10) inserted into (3.18) yields after elementary transformations:

$$\frac{du_S}{d\tau} = \frac{3}{4} \cdot \frac{c_W(Re_{S,\text{rel}})}{d_S} \cdot \frac{\varrho_F}{\varrho_P + j \cdot \varrho_F} \cdot (u_F - u_S)^2 - \frac{\varrho_P - \varrho_F}{\varrho_P + j \cdot \varrho_F} \cdot g \tag{3.20}$$

Definition of the working range:
The relative velocity $u_{\text{rel}} = (u_F - u_S)$ between gas and particle is maximal at the beginning of the acceleration process and amounts to the task under investigation $u_{\text{rel,in}} = u_F = 10\,\text{m/s}$. This leads to Reynolds numbers of

$$Re_{S,\text{rel}} \le \frac{u_{\text{rel,in}} \cdot d_S \cdot \varrho_F}{\eta_F} = \frac{10.0\,\text{m/s} \cdot 10 \cdot 10^{-6}\,\text{m} \cdot 1.20\,\frac{\text{kg}}{\text{m}^3}}{18.26 \cdot 10^{-6}\,\text{Pa s}} = 6.57$$

during the acceleration phase and justifies the application of Stokes' resistance law. $c_W(Re_{S,\text{rel}}) = 24/Re_{S,\text{rel}}$, (3.4). If the simplifications resulting from the present condition $\varrho_P \gg \varrho_F$ are taken into account simultaneously, it follows from (3.20):

$$\frac{du_S}{d\tau} = \frac{1}{\Delta\tau_{\text{Stokes}}} \cdot (u_F - u_S) - g \tag{3.21}$$

with the constant = relaxation time.

$$\Delta\tau_{\text{Stokes}} = \frac{d_S^2 \cdot \varrho_P}{18 \cdot \eta_F}$$

(3.14) shows that in the Stokes range, the relationship

$$w_T = \Delta\tau_{\text{Stokes}} \cdot g$$

applies.

Particle velocity u_S as a function of time τ:
(3.21) can be transformed into the form

$$\frac{du_S}{(u_F - w_T) - u_S} = \frac{d\tau}{\Delta\tau_{\text{Stokes}}}$$

and from $u_S = 0$ to $u_S = u_S$ integrated. This results in:

$$u_S = (u_F - w_T) \cdot \left(1 - \exp\left(-\frac{\tau}{\Delta\tau_{\text{Stokes}}}\right)\right) \tag{3.22}$$

It is evident that the maximum = steady state particle velocity $u_{S,\text{out}} = (u_F - w_T)$ is only reached at $\tau_{\text{out}} \to \infty$. From

$$R = \frac{u_S}{u_F - w_T} = 1 - \exp\left(-\frac{\tau_{\text{out}}}{\Delta\tau_{\text{Stokes}}}\right)$$

however, for a relative approximation $R < 1$ at $u_{S,\text{out}}$ a finite acceleration time τ_{out} can be calculated:

$$\tau_{\text{out}} = -\Delta\tau_{\text{Stokes}} \cdot \ln(1 - R) \tag{3.23}$$

Acceleration distance:
If we denote the distance that the particle travels during time τ as s, then with (3.22):

$$\frac{ds}{d\tau} = (u_F - w_T) \cdot \left(1 - \exp\left(-\frac{\tau}{\Delta\tau_{\text{Stokes}}}\right)\right)$$

By integrating once more over time, we obtain:

$$s = (u_F - w_T) \cdot \left[\tau - \Delta\tau_{\text{Stokes}} \cdot \left(1 - \exp\left(-\frac{\tau}{\Delta\tau_{\text{Stokes}}}\right)\right)\right]$$

Taking into account (3.23) yields the total acceleration distance covered in τ_{out}:

$$s_{\text{out}} = (u_F - w_T) \cdot \Delta\tau_{\text{Stokes}} \cdot (-\ln(1 - R) - R) \tag{3.24}$$

Numerical results:
From the numerical calculations it follows:

$$\begin{aligned} \Delta\tau_{\text{Stokes}} &= \frac{d_S^2 \cdot \varrho_P}{18 \cdot \eta_F} = \frac{(10 \cdot 10^{-6}\,\text{m})^2 \cdot 2700\,\frac{\text{kg}}{\text{m}^3}}{18 \cdot 18.26 \cdot 10^{-6}\,\text{Pa s}} = 8.215 \cdot 10^{-4}\,\text{s} \\ w_T &= \Delta\tau_{\text{Stokes}} \cdot g = 8.215 \cdot 10^{-4}\,\text{s} \cdot 9.81\frac{\text{m}}{\text{s}^2} = 8.06 \cdot 10^{-3}\frac{\text{m}}{\text{s}} \end{aligned}$$

Since $w_T < u_F$ is, the vertical particle transport assumed above takes place. With $R = 0.99$, i.e., the acceleration is considered complete when the particles have reached 99% of their steady state = final velocity $u_{S,\text{out}}$, it follows for the acceleration path:

$$s_{\text{out}} = (u_F - w_T) \cdot \Delta\tau_{\text{Stokes}} \cdot (-\ln(1-R) - R)$$
$$= \left(10.0\,\frac{\text{m}}{\text{s}} - 8.06 \cdot 10^{-3}\,\frac{\text{m}}{\text{s}}\right) \cdot 8.215 \cdot 10^{-4}\,\text{s} \cdot (-\ln(1-0.99) - 0.99)$$
$$s_{\text{out}} = 2.97 \cdot 10^{-2}\,\text{m} = 29.7\,\text{mm}$$

3.2 Flow through Bulk Materials

First, the pressure loss caused by the bulk material on the flowing fluid, here gas, is presented, followed by a discussion of the impacts of specific effects not considered in the calculation approaches. Newtonian behavior is assumed for the gas.

The considered bed packings are assumed to be complete random packings with a constant cross-section along the straight length of flow L_b. The inlet and outlet cross-sections are flat and perpendicular to the flow direction of the fluid. Thus, $L_b = \text{const.}$ applies to all parallel flow channels.

Possible calculation approaches for fluid pressure loss are:

- Model of hydraulic diameter: The fluid flows through the packing through a multitude of parallel, possibly of different diameter, winding tubes, e.g. [10].
- Model of flow-around individual particles: The fluid flows in the packing around a multitude of mutually influencing individual particles, e.g. [11].

Subsequently, the approach based on the hydraulic diameter model analogous to pipe flow according to Ergun [10] is presented as an example. This has also proven successful in extrapolations to high pressures and high temperatures [12]. Alternative approaches can be found in the relevant literature, including [4, 8, 9, 11].

3.2.1 Stationary Bed Packing

The pressure loss of an *incompressible* fluid in a steady state through a bed packing of length L_b is calculated according to Ergun [10] as follows:

$$-\frac{\Delta p}{L_b} = \frac{p_{\text{out}} - p_{\text{in}}}{L_b} = f_1 \cdot v_F + f_2 \cdot \varrho_F \cdot v_F^2 \tag{3.25}$$

$$f_1 = 150 \cdot \frac{(1-\varepsilon_F)^2}{\varepsilon_F^3} \cdot \frac{\eta_F}{(\psi \cdot d_{S,SD})^2}, \tag{3.26}$$

$$f_2 = 1.75 \cdot \frac{(1-\varepsilon_F)}{\varepsilon_F^3} \cdot \frac{1}{\psi \cdot d_{S,SD}} \tag{3.27}$$

with:

$v_F = \dot{m}_F/(A_R \cdot \varrho_F)$	Empty tube gas velocity in the cross-section A_R,
$d^*_{S,SD} = (\psi \cdot d_{S,SD})$	shape-corrected Sauter diameter, see (2.28),
ε_F	average relative void volume of the flowed-through bed, see (2.19).

The first term on the right side of (3.25) is decisive for predominant viscosity influence, the second for dominating inertial forces. The gas flow thus forms analogously to that in particle flow, see Sect. 3.1.1, whereby due to the network-like flow guidance in the bulk material, an earlier dominance of inertial effects can be expected. (3.25) can be simplified if, according to the flow state, either only (3.26) or (3.27) is used.

After introducing a resistance coefficient λ_b can (3.25) be represented in dimensionless form as follows:

$$\left(\frac{-\Delta p}{\varrho_F \cdot v_F^2}\right) \cdot \left(\frac{d_{S,SD}}{L_b}\right) \cdot \psi \cdot \left(\frac{\varepsilon_F^3}{1-\varepsilon_F}\right) = \lambda_b = \lambda_b\left(Re_b = \frac{v_F \cdot d_{S,SD} \cdot \varrho_F}{\eta_F}, \psi, \varepsilon_F\right) \tag{3.28}$$

$$\lambda_b = 150 \cdot \frac{1-\varepsilon_F}{Re_b} + 1.75, \quad \text{Validity range: } 0.4 < Re_b < 1000 \tag{3.29}$$

with:

$Eu_b = -\Delta p/(\varrho_F \cdot v_F^2)$ Euler number = dimensionless pressure loss.

The corresponding pressure loss of an isothermal *compressible* fluid = gas follows from the integration of the differential form of (3.25) and for a given/known pressure p_{out} at the gas outlet of the bed.

$$-\Delta p = (p_{out} - p_{in}) = p_{out} \cdot \left(\sqrt{1 + \frac{2 \cdot (f_1 \cdot v_{F,out} + f_2 \cdot \varrho_{F,out} \cdot v_{F,out}^2) \cdot L_b}{p_{out}}} - 1\right) \tag{3.30}$$

An analogous calculation equation results when a given pressure p_{in}.

3.2.2 Moving Bulk Material

If the bulk material, which is flowed through by a fluid, is itself in motion, e.g., as a plug in a pneumatic conveying line, the empty pipe velocity in (3.25)–(3.30)

$$v_F = \varepsilon_F \cdot u_F = \varepsilon_F \cdot u_{rel} \quad \rightarrow \quad \text{stationary bulk, } u_S = 0 \tag{3.31}$$

must be replaced by

$$\varepsilon_F \cdot (u_F - u_S) = (v_F - \varepsilon_F \cdot u_S) = \varepsilon_F \cdot u_{rel} \quad \rightarrow \quad \text{moving bulk} \tag{3.32}$$

Fig. 3.7 Definition of relative velocity $u_{\mathrm{rel}} = (u_F - u_S)$

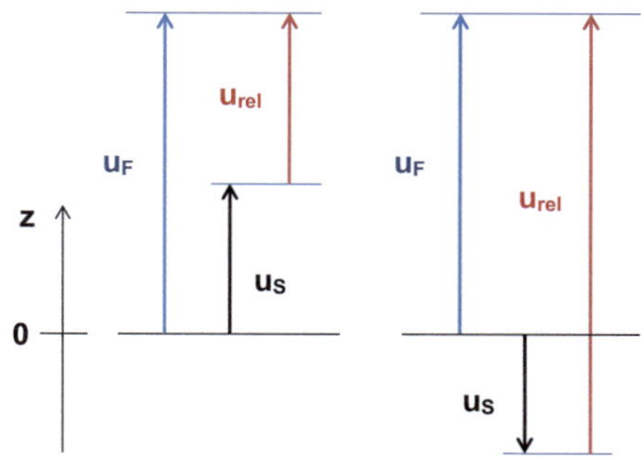

with:

$u_F = \dot{m}_F/(\varepsilon_F \cdot A_R \cdot \varrho_F)$	true average gas velocity,
$u_S = \dot{m}_S/((1-\varepsilon_F) \cdot A_R \cdot \varrho_P)$	true average solid velocity,
$(1-\varepsilon_F) \cdot \varrho_P = \varrho_b$	bulk density of the moving bed

[13]. If the flow direction of the fluid is defined as positive, as shown in Fig. 3.7, for a co-current flow of gas and solid $u_{\mathrm{rel}} = (u_F - u_S)$ while for an opposite flow direction $u_{\mathrm{rel}} = (u_F - (-u_S)) = (u_F + u_S)$ must be used.

3.2.3 Further Influences

When applying the presented calculation approach in practice, additional influencing factors that are not considered in the model may need to be taken into account. Examples include:

- **Edge effect**: The relative void volume ε_F of a bulk material approaches the value $\varepsilon_F \to 1$ ($\to$ if rigid spherical monoparticles) at limiting walls/fixtures in theory. This results in the specific gas volume flow near the wall being greater than that in the core area of the bulk material, given the same pressure gradient, see (3.25). Figure 3.8 shows the effects on the fluid flow profile. The higher fluid-permeable wall zone is generally limited to a wall distance of $y \cong (2{-}4) \cdot d_{S,50}$, depending on the grain shape

Fig. 3.8 Edge effect of a flowed-through bulk layer, according to [4]

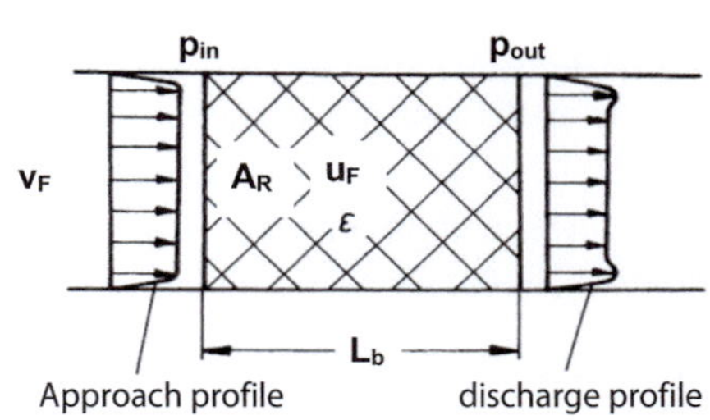

and width of the particle size distribution [14]. The effect is more pronounced the smaller the dimensions of the flow channel are.

- **Sloping of free surfaces**: In horizontal reactors/tubes, solid bulk materials form cones at both ends if they are not mechanically fixed, e.g., by screens. This leads to a decreasing local bulk = flow length L_b with increasing bed height, as a result of which a fluid velocity distribution with a minimum velocity at the lower reactor bottom and a maximum value in the apex area forms across the bulk cross-section. This effect is reinforced by the fact that the bulk material in the apex area is generally more loosely packed than in the core area. In pneumatic plug conveying systems, this can lead to unwanted plug dissolution due to shearing of the upper plug part against the lower one if the plug lengths are too short.
- **Porosity of stationary/moving bulk solids**: The relative void volume of a moving bulk solid $\varepsilon_{F,u}$ is always greater than that of a stationary bulk solid $\varepsilon_{F,0}$ [15]. At the same operating conditions ($u_{\text{rel}}, L_b =$ konst.) a smaller pressure loss occurs over the moving bulk solid than over a stationary one. The determining factors for the difference $(\varepsilon_{F,u} - \varepsilon_{F,0})$ are the grain shape and width of the particle size distribution of the respective bulk solid present.

Note: For the practical applications of the above equations, the porosity ε_F of the bulk material, the Sauter diameter $d_{S,SD}$ characterizing the particle size distribution, and a shape factor ψ characteristic for the entire particle collective must be known. Their determination is only possible with sufficient accuracy through appropriate model experiments. With these results, the design equations are then calibrated. Calculation example 5, Sect. 3.2.4, illustrates the procedure.

3.2.4 Calculation Example 5: Determination of a Characteristic Particle Diameter

A quartz product is flowed through by gases with different values of temperature, pressure, and composition in a process. The resulting pressure losses are to be calculated using the Ergun equation. This must be calibrated with respect to the bulk material characteristics $(\varepsilon_F, d_{S,SD}, \psi)$. The current gas data (ϱ_F, η_F) are known. A representative 50 kg sample is available.

Experimental setup and procedure:
The measurements are carried out in an open-topped vertical glass tube, here: inner diameter $D_R = 290.0\,\text{mm}$, height $H_R \cong 1200\,\text{mm}$, which has a gas-permeable gas distributer at its lower end, to which an adjustable gas flow is supplied. Mesured are the gas volume flow $\dot{V}_{F,0}$ in the intake = ambient condition, the gas pressure p_{in} directly above the gas distributor, the ambient pressure $p_0 = p_{\text{out}}$, and from this, the flow pressure loss of the solid bed $\Delta p = (p_{\text{out}} - p_{\text{in}})$ and their current bed height L_b. The loosely filled bulk material, up

to a chosen (L_b/D_R)-ratio, is carefully leveled in the measuring device and then gradually flowed through with increasing gas flows. In doing so, the above-mentioned measurement data are recorded. Exceeding a limit velocity leads to the formation of a fluidized bed and bed expansion, see Sect. 3.3. The calibration of the Ergun equation is only carried out with the measured values of the fixed bed area, i.e., the resting bulk material.

Experimental setup:

- Bulk material data:
  Sieve residue $R(d_S)$ at grain diameter d_S:
  $R(2.0\,\text{mm}) = 0\,\text{M.\%}$,
  $R(1.2\,\text{mm}) = 0.10\,\text{M.\%}$,
  $R(1.0\,\text{mm}) = 1.49\,\text{M.\%}$,
  $R(0.6\,\text{mm}) = 5.25\,\text{M.\%}$,
  $R(0.315\,\text{mm}) = 56.80\,\text{M.\%}$,
  $R(0.160\,\text{mm}) = 98.80\,\text{M.\%}$,
  $R(0.090\,\text{mm}) = 99.90\,\text{M.\%}$,
  Median diameter: $d_{S,50} = d_S(R = 50\,\text{M.\%}) = 0.350\,\text{mm}$,
  Particle- = solid density: $\varrho_P = 2650\,\text{kg/m}^3$,
  Solid temperature: $T_S = 293\,\text{K}$,
  Filling height: $L_b = 307\,\text{mm}$, $(L_b/D_R) = 307\,\text{mm}/290.0\,\text{mm} = 1.06$,
  Solid weight: $m_S = 32.444\,\text{kg}$.
- Gas data:
  Gas type: Air,
  Gas temperature: $T_F = T_S = 293\,\text{K}$,
  Gas pressure: $p_F = p_{\text{out}} = 1.0\,\text{bar}$,
  from this:
  Gas density, (2.3): $\varrho_F = 1.20\,\text{kg/m}^3$,
  dynamic viscosity, (2.6): $\eta_F = 18.26 \cdot 10^{-6}\,\text{Pa s}$.

Determination of the calculation approach:

Since only small bed pressure losses are expected due to the relatively coarse-grained bulk material and the low bed height, the gas is treated as incompressible according to (3.25). The reference state is the one at the bed outlet: (1.0 bar, 293 K).

(3.25) can be applied to the particle diameter characterizing the particle collective through which the flow passes

$$
\begin{aligned}
d^*_{S,SD} &= (\psi \cdot d_{S,SD}) \\
&= 0.875 \cdot \frac{1-\varepsilon_F}{\varepsilon_F^3} \cdot \frac{\varrho_F \cdot v_F^2}{\left|\frac{\Delta p}{L_b}\right|} \cdot \left(1 + \sqrt{1 + 195.92 \cdot \left(\frac{\varepsilon_F}{v_F}\right)^3 \cdot \frac{\eta_F}{\varrho_F^2} \cdot \left|\frac{\Delta p}{L_b}\right|}\right)
\end{aligned}
\tag{3.33}
$$

rearranged and evaluated with the measurement sets of the flow experiments.

Experimental results:
From the measurement data, it follows:

- Cross-sectional area of the glass tube:

$$A_R = \frac{\pi}{4} \cdot D_R^2 = \frac{\pi}{4} \cdot (0.290\,\text{m})^2 = 0.06605\,\text{m}^2$$

- Empty tube gas velocity:

$$v_F = \frac{\dot{V}_{F,0}}{A_R} = \frac{\dot{V}_{F,0}}{0.06605\,\text{m}^2}$$

- Bulk density:

$$\varrho_b = \frac{m_S}{A_R \cdot L_b} = \frac{32.444\,\text{kg}}{0.06605\,\text{m}^2 \cdot 0.307\,\text{m}} = 1600\,\frac{\text{kg}}{\text{m}^3}$$

- Relative void volume, (2.19), (2.21):

$$\varepsilon_F = 1 - \frac{\varrho_b}{\varrho_P} = 1 - \frac{1600\,\frac{\text{kg}}{\text{m}^3}}{2650\,\frac{\text{kg}}{\text{m}^3}} = 0.396$$

- Measured dependency $\Delta p(v_F)$:
  This is shown in Fig. 3.9. The red circles denote the measured points determined by experiment. Measurements were taken up to the fluidized bed area. The subsequent evaluation is based only on the measured values from the fixed bed area, $L_b = \text{const.}$.

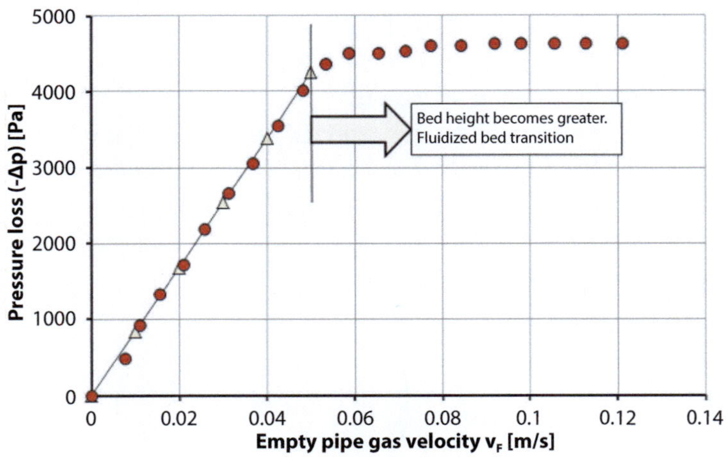

Fig. 3.9 Dependency of bed pressure loss $(-\Delta p)$ on the flow velocity v_F, explanations in the text

Table 3.1 Excerpt from the experimental data

Row	v_F [m/s]	$(-\Delta p)$ [Pa]	$d^*_{S,SD}$ [µm]
1	0.0154	1334	239.4
2	0.0209	1736	244.7
3	0.0313	2668	241.9
4	0.0424	3551	244.5

Determination of $d^*_{S,SD}$:
Table 3.1 shows some of the evaluated experimental data from the fixed bed area of Fig. 3.9 as an example. The evaluation of row 3 is shown as an example:

$$\begin{aligned} d^*_{S,SD} &= 0.875 \cdot \frac{1-0.396}{0.396^3} \cdot \frac{1.20\,\frac{\text{kg}}{\text{m}^3} \cdot \left(0.0313\,\frac{\text{m}}{\text{s}}\right)^2}{\left|\frac{2668\,\text{Pa}}{0.307\,\text{m}}\right|} \\ &\cdot \left(1 + \sqrt{1 + 195.92 \cdot \left(\frac{0.396}{0.0313\,\frac{\text{m}}{\text{s}}}\right) \cdot \frac{18.26 \cdot 10^{-6}\,\text{Pa s}}{\left(1.20\,\frac{\text{kg}}{\text{m}^3}\right)} \cdot \left|\frac{2668\,\text{Pa}}{0.307\,\text{m}}\right|}\right) \\ &= 241.9 \cdot 10^{-6}\,\text{m} \end{aligned}$$

The average value of all measured diameters is: $\overline{d}^*_{S,SD} = 242.6\,\mu\text{m}$

In Fig. 3.9, the dependence calculated with (3.25) and the averaged $\overline{d}^*_{S,SD} = 242.6\,\mu\text{m}$ is represented by gray-shaded triangles and a fitting curve. The agreement with the measurements is good.

Since the Sauter diameter $d_{S,SD}$ can only be calculated with difficulty from the particle size distribution, a grain shape factor based on the median diameter $d_{S,50} = 350\,\mu\text{m}$ can be defined. For this, for example, applies:

$$\varphi = \frac{\overline{d}^*_{S,SD}}{d_{S,50}} = \frac{242.6\,\mu\text{m}}{350\,\mu\text{m}} = 0.693$$

When $\overline{d}^*_{S,SD}$ has been determined for a particle collective, a division into form factor and reference particle diameter is not necessary: $\overline{d}^*_{S,SD}$ can directly used for calculations of pressure losses.

Note: In the manner described above, the measured values of the fluidized bed area could also be evaluated. However, the ε_F values change here with the gas velocity v_F. Due to the onset of bed expansion and the possible bubble formation associated with it, see Sect. 3.3, the bed heights L_b and thus ε_F can only be determined relatively inaccurately.

3.3 Fluidized Beds

As indicated in Calculation Example 5, Sect. 3.2.4, Fig. 3.9, the aeration test can be carried out with such high gas velocities v_F that an expansion of the bulk material bed and thus the formation of a fluidized bed occurs. Due to the bed expansion, the solid particles in this state exhibit an increasing mobility with incrasing gas velocity, which results in a "liquid-like" behavior with only low internal friction/viscosity. In the following, gas/solid fluidized beds are considered. Deviations will be pointed out.

3.3.1 Fluidization Velocity

As can already be seen in Fig. 3.9, the pressure loss $(-\Delta p)$ of the vertically flowed through bulk material above a critical gas velocity $v_{F,L}$ remains constant. This velocity is referred to as minimum fluidization velocity $v_{F,L}$. From the so-called fluidization or loosening point and over the entire fluidized bed area, the bed weight is supported by the applied pressure difference $(-\Delta p) = (-\Delta p_{WS})$. Thus, it holds:

$$(-\Delta p_{WS}) \cdot A_R = F_G - F_A = A_R \cdot L_{WS} \cdot (1 - \varepsilon_{WS}) \cdot (\varrho_P - \varrho_F) \cdot g = \text{const.}$$
$$\rightarrow \quad v_F \geq v_{F,L} \tag{3.34}$$

Bed height L_{WS} and relative void volume ε_{WS} increase with increasing gas velocity $v_F > v_{F,L}$. With the porosity ε_L and the bed height $L_L \cong L_b$ at the loosening point, it follows from (3.34) for gases ($\rightarrow \varrho_P \gg \varrho_F$):

$$(-\Delta p_{WS}) \cong \frac{\Delta m_S \cdot g}{A_R} = \frac{G_S}{A_R} = L_{WS} \cdot (1 - \varepsilon_{WS}) \cdot \varrho_P \cdot g = L_L \cdot (1 - \varepsilon_L) \cdot \varrho_P \cdot g \tag{3.35}$$

The loosening velocity $v_{F,L}$ can be determined by equating the specific pressure loss of the fluidized bed $(-\Delta p_{WS})/L_L = (1 - \varepsilon_L) \cdot \varrho_P$, (3.35) with that of the stationary solid bed, (3.25) ($\rightarrow$ intersection of the fixed bed and fluidized bed characteristic lines). This results in:

$$v_{F,L} = 42.9 \cdot \frac{1 - \varepsilon_L}{d_{S,SD}^*} \cdot \frac{\eta_F}{\varrho_F} \cdot \left(\sqrt{1 + \frac{g}{3214} \cdot \frac{\varepsilon_L^3 \cdot}{(1 - \varepsilon_L)^2} \cdot (d_{S,SD}^*)^3 \cdot \frac{(\varrho_P - \varrho_F) \cdot \varrho_F}{\eta_F{}^2}} - 1 \right) \tag{3.36}$$

and in dimensionless form:

$$Re_L = 42.9 \cdot \frac{1 - \varepsilon_L}{\psi} \cdot \left(\sqrt{1 + \frac{\psi^3}{3214} \cdot \frac{\varepsilon_L^3 \cdot}{(1 - \varepsilon_L)^2} \cdot Ar} - 1 \right) \tag{3.37}$$

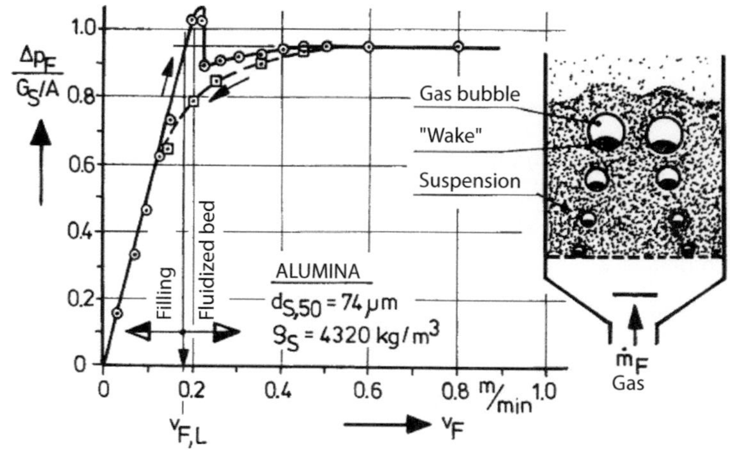

Fig. 3.10 Fluidization of a solid bulk material; Fluidizing gas: air

The Reynolds number Re_L is calculated with $v_{F,L}$ and $d_{S,SD}$, the Archimedes number Ar with $d_{S,SD}$ formed. Figure 3.10 shows the above-described facts using the example of the measured pressure loss characteristics of a "sandy" alumina. The relative pressure loss $(\Delta p/\Delta p_{WS}) = (\Delta p/(G_S/A_R))$ is plotted on the ordinate against the empty pipe velocity v_F of the gas on the abscissa.

When initially starting up the bulk material bed with increasing gas velocity v_F, breaking of particle interactions, and possibly entanglements/compressions, leads to a pressure peak that does not occur when shutting down with decreasing v_F. The relative pressure loss in the present case does not settle at the theoretical value of 1, but only at $(\Delta p/(G_S/A_R)) \cong 0.95$. The proportion $(1 - 0.95)$ of the bulk material weight is thus supported by the distributor floor and is not carried by the gas. It is also evident that the loosening velocity, which is formally defined as the intersection of the fixed bed and fluidized bed characteristics, assumes the value $v_{F,L} \cong 0.18\,\text{m/min}$, but the practical loosening occurs over a larger velocity range. This is influenced by grain shape and width of the particle size distribution. For the indicated reasons, it is recommended to determine loosening velocities by measurement.

While liquid/solid fluidized beds expand largely homogeneously with increasing fluid velocity, gas/solid fluidized beds exhibit deviating behavior: Above the loosening point, the fluidizing gas stream separates into a suspension and a bubble or channel phase. In the latter case, it involves solid-free/-poor channels that extend over the entire or a large area of the bed height. Figure 3.10 shows an example of bubble formation. The behavior that occurs in individual cases is largely determined by the respective particle properties. Based on this, a classification scheme for bulk materials has been developed in [16], which will be discussed in Sect. 3.3.2.

The working range of fluidized beds theoretically extends from the loosening velocity $v_{F,L}$ to the settling = terminal velocity w_T of the particle inventory. With the inhomogeneous gas/solid systems considered here, fluidization states far above w_T can also be realized in so-called circulating fluidized beds.

3.3.2 Geldart Diagram/Classification

Geldart [16] has developed a classification system based on his own and external measurements, which divides dry bulk materials with regard to their fluidization behavior in gas/solid fluidized beds into four groups - A, B, C, D. Figure 3.11 shows this classification scheme. On the ordinate, the difference between the particle density $\varrho_P \cong \varrho_S$ and the gas density ϱ_F, on the abscissa a characteristic particle diameter, in the original work $d_{S,SD}$, here simplifying and with sufficient accuracy $d_{S,50}$, is plotted. Geldart defines the behavior of the four bulk material classes as described below [11, 16]. For comparison purposes, a reference bulk material with $\varrho_P = \varrho_S = 2600\,\text{kg/m}^3$ (→ quartz particles) in different grinding is introduced.

Group A Bulk materials consisting of fine-grained particles and/or low particle density. Fluidized beds of these materials expand homogeneously to noticeably above the minimum fluidization before bubble formation begins. The bubble size remains relatively small. If the gas supply is suddenly switched off, the bed collapses very slowly at a speed

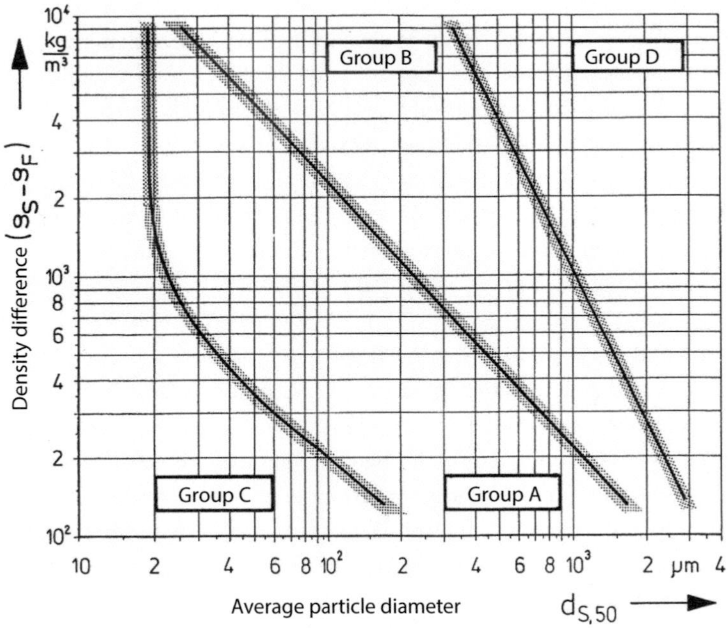

Fig. 3.11 Solid classification according to Geldart

that corresponds approximately to the gas-empty pipe speed in the suspension phase. All gas bubbles rise faster than the interstitial gas in the suspension phase. The average bubble size can be reduced in two ways: by a wide particle size distribution and/or a small average particle size. There is a maximum bubble size. Particle size range of the reference bulk material: $d_{S,50} \cong (20-90)\,\mu\text{m}$.

Group B This group contains most materials in the range of medium particle sizes and densities. In contrast to Group A, bubble formation begins directly above the minimum fluidization for Group B bulk materials. The bed expansion is low, and when the gas supply is suddenly switched off, the bed collapses very quickly. The bubbles rise significantly faster than the interstitial gas. There is no upper limit for bubble growth; it is generally limited by the dimensions of the apparatus. Particle size range of the reference bulk material: $d_{S,50} \cong (90-650)\,\mu\text{m}$.

Group C This group includes bulk materials that are cohesive in some way. "Normal" fluidization of these materials is extremely difficult. In small, smooth tubes, the bulk material is lifted as a whole/as a plug, see Fig. 3.12, or the gas merely blows free individual channels reaching from the inflow bottom to the bed surface (→ "rat holes"). The reason for this is that the adhesive forces acting between the particles are noticeably greater than the forces exerted by the gas on the particles. Only by using mechanical stirrers can more or less good fluidization be enforced. Particle size range of the reference bulk material: $d_{S,50} \leq 20\,\mu\text{m}$.

Group D These are bulk materials made of large and/or very heavy particles. The gas velocity in the suspension phase is comparatively large. Most gas bubbles rise at a lower speed than the interstitial gas. As a result, the suspension gas flows through the gas bubbles from bottom to top. This results in a gas exchange mechanism between the suspension and bubble phases that differs from that of Group A and B bulk materials. Group D materials can form so-called *spouted beds*. Particle size range of the reference bulk material: $d_{S,50} \geq 650\,\mu\text{m}$.

The transitions between the individual groups are fluid. The Geldart diagram implicitly describes the influence of two summary bulk material characteristics: gas holding capacity and gas permeability. of a particle assembly. The gas holding capacity can be characterized by the venting time $\Delta\tau_E$ of a fluidized bed suddenly separated from the gas supply, measured in a standardized apparatus under standardized conditions [17]. Figure 3.12 shows a possible device (→ cross-sectional area: $A_R = 1.0\,\text{dm}^2$, filling mass: $\Delta m_S = 2.0\,\text{kg}$), while Fig. 3.13 exemplarily compares the venting behavior of three different bulk materials determined in this apparatus. Details on the measurement methods can be found in, among others, [17, 18]. The gas permeability is a measure of the specific flow resistance of a bulk material. It is defined as the flow through the bulk material per unit area A_R due to an applied unit pressure gradient $(-\Delta p/L_L)$ by the flow-

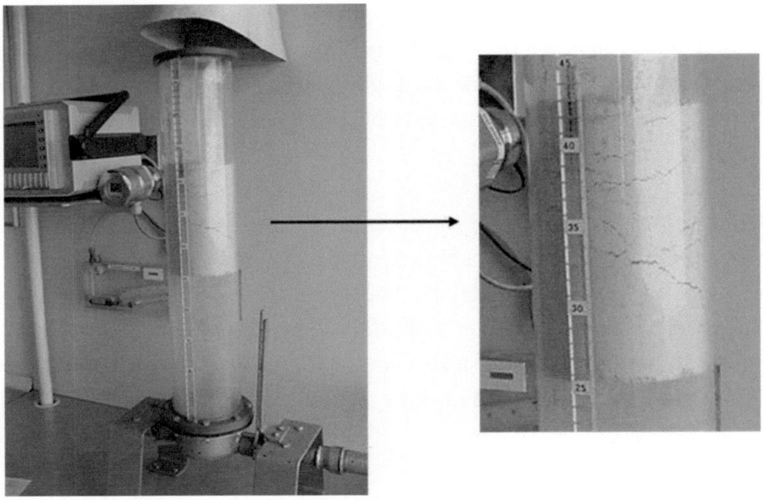

Fig. 3.12 Plug formation of a Group C bulk material

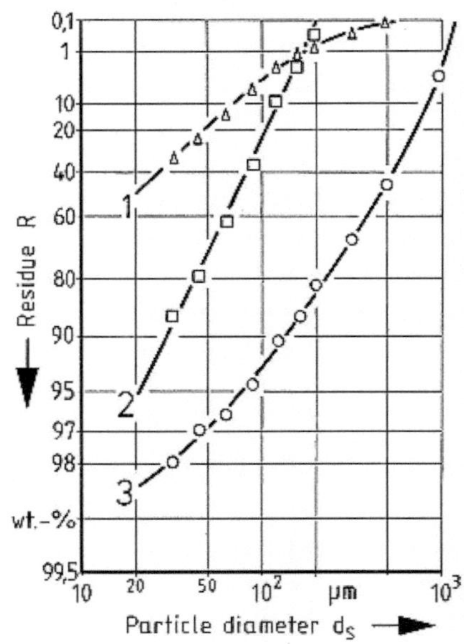

Curve	1	2	3
Bulk solid	Fly ash	Alumina	Coarse limestone
Geldart group	A	A/B	B/D
Bulk density [kg/m³]	1090	1040	1600
Vibrated bulk density [kg/m³]	1560	1160	1700
Particle density [kg/m³]	2480	4320	2730
Deaeration test:			
Quantity [kg]	2	2	2
Deaeration time [s]	191	20	1

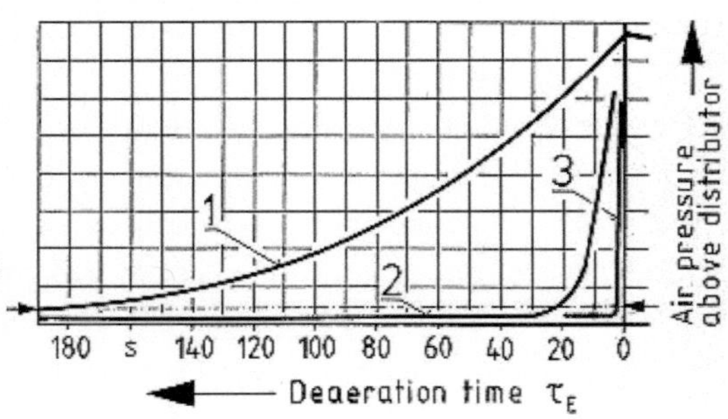

Fig. 3.13 Comparison of the venting behavior of different bulk materials

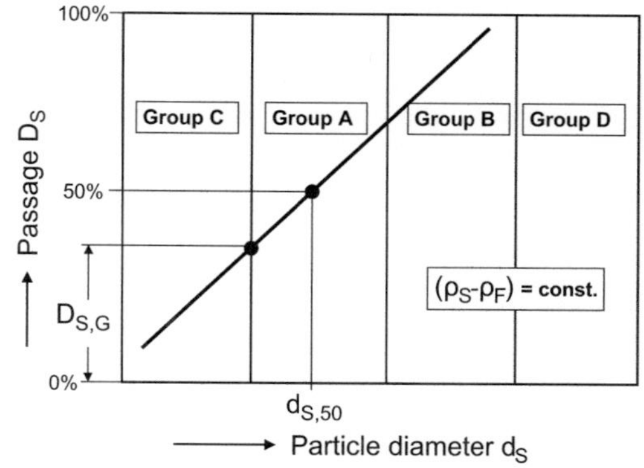

Fig. 3.14 Procedure for wide grain size distributions

ing gas volume flow $\dot{V}_F$. The larger $\dot{V}_F$ is, the greater the gas permeability. Gas holding capacity and permeability are coupled: increasing permeability results in decreasing gas holding capacity and vice versa. Gas permeability increases with increasing particle diameter $d_{S,50}$, i.e., from left to right in the Geldart diagram, while the gas holding capacity decreases in the same direction. From left to right in the diagram, the loosening velocity $v_{F,L}$ required for the generation of a fluidized bed increases from a few mm/s to the range of m/s.

The influence of the width of the particle size distribution of a bulk material on its fluidization behavior is only inadequately captured by the Geldart diagram. In the case of wide particle size distributions (→ for example, scattering parameters of the RRSB line $n \leq 1.5$) should not be classified solely based on the sizes $(\varrho_P, d_{S,50})$, but rather through additional venting tests. If these are not possible, the following procedure can be used, see Fig. 3.14: Based on the grain size distribution, it is checked whether a passage of more than $D_{S,G} \cong 40\,\text{M.\%}$ of the bulk material of a Geldart group is to be assigned to the left of the medium grain $d_{S,50}$. If this is the case, this proportion of the grain size distribution is set to 100 M.% and reclassified, i.e., the "coarse fraction" is discarded as having no influence. Justifications for this approach can be found in, among others, [19–22]. For bulk materials with a wide grain size distribution, the gas holding capacity increases with increasing fine fraction, while the gas permeability decreases. The fine fraction fills the void volume of the coarse fraction, i.e., the entire relative void volume ε_F decreases and accordingly (2.20), $(k \cdot \varepsilon_F) \cong \pi$, the number k of adhesion force-enhancing contact points between the particles increases.

The classification according to Geldart not only allows the assessment of the expected behavior of a bulk material during its fluidization, but also allows estimates of its general handling behavior whenever it is in a loosened state. This is, among other things, the case for pneumatic conveying, especially dense phase conveying, see Sect. 4.3, for rapidly flowing bulk materials, e.g., on chutes, behind belt feeders, when filling containers, or when collapsing bridges, funnels, and weirs in silos. Details on this can be found, among others, in [17]. A physical explanation of the bulk material behavior and the position of the boundary curves is given, among others, in [23].

3.3.3 Operation of Fluidized Beds

Subsequently, the effects/consequences of the influences described in Sect. 3.3.2 on fluidized bed operation are summarized. This is also necessary because gas/solid fluidized beds can be converted into pneumatic vertical conveyors by increasing the gas velocity v_F and thus comparable effects can be expected there:

- **Group C materials**: These can only be fluidized stably with external aids, e.g., agitators that destroy the internal gas flow channels, or possibly very high gas velocities. They then behave like Group A materials. Preliminary tests on appropriate test devices are absolutely necessary.
- **Bubble formation**: At the gas velocity

$$v_{F,B} = 2.07 \cdot \exp(0.76 \cdot g(45\,\mu\text{m})) \cdot \frac{d_{S,50} \cdot \varrho_F^{0.06}}{\eta_F^{0.347}} \tag{3.38}$$

with:

$g(45\,\mu\text{m})$ Weight fraction of the particle fraction with grain sizes $d_S < 45\,\mu\text{m}$,
$v_{F,B}$ [m/s]: if: $d_{S,50}$ [m], ϱ_F [kg/m^3], η_F [Pa s],

bubble formation begins [24]. The bubble formation speed $v_{F,B}$ normalized with the loosening speed $v_{F,L}$ (3.36), can take the following values:

$$\frac{v_{F,B}}{v_{F,L}} = K_{L/B} \geq 1 \tag{3.39}$$

- **Group A materials**: Their $(v_{F,B}/v_{F,L})$ ratios generally have values $K_{L/B} > 1$, which approach $K_{L/B} = 1$ as they approach the boundary curve to Group B. A-materials thus expand in the gas velocity range $(v_{F,L} \leq v_F \leq v_{F,B})$ homogeneous, i.e., the solid inventory is (statistically speaking) evenly distributed over the bed volume. They are transformed into a liquid-like state by the introduction of even small amounts of gas, which persists even after the gas supply is interrupted, see curve 1, fly ash, in Fig. 3.13. Above $v_{F,B}$, the fluidizing gas stream is divided into a bubble and a suspension phase. The suspension phase remains in the state of initial bubble formation $(\varepsilon_B, v_{F,B})$. The volume of the bubbles passing through the bed increases due to gas expansion and bubble coalescence over its height. If a volume-equivalent limit/maximum diameter $D_{B,\max}$ is exceeded ($\rightarrow$ lies in the centimeter range), the bubbles break up into smaller units. Group A materials are characterized by high gas holding capacity and correspondingly low gas permeability and show a strong tendency to flooding = uncontrolled flow. The group A behavior has a positive/supporting effect on the functions of aerated channel and horizontal pneumatic strand conveyors.
- **Group B and D materials**: In these cases, bubble formation starts directly above the loosening speed: $v_{F,B} = v_{F,L}, K_{L/B} = 1$. The suspension phase remains in the state of minimal fluidization $(\varepsilon_L, v_{F,L})$. There is no size limitation for the bubble growth,

which is essentially caused by coalescence. In reactors with a small diameter D_R and/or a large height/diameter ratio (L_{WS}/D_R), bubbles can fill the entire reactor cross-section (→ *slugs*). This leads to the formation of a successive series of gas bubbles and bulk material plugs, characterized by strong pressure pulsations and referred to as pulsating fluidized bed. Since the rising speed u_B of a bubble with its diameter D_B increases, $u_B \propto \sqrt{g \cdot D_B}$, a fluidized bed of Group B or D bulk materials collapses almost instantaneously after the gas supply to the bed is switched off due to the bypass effect of the relatively large bubbles, i.e., the gas holding capacity of these bulk materials is low, the gas permeability, however, is high and increasing further towards Group B after D. Flooding is only possible to a limited extent with Group B bulk materials and not possible with D materials. Bulk materials to the right of the boundary curve A/B in the Geldart diagram can be classified as free-flowing with a sufficiently narrow particle size distribution.

- **Bed expansion**: The expansion of bubble-forming heterogeneous fluidized beds is, due to the decreasing bubble residence time in the fluidized bed with increasing bubble size, less than that of homogeneously expanding fluidized beds.

If a fluidized bed can be operated at a given gas flow $\dot{m}_F$ with higher gas velocity v_F, this, among other things, reduces the reactor diameter and thus its investment costs. This has led to corresponding investigations and modifications of fluidized bed systems, which are briefly summarized below:

- ***Turbulent fluidization***: With a slow increase in gas velocity v_F in a given system with a predetermined bulk material, the operating states discussed earlier are successively passed through. If these are characterized, for example, by the pressure fluctuations $K_{\Delta p}$ related to the respective average bed pressure loss, this parameter increases with increasing v_F as expected (→ increasing bubble size, possibly formation of a slugging fluidized bed), and then with further v_F-increase, it drops again to a significantly smaller, approximately constant value, i.e., the pulsations of the bed are damped and the bed is transferred to a "more homogeneous" state with more intense interactions between gas and solid. The gas velocity $v_{F,tf}$ associated with the onset of this state is at the beginning of the $K_{\Delta p}$-plateau. A recognized definition for determining this has not yet been established. Measurement results [25] show that a normalization of $v_{F,tf}$ with the settling velocity w_T—this is calculated with the respective average particle diameter $d_{S,SD}$ of the examined particle collective—for fine-grained group A bulk materials to values $(v_{F,tf}/w_T) \cong 10$, for coarse-grained D materials, however, to $(v_{F,tf}/w_T) \leq 1$ leads. Visual observations illustrate that due to the high fluidization velocities $v_F \geq v_{F,tf}$ large gas bubbles and *slugs* in the bed break down into smaller ones, gas bubbles are deformed into elongated gas spaces, and particle clusters and strands form. The fact that values $(v_{F,tf}/w_T) \gg 1$ can be realized for fine-grained bulk materials is explained by the fact that bulk material conglobations and strands, whose "apparent" terminal velocity is significantly greater than that of individual particles, are discharged from the bed and fall back into it before they have dissolved into individual particles. A limited amount of individual particles is still discharged and may need to be returned to the bed. The transition of the turbulent fluidized bed into the free space above is not sharply delimited but has a diffuse transition area. The existence of *turbulent fluidization* ends with the so-called transport velocity $v_{F,tr}$, which in turn marks the beginning of *fast fluidization*.

- ***Fast fluidization***: When the transport velocity $v_{F,tr}$, which is determined by the empirical equation

$$Re_{tr} = \frac{v_{F,tr} \cdot d_S \cdot \varrho_F}{\eta_F} = 2.28 \cdot Ar^{0.419} \tag{3.40}$$

is exceeded [26, 27], see Fig. 3.15, there is an extreme increase in the solid mass flow carried out of the bed, which would lead to it emptying completely within a very short time. By installing a solid return system (→ gas/solid separator, solid return in a *downcomer* and introduction against the overpressure at the base of the reactor), this is prevented and stable operation is ensured. Such systems are called circulating fluidized beds. In Fig. 3.15, in addition to (3.40), the course of the *Re* number $Re_{S,T}$ of the corresponding single-particle terminal velocity w_T, (3.12), as well as the limits of the various Geldart groups are plotted. The comparison of (3.40) and (3.12) shows the effect already described in *turbulent fluidization*, that the discharge speed $v_{F,tr}$ of

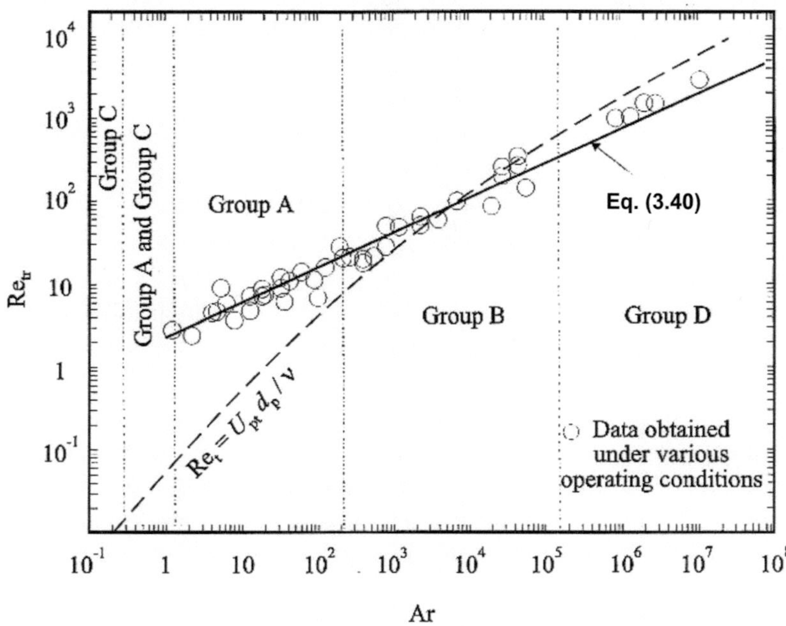

Fig. 3.15 Empirical basis of (3.40) [27]

fine-grained bulk materials is a multiple of their settling velocity w_T, while for coarser products it approximately matches this value.

The operating behavior of circulating fluidized beds is influenced by the chosen operating velocity $v_F > v_{F,tr}$, but also by the size of the circulating specific solid mass flow $(\dot{m}_S/A_R)$, which increases with v_F [27–31]. The resulting axial distribution of solid volume concentration is due to the combined effect of both factors. $\varepsilon_S = (1 - \varepsilon_F)$, which typically has a denser soil area with $\varepsilon_S \cong (0.2-0.4)$ near the bottom, a subsequent, variably high region with $\varepsilon_S \cong 0.2$ and a transition area with lower ε_S values. The current reactor height also influences this distribution. The solid moves in the form of particle clusters and strands of varying density. Uneven solid and gas distribution profiles are established across the reactor cross-section A_R: lower solid concentrations and thus higher gas velocities in the central area, leading to a vertical upward transport of the solid, and higher solid concentrations and lower gas velocities in the wall area, resulting in a downward-moving solid fraction against the direction of the gas flow. Increasing v_F reduces this fraction. An upper operating limit for *fast fluidization* is defined as the transition to pneumatic vertical conveying, which is determined by the empirical equation

$$\frac{v_{F,pf}}{\sqrt{g \cdot d_S}} = 21.6 \cdot \left(\frac{\dot{m}_S}{A_R \cdot \varrho_F \cdot v_{F,pf}} \right)^{0.542} \cdot Ar^{0.105} \tag{3.41}$$

with:

$v_{F,pf}/\sqrt{g \cdot d_S} = \sqrt{Fr_{pf}}$ $\quad Fr_{pf} =$ Froude number of the process,
$\dot{m}_S/(A_R \cdot \varrho_F \cdot v_{F,pf}) = \mu_{pf}$ $\quad \mu_{pf} =$ corresponding solid/gas loading,

can be described [26, 29] and represents the operating condition at which no solid backflow can be observed [29]. This will be discussed again in Sect. 4.6.3.

The pressure losses of *turbulent* and *fast fluidization* in the actual reactor can be estimated with (3.35) if the now existing, but compared to the static pressure of the bed weight $(\Delta m_S \cdot g/A_R)$ low acceleration and friction pressure losses can be taken into account globally:

$$(-\Delta p_{WS}) \cong (1.1-1.2) \cdot \left(\frac{\Delta m_S \cdot g}{A_R} \right) \tag{3.42}$$

For Δm_S the mass of the entire solid inventory currently in the reactor must be used.

Note For a direct scale-up of the measurement results from experimental fluidized beds to operating plants, the diameter of the experimental fluidized bed $D_R \geq 0.50\,m$ should be executed. For smaller D_R, a pronounced additional diameter dependency of the results must be taken into account [32].

3.4 Particle Swarm in Gas Flow

Bulk materials are predominantly moved and flowed through as particle swarms in process engineering processes. Fluidized beds can be considered as particle swarms in this sense. In comparison to isolated individual particles, additional forces/effects act on the particles bound in these particle collectives, which depend significantly on the current solid volume concentration $\varepsilon_S = (1 - \varepsilon_F)$. Among other things, this leads to strong changes in the gas velocity field, increased momentum exchange between gas/solid and particle/particle (→ swarm turbulence) and, as already shown in Sect. 3.3.3, also to gas/solid segregation. Therefore, a distinction must be made between (statistically) homogeneous and heterogeneous distribution of the solid in the considered volume.

- **Homogeneous Distribution**: For particle distances $x > 6 \cdot d_S$, corresponding to $\varepsilon_F > 0.9975$, the mutual influence is negligibly small: The particles in the swarm behave like isolated individual particles, i.e., the swarm resistance $F_{W,\varepsilon}$ corresponds to the sum of the resistances F_W of the individual particles calculated with (3.3) and the swarm terminal velocity $w_{T,\varepsilon}$ of that w_T of an isolated individual particle.
  In the range of less strongly loosened particle accumulations, $\varepsilon_F < 0.9975$, down to pourings of contacting particles, $\varepsilon_F \gtrsim 0.40$, there is a reduction in the swarm terminal velocity $w_{T,\varepsilon}$ ($= v_F$ in fluidized beds) compared to that of isolated single particles w_T, i.e. $w_{T,\varepsilon}/w_T < 1$, or $v_F/w_T < 1$ in fluidized beds. For the calculation of the dependencies $w_{T,\varepsilon}(\varepsilon_F)$ or $v_F(\varepsilon_F)$ there are a number of approaches with respective limiting boundary conditions, see, among others, [4, 11, 14, 33]. However, such models are hardly applicable in the context of the pneumatic conveying, as this, especially in the dense flow range, is characterized by highly segregated gas/solid flows.
- **Heterogeneous Distribution**: Is caused by the spontaneous formation of gas bubbles, particle aggregations/balls/clusters, and strands within the swarm and is only calculable in individual cases. As an example, consider a vertical conveying pipe through which gas and solid flow, in which the solid has condensed into a local bulk material aggregation/solid ball. The (porous) conglobation, due to its higher volume concentration of particles, has a greater flow resistance for the conveying gas than the surrounding, particle-depleted gas/solid flow. This leads to the conveying gas partially bypassing the conglobation and flowing around it at a correspondingly higher speed. At the same time, particles from the less dense outer flow collide with the conglobation and additionally drive it. By absorbing colliding/following particles (→ slipstream effect), it can increase its volume. Overall, the conglobation behaves like a "large" particle with a correspondingly large settling velocity. If this is greater than the inflow velocity, the ball falls downward against the gas flow direction and dissolves, so that its individual particles can be flowed around again and transported a distance vertically upward before a new separation occurs. This "climbing" of successively forming and dissolving balls is a phenomenon frequently observed in vertical

pneumatic dense phase conveying. The described process can easily be converted into a mathematical model. However, its applicability is limited by the fact that there are currently no reliable findings on why, when, where, and how (→ triggering mechanism, size, density, etc.) such aggregations form. An exception to this is the bubble formation in stationary fluidized beds described above. The fact that the structural design of a reactor/transport system also provides additional triggers for separations is illustrated by a simple flow deflection: The solid, evenly distributed over the transport cross-section before this, leaves the deflector as a strand adhering to its outer wall.

References

1. Merker, G.P., Baumgarten, C.: Fluid- und Wärmetransport: Strömungslehre. Teubner, Stuttgart (2000)
2. Kaskas, A.A.: Schwarmgeschwindigkeit in Mehrkornsuspensionen am Beispiel der Sedimentation. TU Berlin, Berlin (1970). Dissertation
3. Martin, H.: Wärme- und Stoffübertragung in der Wirbelschicht. Chem.-Ing.-Tech. **52**(3), 199–209 (1980)
4. Schubert, H. (Hrsg.): Handbuch der Mechanischen Verfahrenstechnik Bd. 1. Wiley-VCH, Weinheim (2003)
5. Haider, A., Levenspiel, O.: Drag coefficient and terminal velocity of spherical and non-spherical particles. Powder Technol. **58**, 63–70 (1989)
6. Crowe, C., Sommerfeld, M., Tsuji, Y.: Multiphase flows with droplets and particles. CRC Press, Boca Raton (1998)
7. Sawatzki, O.: Über den Einfluss der Rotation und der Wandstöße auf die Flugbahnen kugeliger Teilchen im Luftstrom. Univ. Karlsruhe, Karlsruhe (1961). Dissertation
8. Schubert, H., Heidenreich, E., Liepe, F., Neeße, T.: Mechanische Verfahrenstechnik I. VEB Deutscher Verlag für Grundstoffindustrie, Leipzig (1977)
9. Löffler, F., Raasch, J.: Grundlagen der Mechanischen Verfahrenstechnik. Vieweg, Braunschweig, Wiesbaden (1992)
10. Ergun, S.: Fluid flow through packed columns. Chem. Engng. Progr. **48**(2), 89–94 (1952)
11. Molerus, O.: Fluid-Feststoff-Strömungen. Springer, Berlin (1982)
12. Werther, J.: Strömungsmechanische Grundlagen der Wirbelschichttechnik. Chem.-Ing.-Tech. **49**(3), 193–202 (1977)
13. Yoon, S.M., Kunii, D.: Gas flow and pressure drop through moving beds. Ind. Eng. Chem. Process. Des. Dev. **9**(4), 559–565 (1970)
14. Brauer, H.: Grundlagen der Einphasen- und Mehrphasenströmungen. Sauerländer, Aarau (1971)
15. Jenike, A.W.: Storage and flow of solids. Bull. No. 123, Utah Engng. Exp. Station. Univ. of Utah, Salt Lake City (1964)
16. Geldart, D.: Types of gas fluidization. Powder Technol. **7**, 285–292 (1973)
17. Hilgraf, P.: Einfluss von Schüttgütern im Hinblick auf ihre Handhabung. VDI-Berichte 1918., S. 63–93 (2006)
18. McGlinchey, D. (Hrsg.): Characterization of Bulk Solids. Blackwell, Oxford (2005)
19. Schönlebe, K., Seewald, H.: Fließverhalten binärer Kohlemischungen. Aufbereitungs Tech. **32**(7), 335–343 (1991)

20. Husemann, K., Höhne, D., Schünemann, U.: Der Einfluss des Feinkornanteils auf die Fließeigenschaften von groben Schüttgütern. Aufbereitungs Tech. **35**(2), 61–70 (1994)
21. Hilgraf, P.: Handling of bulk material mixtures. ZKG Int. **61**(11), 46–59 (2008)
22. Molerus, O.: Schüttgut-Mechanik. Springer, Berlin (1985)
23. Molerus, O.: Interpretation of Geldart's type A, B, C and D powders by taking into account interparticle cohesion forces. Powder Technol. **33**, 81–87 (1982)
24. Abrahamsen, A.R., Geldart, D.: Behavior of gas fluidized beds of fine powders, part I: homogeneous expansion. Powder Technol **26**, 35 (1980)
25. Yerushalmi, J.: High velocity fluidized beds. Chapter 7. In: Geldart, D. (Hrsg.) Gas fluidization technology. John Wiley & Sons, New York (1986)
26. Bi, H.T., Fan, L.-S.: Regime transitions in gas-solid circulating fluidized beds. AIChE Annual Meeting, Los Angeles. Paper # 101e., S. 17–22 (1991)
27. Fan, L.-S., Zhu, C.: Principles of gas-solid flows. Cambridge University Press, New York (1998)
28. Kunii, D., Levenspiel, O.: Fluidization engineering, 2. Aufl. Butterworth-Heinemann, Boston (1991)
29. Bi, X.: Flow regime transitions in gas-solid fluidization and transport. University of British Columbia, Vancouver (1994). Ph.D. Thesis,
30. Bi, H.T., Grace, J.R., Zhu, J.X.: Types of choking in vertical pneumatic systems. Int. J. Multiph. **19**(6), 1077–1092 (1993)
31. Wirth, K.-E.: Zirkulierende Wirbelschichten. Springer, Berlin (1990)
32. Werther, J.: Zur Maßstabsvergrößerung von Gas/Feststoff-Wirbelschichten. Aufbereitungs-Technik **15**(12), 670–677 (1974)
33. Gibilaro, L.G.: Fluidization-dynamics. Butterworth-Heinemann, Oxford (2001)

4 Basics of Pneumatic Conveying

First, the conveying/state diagram of the PF and the flow patterns that occur in a PF, as well as their dependence on the properties of the conveyed bulk material, are presented. A discussion of the calculation bases required for system dimensioning and suggestions for system design follows.

4.1 Conveying Diagram

State or conveying diagrams are used to describe conveying technical relationships.

4.1.1 Standard Representation

Figure 4.1 shows schematically the usual plotting of such a diagram. On the ordinate, the pressure difference $\Delta p_R = (-\Delta p) = (p_{\text{in}} - p_{\text{out}})$ at the conveying line, on the abscissa a characteristic gas velocity v_F, usually the one at the line end $v_{F,E}$ (Index E), plotted. The parameter is the solid throughput $\dot{m}_S$. Such a state diagram is only valid for the respective investigated combination of conveyed material and specific conveying line. Each operating point in the diagram corresponds to a conveying with the specific operating data ($\dot{m}_S$, Δp_R, v_F). In Fig. 4.1 two limit curves are plotted, between which a conveying operation can be realized [1, 2, 3]:

- Curve A = solid-free conveying pipe: Curve A corresponds to the pressure loss characteristic of the conveying gas flow through the empty conveying pipe. Calculation of Curve A with (2.79). The power input by the gas flow is completely used to overcome its own flow resistance. To the right of Curve A, no solid transport is possible.

P. Hilgraf, *Pneumatic Conveying*, https://doi.org/10.1007/978-3-662-67223-5_4

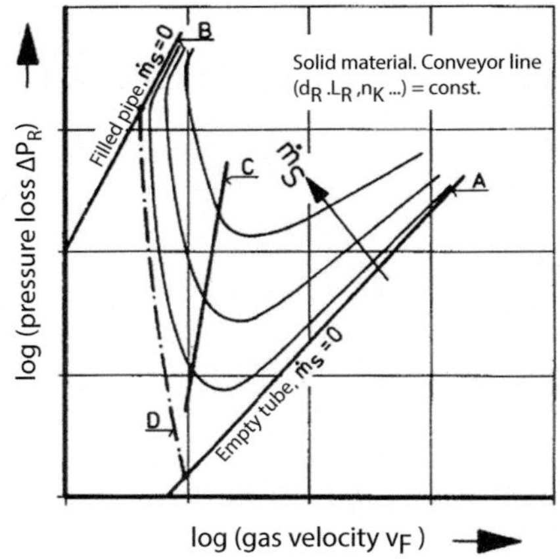

Fig. 4.1 Conveying diagram, schematic [1]

- Curve B = completely filled with stationary solid material conveying pipe: Curve B corresponds to the pressure loss characteristic of the conveying gas flow through the conveying pipe filled with solid material. Calculation of Curve B with (3.30). The power input by the gas flow is completely used to overcome its own flow resistance. To the left of Curve B, no solid transport is possible.

Additional resistances, e.g., deflections and/or changes in height, must be considered separately in (2.79) and (3.30). All possible conveying states lie between the boundary curves A and B. From Fig. 4.1, it is already apparent that, for a given conveying line pressure difference Δp_R, the bulk material must be in a "loosened/diluted" state to the right of the boundary curve B in order to be transported.

The lines of constant solid mass flow $\dot{m}_S$ in Fig. 4.1 shift towards larger pressure losses Δp_R with increasing throughput. All pass through a pronounced pressure loss minimum. For each solid throughput $\dot{m}_S$, there are two operating points for a given Δp_R: one at the higher gas velocities v_F to the right of the pressure loss minimum, another at lower gas velocities to the left of it. Operating points to the right of the connecting line of the individual Δp_R minima are often assigned to dilute phase conveying, and those to the left to the so-called dense phase conveying.

If the conveying diagram, Fig. 4.1, is recorded on a conveyor line with horizontal and vertical sections as well as deflections, there is the problem of transferring the results to other pipeline geometries, e.g., that of an operating system. The individual pipe elements have different specific flow resistances. They influence each other and also the flow states that develop in the conveyor pipe. A direct scale-up of the measurements is only possible for geometrically similar conveyor lines with the same pipe diameter D_R . Pilot plants for creating conveying diagrams are typically subject to structural restrictions regarding pipeline routing. To obtain the necessary information about the behavior of the various pipe elements, the test sections should contain at least one unaffected horizontal and one vertical measuring section each. Unaffected means: no influences of upstream and downstream pipe elements on the measurement result. The additional measurement data (→ i.e., general pressure measurements, possibly visual observations through glass pipe elements) allow the construction (of different) horizontal and vertical state diagrams and the determination of the associated resistance coefficients. This will be discussed later. The following considerations are carried out using the example of a straight horizontal conveyor line. Deviations will be pointed out.

Curve C in Fig. 4.1 is referred to as the "clogging limit". A uniform definition for this is lacking. The clogging limit generally describes the transition from stable steady state flow forms to unsteady and thus more clogging-prone conveying states in the conveyor line. In horizontal lines, this corresponds to the transition from strand to dune, bale, or plug conveying, see Sect. 4.2. The clogging limit is therefore not a physical conveying limit. In practical operation, it depends not only on the flow conditions in the pipe but also on the conveying/delivery characteristic $\Delta p(\dot{V}_{F,0})$ of the pressure generator used. If its characteristic curve is sufficiently steep, i.e., the sucked gas volume flow $\dot{V}_{F,0}$ changes only slightly with increasing pressure difference Δp, then safe conveying operation can be maintained even to the left of the clogging limit, i.e., with lower gas velocities. Many dense phase conveying systems operate in this area.

The boundary curves A and B intersect at the coordinate origin, see Fig. 4.1. Since a differential velocity $u_{\text{rel}} = (u_F - u_S)$ between gas and solid is required for solid propulsion, an arbitrary reduction of the conveying gas velocity v_F is not possible. Thus, there must exist another boundary curve D, connecting curves A and B, on which $\dot{m}_S = 0$ will exist. This curve is crucial for practical plant design and can only be determined experimentally. Curve D ultimately describes the minimum conveying gas velocities $v_{F,\min}$ required for material transport in highly segregated gas/solid flows. This will be discussed in more detail in Sect. 4.6.2.

While with fine-grained bulk materials (→ $d_{S,50} \lesssim 100\,\mu\text{m}$) any combination of $\Delta p_R, v_F$ and $\dot{m}_S$ can be realized in the conveying diagram, unstable areas occur with coarser bulk materials with a relatively narrow particle size distribution, which are unsuitable for practical operation. This leads to pronounced pressure pulsations, shocks, and strong noise development (→ rattling) in the conveying line as well as periodic and uncontrollable sudden line blockages. This will be discussed in more detail in Sect. 4.13.4.

The conveying diagram of a vertical conveying line is largely identical to that of horizontal transport routes. The intersection of curves D and B can be identified (theoretically) as minimum fluidisation velocity $v_{F,L}$, that of D and A approximately as terminal velocity w_T of the characteristic individual particle of the solid. The explanations in Sect. 3.3.3 illustrate that the conveying behavior of the various bulk material classes is more complex, e.g., (3.40) must be considered in the diagram. Along curve B, due to gas expansion in the upper area of the vertical section, fluidization and bed expansion can already occur before the loosening speed $v_{F,L}$ is exceeded at the base of the line, etc.

4.1.2 Modified Representations

The state diagram shown in Fig. 4.1 is often used in a slightly modified form: On the ordinate, instead of the pressure difference Δp_R its value related to the conveying distance L_R $(\Delta p_R/L_R)$ and/or as a parameter instead of the solid throughput $\dot{m}_S$ the corresponding loading $\mu = \dot{m}_S/\dot{m}_F$ and/or on the abscissa instead of the gas exit velocity $v_{F,E}$ the gas mass flow $\dot{m}_F$ or its specific value $(\dot{m}_F/A_R)$ is used. Further variants are possible.

More importantly, the type of plotting of a state diagram may need to be adapted to the currently available experimental technique or to clarify certain operating parameters. The described standard representation is easy to implement if the bulk material is fed into the conveying line in a controlled manner, e.g., by means of a rotary valve. By adjusting its revolution speed, a constant solid mass flow can be maintained. $\dot{m}_S$ is set, with various gas velocities v_F conveyed and the resulting pressure losses Δp_R are measured. Then, $\dot{m}_S$ is adjusted to a new value and the whole process is repeated, i.e., the dependency $\Delta p_R(\dot{m}_S, v_F)$ is determined. When introducing the solid material using a pressure vessel, the gas velocity v_F is varied at a fixed conveying pressure difference Δp_R and the resulting solid material flow rates $\dot{m}_S$ are measured, thus: $\dot{m}_S(\Delta p_R, v_F)$. Figure 4.2 shows a typical plot of such measurements in practice. The individual curves $\Delta p_R = \text{const.}$ corresponds to horizontal cuts through the standard diagram shown in Fig. 4.1, while the ordinate intersections of the lower $\dot{m}_S = 0$ extrapolated Δp_R-curve branches represent gas velocities $v_{F,\min}$ on the boundary curve D. As shown in [4], among other things, a correctly designed and shaped conveying line will always become unstable at the beginning of the line, i.e., at the lowest gas velocity, when subjected to externally imposed disturbances. It is therefore reasonable to use the gas velocity at the beginning of the conveying line $v_{F,A}$ (Index A) as the characteristic gas velocity in the state diagram shown in Fig. 4.2. The velocities determined in this way $v_{F,A,\min}$ represent the absolute lower operating limit of the considered conveying system [5]. Practical designs must maintain an appropriate safety margin to this. Details in Sect. 4.6.2.

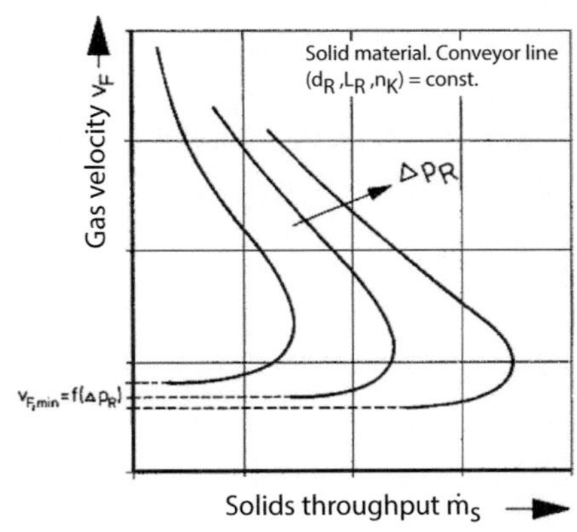

Fig. 4.2 Modified conveying diagram, schematic [1]

Since the characteristic values and operating behavior of pneumatic conveyors, among other things, are also influenced by the size of the pipe diameter D_R, measurements on at least two conveying lines with significantly different diameters D_R are required for a safe scale-up.

Dimensionless representations of the conveying diagram have not yet become established.

Note: Prerequisites for usable measurement results from the investigations described above are correct pipeline routing in the test facilities, i.e., sufficiently long acceleration or re-acceleration sections at the beginning of the conveying line or behind/between deflections, etc., and an accordingly unaffected positioning of the measuring points.

4.2 Flow Forms

Initially, straight horizontal conveying lines are considered again. Depending on the size of the conveying gas velocity v_F, different flow states occur in the conveying line, see Fig. 4.3. With sufficiently high gas velocities v_F, the solid flows through the conveying pipe relatively evenly distributed across its cross-section. Gradual reduction of v_F with constant mass flow $\dot{m}_S$ leads to an increasingly stronger enrichment of the flow with solid in the lower pipe cross-section (→ consequence of the gravitational force acting perpendicular to the flow direction). The solid concentration profile across the pipe height

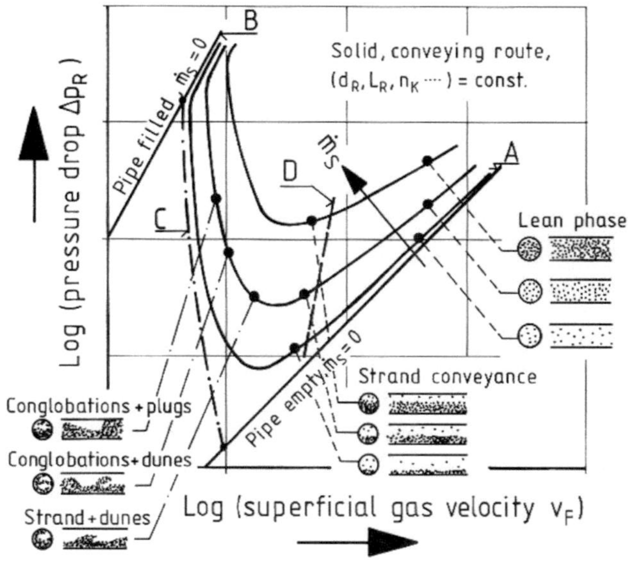

Fig. 4.3 Conveyance diagram with flow forms [7]

becomes increasingly asymmetric. Because each conveying gas can only carry a limited amount of solid in flight, when falling below a critical gas velocity v_F, usually referred to as *saltation velocity*, the solid content exceeding the associated limit load μ_{strand} falls out from the gas stream and must be transported in the form of a strand sliding over the tube bottom. The drive is mainly provided by the impulse of impacting particles from the pneumatic conveying remaining above the strand. Strand formation often occurs at gas velocities v_F, which are still to the right of the connecting line of the pressure loss minima of the curves $\dot{m}_S = \text{const.} \rightarrow$ curve D in Fig. 4.3. An estimation of the limit load μ_{strand} can be made for fine-grained solids ($\rightarrow d_S \lesssim 100\,\mu\text{m}$) using the equation [6, 7]

$$\mu_{\text{strand}} = K_0 \cdot \frac{v_F^3}{g \cdot w_T \cdot D_R} = K_0 \cdot \frac{Fr_R^{3/2}}{Fr_T^{1/2}}, \tag{4.1}$$

with:

$Fr_R = v_F^2/(g \cdot D_R)$ Froude number of conveyance,
$Fr_T = w_T^2/(g \cdot D_R)$ Froude number of terminal velocity.

K_0 must be determined specifically for the bulk material. As an approximation for the beginning of solid fallout, $K_0 \cong 3 \cdot 10^{-5}$ can be used.

Further reduction in gas velocity increases the solid content conveyed as a strand. (4.1) can be applied to the pneumatic conveying in the remaining free pipe cross-section $(\varphi \cdot A_R)$ above the strand, if (v_F/φ) is used instead of v_F and $(D_R \cdot \sqrt{\varphi})$ instead of D_R. K_0 must be determined anew. The beginning and course of solid separation are influenced, as (4.1) exemplarily illustrates, by the size of the conveying pipe diameter D_R, i.e., the respective plant design, and by all those solid properties that are implicitly described by the settling velocity w_T of individual particles, including their density, size, grain shape, and surface structure. With increasing strand content, those solid characteristics that result from the integration of individual particles into a bulk material structure increasingly influence the conveying process. These include, for example: internal friction angle, wall friction angle, gas permeability, gas holding capacity, etc.

Further reduction of the gas velocity causes the solid strand to flow increasingly irregularly and eventually leads to the formation of waves/dunes, which run over the strand, grow in height, and can temporarily fill the entire pipe cross-section locally. Triggers: increased gas and thus solid velocities in the free space, more intense particle bombardment of the strand from the upper space, larger gas shear stresses on the strand surface, etc. Converging dunes form conglobations, which in turn can be compressed into closed plugs. When exceeding a critical plug length dependent on the available conveying pressure difference Δp_R, the gas mass flow $\dot{m}_F$ and the properties of the respective solid, the pipeline becomes clogged by "wedging". In this case, the radial support of the axial forces acting on the plug on the pipe wall increases the wall friction so much that the usable pressure gradient is no longer sufficient for plug propulsion, see Sect. 2.2.1. The existence ranges of the individual flow forms and their effects on practical conveying operation are significantly influenced by the type or the characteristics of the conveyed solid. This is discussed in Sect. 4.5.

As a result of the pressure difference applied to the conveying line, the conveying gas expands in the direction of flow. The flow forms shown above are thus realized more or less pronounced one after the other along a conveying route. Additional influences include, among others, the type of entry device and the current pipeline routing. Figure 4.4 considers the conveying route as a series connection of pipe sections with approximately constant gas velocity v_F, in which the flow forms described above occur. The states run along a curve $\dot{m}_S$ = konst. and are shown in Fig. 4.3 only for correspondingly short pipelines. In implemented conveying systems, such state changes only cover a limited partial area of the curve $\dot{m}_S$ = const.

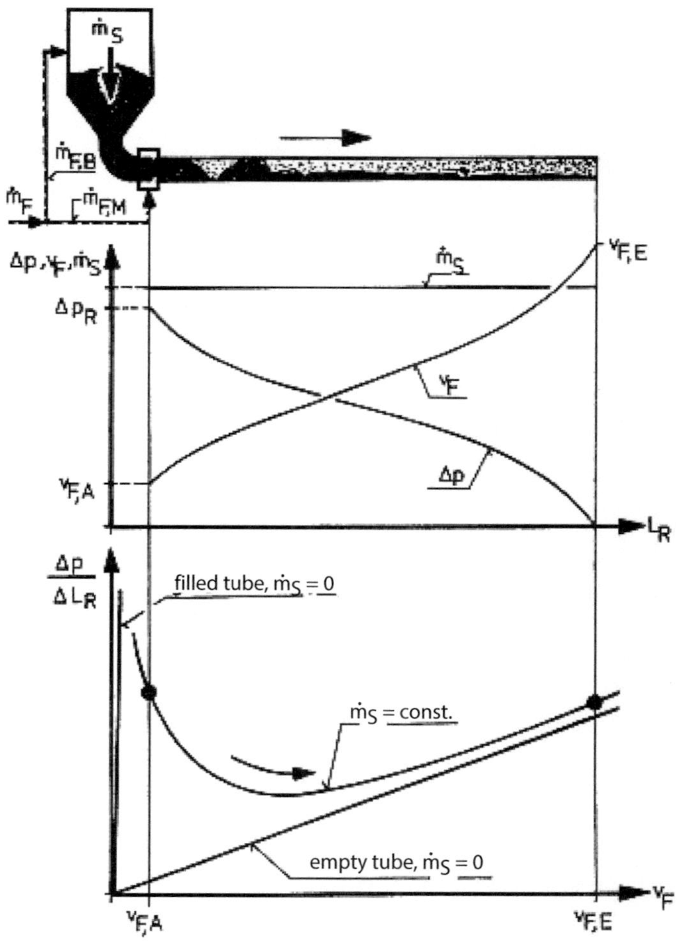

Fig. 4.4 Flow forms along a conveying route

The forces driving or counteracting the solid matter change with the flow pattern. Figure 4.5 shows this for three different flow patterns. In the lean phase conveying area, the driving force is provided by the drag forces F_W of the individual particle flow. With decreasing gas velocity = transition to the dense phase area, pressure difference forces $\Delta p \cdot A_R \cdot (1 - \varepsilon_F)$ gain importance. The drive of a strand is also supported by particle bombardment from the flight phase. Plugs are completely flowed through by the gas. The processes at the plug are discussed in more detail in Sect. 4.4.

The pressure loss minima of the individual $\dot{m}_S$-curves result from the competition of two different resistance mechanisms. At high gas velocities, the conveying resistance results from particle/wall and particle/particle collisions (→ deceleration and

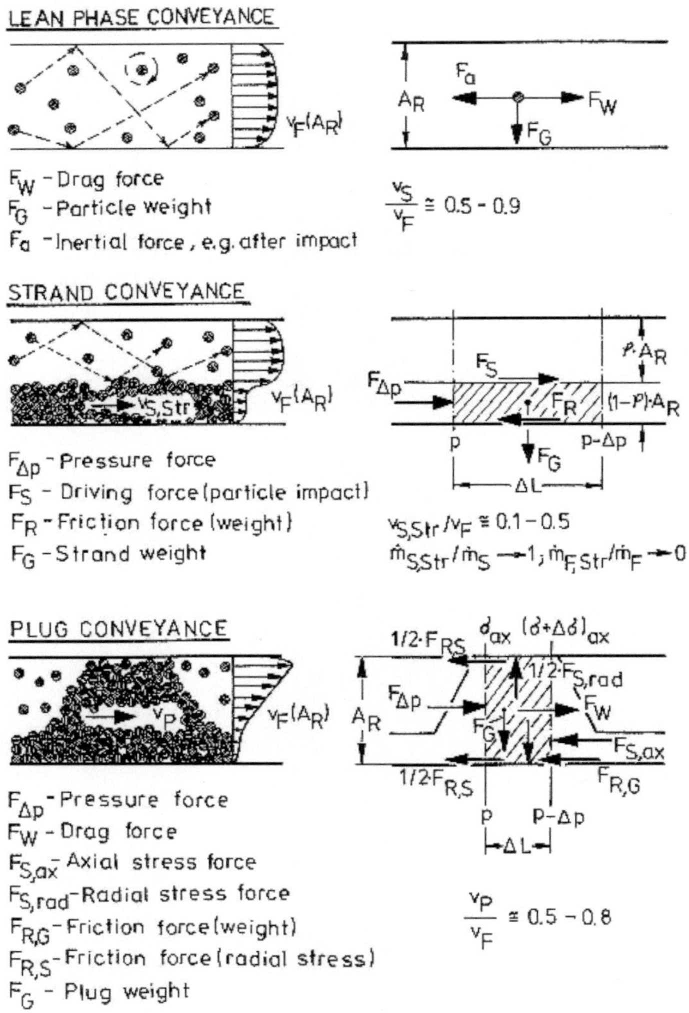

Fig. 4.5 Forces on the solid matter during lean phase, strand, and plug conveying

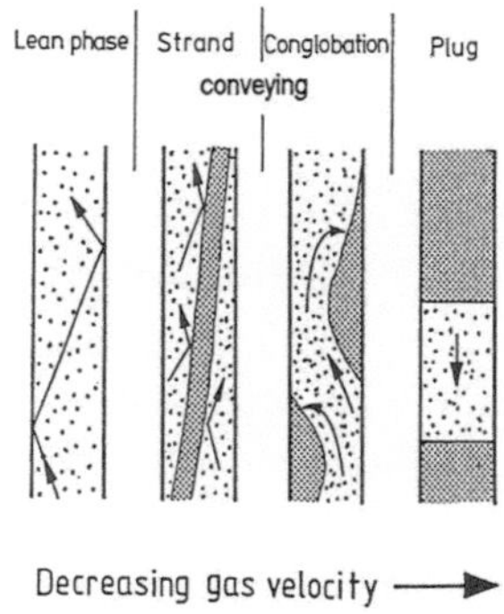

Fig. 4.6 Flow patterns in vertical pneumatic conveying

re-acceleration) and at correspondingly low gas velocities from the sliding friction (→ Coulomb friction) between bulk material and pipe wall. At the pressure loss minimum, the sum of the two overlapping effects is the lowest.

A so-called thrust conveying, in which the entire conveying pipe is filled with solid matter, can only be realized due to gas expansion by installing a counteracting force at the conveying pipe end. Uneconomically high conveying pressure differences are required. Flow deflections, i.e., deflecting bends or elbows in the conveying line, support the wedging of the solid matter. Thrust conveying can therefore only be realized in special cases in short straight lines.

The conveying states in vertical transport sections are analogous to those in horizontal lines, see Fig. 4.6. With decreasing gas velocity, strands and conglobations are formed, whose "apparent" terminal velocity is significantly greater than that of the individual particles. As a result, local backflows of the solid matter are possible in these flow patterns. Due to the constant dissolution and reformation of the strands or conglobations, however, the solid particles are always individually approached and conveyed upwards. On average, vertical transport takes place. In the case of plug conveying, which is generally only stably achievable with coarser bulk material with a relatively narrow particle size distribution, e.g., plastic granulate, the individual plugs lose solid particles at their lower side, which are picked up again by the following plug.

In conveying systems with horizontal, vertical sections and deflections, the routing of the pipeline also influences the flow patterns that occur. Example: Gas/solid separation in bends, back-mixing in the respective subsequent pipe sections, etc.

4.3 Introduction to Dense Phase Conveying

As already indicated in Sect. 1.3, current developments in the field of pneumatic conveying are particularly concerned with reducing the drive power requirements through the use of so-called dense phase conveying methods, which are offered in various forms. Dense phase conveying is characterized by very low conveying gas velocities and, as a result, high loadings μ. The following is a brief introduction to the topic, which will be explored in more depth in later chapters.

4.3.1 Delimitation of Dilute and Dense Phase Conveying

There is no uniform, generally accepted definition of the term dense phase conveying. Various criteria exist for distinguishing dense phase conveying (DPF) from dilute phase conveying (LPF). Since some of the special features of DPF can be illustrated using these definitions, they will be briefly discussed. A DPF is present when:

a) the loading μ exceeds a typical value, in general $\mu \cong 30\,\frac{\text{kg S}}{\text{kg F}}$, e.g., [2],
b) the solid volume concentration $\varepsilon_S = (1 - \varepsilon_F)$ is greater than $\varepsilon_S \cong (0.03\text{–}0.04)\,\frac{\text{m}^3\,\text{S}}{\text{m}^3\,\text{F}}$, e.g., [2, 9],
c) the bulk material completely fills the conveying pipe cross-section at one or more points along the conveying path, e.g., [10],
d) the operating point of the conveying system in the state diagram, see Fig. 4.3, is to the left of the pressure loss minimum of the associated solid flow rate $\dot{m}_S = \text{const.}$, i.e., at lower gas velocities, e.g., [11],
e) the conveying gas is no longer able to carry the entire solid flow in flight, e.g., [12, 13].

Each of the mentioned definitions can be criticized, as will be shown below [14]:

Regarding a) The simplified energy balance of a pneumatic conveying line, see Box 4.1, illustrates that the loading μ decreases with increasing conveying distance L_R under otherwise identical boundary conditions, (4.2) in Box 4.1. As a consequence, for example, the fly ash transport No. 1 (→ Table in Box 4.1) with $\mu = 81.5\,\text{kg}\,S/\text{kg}\,F$ of the DPF, the conveying No. 2 with $\mu = 11.0\,\text{kg}\,S/\text{kg}\,F$ would, however, be assigned to the LPF, although identical flow conditions in the conveying pipe are established in both cases due to the identical operating conditions ($\rightarrow \Delta p_R, D_R, v_{F,A}, v_{F,E} = \text{konst.}$). The same energy input ($\Delta p_R \cdot \dot{V}_F$) is used in conveying No. 1 to transport a large solid mass flow $\dot{m}_S$ over the short resistance distance $L_R = 152\,\text{m}$, with conveyor No. 2 having a correspondingly smaller mass flow $\dot{m}_S$ over the considerably longer resistance distance $L_R = 796\,m$ to transport. In the associated state diagrams, both conveyors are clearly classified as DPF. The use of the loading μ as a distinguishing criterion between LPF and DPF is thus only meaningful when referring to a reference length, e.g., $L_R = 100\,m$. The comparison of conveyors 1 and 3 in Box 4.1 shows the influence of the driving pressure difference Δp_R described by (4.2).

Box 4.1: Dependencies of loading [14]

Simplified energy balance: Energy released by the gas = transport energy absorbed by the solid

$$\Delta E_F = \Delta E_S \rightarrow P_F \cdot \frac{L_R}{v_F} = P_S \cdot \frac{L_R}{u_S}$$

$$\Delta p_R \cdot \dot{V}_F = \dot{m}_S \cdot g \cdot \beta_R \cdot L_R \cdot \left(\frac{v_F}{u_S}\right)$$

with:

$P_{F/S}$ Power input,
β_R Resistance coefficient.

Inserting

$$\dot{V}_F = \frac{\dot{m}_F}{\varrho_F}, \quad \mu = \frac{\dot{m}_S}{\dot{m}_F}$$

results in:

$$\mu = \left(\frac{u_S/v_F}{g \cdot \beta_R \cdot \varrho_F}\right) \cdot \frac{\Delta p_R}{L_R} \tag{4.2}$$

Thus:

$$\mu \propto \frac{1}{L_R}, \quad \mu \propto \Delta p_R, \quad \mu \propto \frac{1}{\varrho_F}$$

Practical measurement values: Fly ash/air, counter pressure $p_E = 1.0\,\text{bar}$

No.	Δp_R [bar]	$v_{F,A}$ [m/s]	$v_{F,E}$ [m/s]	D_R [mm]	L_R [m]	μ [kg/kg]
1	2.0	4.0	12.0	82.5	152	81.5
2	2.0	4.0	12.0	82.5	796	11.0
3	1.0	4.0	8.0	82.5	152	41.2

Note: (4.2) can also be determined from a simple force balance ($\rightarrow \Delta p_R \cdot A_R = \Delta m_S \cdot g \cdot \beta_R$, Δm_S = instantaneous solid mass in the conveying pipe). It is suitable for estimating the effects of changes/modifications to an existing conveying system, e.g., an extension of the conveying route, based on the operating results of the existing system, see calculation example 6, Sect. 4.3.3.

Regarding b): When using the solid volume concentration ε_S, analogous problems arise. (4.3) in Box 4.2 shows that this also contains the loading and changes additionally along the transport path. The same applies to the bulk density ϱ_b, (4.4). For comparisons, both a reference length and a reference position along the conveyor line are required. The quantities ε_S and ϱ_b have the advantage that they correctly describe the spatial approach of the bulk material particles at high system pressures, e.g., when feeding a reactor under overpressure. This is not provided by the loading.

Box 4.2: Dependencies of solid volume concentration and bulk density

Local solid volume concentration: This amounts to the pipe element ΔV_R:

$$\varepsilon_S = (1-\varepsilon_F) = \frac{\Delta V_S}{\Delta V_R} = \frac{\dot{m}_S \cdot \Delta\tau_S}{\varrho_P \cdot A_R \cdot \Delta L_R} = \frac{\dot{m}_S \cdot \Delta L_R}{\varrho_P \cdot u_S \cdot A_R \cdot \Delta L_R}$$

With $A_R = \dot{m}_F/(\varrho_F \cdot v_F)$ follows:

$$\varepsilon_S = \mu \cdot \frac{\varrho_F}{\varrho_P} \cdot \frac{v_F}{u_S} \tag{4.3}$$

(2.21) provides the corresponding *local bulk density*:

$$\varrho_b = \varepsilon_S \cdot \varrho_P = \mu \cdot \varrho_F \cdot \frac{v_F}{u_S} \tag{4.4}$$

Application of equations (4.3) and (4.4) to conveyor No. 1 in Box 4.1. Additional operating data: speed ratio: $u_S/v_F \cong 0.80$, mixing temperature: $T_M = 293\,\text{K}$, gas density at the end of the pipeline: $\varrho_{F,E} = 1.20\,\text{kg/m}^3$, gas density at the beginning of the pipeline: $\varrho_{F,A} = \varrho_{F,E} \cdot p_A/p_E = 1.20\,\text{kg/m}^3 \cdot 3.0\,\text{bar}/1.0\,\text{bar} = 3.60\,\text{kg/m}^3$, particle density:$\varrho_P = \varrho_S = 2300\,\text{kg/m}^3$. From this, it follows for the end of the conveying line:

$$\varepsilon_{S,E} = 81.5\,\frac{\text{kg S}}{\text{kg F}} \cdot \frac{1.20\,\frac{\text{kg}}{\text{m}^3}}{2300\,\frac{\text{kg}}{\text{m}^3}} \cdot \frac{1}{0.80} = 0.0532$$

$$\varrho_{b,E} = 0.0532 \cdot 2300\,\frac{\text{kg}}{\text{m}^3} = 122.3\,\frac{\text{kg}}{\text{m}^3}$$

and the beginning of the line ($\rightarrow$ after the completion of the initial acceleration):

$$\varepsilon_{S,A} = 0.1595$$

$$\varrho_{b,A} = 366.8\,\frac{\text{kg}}{\text{m}^3}$$

The loose bulk density of the fly ash was measured with $\varrho_{SS} \cong 800\,\text{kg/m}^3$.

Example: If the conveyance No. 1 in Box 4.1 were operated with the same pressure difference $\Delta p_R = 2.0\,\text{bar}$, but with a halved initial gas velocity $v_{F,A}^* = 2.0\,\text{m/s}$, against a pressure of $p_E^* = 4.0\,\text{bar}$ at the end of the line instead of $p_E = 1.0\,\text{bar}$, the same loading would result in both conveyances $\mu = \mu^* = 81.5\,\text{kg}\,S/\text{kg}\,F$ adjust. However, the solid volume concentration and the bulk density at the end of the pipeline increase from $\varepsilon_{S,E} = 0.0532$ or $\varrho_{b,E} = 122.3\,\text{kg/m}^3$ to $\varepsilon_{S,E}^* = 0.2126$ or $\varrho_{b,E}^* = 489.0\,\text{kg/m}^3$ and at the beginning of the pipeline from $\varepsilon_{S,A} = 0.1595$ or $\varrho_{b,A} = 366.8\text{kg/m}^3$ to $\varepsilon_{S,A}^* = 0.3189$ or $\varrho_{b,A}^* = 733.5\,\text{kg/m}^3$. At the beginning of the conveying line, the bulk density $\varrho_{b,A}^*$ thus remains only slightly below the possible maximum value of $\varrho_{SS} \cong 800\,\text{kg/m}^3$, see also Box 4.2.

Regarding c): The suggestion that a DPF is present when the bulk material completely fills the conveying pipe cross-section in some areas can, as will be shown, only be applied to bulk materials with high gas permeability, i.e., to coarse-grained products with a narrow particle size distribution, e.g., plastic granules. For fine-grained solids, this generally leads to pipe blockages.

Regarding d): Classifying a pneumatic conveying system as DPF when its operating point in the associated state diagram is to the left of the connecting line of the pressure loss minima of the curves $\dot{m}_S =$ konst. appears to be the most sensible method for DPF delimitation. Since the state diagram is only valid for the respective present/examined combination of bulk material and specific conveying line, the above DPF definition automatically captures all influencing parameters on the current conveying state. These include beside conveying gas velocity v_F and solid throughput $\dot{m}_S$ in particular: type of bulk material, conveying distance L_R, conveying pipe diameter D_R. Definition d) is used in the present text.

Regarding e): The beginning of the conveying gas/solid separation, i.e., the formation of a bulk material strand at the bottom of the (horizontal) conveying pipe, is very difficult to measure and also occurs in the area of pressure loss minima. Definition e) can therefore be equated with definition d).

In pneumatic dense phase conveying, due to the low transport velocities, there are always strongly separated two-phase flows, see Fig. 4.3. The emerging flow patterns (→ strand, dune, conglobation, plug, flow conveying) are determined by the properties of the specific bulk material and the chosen operating conditions. Their behavior covers the range of quasi-steady state to non-steady state flows. The transitions are fluid.

An alternative discussion of various DPF definitions can be found in [15].

4.3.2 Advantages and Disadvantages

Dense phase conveying is operated close to the minimum possible conveying gas velocity. The size and course of this velocity are influenced, among other things, by the properties of the solid material and operating conditions and must be known for a reliable plant design. For the solid material feeding device, minimum requirements regarding the avoidance of uncontrolled gas losses/leakages at the planned conveying pressures result from the proximity to the conveying limit. The usual solids feeder for DPF is therefore the pressure vessel. The safety distance to be maintained in individual cases compared to the conveying limit must be determined taking into account possible fluctuations/changes in product properties, possibly incomplete knowledge of the conveying behavior of the bulk material at hand, its specific conveying behavior itself, and including the delivery characteristics of the intended pressure generator. Favorable conditions allow an approach to the respective limit velocity at the beginning of the pipeline up to approx. $\Delta v_{F,A} \cong (1.0-1.5)\,\mathrm{m/s}$.

For the reasons mentioned above, a number of different conveying principles are offered for the practical implementation of dense phase conveying, which can be roughly divided into conventional and process-stabilizing/supporting methods. While in conventional DPF, the transport takes place with the minimum possible conveying gas velocity for safe operation without additional auxiliary devices through smooth pipes, this is supported or monitored in stabilizing methods by various measures, e.g., pointwise gas addition along the conveying route from an accompanying bypass line. None of the commercially available DSF methods is universally applicable. The general goal of each conveying principle must be to support the natural conveying behavior of the considered bulk material in a suitable manner.

The operation of a DPF system is influenced to a much greater extent by the bulk material/solid properties than that of a dilute phase pneumatic conveying system. For system designs, these properties must therefore be determined and suitably characterized or classified. In general, dense phase conveying does not require larger conveying pressure differences Δp_R than a comparable pneumatic conveying. DPF systems are generally designed as positive pressure conveying systems.

Compared to dilute phase conveying, dense phase conveying is distinguished by the following advantages and disadvantages:

Advantages:

- More gentle product conveying, i.e., overall low mechanical stress on the conveyed material, little abrasion and/or grain breakage,
- minimal wear of conveying lines,
- partially drastically reduced energy consumption; this is especially true for fine-grained bulk materials ($\rightarrow d_{S,50} \lesssim 50\,\mu\mathrm{m}$),

- smaller dimensions of plant components, e.g., compressors, receiving filters, conveying line diameters,
- economical operation of conveying systems under protective/inert gas.

Disadvantages:
- Distinctly solid-specific "tailor-made" conveying systems, i.e., with a selected operating setting, only a limited range of solids can be transported,
- high design accuracy and thus detailed knowledge of the conveyed material and task as well as operating conditions required,
- more complex locking devices for bulk material entry into the conveying line necessary,
- often higher investment costs, especially for processes that stabilize and/or support the conveying process through additional measurement technology and equipment measures.

Since not all pneumatic conveying tasks can be solved by DPF methods, dilute phase conveying will continue to have its areas of application.

4.3.3 Calculation Example 6: Expansion of a Conveying System

An existing pneumatic conveying system transports a fine-grained mineral over a conveying distance of $L_R = 100\,\mathrm{m}$ through a steel pipe with an inner diameter of $D_R = 127.1\,\mathrm{mm}$ ($\rightarrow$ pipe $\varnothing 139.7\,\mathrm{mm} \times 6.3\,\mathrm{mm}$). The following operating data are known:

- Conveying gas: Air,
- Solid mass flow: $\dot{m}_S = 50\,\mathrm{t/h}$,
- Pressure at the end of the conveying line: $p_E = 1.00\,\mathrm{bar}$,
- Pressure difference in the conveying line: $\Delta p_R = 1.50\,\mathrm{bar}$,
- Gas velocity at the beginning of the conveying line: $v_{F,A} = 7.0\,\mathrm{m/s}$,
- Gas/solid mixture temperature: $T_M = 293\,\mathrm{K}$.

The conveying ends in a storage silo. To increase the storage capacity, a new silo of the same size should be installed in the conveying direction behind the existing one. The existing conveying line will be extended via a two-way diverter on the existing silo roof to the new silo: $D_{R,\mathrm{new}} = D_R$. The total conveying distance to the new storage silo is: $L_{R,\mathrm{new}} = 135\,\mathrm{m}$.

The following should be checked in advance for the plant planning:

a) What solid throughput $\dot{m}_{S,\mathrm{new}}$ can be achieved with the existing conveying gas supply ($\rightarrow$ same conveying pressure and same conveying gas flow as before)?

b) What conveying pressure difference $\Delta p_{R,\text{new}}$ is required to transport the previous solid throughput $\dot{m}_S$ with the existing gas flow over the distance $L_{R,\text{new}}$? Boundary condition to be maintained: The initial gas velocity must not fall below a value of $v_{F,\text{limit}} = 6.0\,\text{m/s}$ (→ negative operating experiences of the operator).
c) What gas flow rate is required if $v_{F,\text{limit}}$ is undercut?

Calculation Approach
Since the deviations from the original operating state are not very large, the calculations are carried out with (4.2) from Box 4.1 in the form

$$\mu = \frac{\dot{m}_S}{\dot{m}_F} = C_b \cdot \frac{\Delta p_R}{L_R} \quad \rightarrow \quad C_b \cong \text{const.} \tag{4.5}$$

performed.

Regarding point a):
Application of (4.5) to the existing and the new conveyor results in their ratio with $\dot{m}_{F,\text{new}} = \dot{m}_F$, $\Delta p_{R,\text{new}} = \Delta p_R$:

$$\dot{m}_{S,\text{new}} = \dot{m}_S \cdot \frac{L_R}{L_{R,\text{new}}} = 50\,\frac{\text{t}}{\text{h}} \cdot \frac{100\,\text{m}}{135\,\text{m}}$$

$$\dot{m}_{S,\text{new}} = 37.0\,\frac{\text{t}}{\text{h}}$$

This throughput is too low for continuous operation of the upstream plant.

Regarding point b):
Reapplying (4.5) to the existing and the new conveyor results in their ratio with $\dot{m}_{F,\text{new}} = \dot{m}_F, \dot{m}_{S,\text{new}} = \dot{m}_S$:

$$\Delta p_{R,\text{new}} = \Delta p_R \cdot \frac{L_{R,\text{new}}}{L_R} = 1.50\,\text{bar} \cdot \frac{135\,\text{m}}{100\,\text{m}}$$

$$\Delta p_{R,\text{new}} = 2.025\,\text{bar}$$

Checking the initial gas velocity $v_{F,A,\text{new}}$ at $\dot{m}_{F,\text{new}} = \dot{m}_F$. From the continuity equation and the ideal gas law, we get:

$$\frac{\dot{m}_{F,\text{new}}}{\dot{m}_F} = 1 = \frac{v_{F,A,\text{new}} \cdot \varrho_{F,A,\text{new}}}{v_{F,A} \cdot \varrho_{F,A}} = \frac{v_{F,A,\text{new}}}{v_{F,A}} \cdot \frac{(p_E + \Delta p_{R,\text{new}})}{(p_E + \Delta p_R)}$$

$$v_{F,A,\text{new}} = v_{F,A} \cdot \frac{(p_E + \Delta p_R)}{(p_E + \Delta p_{R,\text{new}})} = 7.0\,\frac{\text{m}}{\text{s}} \cdot \frac{(1.00 + 1.50)\,\text{bar}}{(1.00 + 2.025)\,\text{bar}}$$

$$v_{F,A,\text{new}} = 5.79\,\frac{\text{m}}{\text{s}} < v_{F,\text{limit}} = 6.0\,\frac{\text{m}}{\text{s}}$$

Regarding point c):
In order to maintain the initial gas velocity $v_{F,A,\text{new}} \geq v_{F,\text{limit}} = 6.0\,\text{m/s}$ under the operating conditions determined for point b), the continuity equation and ideal gas law apply at the pressure $p_{A,\text{new}} = (p_E + \Delta p_{R,\text{new}})$ at the beginning of the delivery line:

$$\frac{\dot{m}_{F,\text{limit}}}{\dot{m}_F} \geq \frac{v_{F,\text{limit}}}{v_{F,A,\text{new}}} = \frac{6.0\,\frac{\text{m}}{\text{s}}}{5.79\,\frac{\text{m}}{\text{s}}} = 1.04$$

$$\dot{m}_{F,\text{new}} \geq \dot{m}_{F,\text{limit}} = 1.04 \cdot \dot{m}_F$$

The gas mass flow must therefore be increased by at least 4% based on the existing value.

(4.5) allows for a simple estimation of the changes in operating data resulting from a planned plant conversion.

4.4 Processes at the Plug

For each combination of available pressure difference Δp_R, gas flow $\dot{m}_F$ and solid to be conveyed, there is a critical, just manageable plug length L_P, whose exceeding leads to a blockage of the conveying line. The formation of such plugs must be prevented by appropriate design and/or suitable auxiliary devices, see Sects. 6.1 to 6.6. The forces on the solid shown in Fig. 4.5 for three different conveying states show that due to the external stresses imposed on the individual plug, mechanical axial compressive stresses $\sigma_{S,\text{ax}}$ build up in the bulk material, transmitted from particle contact to particle contact. These cause radial stresses $\sigma_{S,\text{rad}}$ perpendicular to the surrounding pipe wall. For the ratio of the two stresses, the following applies analogously (2.15):

$$K = \frac{\sigma_{S,\text{rad}}}{\sigma_{S,\text{ax}}} \tag{4.6}$$

The stress transfer coefficient K is similar to the well-known static pressure coefficient in silo technology K_0. The stress $\sigma_{S,\text{rad}}$, or force $F_{S,\text{rad}} = \sigma_{S,\text{rad}} \cdot D_R \cdot \pi \cdot L_P$, acting normally on the pipe wall, results in an additional friction force $F_{R,S} = \beta_R \cdot F_{S,\text{rad}}$, which is superimposed on the friction force $F_{R,G}$ caused by the weight of the material. While $F_{R,S}$ acts over the entire pipe circumference, this is only the case for $F_{R,G}$ in the lower half of the pipe. By "wedging" of the solid, the friction forces to be overcome by the driving conveying gas ($F_{R,G} + F_{R,S}$) can increase many times compared to the friction caused by the weight of the material.$F_{R,G}$ increase [16].

To move a single plug of length L_P in a conveying line of diameter D_R, a certain pressure difference, referred to as displacement pressure, Δp_P is required. This must be generated by the conveying gas when flowing through the plug. (3.25) and (3.30) show that a minimum gas velocity $v_{F,P,\min}$ and thus a minimum gas flow $\dot{V}_{F,P,\min}$ is required. The three quantities ($\Delta p_P, L_P, v_{F,P,\min}$) are assigned to each other.

Results of displacement pressure measurements in horizontal conveying pipes are exemplarily shown in Figs. 4.7, 4.8 and 4.9. Figure 4.7 shows displacement pressures Δp_P, which were obtained with fine-grained, extremely cohesive titanium dioxide at different pre-compressions Δp_V were measured [17], Fig. 4.8 those of a alumina with a very broad particle size distribution ($d_S \leq 250\,\mu\text{m}$, $d_{S,50} \cong 66\,\mu\text{m}$) without pre-compaction [16]. Pre-compaction means that the bulk material in the test setup was pre-compacted by a gas pressure difference Δp_V before the actual displacement pressure measurement. This is intended to simulate the conveying pressure difference locally applied to the plug in the conveying line. Both investigations were carried out on stationary plugs and at a counter pressure of $p_E = 1.0\,\text{bar}$. They illustrate that the displacement pressure Δp_P increases progressively with the plug length L_P for the relatively fine-grained bulk materials studied and that the plug lengths that can be conveyed with a pre-compaction Δp_V and an available pressure difference are drastically shortened. The displacement pressures Δp_P increases rapidly towards infinity with increasing plug length ΔL_P.

Figure 4.7 confirms practical experience, which states that a pipe blockage should not be eliminated by increasing the conveying pressure, but initially by gas suction and loosening of the plug. Only after sufficient loosening should the gas flow be increased slowly. The implementation of this procedure requires appropriate measures to be planned during the system installation.

Figure 4.8 shows the influence of the pipe diameter D_R on Δp_P: With the same displacement pressure Δp_P, increasingly longer plugs D_R can be conveyed in pipes with increasing diameter. The cause is the increasing ratio (A_R/U_R), $U_R =$ pipe circumference, which allows longer plugs $\Delta p_P =$ friction lengths L_P with the same driving pressure difference. This is shown in Box 4.3.

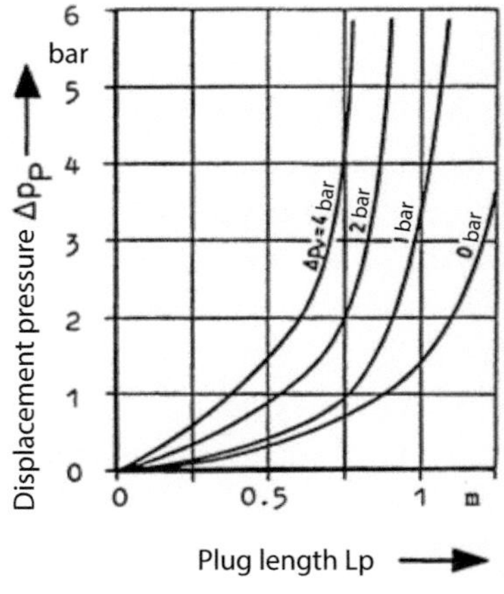

Fig. 4.7 Displacement pressures Δp_P as a function of plug length L_P and pre-compression pressure difference Δp_V, $D_R = 80\,\text{mm}$, bulk material: TiO_2, type Kronos AD, $d_S \leq 1\,\mu\text{m}$ [17]

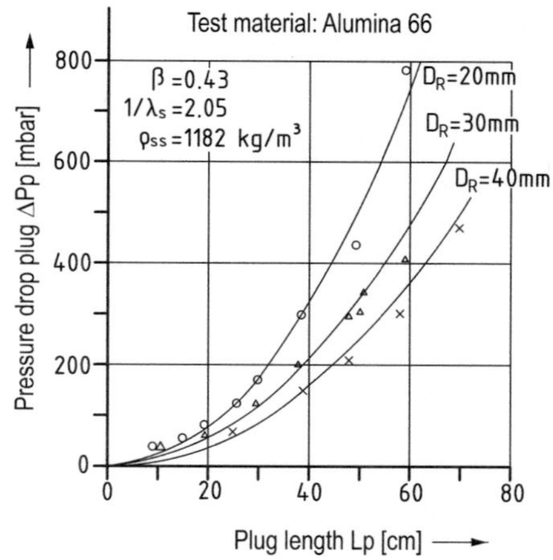

Fig. 4.8 Displacement pressures Δp_P depending on plug length L_P and pipe diameter $D_R = (20, 30, 40)$ mm, no pre-compaction, bulk material: Al_2O_3, alumina, $d_{S,50} \cong 66\,\mu$m [16]

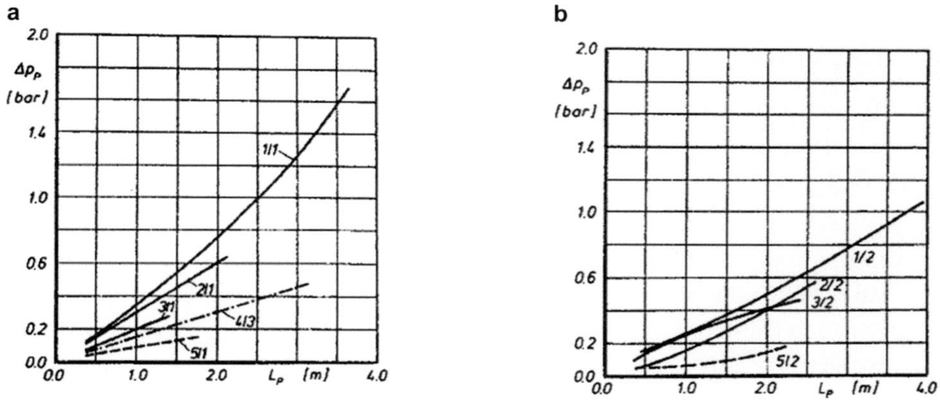

Fig. 4.9 Displacement pressures Δp_P depending on plug length L_P and pipe diameter $D_R = (40, 65)$ mm, no pre-compaction, various bulk materials [18]. Curve designation:$x/y \rightarrow$ $x =$ Bulk material number, $y =$ Pipe number. **a** $D_R = 40$ mm: Steel pipe 1, acrylic glass pipe 3, **b** $D_R = 65$ mm: Steel pipe 2. Bulk materials: 1 – Sand with $d_{S,50} = 0.51$ mm, 2 – Sand with $d_{S,50} = 1.68$ mm, 3 – Sand with$d_{S,50} = 2.83$ mm, 4 – Polyethylene granulate with $d_{S,50} = 3.19$ mm, 5 – Bitter lupines with $d_{S,50} = 5.83$ mm

Box 4.3: Plug lengths in pipes with different diameters

The different plug lengths ($L_{P,1}, L_{P,0}$), which occur at the same displacement pressure Δp_P in two lines with the diameters $D_{R,1} > D_{R,0}$ should be estimated. From a highly simplified force balance on the plug, it follows:

$$\Delta p_p \cdot \frac{\pi}{4} \cdot D_R^2 = \tau_W \cdot \pi \cdot D_R \cdot L_P \tag{4.7}$$

with:

τ_W Wall shear stress.

Application of (4.7) to both pipe diameters, assuming that $\tau_{W,1} = \tau_{W,0}$ applies, for their ratio:

$$\frac{D_{R,1}^2}{D_{R,0}^2} = \frac{D_{R,1} \cdot L_{P,1}}{D_{R,0} \cdot L_{P,0}}$$

Thus:

$$L_{P,1} = L_{P,0} \cdot \frac{D_{R,1}}{D_{R,0}} \tag{4.8}$$

The plug length that can be controlled with a given pressure difference increases with increasing pipe diameter. A more detailed analysis, taking into account, for example, the bulk material's own weight and its acceleration, results in smaller plug extensions than with the above approach.

Linear displacement pressure increases $\Delta p_P(L_P)$ are measured for coarse-grained bulk materials with a narrow particle size distribution ($\rightarrow d_{S,50} \gtrsim 500\,\mu\text{m}$). Figure 4.9 shows corresponding characteristic curves from [18], which were determined on conveyed plugs and two different pipe diameters. As expected, longer plugs can be conveyed with the same pressure difference in the larger pipe.

The height of the displacement pressure Δp_p is a measure of the "wedging forces" acting on the plug. If one forms the ratio between Δp_p and the pressure loss caused by weight friction alone $\Delta p_{R,G} = (\beta_R \cdot \varrho_{SS} \cdot g \cdot L_P)$, i.e. [1]

$$\frac{\Delta p_P}{\Delta p_{R,G}} = \frac{\Delta p_P}{\beta_R \cdot \varrho_{SS} \cdot g \cdot L_P} = \frac{\beta_R^*}{\beta_R} = \kappa \tag{4.9}$$

so the $\kappa = 1$ exceeding portion is attributable to wedging forces. Fine-grained bulk materials result in amplification ratios up to $\kappa \cong 25$ and larger [16, 17], coarse-grained products only up to $\kappa \cong 6$ [18]. β_R^* is a total friction coefficient related to the plug weight, which generally dominates over β_R.

Figure 4.10a shows the blowing out of a stationary bulk material plug from a (clogged) pipe in various stages [19]. The conveyed material is a spherical plastic granulate with a $d_{S,50} = 3.0\,\text{mm}$ particle diameter ($\rightarrow$ Monograin). The "shifting" of the plug occurs here as a sliding of a material layer in the area of the pipe apex over an underlying, slower-moving strand. In Fig. 4.10b, the associated displacement pressures are shown. For these, amplification ratios of $\kappa < 2$. With a correspondingly increased conveying gas flow, it is possible to transport such plugs as compact units. This indicates practical problems with the displacement pressure measurements and also a mini-

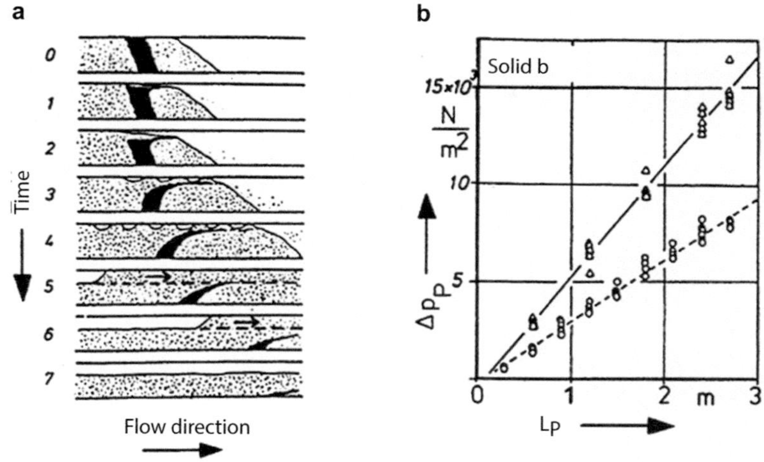

Fig. 4.10 Blowing a plug out of a pipe. **a** flow profile, **b** displacement pressures, —— $D_R = 27.0\,\text{mm}$, – – – $D_R = 50.3\,\text{mm}$, according to [19]

mum plug length for the realization of a stable plug conveying (→ Recommendation: $(L_R/D_R) > 5)$): The cause is the larger relative gap volume ε_F in the upper area of the conveying pipe cross-section as well as the sloping of the plug ends, i.e., different gas flow paths with different specific resistance coefficients are formed. The result is shown in Fig. 4.5 schematically shows the gas velocity distribution across the conveying pipe cross-section during plug conveying with a maximum in the apex area. This leads to shear forces from the gas on the solid. Figure 4.10 shows that $\Delta p_P(L_P)$ increases linearly and larger pipe diameters D_R lead to longer manageable plug lengths L_P.

In Fig. 4.11, the maximum lengths L_P of individual plugs, which can be conveyed/shifted in a conveying pipe $D_R = 40\,\text{mm}$ with a pressure difference of $\Delta p_R = 1.0\,\text{bar}$, are plotted against the mean particle diameters $d_{S,50}$ of different bulk materials. Measurement results from Figs. 4.7, 4.8 and 4.9 were used, without considering the different widths of particle size distributions, various solid densities, grain shapes, etc. Also included are the existence ranges of bulk material classes according to Geldart. Figure 4.11 illustrates that the controllable plug lengths with a given conveying pressure increase with the particle size of the conveyed material. This statement is generalized in Sect. 4.5: The greater the gas permeability of the bulk material to be conveyed, the longer the plugs that can be conveyed or controlled, and vice versa.

As long as a plug is so short that the required displacement pressure Δp_P is less than the available conveying pressure difference Δp_R, it will be conveyed. The line becomes clogged when longer plugs form and $\Delta p_P \geq \Delta p_R$ occurs. A basic method for removing/preventing plugs is shown in Fig. 4.12. Via a bypass line, conveying gas is passed around the plug and reintroduced at a point where the conveying pressure difference Δp_R is greater than the displacement pressure Δp_P of the plug section behind the input point. This part is then conveyed.

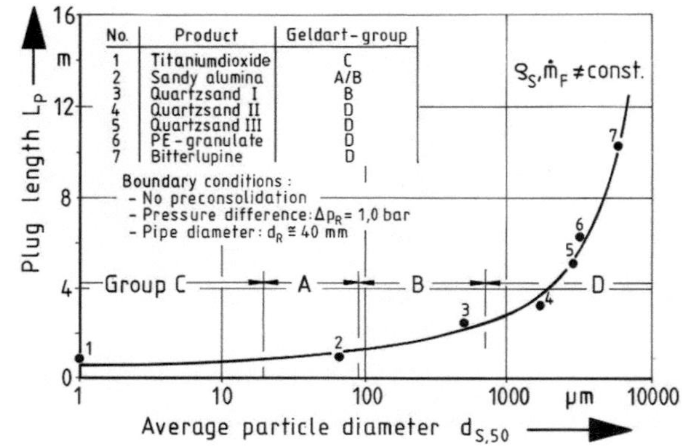

Fig. 4.11 Maximum plug lengths of various bulk materials at a displacement pressure of $\Delta p_P = 1.0$ bar

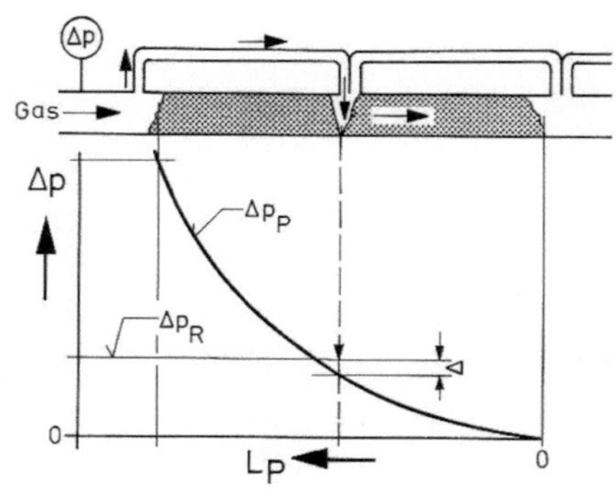

Fig. 4.12 Dissolving a plug by targeted gas supply

Details on the DPF methods and calculation approaches for plug conveying are discussed in Sects. 4.6 and 5.2.

4.5 Bulk Material Influences on Conveying Behavior

The multitude of bulk material properties influencing the conveying process requires characteristic values that capture essential influencing variables in a summary manner. In the dilute phase pneumatic conveying range, where the properties of the relatively isolated flowing individual particles are relevant for the conveying behavior, this is achieved using the terminal velocity w_T of a particle characterizing the bulk material (→ captures grain size, density, shape, surface structure) or a Froude number formed with w_T and a characteristic length L, e.g., the particle or pipe diameter, $Fr_T = w_T^2/(g \cdot L)$. These are used, among other things, for the correlation of resistance coefficients or for estimating the particle impact angle and the frequency of a particle's impact on the surrounding pipe wall.

Due to the pronounced gas/solid separation, in the area of dense phase conveying, those material properties that influence the conveying process are preferred, which result from the integration of individual particles into a higher-level bulk material structure, e.g., into a strand or a plug. For DPF, characteristic values are required that describe the behavior of this bulk material structure as a whole, which is flowed through and/or overflowed by conveying gas. The gas holding capacity and the gas permeability of a solid bed have proven to be particularly suitable for this purpose [13, 20, 21, 7]. Both capture as overarching characteristic values the cumulative effects of a multitude of individual characteristic values (→ among others, particle size distribution, density, shape, structure/composition of the bulk material structure), which influence the conveying behavior of the considered product with possibly varying weighting. Gas holding capacity and permeability are reciprocally coupled and are qualitatively captured by the solid classification scheme developed by Geldart [22]. In the following, the gas holding capacity is characterized by the venting time $\Delta\tau_E$ of a suddenly separated fluidized bed measured under standardized conditions, while the gas permeability is defined as the gas volume flow $\dot{V}_F$ flowing through the bed per unit area A_R due to an applied unit pressure gradient $(-\Delta p/L_L)$. The larger $\dot{V}_F/A_R = v_F$ is, the greater the gas permeability. It increases with increasing particle diameter, see Sects. 3.2.1 and 3.3.2.

The position of a bulk material in the Geldart diagram should therefore provide conclusions about its suitability for conventional dense phase conveying and the expected conveying behavior. Since the fluidization behavior described in the diagram is not necessarily identical to the DPF conveying behavior, the existing classification of fluidization behavior must be extended by class boundaries for different conveying behavior. This is shown in Fig. 4.13 [7].

Figure 4.13 shows an extended Geldart diagram with two hatched areas added, dividing the diagram area into regions of different conveying behavior. These allow an assessment of the basic suitability of a given bulk material for a DPF and at the same time provide information on the flow forms that occur and the pneumatic conveying methods to be used. In the diagram, the gas holding capacity increases from right to left, cf. Fig. 3.13, while the gas permeability decreases in the same direction, cf. Fig. 4.11. The changes in both parameters along the $d_{S,50}$-axis are non-linear. Bulk materials that are ideally suited for conventional DPF are located in the two hatched areas. The conveying gas initial velocities $v_{F,A}$ mentioned in the following paragraphs refer to conveying pipe diameters in the range of $D_R \cong 100\,\text{mm}$.

Group 1: The fine dust bulk materials in the left hatched area of Fig. 4.13 allow conveying gas velocities at the beginning of the pipeline down to $v_{F,A} \cong (3\text{–}6)\,\text{m/s}$. Even a small amount of gas mixing leads to fluidization. Due to the very high gas holding capacity of the bulk materials, this state is maintained for a long time even after interruption/reduction of the gas supply. The behavior is similar to that of a liquid, the angle of internal friction is greatly reduced and significantly smaller than the associated wall friction angle. The conveyed material is transported in the form of a more or less

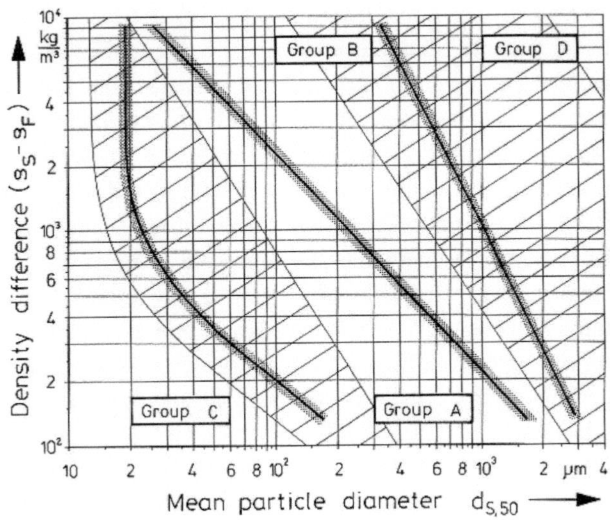

Fig. 4.13 Extended Geldart Diagram

strongly loosened solid strand at the bottom of the (horizontal) conveying line. Dunes, onglobations, or plugs that build up dissolve/liquefy immediately under these conditions. If a plug filling the pipe cross-section should still form, the low gas permeability and the required high displacement pressure cause even short bulk material plugs (→ Lengths $L_P \lesssim 1\,\text{m}$), see Fig. 4.11, lead to a sudden pipeline blockage [2, 7]. The reason for the DPF suitability of the considered bulk material group is their very high gas holding capacity. The required gas/solid mixing takes place at the beginning of the conveying line: The bulk material leaving the solid lock falls into the conveying gas stream. The developing strand vents into the free space above the strand, but since the venting times $\Delta\tau_E$ of this product group are long, the fluidized state is maintained over longer conveying distances. Example: The venting time of 2.0 kg of the fly ash shown in Fig. 3.13 is $\Delta\tau_E = 191\,\text{s}$ (→ Group 1 material). Assuming a (realistic) average solid velocity of $u_S = 10\,\text{m/s}$ along the conveying line, this fly ash could be transported over almost $L_R = u_S \cdot \Delta\tau_E = 10\,\text{m/s} \cdot 191\,\text{s} \cong 1900\,\text{m}$ in the advantageous fluidized state.

Group 2: With the coarse-grained bulk materials in the right hatched area of Fig. 4.13, conveying gas initial velocities of $v_{F,A} \cong (2\text{–}5)\,\text{m/s}$ can also be realized if at the same time a relatively narrow particle size distribution (→ inclination of the RRSB line $\alpha_{RRSB} \geq 60°$) is present. These bulk materials fall out of the conveying gas stream under dense flow conditions, accumulate on the pipe bottom, and are compressed into conglobations and plugs that fill the entire pipe cross-section. The plugs pass through the conveying section more or less as compact units. They are flowed through by the conveying gas. Due to the high gas permeability of the bulk material group considered here, very long and thus stable plugs (→ $L_P/D_R \gg 10$) can be safely conveyed, see Fig. 4.11. The fact that the internal friction angle is generally larger for coarse-grained products than the corresponding wall friction angle further increases plug stability. The natural plug formation can be controlled by simple equipment measures. The reason for the DPF suitability of this bulk material group is thus their very high gas permeability.

Group 3: Between the two hatched fields in Fig. 4.13 lies an area of extremely difficult-to-convey bulk materials. The already insufficient gas holding capacity is not yet met by a sufficiently high gas permeability and vice versa. Both properties are lacking and must be compensated for by high conveying gas velocities. Required are $v_{F,A} \gtrsim 10\,\mathrm{m/s}$. Conventional DPF cannot be realized with the bulk materials of this group. A reduction of the gas velocities into the dense flow/slow conveying range is only possible through the use of special bypass conveying methods. This results in a combined conglobation/plug conveying in the conveying section, whose unstable behavior would lead to sudden pipe blockages in a conventional DPF.

The relationships discussed so far have been confirmed by measurements [13, 20, 21, 22, 23, 24]. Figure 4.14 shows measured values of minimal required conveying gas velocities for various bulk materials and stable operation as a function of their gas retention capacity [20], confirming the behavior described above. The gas retention capacity is characterized here by a *vibrated deaeration constant* K_τ. This is formed in analogy to the diffusion constant of a Fick's diffusion process. It applies: With increasing K_τ, the deaeration time $\Delta\tau_E$ becomes smaller. The hatched areas in Fig. 4.13 correspond to the shaded areas in Fig. 4.14, both with bulk material group 3 in between. As part of the investigations presented in Fig. 4.14, 24 different bulk materials were systematically conveyed and analyzed through conventional conveying lines on an industrial scale ($\rightarrow D_R = 53/81\,\mathrm{mm}$) [20, 21].

Group 4: Bulk materials that lie to the left of the left hatched area in Fig. 4.13 are extremely cohesive and unsuitable for dense phase conveying. As pronounced Group C materials, they are difficult to fluidize. They form channels/rat holes through which the introduced gas can escape or bypass the bulk material. Their venting times $\Delta\tau_E$ are

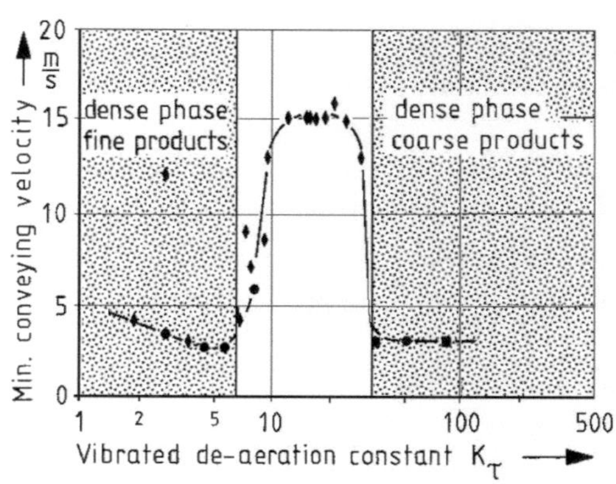

Fig. 4.14 Minimum conveying gas velocities of different bulk materials as a function of their deaeration behavior, according to [20]

therefore short, i.e., the gas holding capacity is low. Bulk materials of this group form stable agglomerates and wall deposits in both the solids lock and the conveying line, which can lead to operational problems. To overcome the interparticle and particle/pipe wall adhesion forces, high conveying gas velocities, $v_{F,A} \gtrsim 10\,\mathrm{m/s}$, are required. A common problem in the transport of such products is the entry of bulk material from the upstream plant (→ e.g., from a decoupling storage container) via the solids lock into the conveying line. This leads to discharge problems in both the pre-container and the lock due to bridging/adhesion/compression. To what extent the wall deposits that may build up in the conveying line are limited or reduced in thickness by the high gas velocities required for this product group cannot be predicted.

Group 5: Bulk materials that lie to the right of the right hatched area in Fig. 4.13 must be conveyed at high gas velocities due to the large individual particle masses and the large flow forces required for their propulsion, i.e., in dilute phase. Dense phase conveying is generally not possible. Example: Conveying of iron sponge pellets, solid density: $\varrho_S = 7790\,\mathrm{kg/m^3}$, bulk density: $\varrho_{SS} = 1690\,\mathrm{kg/m^3}$, mean particle diameter: $d_{S,50} = 12.0\,\mathrm{mm}$, maximum grain diameter: $\hat{d}_S \cong 20.0\,\mathrm{mm}$, inclination of the RRSB line: $\alpha_{\mathrm{RRSB}} = 76.0°$. Attempts were made to convey the iron sponge in the form of discrete plugs. This failed; there were pipeline blockages. Conveyances with initial gas velocities of $v_{F,A} \gtrsim 17\,\mathrm{m/s}$ could be realized without problems. The gas velocities at the pipeline end were on the order of $v_{F,E} \cong 40\,\mathrm{m/s}$ and thus in the range of single-particle terminal velocities. It was conveyed through pipes with $D_R = 82.5\,\mathrm{mm}$ over distances up to $L_R = 472\,\mathrm{m}$. The pipeline wear, especially that of the last deflections, and the noise pollution caused by flow noises were enormous.

In connection with the extended Geldart diagram, Fig. 4.13, the bulk material grouping described above allows a simple assessment of the suitability of a given material for conventional dense phase conveying and the resulting flow form. However, since the Geldart diagram does not adequately take into account the width of the particle size distribution or the grain shape of the bulk material, the classification must be secured by appropriate measurements. For bulk materials with a wide grain distribution, for example, the fine fraction fills the free gap volume between the coarser particles. The result is a reduced gas permeability and a generally increased gas holding capacity. The same applies to grain shapes that deviate from that of a sphere.

A generally accepted standard test for measuring the gas holding capacity and gas permeability of a bulk material does not currently exist. Depending on the user, different methods (*) with varying setups, different apparatus sizes, and setting parameters are used. For the assignment of the characteristic values determined by these methods, venting time $\Delta\tau_E^*$ and gas permeability v_F^* to the individual bulk material groups applies:

- Group 1: The venting time $\Delta\tau_E^*$ of the bulk materials must exceed a critical value $\Delta\tau_{E,\mathrm{crit}}^*$, $\Delta\tau_E^* > \Delta\tau_{E,\mathrm{crit}}^*$. To the left and right of the left hatched area in the extended Geldart diagram, $\Delta\tau_E^*$ decreases to smaller values.

- Group 2: The gas permeability v_F^* of the bulk materials must be above a critical value $v_{F,\text{crit}}^*$, $v_F^* > v_{F,\text{crit}}^*$. To the left of the right hatched area in the extended Geldart diagram, v_F^* decreases to smaller values.
- Group 3: For these materials located in the middle range of the extended Geldart diagram, the following applies: $\Delta\tau_E^* < \Delta\tau_{E,\text{crit}}^*$ and $v_F^* < v_{F,\text{crit}}^*$.
- Group 4: The following applies: $\Delta\tau_E^* < \Delta\tau_{E,\text{crit}}^*$ and $v_F^* \to 0 \ll v_{F,\text{crit}}^*$. Additionally, the extreme cohesiveness of these bulk materials must be taken into account.
- Group 5: Characteristic for these conveyed materials is: $\Delta\tau_E^* \to 0 \ll \Delta\tau_{E,\text{crit}}^*$ and $v_F^* \gg v_{F,\text{crit}}^*$. Due to the very high gas permeability v_F^* high gas velocities are required to generate the driving forces for the large/heavy particles of this bulk material group.

The specific values of $\Delta\tau_{E,\text{crit}}^*$ and $v_{F,\text{crit}}^*$ depend on the chosen measurement method. As a simple measurement method, the survey of the flow through and fluidized bed characteristics of the respective bulk material described in Fig. 3.10 is suitable, which is determined by the measurement of the venting time.$\Delta\tau_E^*$ ($\to$ sudden shut-off of the fluidizing gas flow) is completed. v_F^* can be determined from the measurement data of the gas flow through bulk. At the same time, the method offers the possibility to determine a particle diameter characterizing the present particle collective $d_{S,SD}^* = (\psi \cdot d_{S,SD})$, see Sect. 3.2.4, calculation example 5.

Instead of the direct ($\Delta\tau_E^*$, v_F^*)-values, as for example Fig. 4.14 with K_τ shows, derived characteristics can also be used to describe gas holding capacity and gas permeability. Examples:

- The gas holding capacity is often represented by a venting constant K_E, which is represented by the empirical venting equation

$$K_E = \left(\frac{-\Delta p}{L}\right)_{WS} \cdot \Delta\tau_E \tag{4.10}$$

with:

$(-\Delta p/L)_{WS}$; Pressure gradient across fluidized bed [24].

(4.10) describes the venting behavior only approximately: measurement strips clearly show the faster escape of the bubble phase and the slower process of the subsequent outflow of the suspension phase of the gas from the collapsing fluidized bed.

- As a measure of gas permeability, the permeability constant

$$K_P = \frac{v_F}{(-\Delta p/L)_b} \tag{4.11}$$

with:

$(-\Delta p/L)_b$; Pressure gradient across bulk material, for example, [20].

The approach according to (4.11) corresponds to the viscosity-dominated = Stokes's region of the Ergun equation.

In a series of further works, alternative/complementary approaches to classifying a bulk material with regard to its expected behavior in pneumatic conveying lines are proposed. A brief summary of the state of the art is contained in [25].

4.6 Characteristic Gas Velocities

The initial conveying gas velocity to be selected in the case of a plant design is determined, among other things, by the specific task at hand. It is clear that this must be above the current - and known - conveying limit velocity. Nevertheless, different gas velocities are required for the task of "pulsation-free feeding of a reactor" than for a "(A to B) conveying with minimal energy input". In the first case, the gas velocity should be at least above that of the so-called clogging limit (→ largely pulsation-free conveying), while in the second example it should be chosen so that the loading μ is set to a maximum (→ energetically optimal operating point). In the following, approaches for determining such gas velocities are presented and discussed.

4.6.1 Interaction of Pressure Generator and Conveying Line

The delivery characteristic $\Delta p(\dot{V}_F)$ of the pressure generator is crucial for the feasibility and stability of the chosen operating point of a conveyor system. Figure 4.15 shows the state diagram of a conveyor system (→ system characteristics), in which the characteristics of two different pressure generators are plotted. The solid mass flow $\dot{m}_{S,2}$ is to be conveyed. With the flat pressure generator characteristic "a", e.g., of a high-pressure fan, two operating points A and B are possible. When the conveying gas flow is reduced from operating point A, the necessary conveying pressure loss (→ along the curve $\dot{m}_{S,2}$) remains smaller than the pressure difference provided by the pressure generator "a" (→ along curve "a"), until the operating point B is reached. A further reduction of the gas

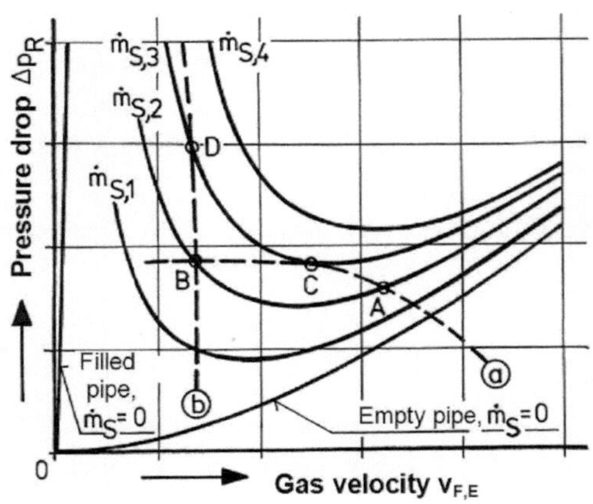

Fig. 4.15 Stable and unstable operating points of different pressure generator systems

flow suddenly clogs the conveying line since the required pressure difference becomes greater than the one provided.

In addition to changes in the gas flow, fluctuations in the bulk material throughput $\dot{m}_S$ are possible, e.g., due to different filling levels of an inflowing rotary valve. Starting again from characteristic curve "a" and operating point A, the conveying is stable when the solid throughput is increased up to operating point C. If the throughput corresponding to C, $\dot{m}_{S,3}$, is exceeded, the conveying line becomes clogged, as the pressure generator "a" no longer provides a sufficient conveying pressure difference. On the branch of curve "a" between C and B, mass flows $\dot{m}_{S,2} \leq \dot{m}_S \leq \dot{m}_{S,3}$ can be realized.

The associated gas volume flow ($\rightarrow \propto v_{F,E}$) adjusts along curve "a" accordingly between C and B. These operating points are unstable. To the left of B, operation with $\dot{m}_S \geq \dot{m}_{S,2}$ is not possible. Taking into account all conceivable disturbances, it becomes clear: With the flat pressure generator characteristic curve "a", only operating points A to the right of the connecting line of the individual $\dot{m}_S$-pressure loss minima can be operated stably.

From Fig. 4.15, it is directly apparent that with the very steep pressure generator characteristic "b", operating points B to the left of the $\dot{m}_S$-pressure loss minima can be practically utilized. For dense phase conveying systems operating in this diagram range, sufficiently steep, i.e., counter-pressure independent, pressure generator characteristics $\Delta p(\dot{V}_F)$ are required in any case. These must be steeper than the corresponding system characteristics. Suitable units for this purpose are, for example, screw compressors and, for smaller pressures, rotary lobe blowers.

The installation of an gas receiver between the pressure generator and the conveying system usually leads to a flattening of the delivery characteristic. This can be counteracted by raising the gas receiver pressure to a level significantly higher than the conveying pressure and gas extraction via a Laval nozzle. In the same way, extraction from a factory network (= very large air receiver with multiple extraction) must be used. The gas flow rate through a Laval nozzle is constant over a wide range of counter-pressures when properly designed [1, 26, 3].

4.6.2 Horizontal Minimum Conveying Gas Velocity

The initial velocities of the conveying gas mentioned in the previous sections are influenced in individual cases not only by the respective conveyed material but also by the specific operating conditions. An exact determination of the absolute conveying limit, curve D in Fig. 4.1, is required, especially in the dense phase range. This can be taken from representations of the state diagram according to Fig. 4.2. The measured diagrams in Fig. 4.16a,b show this exemplarily. [27, 28, 5]. The quartz powder conveyed here is a group 1 bulk material, i.e., it is located in the left hatched area of the extended Geldart diagram, Fig. 4.13, and is ideally suited for conventional dense phase conveying. There, a fluidized strand conveying is established, which was also observed.

As a characteristic conveying gas velocity, in Fig. 4.16a,b the one at the beginning of the line, $v_{F,A}$, is used. The reference state at the beginning of the pipeline was chosen because pneumatic conveying – assuming correct pipeline routing – becomes unstable here first. It can be seen that each pressure loss curve $\Delta p_R = \text{const.}$ has a smallest possible initial conveying gas velocity $v_{F,A,\min}$ assigned to it (→ extrapolation of the lower curve branches to the ordinate at $\dot{m}_S = 0$), whose size depends on both Δp_R and the conveying pipe diameter D_R. Further investigations at increased back pressures at the pipeline end, $p_E \gg 1.0\,\text{bar}$, show that not the pressure difference Δp_R, but the absolute pressure p_A at the beginning of the pipeline determines the size of the minimum initial gas velocity $v_{F,A,\min}$. The dependency $v_{F,A,\min}(D_R, p_A)$ can be described with sufficient accuracy for a given bulk material by the empirical power approach

$$v_{F,A,\min} = K_v^* \cdot \frac{D_R^k}{p_A^l} \tag{4.12}$$

with:

K_v^*, k, l solid-specific constants/exponents,

Since (4.12) contains no influences of gas type, temperature, etc., but these have proven effects on the size of $v_{F,A,\min}$, it can only be partially correct. From the explanations of

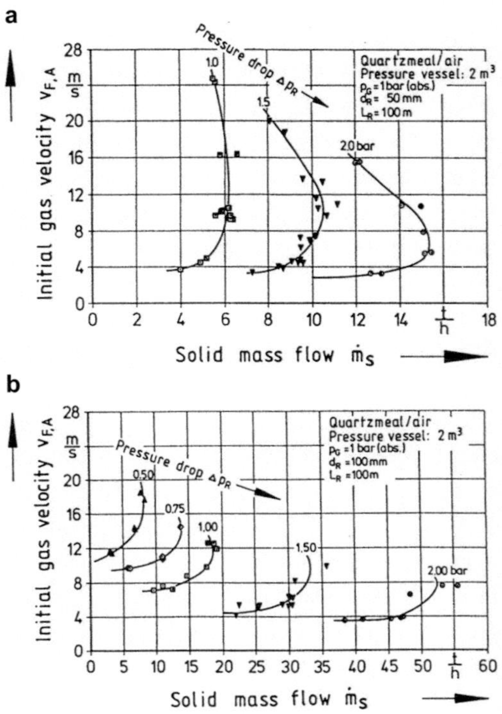

Fig. 4.16 Measured conveying diagrams: Quartz powder/air, $d_{S,50} = 39\,\mu\text{m}$, $\varrho_S = 2610\,\text{kg/m}^3$, $\varrho_{SS} = 1210\,\text{kg/m}^3$. **a** $D_R = 50\,\text{mm}$, $L_R = 100\,\text{m}$, **b** $D_R = 100\,\text{mm}$, $L_R = 100\,\text{m}$, parallel conveying line layout

Chap. 3, it follows that for the drive of the bulk material, the gas density ϱ_F is essential. Its size is determined by the ideal gas law, (2.3), (2.4), coupled to the gas type, temperature, and local pressure. It is obvious that in (4.12) instead of the pressure p_A the gas density $\varrho_{F,A}$ must be used. Thus, the following applies:

$$v_{F,A,\min} = K_v \cdot \frac{D_R^k}{\varrho_{F,A}^l} = K_v \cdot \frac{D_R^k}{p_A^l} \cdot (R_F \cdot T_A)^l = K_v \cdot (\frac{p_0}{T_0})^l \cdot \frac{D_R^k}{p_A^l} \cdot (\frac{T_A}{\varrho_{F,0}})^l \tag{4.13}$$

with:

R_F specific gas constant,

T_A gas temperature at the beginning of the pipeline $\cong$ mixing temperature T_M, (2.14),

$\varrho_{F,0}$, p_0, T_0 gas density, pressure, temperature at reference state "0".

Systematic conveying tests with ground iron ore and mixing temperatures of $t_M = (20-300)\,°\mathrm{C}$ confirm (4.13) [29].

So far, measured diameter exponents range between $k = (0-0.80)$, gas density exponents in the range $l = (0-1.30)$, i.e., $v_{F,A,\min}$ increases with increasing delivery pipe diameter D_R and decreasing gas density $\varrho_{F,A}$. The following bulk material dependencies can be observed:

- fine-grained bulk materials with high gas holding capacity, ball-like shape, and low cohesion: $k \to 0$, i.e., low influence of D_R on $v_{F,A,\min}$,
- coarse-grained bulk materials with low gas holding capacity: $l \to 0$, i.e., low influence of $\varrho_{F,A}$ on $v_{F,A,\min}$.

Bulk materials between these extremes can generally be described by exponents of size k or $l = (0.4–0.8)$. Figure 4.17a, b shows, as an example, the dependency $v_{F,A,\min}(D_R, p_A)$ using the conveying diagrams of a coarser limestone. Due to its relatively broad particle size distribution, it is a Group 3 bulk material. Its density exponent is $l \cong 0$.

A summary dimensionless representation of the $v_{F,A,\min}$-dependencies of all bulk materials studied so far using the approach in Calculation Example 3, Sect. 2.4.3,

$$\begin{aligned} Fr_{R,\min} &= F\left(\frac{D_R}{d_{S,50}}, \frac{\varrho_{F,A}}{\varrho_P}, Ar, n_{\mathrm{RRSB}}, \psi, \xi_S\right) \\ &= K_v^{**} \cdot \left(\frac{D_R}{d_{S,50}}\right)^{a1} \cdot \left(\frac{\varrho_{F,A}}{\varrho_P}\right)^{a2} \cdot Ar^{a3} \cdot n_{\mathrm{RRSB}}{}^{a4} \cdot \psi^{a5} \cdot \xi_S{}^{a6} \end{aligned} \tag{4.14}$$

lead to deviations of up to $\pm 30\,\%$ of the measured data and is therefore too inaccurate for practical designs. Therefore, the more accurate solid-specific equation (4.13) is used here. (4.14) is suitable for initial estimates of the expected behavior of a new product.

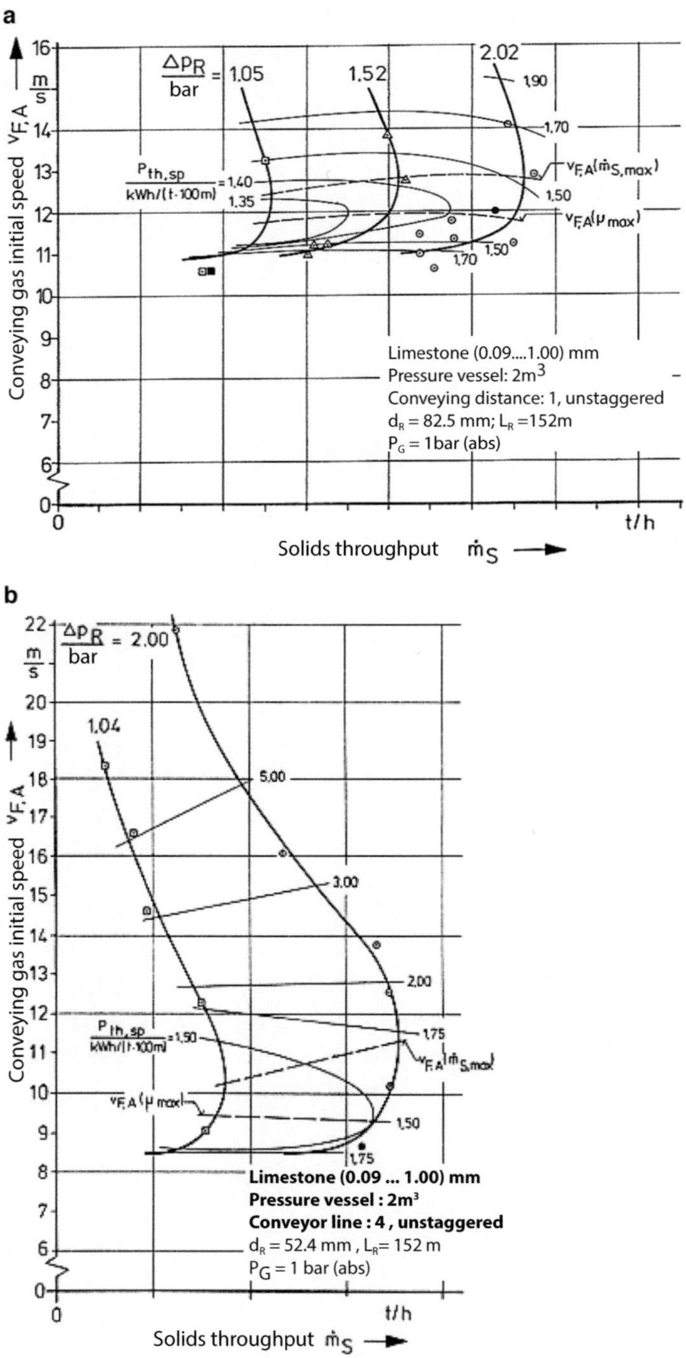

Fig. 4.17 Measured conveying diagrams: Limestone grit/air, $d_{S,50} = 460\,\mu\text{m}$, $\varrho_S = 2750\ \text{kg/m}^3$, $\varrho_{SS} = 1320\,\text{kg/m}^3$. **a** $D_R = 82.5\,\text{mm}$, $L_R = 152\,\text{m}$, **b** $D_R = 52.4\,\text{mm}$, $L_R = 152\,\text{m}$, parallel conveying line layout

Another method for a preliminary estimate of the size of the minimum velocity is shown in Fig. 4.18 [5]: $v_{F,A,\min}$ is plotted against the reciprocal value of the volume-related separation performance

$$P_{E,sp} = \frac{g \cdot (1 - \varepsilon_F) \cdot \varrho_P \cdot d_{S,50}}{\Delta\tau_E} \tag{4.15}$$

with:

ε_F relative void volume of the loosely poured bulk material,

for a reference pressure $p_A = \text{const.}$. The parameter is the conveying pipe diameter D_R. The separation performance is defined as the work to be performed by the gas to separate two particle layers by the diameter $d_{S,50}$ based on the duration of this separation.→ time duration of reduced internal friction). $P_{E,sp}$ is, for example, smaller the longer the venting time $\Delta\tau_E$ of a fluidized bed is. Therefore, $\Delta\tau_E$ is used as the reference time. The characteristic number $P_{E,sp}$ thus describes the work done during the venting duration $\Delta\tau_E$ of a fluidized bed separated from the gas supply by the escaping gas on the solid particles in the bulk volume to maintain the fluidization state, i.e., the spatial separation of the particles. From Fig. 4.18 it can be seen, among other things, that with increasing $\Delta\tau_E$, or $(1/P_{E,sp})$, the influence of the pipe diameter is smaller, but with decreasing $\Delta\tau_E$ it becomes larger. This corresponds to the above-mentioned tendency of the experimentally determined exponents k for fine and coarse particles. The systematic plotting of measured minimum velocities $v_{F,A,\min}$ against $(1/P_{E,sp})$ allows estimating the $v_{F,A,\min}$ of a bulk material based on fewer, laboratory-determined parameters without conveying tests.

Conveying must be designed with a distance $\Delta v_{F,A}$ to the minimum velocity $v_{F,A,\min}$. The operating speed $v_{F,A}$ is thus obtained as follows:

$$v_{F,A} = v_{F,A,\min} + \Delta v_{F,A} \tag{4.16}$$

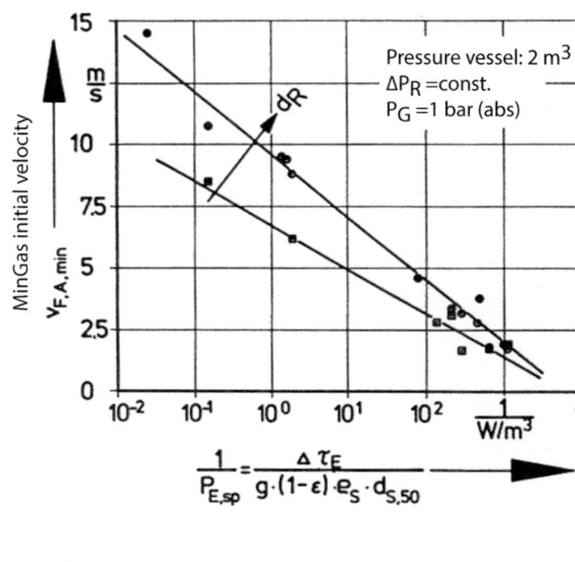

Fig. 4.18 To estimate the minimum velocity $v_{F,A,\min}$

The size of $\Delta v_{F,A}$ depends on the type of the solids feeder and the task, see Sect. 4.3.2. With the true relative velocity between gas and solid at the beginning of the pipeline $u_{A,\text{rel}} = (u_{F,A} - u_{S,A}) = u_{F,A,\text{min}}$ and the associated voidage $\varepsilon_{F,A}$ follows from (4.16):

$$\Delta v_{F,A} = v_{F,A} - v_{F,A,\text{min}} = \varepsilon_{F,A} \cdot u_{S,A} \tag{4.17}$$

Assuming that the local solid velocity u_S along the (horizontal) conveying distance is determined by the approach $u_S = (v_F - v_{F,A,\text{min}})/\varepsilon_F$, see (4.17), with: v_F, ε_F – local gas velocity, local void fraction, $v_{F,A,\text{min}} = \text{const.}$, can be described, the velocity ratio follows:

$$C_h = \frac{u_S}{u_F} = \frac{(v_F - v_{F,A,\text{min}})/\varepsilon_F}{v_F/\varepsilon_F} = 1 - \frac{v_{F,A,\text{min}}}{v_F} \tag{4.18}$$

C_h changes along the conveying distance. It is required in the later presented approaches for calculating the pressure loss, see Sect. 4.7.

Note 1: Compare the note at the end of Sect. 4.1.2. In our own experiments, the length of the straight undisturbed horizontal section behind the bulk material feed at the beginning of the conveying line was always $(L_{A,h}/D_R) \gtrsim 60$.

Note 2: The above explanations, especially Figs. 4.16 and 4.17, once again emphasize the need for systematic measurements on at least two conveying lines of different diameters D_R. Without such investigations, a safe scale-up to other tasks is problematic. Example: For the quartz powder of Fig. 4.16, a conveying system with the data: $\dot{m}_S = 25\,\text{t/h}$, $L_R = 100\,\text{m}$, $\Delta p_R \leq 1.0\,\text{bar}$, $p_E = 1.0\,\text{bar}$, is to be created. The offering company carries out tests on its test facility, dimensions: $D_R = 50\,\text{mm}$, $L_R = 100\,\text{m}$, with the quartz powder (→ Fig. 4.16a). Since the bulk material is extremely abrasive, a dense phase conveying system with the lowest possible gas velocities should be designed. The conveying limit for $\Delta p_R = 1.0\,\text{bar}$ is determined in the experiments with $v_{F,A,\text{min}} \cong 4.0\,\text{m/s}$ and, based on this, the initial conveying gas velocity of the planned system is set to $v_{F,A} = 6.0\,\text{m/s}$. For the throughput of the operating system, $\dot{m}_S = 25\,\text{t/h}$, however, a conveying line diameter $D_R > 100\,\text{mm}$, see Fig. 4.16a, b, is required. Its conveying limit velocity is at $v_{F,A,\text{min}} > 7.0\,\text{m/s}$ (→ Fig. 4.16b), i.e., when the operating system is started for the first time, there will immediately be massive blockages in the pipeline at the beginning of the conveyor pipe. The system is not operational.

4.6.3 Vertical Minimum Conveying Gas Velocity, *Choking*

To narrow down this limit velocity, (a) a fluidized bed whose operating velocity is successively increased, and (b) a pneumatic conveying whose conveying gas velocity is gradually reduced, can be considered.

Procedure (a): This has already been discussed in Sect. 3.3.3, *fast fluidization*: If the empty pipe gas velocity v_F exceeds the value defined by (3.40) $v_{F,tr}$, a strong increase in the solid mass flow carried out of the bed begins. If the current mass flow $\dot{m}_S(v_F)$ is known/calculable and at the same time exactly this solid mass flow is fed to the system via a bulk material lock, then a pneumatic vertical conveying with gas velocities down to $v_{F,tr}$ would be feasible. In a circulating fluidized bed this dosing is self-regulating through the installed solid return system, while for pneumatic conveyors complex control loops are required. By increasing the gas velocity for a given $\dot{m}_S$ on $v_F \geq v_{F,pf}$, (3.41), a state of operation is established in the fluidized bed in which no solid backflow can be observed across the reactor cross-section, i.e., there is a pure upward movement of the solid. Accumulation of bulk material in the system due to differently sized inflow and outflow solid streams can thus be avoided. $v_{F,pf}$ describes the transition from *fast fluidization* to pneumatic vertical conveying. The resolution of (3.41) for $v_{F,pf}$ leads to:

$$v_{F,pf} = (21.6 \cdot \sqrt{g \cdot d_S} \cdot Ar^{0.105})^{0.6485} \cdot \left(\frac{\dot{m}_S}{A_R \cdot \varrho_F} \right)^{0.3515} \tag{4.19}$$

Procedure (b): Pneumatic vertical conveying tends to choke at low transport speeds, a phenomenon known as *choking* [30, 31, 32, 33]. The term *choking* is relatively loosely defined. It describes unstable, possibly plug-forming, or extremely strongly pulsating processes on the left branch of the state diagram of the respective vertical conveying section. Such states occur when the gas velocity falls below a limit speed, $v_{F,ch}$. In the conveying diagram $\Delta p_R(v_{F,E}, \dot{m}_S)$ in Fig. 4.15, which also describes the processes in vertical sections, the individual $\dot{m}_S$-curves pass through a pressure loss minimum. On the curve branches to the right of the minimum, the pressure loss resulting from the frictional resistance of the bulk material dominates ($\rightarrow \Delta p_{R,\text{fric}} \propto v_F$), while to the left of the minimum, the additional lifting pressure loss, which carries the instantaneous weight of the bulk material in the vertical pipe ($\rightarrow \Delta p_{R,\text{lift}} \propto 1/v_F$), is decisive. The left curve branch rises extremely steeply with decreasing gas velocity: small velocity changes lead to very large pressure changes and, as a result, to the above-described *choking* effects, see also Sect. 4.2. The *choking* phenomena discussed in the literature can be classified into three types with different triggers [31, 32]:

- Type C-choking: When the critical empty pipe gas velocity $v_{F,ch,C}$ is undercut, bulk material plugs and gas cushions alternately form in the conveying section, each filling the entire pipe cross-section. This leads to strong pressure pulsations and shocks in the pipeline ($\rightarrow$ classical *choking* definition). Type C-choking preferably occurs with coarser bulk materials and in pipes with smaller diameters. Whether Type C-choking is possible in a transport system can be estimated, among other things, with (4.20) [33]. No *choking* occurs when the condition

$$\frac{w_T}{\sqrt{g \cdot D_R}} \leq 0.35 \tag{4.20}$$

is fulfilled. Fig. 4.19 shows an example of the evaluation of this equation for fine-grained bulk materials, whose terminal velocity w_T was calculated with (3.14) for the Stokes range (→ conveying gas = air, temperature $t_F = t_S = 20\,°C$). Type C-choking does not occur for particle diameters d_S below the corresponding particle density curve ϱ_P, but it is possible above. From Fig. 4.19 it follows that in the pneumatic vertical conveying of solids of Geldart groups C and A, Type C-choking usually does not occur. Although gas/solid separation, e.g., strand formation, can also occur there, it only causes minor pressure pulsations. Other types of *choking* are possible, however. Figure 4.19 confirms the statements of the Geldart classification that stable large gas bubbles cannot form in material classes A and C, but they can in the coarser classes B and D.

- Type B-choking: This results from limitations of the plant equipment. There is no plug formation, but the conveying collapses abruptly. In the pneumatic conveying considered here, this unstable state can be caused by an incorrectly chosen pressure generator characteristic curve: Details can be found in Sect. 4.6.1. Falling below the gas velocity $v_{F,ch,B}$, triggered, for example, by pressure/flow rate fluctuations, leads to a collapse of the conveying. Other causes can be solid locks with uncontrolled high gas leakages.
- Type A-choking: In this process, also referred to as accumulating *choking*, when the empty pipe gas velocity $v_{F,ch,A}$ is undercut, bulk material backflow begins at the pipe wall, i.e., a solid velocity profile is established across the pipe cross-section, which has the desired upward flow in the core area and a downward movement in the wall area. This leads to an enrichment/accumulation of bulk material in the foot-point area of the vertical section, there to the formation of a dense fluidized bed and to a reduction of the design solid throughput $\dot{m}_{S,0}$. The solid mass flow $\dot{m}_S(v_{F,ch,A})$

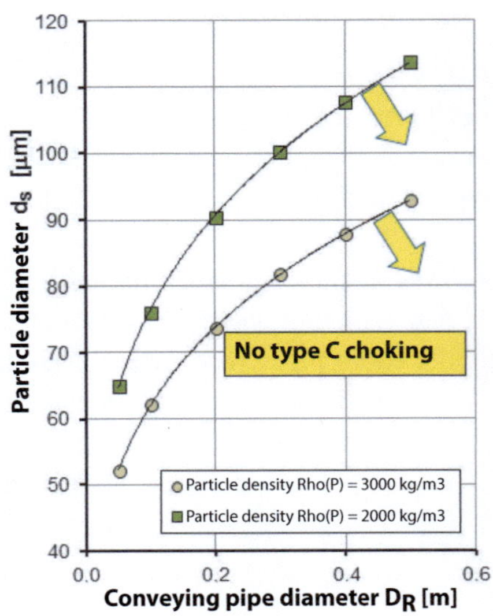

Fig. 4.19 Boundary curves for Type C-choking, (4.20), fine-grained bulk materials of different particle densities

Fig. 4.20 Limiting velocities for Type A-choking, (4.19), fine-grained bulk materials of different particle densities

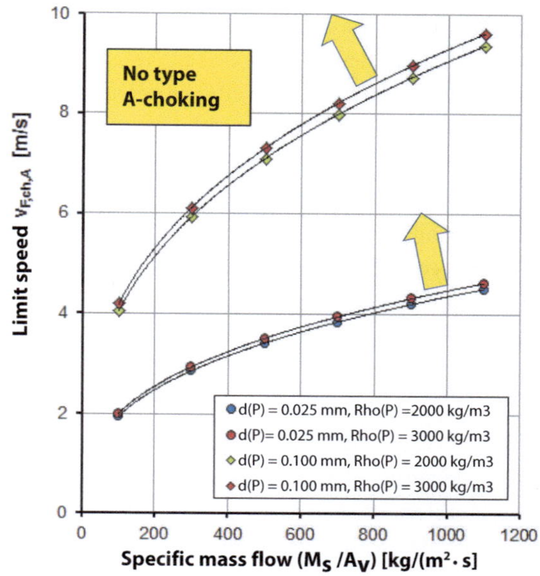

discharged at $v_{F,ch,A}$ is the maximum achievable throughput without accumulation. If the conveying line is still supplied with a mass flow by the installed solid lock, $\dot{m}_{S,0} < \dot{m}_S(v_{F,ch,A})$, it leads to a permanent solid enrichment or expansion of the dense zone and, as a result, a continuous pressure increase. This then leads to a failure/malfunction of the pressure generator. The velocity $v_{F,ch,A}$ can be calculated with (4.19), i.e. $v_{F,ch,A} = v_{F,pf}$. Fig. 4.20 shows an example of the evaluation of (4.19) for the particle diameters $d_S = (25\,\mu\text{m}, 100\,\mu\text{m})$ and the particle densities $\varrho_P = (2000\,\text{kg/m}^3, 3000\,\text{kg/m}^3)$. Further boundary conditions: conveying gas = air, temperature $t_F = t_S = 20\,°\text{C}$, pressure at the base of the vertical distance $p_R = 2.0\,\text{bar}$. Plotted is $v_{F,ch,A}$ over the specific solid mass flow rate related to the pipe cross-sectional area $(\dot{m}_S/A_R)$. The limiting velocity $v_{F,ch,A}$ increases with increasing values of $(\dot{m}_S/A_R)$ and d_S, the density ϱ_P has a minor influence. At gas velocities below the individual curves, Type A-choking occurs.

Various measurements/investigations, e.g. in [31], illustrate that the velocities $v_{F,ch}$ of the various *choking* types occur in the order

$$v_{F,ch,C} < v_{F,ch,B} < v_{F,ch,A} \tag{4.21}$$

(→ the classification of $v_{F,ch,B}$, which is also influenced by A- and C-choking phenomena, appears problematic). For the vertical pipelines in pneumatic conveying systems considered here, $v_{F,ch,A}$ is the critical gas velocity at the foot of the vertical section that must not be undercut for stable, jam-free operation. $v_{F,ch,A} = v_{F,pf}$ thus corresponds to the "clogging limit" of horizontal conveyances, referred to as curve C in Fig. 4.1, see Sect. 4.1.1. As described there, conveying operation is also possible at lower gas velocities if the equipment and control technology of the system are adapted to this application. The absolute lower limit velocity for this is $v_{F,tr}$. Reference is also made to the information in Fig. 3.12, Sect. 3.3.3, pointed out the fact that the gas velocity $v_{F,tr}$ for dense transport of fine-grained bulk materials is a multiple of the individual particle settling velocity w_T, while it approximately matches this for coarser solids. Such behavior is also expected for gas velocities $v_F > v_{F,tr}$.

In the relevant literature, e.g. [30, 31, 32, 34, 35, 36], alternative empirical equations are proposed to describe the *choking* problem. The above selection is based on the author's positive experiences with the described approaches.

Note: The relationships discussed above are more complex in detail than presented: In *fast fluidization* states with gas velocities v_F above the transport velocity $v_{F,tr}$, for example, there are two limiting velocity curves at a given and constant specific mass flow $(\dot{m}_S/A_R)$. These run together at the operating point $(v_F, (\dot{m}_S/A_R))_{tr}$, see Fig. 4.21. The right limiting curve $v_{F,pf}$ describes the transition to accumulation-free pneumatic vertical conveying and can be calculated using (4.19), while when the gas velocity falls below the left boundary curve $v_{F,dp}$ a "dense" fluidized bed with a changed solid concentration profile over the reactor height and internal solid circulation is established. This boundary curve can be determined by the empirical equation

$$\frac{v_{F,dp}}{\sqrt{g \cdot d_S}} = 39.8 \cdot \left(\frac{\dot{m}_S}{A_R \cdot \varrho_F \cdot v_{F,dp}}\right)^{0.311} \cdot \frac{1}{Re_T^{0.078}} \qquad (4.22)$$

$$\rightarrow \quad D_R < 0.3\,\text{m},\ \textbf{Geldart}\ \text{"}\mathbf{A}\text{" and "}\mathbf{B}\text{"}$$

with:

Re_T $(= w_T \cdot d_S \cdot \varrho_F/\eta_F)$, Reynolds number formed with w_T and d_S,

[30]. It is applicable to systems with and without *choking*. Transformation to $v_{F,dp}$ results in:

$$v_{F,dp} = (39.8 \cdot \sqrt{g \cdot d_S} \cdot Re_T^{-0.078})^{0.763} \cdot \left(\frac{\dot{m}_S}{A_R \cdot \varrho_F}\right)^{0.237} \qquad (4.23)$$

From Fig. 4.21, the values for different operating modes ($\rightarrow$ e.g. $v_F = \text{const.}$, $(\dot{m}_S/A_R) = \textit{variable}$ or $v_F = \textit{variable}$, $(\dot{m}_S/A_R) = \text{const.}$) adjusting operating conditions can be taken from.

Fig. 4.21 Boundaries of *fast fluidization*

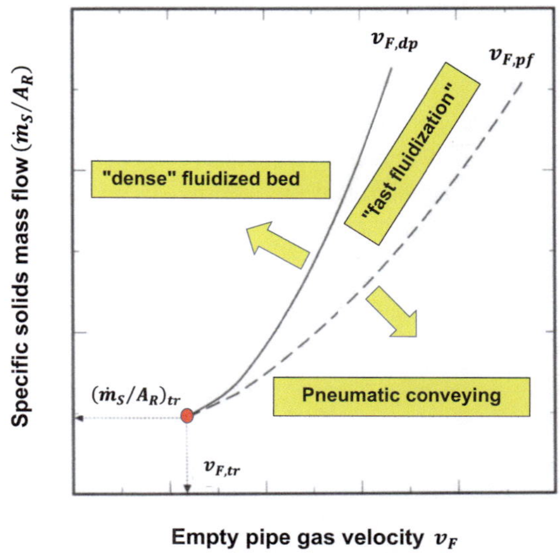

4.6.4 Characteristic Horizontal Conveying Gas Velocities

The following exemplary calculation approaches allow an estimation of those gas velocities at which the transition of a characteristic conveying/flow state into another takes place when exceeded or fallen below. Considered are: clogging limit velocity $v_{F,st}$, gas velocity at pressure loss minimum $v_{F,mp}$, *saltation* velocity $v_{F,sa}$ and *pickup* velocity $v_{F,pu}$. The velocities $(v_{F,sa}, v_{F,mp}, v_{F,st})$ describe with decreasing values the transition from flight to strand conveying and from a low-pulsation strand conveying to an unsteady, strongly pulsating dune/conglobation conveying. At gas velocities $v_F \geq v_{F,pu}$ the conveying gas stream picks up the solid from a resting strand and carries it away. Due to the non-standardized measurement methods, see, for example, [37], all these methods contain a "subjective" component (→ on which strand thickness/cross-sectional proportion is the *saltation* velocity $v_{F,sa}$, on which amplitude of pressure pulsations is $v_{F,st}$ based on, etc.?). All mentioned limiting velocities are in the range of the pressure loss minima of the standard conveying diagram, see Fig. 4.1. This leads to the fact that they are used interchangeably by the individual authors and are also directly compared with each other in the literature [38]. The following approaches provide reliable statements from the author's point of view:

a) Clogging velocity $v_{F,st}$**:** This is described by curve C in Fig. 4.1 and defines the transition from a low-pulsation conveying state with respect to the conveying pressure to an unsteady, strongly pulsating conveying state. As already emphasized several times, conveying operation at gas velocities $v_F < v_{F,st}$ is possible with a correspondingly steep delivery curve of the pressure generator. However, if special reactors are pneumatically charged, including coal dust burners and blast furnaces, a pulsation-free/-low solid entry is required for their economical and safe operation. In the case of coal dust burners, for example, strong flow pulsations can lead to incomplete combustion and thus to the formation of the explosive gas CO. Its concentration is monitored and triggers the shutdown of the plant when a limit value is exceeded. In such cases, the condition $v_F > v_{F,st}$ must be met.

The dependence of the loading μ_{st} on the Froude number $Fr_{D,st}$ at the choking limit can be described according to [39] by the approach

$$\mu_{st} = K_{st} \cdot Fr_{R,st}^{n} \tag{4.24}$$

with:

μ_{st} $(= \dot{m}_S/(A_R \cdot \varrho_F \cdot v_{F,st}))$, clogging limit load,
$Fr_{R,st}$ $(= v_{F,st}^2/(g \cdot D_R))$, Froude number of the clogging limit can be represented.

(4.24) generalizes an approach from [40], which, based on investigations on coarser millimeter-sized solids, suggests an exponent $n = 2$. K_{st} and n must be adapted to the properties of the respective bulk material, especially its particle size. For a given limit load μ_{st}, the minimum permissible gas velocity for a low-pulsation operation is determined.$v_{F,st}$ directly from (4.24):

$$v_{F,st} = \sqrt{g \cdot D_R} \cdot \left(\frac{\mu_{st}}{K_{st}}\right)^{\frac{1}{2 \cdot n}} \tag{4.25}$$

The solid mass flow corresponding to μ_{st} is then calculated as: $\dot{m}_{S,st} = \mu_{st} \cdot A_R \cdot \varrho_F \cdot v_{F,st}$. For the usual design case, where the solid throughput $\dot{m}_S$ and not the loading μ_{st} is specified, it follows from (4.24):

$$v_{F,st} = \left(\frac{1}{K_{st}} \cdot \frac{(g \cdot D_R)^n}{\varrho_F} \cdot \frac{\dot{m}_S}{A_R}\right)^{\frac{1}{2 \cdot n+1}} \tag{4.26}$$

In [4] the approach developed based on a stability analysis and verified by measurements and comparisons with literature data

$$\mu_{st} = 0.018 \cdot \frac{1}{\beta_R^2} \cdot \frac{\varrho_F}{\varrho_{b,\mathrm{Str}}} \cdot Fr_{R,st}^2 \tag{4.27}$$

with:

β_R Sliding friction coefficient bulk material strand-pipe wall,
$\varrho_{b,\mathrm{Str}}$ Bulk material density of the strand

is recommended.

(4.27) was applied to bulk materials with particle diameters in the range of $d_{S,50} = (90\text{-}3300)\ \mu\text{m}$, particle densities of $\varrho_P = (1280\text{-}2850)\ \text{kg/m}^3$ and sliding friction coefficients of $\beta_R = (0.23\text{-}0.57)$ tested and provides reliable results.

(4.24), and thus also (4.27), cannot be applied to the conveyance of fine dust particles [8]. For very large loadings μ, the clogging limit proves to be independent of the conveying gas velocities v_F and is determined only by the friction coefficient β_R between the solid and the pipe wall [41]. For practical estimates of the clogging limit velocity $v_{F,st}$, the following calculation equations are proposed in [42, 8]. For coarse-grained materials with $d_{S,50} > 1$ mm applies:

$$v_{F,st} \cong 1.3 \cdot w_T \cdot \mu_{st}^{1/4} \quad \rightarrow \quad \mu_{st} < 15 \tag{4.28}$$

For fine-grained solids with $d_{S,50} < 100\,\mu\text{m}$ can be applied:

$$v_{F,st} \cong 0.25 \cdot \frac{\sqrt{g \cdot D_R}}{\sqrt{\varrho_F / \varrho_{b,\text{Str}}}} \quad \rightarrow \quad \mu \cdot \frac{\varrho_F}{\varrho_{b,\text{Str}}} < 0.75 \tag{4.29}$$

Both equations provide satisfactory, rather somewhat conservative values.

b) Saltation velocity $v_{F,sa}$: It describes the onset of bulk material particles falling out of the conveying gas stream. These particles are supported on the pipe bottom and form a moving or stationary strand there. $v_{F,sa}$ is thus the lower limit velocity for dilute phase pneumatic conveying. For coarse-grained solids, *saltation* velocity $v_{F,sa}$ and gas velocity $v_{F,mp}$ at the pressure loss minimum are largely identical, while fine-grained products already fall out of the gas stream at higher gas velocities $v_{F,sa} > v_{F,mp}$, see curve D in Fig. 4.3.

The extensive measurement-based investigations in [43] show that fine and coarse-grained bulk materials behave differently with respect to $v_{F,sa}$: The *saltation* velocity of coarse-grained products decreases as expected with decreasing particle diameter, but then increases again for fine-grained solids. The boundary diameter d_S^* between coarse and fine-grained is determined by

$$\frac{d_S^*}{D_R} = 1.39 \cdot \left(\frac{\varrho_P}{\varrho_F}\right)^{-0.74} \tag{4.30}$$

For $d_S \geq d_S^*$ applies

$$\mu_{sa} = 0.373 \cdot \left(\frac{\varrho_P}{\varrho_F}\right)^{1.06} \cdot \left(\frac{\sqrt{Fr_{S,T}}}{10}\right)^{-3.7} \cdot \left(\frac{\sqrt{Fr_{R,sa}}}{10}\right)^{3.61} \tag{4.31}$$

and for $d_S < d_S^*$

$$\mu_{sa} = 5.56 \cdot 10^3 \cdot \left(\frac{d_S}{D_R}\right)^{1.43} \cdot \left(\frac{\sqrt{Fr_{R,sa}}}{10}\right)^{4} \tag{4.32}$$

with:

μ_{sa} $(= \dot{m}_S / (A_R \cdot \varrho_F \cdot v_{F,sa}))$, loading at the beginning of *saltation*,
$Fr_{R,sa}$ $(= v_{F,sa}^2 / (g \cdot D_R))$, Froude number at the beginning of *saltation*,

$Fr_{S,T}$ $(= w_T^2/(g \cdot d_S))$, Froude number formed with single particle settling velocity w_T and characteristic particle diameter d_S.

(4.30)–(4.32) are based on measurements of bulk materials with particle diameters in the range of $d_{S,50} = (20\text{-}1640)\,\mu\text{m}$ and particle densities of $\varrho_P = (2500, 8700)\,\text{kg/m}^3$. $v_{F,sa}$ must be calculated analogously to (4.25) and (4.26).

The different behavior of coarse and fine particles can be explained as follows: Lateral movements of coarse particles in the conveying tube occur largely unaffected by gas turbulence. After a wall impact, the particles bounce back into the accelerating conveying gas stream. This self-motion is determined by the shape and hardness of the particles. Fine particles, on the other hand, are forced into lateral movements by gas turbulence and are also aerodynamically decelerated during wall impacts in the boundary layer, i.e., a rebound into the core flow does not occur due to the small rebound distances. Already at higher gas velocities, an accumulation of fine particles in the lower pipe wall area is supported by gravity. Adhesion/cohesion forces reinforce this effect.

In [44], *saltation* measurements with coarse-grained bulk materials ($\rightarrow$ $d_{S,50} \geq 450\,\mu\text{m}$) are reported, in which the *saltation* velocities $v_{F,sa}$ are about 2 m/s below the gas velocities $v_{F,mp}$ at the pressure loss minimum.

c) Velocity $v_{F,mp}$ at the pressure loss minimum: A given solid mass flow $\dot{m}_S$ is transported at the associated pressure loss minimum of the conveying diagram, see Fig. 4.3, with the smallest possible pressure difference Δp_R, i.e., in an energetically advantageous conveying state. Due to the relatively low gas velocity, such conveying is also gentle on the product and pipeline: grain destruction and pipe wear are low. The pressure loss minimum thus represents a preferred/target operating point. Details on energy consumption can be found in Sect. 4.12.

Based on extensive own measurements and data from the literature, the dependency of the loading μ_{mp} in [45] at the minimum pressure loss from the associated gas velocities $v_{F,mp}$ has been determined as follows:

$$\mu_{mp} = \frac{1}{10^\delta} \cdot Fr_{R,mp}^\kappa \quad \rightarrow \quad \delta = 1.44 \cdot \frac{d_S}{\text{mm}} + 1.96, \quad \kappa = 0.55 \cdot \frac{d_S}{\text{mm}} + 1.25 \tag{4.33}$$

with:

μ_{mp} $(= \dot{m}_S/(A_R \cdot \varrho_F \cdot v_{F,mp}))$, loading at the pressure loss minimum,
$Fr_{R,mp}$ $(= v_{F,mp}^2/(g \cdot D_R))$, Froude number at the pressure loss minimum.

The approach corresponds to that of (4.24). $K_{mp} = 1/10^\delta$ and $\kappa = n$ are represented as dependent only on the characteristic particle size d_S. Since the bulk materials investigated in [45] are predominantly relatively coarse-grained plastics, $d_S \cong (0.7\text{-}6.0)\,\text{mm}$, is concerned, their particle density ϱ_P is relatively low compared to those of mineral substances. This leads to the application of (4.33) to specifically heavier solids resulting in exaggerated $v_{F,mp}$ speeds compared to existing measurements, see e.g. [44]. The introduction of a correction term $f(\varrho_P/\varrho_{P,\text{ref}})$ or $f(\varrho_P/\varrho_F)$ in (4.33) would improve its accuracy. (4.33) is based on measured values for conveying pipe diameters of $D_R = (50\text{-}400)$ mm.

In the case of fine-grained bulk materials, i.e. $d_S \to 0$, (4.33) reduces to $\mu_{mp} = 1/10^{1.96} \cdot Fr_{R,mp}^{1.25}$. The Froude number exponent $\kappa = 1.25$ then corresponds approximately to that of (3/2) from (4.1), which describes the *saltation* onset of fine-grained bulk materials. $v_{F,mp}$ must be calculated according to (4.25) and (4.26).

d) *Pickup* speed $v_{F,pu}$: Conveying gas flows with speeds $v_F \geq v_{F,pu}$ pick up the solid from a resting strand/deposit in the horizontal pipe and carry it away. The faster the removal, the greater the distance from v_F to the limit speed $v_{F,pu}$ is chosen. Removal of deposits is, for example, absolutely necessary before a pure product change. Especially in dense phase = slow velocity conveyance, this is sometimes not possible with the available conveying gas flow, as it may only allow speeds $v_F < v_{F,pu}$, i.e., for cleaning the pipeline, a correspondingly larger gas volume flow must be temporarily provided.

The dependence of the gas speed $v_{F,pu}$ on the characteristic particle diameter d_S has a minimum at $d_S \cong (70\text{-}150)\,\mu\text{m}$: For smaller particle sizes d_S the value of $v_{F,pu}$ increases to values up to $v_{F,pu} \cong 15\,\text{m/s}$ at $d_S \cong 5\,\mu\text{m}$ and $D_R = 52\,\text{mm}$ [37, 46]. The cause is, among other things, the friction and cohesion forces between the fine particles on the surface of the deposit, where the local gas velocity also approaches zero due to the adhesion condition. Above the minimum, $v_{F,pu}$ increases with increasing particle diameter d_S as expected. This behavior is analogous to that of the previously described *saltation* velocity, although the responsible mechanisms are completely different [43]. For bulk materials with particle diameters$d_S > 100\,\mu\text{m}$ is used in [37, 44] the calculation equation based on a dimensional analysis

$$\sqrt{Fr_{S,pu}} = 0.0428 \cdot Re_{S,pu}^{0.175} \cdot \left(\frac{D_R}{d_S}\right)^{0.25} \cdot \left(\frac{\varrho_P}{\varrho_F}\right)^{0.75} \tag{4.34}$$

with:

$Fr_{S,pu}$ $(= v_{F,pu}^2/(g \cdot d_S))$, Froude number formed with *pickup* speed $v_{F,pu}$ and particle diameter d_S,

$Re_{S,pu}$ $(= v_{F,pu} \cdot d_S \cdot \varrho_F/\eta_F)$, Reynolds number formed with $v_{F,pu}$ and d_S, validity range: $25 \leq Re_{S,pu} < 5000, 8 \leq (D_R/d_S) < 1340, 700 \leq (\varrho_P/\varrho_F) < 4240$

proposed.

(4.34) provides a constant value for $v_{F,pu}$ for a given solid, conveying gas, gas state and pipe diameter. The *saltation* velocity $v_{F,sa}$ for example, would increase further under the mentioned boundary conditions with increasing load μ. Measurements in [37, 44] show that $v_{F,pu}$ could represent an upper limit for $v_{F,sa}$. At loads $\mu \gtrsim 10$ both would be identical, below that $v_{F,pu} > v_{F,sa}$.

In [6] it is pointed out that by adding a small amount of bulk material to the purging gas stream, the removal of stationary deposits is accelerated by the more intensive momentum exchange at their surface. This can be practically implemented.

Note 1: The calculation approaches presented in subsections a)–d) must, like those of Sects. 4.6.2 and 4.6.3, be applied to the critical operating condition at the beginning of the conveying line, see Sect. 4.1.2.

Note 2: Rabinovich and Kalman have subjected the characteristic gas velocities described in Sects. 4.6.3 and 4.6.4, among others, to a critical review and summary in a series of publications [47, 48, 49, 50, 51, 52, 53] based on their own and available literature measurement results. The various limit velocities can thus be represented with a maximum deviation of approx. ±30 % by power functions of the form

$$Re_S^* = a \cdot (Ar^*)^b \quad \rightarrow \quad Re_S^* = \frac{Re_S}{K_{Re}}, \quad Ar^* = Ar \cdot K_{Ar} \tag{4.35}$$

with:

Re_S	$(= v_{F,x} \cdot d_S \cdot \varrho_F/\eta_F)$, Reynolds number formed with the respective gas velocity $v_{F,x}$ and the particle diameter d_S.
Ar	$(d_S^3 \cdot g \cdot \varrho_F \cdot (\varrho_P - \varrho_F)/\eta_F^2)$, Archimedes number,
K_{Re}, K_{Ar}	empirical correction functions for Re_S and Ar,
a, b	empirical constants, exponents

as master or reference curves.

K_{Re} and K_{Ar} show, depending on the currently considered speed $v_{F,x}$, different structures. To determine a $v_{F,x}$, the modified Archimedes number Ar^* from (4.35) is used to calculate the modified Reynolds number Re_S^* ($\rightarrow Ar^* = Ar \cdot K_{Ar}$ as well as a, b and K_{Re} are known from the task description). From Re_S^*, the Reynolds number describing the current problem $Re_S = K_{Re} \cdot Re_S^*$ and from this, the desired $v_{F,x}$ can be determined. In calculation example 7, the method described above is applied. The results, which are only hinted at here, from [47, 48, 49, 50, 51, 52, 53] allow predictions of the expected behavior of fluid/solid mixtures in different operating situations and should be considered in critical plant designs.

4.6.5 Calculation Example 7: Calculation of Characteristic Conveying Speeds

The new product "limestone chippings" is to be pneumatically transported through an existing horizontal and a vertical conveying section with different lengths. The conveying pipe diameters of the lines are identical. With the existing pressure generators (→

single-stage screw compressors) a gas initial velocity of $v_{F,A} = 14.50\,\mathrm{m/s}$ can be realized in both lines at a conveying pressure of 2.0 bar at the beginning of the line. It is necessary to check which conveying states will occur in the planned conventional conveyances and the given bulk material using the calculation approaches provided in Sects. 4.6.2, 4.6.3 and 4.6.4.

Relevant Design Data:

- Operating data of both systems:
  Solid throughput: $\dot{m}_S = 20\,\mathrm{t/h} = 5.556\,\mathrm{kg/s}$,
  Conveying pipe diameter: $D_R = 82.5\,\mathrm{mm}$, Cross-sectional area:
  $A_R = 5.3456 \cdot 10^{-3}\,\mathrm{m}^2$,
  Conveying pressure at the beginning of the pipeline: $p_A = 2.0\,\mathrm{bar}$,
  Pressure at the end of the pipeline: $p_E = 1.0\,\mathrm{bar}$,
  Design/mixing temperature $\cong$ Solid temperature: $T_M = 293\,\mathrm{K}$.
- Bulk material data:
  Particle size distribution: $d_S \cong (0.03\text{-}1.70)\,\mathrm{mm}$,
  Median diameter: $d_{S,50} = 0.450\,\mathrm{mm}$,
  Particle- = solid density: $\varrho_P = 2750\,\mathrm{kg/m}^3$,
  Bulk density: $\varrho_{SS} = 1600\,\mathrm{kg/m}^3$,
  preliminary conveying tests and bulk material investigations are available.
- Gas data:
  Gas type: Air,
  Gas density at the beginning of the pipeline: $\varrho_{F,A} = 2.40\,\mathrm{kg/m}^3$,
  Gas density at the end of the pipeline: $\varrho_{F,E} = 1.20\,\mathrm{kg/m}^3$,
  Dynamic viscosity: $\eta_F = 18.26 \cdot 10^{-6}\,\mathrm{Pa\,s}$.

a) Horizontal conveying line

- **Conveying limit= Minimum speed** $v_{F,A,\min}$:
  There are conveying tests available, see Fig. 4.17, which show that $v_{F,A,\min}$ depends on the diameter of the conveying pipe D_R but not on the conveying pressure p_A, see Sect. 4.6.2. From Fig. 4.17a, the following applies to the operating case considered here:

$$v_{F,A,\min} = 11.0\,\frac{\mathrm{m}}{\mathrm{s}}$$

- **Clogging limit speed** $v_{F,st}$:
  From (4.27) and taking into account (4.26) for the beginning of the conveying line:

$$v_{F,st} = \left(\frac{\varrho_{b,\mathrm{Str}}}{0.018} \cdot \left(\frac{\beta_R \cdot g \cdot D_R}{\varrho_{F,A}} \right)^2 \cdot \frac{\dot{m}_S}{A_R} \right)^{1/5} \tag{4.36}$$

With $\beta_R = \tan \varphi_W = \tan 20° = 0.364$ and
$\varrho_{b,\mathrm{Str}} \cong 0.90 \cdot \varrho_{SS} = 0.90 \cdot 1600\,\mathrm{kg/m}^3 = 1440\,\mathrm{kg/m}^3$ applies:

$$v_{F,st} = \left(\frac{1440\,\frac{\text{kg}}{\text{m}^3}}{0.018} \cdot \left(\frac{0.364 \cdot 9.81\,\frac{\text{m}}{\text{s}^2} \cdot 0.0825\,\text{m}}{2.40\,\frac{\text{kg}}{\text{m}^3}} \right)^2 \cdot \frac{5.556\,\frac{\text{kg}}{\text{s}}}{5.3456 \cdot 10^{-3}\,\text{m}^2} \right)^{0.20}$$

$$v_{F,st} = 16.58\,\frac{\text{m}}{\text{s}}$$

- **Saltation velocity** $v_{F,sa}$:
  The approach according to (4.30)–(4.32) is used. From (4.30), it follows:

$$d_S^* = 1.39 \cdot D_R \cdot \left(\frac{\varrho_{F,A}}{\varrho_P} \right)^{0.74} = 1.39 \cdot 0.0825\,\text{m} \cdot \left(\frac{2.40\,\frac{\text{kg}}{\text{m}^3}}{2750\,\frac{\text{kg}}{\text{m}^3}} \right)^{0.74} = 6.25 \cdot 10^{-4}\,\text{m}$$

$$d_S^* = 0.625\,\text{mm} > d_{S,50} = 0.450\,\text{mm} \quad \rightarrow \quad \text{continue with}$$

(4.32) considering (4.26) results in:

$$v_{F,sa} = \left(\frac{1}{0.556} \cdot \left(\frac{D_R}{d_{S,50}} \right)^{1.43} \cdot (g \cdot D_R)^2 \cdot \frac{\dot{m}_S}{A_R \cdot \varrho_{F,A}} \right)^{1/5}$$

$$= \left(\frac{1}{0.556} \cdot \left(\frac{82.5\,\text{mm}}{0.450\,\text{mm}} \right)^{1.43} \cdot \left(9.81\,\frac{\text{m}}{\text{s}^2} \cdot 0.0825\,\text{m} \right)^2 \cdot \frac{5.556\,\frac{\text{kg}}{\text{s}}}{5.3456 \cdot 10^{-3}\,\text{m}^2 \cdot 2.40\,\frac{\text{kg}}{\text{m}^3}} \right)^{0.2}$$

$$v_{F,sa} = 15.45\,\frac{\text{m}}{\text{s}}$$

- **Velocity** $v_{F,mp}$ **at the pressure loss minimum**:
  It is calculated with (4.33). From this, taking into account (4.26):

$$v_{F,mp} = \left(10^{\delta} \cdot (g \cdot D_R)^{\kappa} \cdot \frac{\dot{m}_S}{A_R \cdot \varrho_{F,A}} \right)^{1/(2\cdot\kappa+1)}$$

In the present case, the following applies:

$$\delta = 1.44 \cdot \frac{d_S}{\text{mm}} + 1.96 = 1.44 \cdot \frac{0.450\,\text{mm}}{\text{mm}} + 1.96 = 2.608$$

$$\kappa = 0.55 \cdot \frac{d_S}{\text{mm}} + 1.25 = 0.55 \cdot \frac{0.450\,\text{mm}}{\text{mm}} + 1.25 = 1.498$$

$$\frac{1}{2 \cdot \kappa + 1} = \frac{1}{2 \cdot 1.498 + 1} = 0.250$$

$$v_{F,mp} = \left(10^{2.608} \cdot \left(9.81\,\frac{\text{m}}{\text{s}^2} \cdot 0.0825\,\text{m} \right)^{1.498} \cdot \frac{5.556\,\frac{\text{kg}}{\text{s}}}{5.3456 \cdot 10^{-3}\,\text{m}^2 \cdot 2.40\,\frac{\text{kg}}{\text{m}^3}} \right)^{0.250}$$

$$v_{F,mp} = 18.91\,\frac{\text{m}}{\text{s}}$$

- **Alternative: Velocity $v_{F,mp}$ at the pressure loss minimum**:
  Since the present problem is outside the data basis of (4.33),→ $d_{S,50} < 0.7\,\mathrm{mm}$) and since these generally lead to too high velocities for larger particle densities $v_{F,mp}$, an alternative approach based on a more extensive dataset from [49] is used for comparison. From (4.35) it follows

$$Re_S^* = \frac{Re_S}{K_{Re}} = a \cdot (Ar^*)^b = a \cdot (Ar \cdot K_{ar})^b$$

$$Re_S = a \cdot (Ar \cdot K_{ar})^b \cdot K_{Re} \tag{4.37}$$

with the necessary empirical adjustments according to [49]:

$$\begin{aligned} K_{Re} &= 2.7 - 3.1 \cdot \exp\left(-\frac{D_R/D_{R,50}}{1.6}\right) \\ &= 2.7 - 3.1 \cdot \exp\left(-\frac{82.5\,\mathrm{mm}/50\,\mathrm{mm}}{1.6}\right) = 1.5947 \\ K_{Ar} &= \left(\left(\frac{1-e}{e}\right)^{0.1} + 30 \cdot C_V^{0.35}\right)^{2.33} \end{aligned}$$

The volume flow ratio C_V is defined as follows:

$$\begin{aligned} C_V &= \frac{\dot{V}_S}{\dot{V}_S + \dot{V}_F} = \frac{\dot{m}_S/\varrho_P}{\dot{m}_S/\varrho_P + v_{F,mp} \cdot A_R} \\ &= \frac{5.556\,\frac{\mathrm{kg}}{\mathrm{s}}/2750\,\frac{\mathrm{kg}}{\mathrm{m}^3}}{5.556\,\frac{\mathrm{kg}}{\mathrm{s}}/2750\,\frac{\mathrm{kg}}{\mathrm{m}^3} + v_{F,mp} \cdot 5.3456 \cdot 10^{-3}\,\mathrm{m}^2} \\ &= \frac{1}{1 + 2.6459\,\frac{\mathrm{s}}{\mathrm{m}} \cdot v_{F,mp}} \end{aligned}$$

The restitution coefficient is estimated with $e \cong 0.70$ thus:

$$\begin{aligned} K_{Ar} &= \left(\left(\frac{1-0.70}{0.7}\right)^{0.1} + 30 \cdot \left(\frac{1}{1 + 2.6459\,\frac{\mathrm{s}}{\mathrm{m}} \cdot v_{F,mp}}\right)^{0.35}\right)^{2.33} \\ &= \left(0.9188 + 30 \cdot \left(\frac{1}{1 + 2.6459\,\frac{\mathrm{s}}{\mathrm{m}} \cdot v_{F,mp}}\right)^{0.35}\right)^{2.33} \end{aligned}$$

Furthermore, the following applies:

$$Ar = \frac{d_{S,50}^3 \cdot g}{\eta_F^2} \cdot \varrho_{F,A} \cdot (\varrho_P - \varrho_{F,A})$$
$$= \frac{(0.450 \cdot 10^{-3}\,\mathrm{m})^3 \cdot 9.81\,\frac{\mathrm{m}}{\mathrm{s}^2}}{(18.26 \cdot 10^{-6}\,\mathrm{Pa\,s})^2} \cdot 2.40\,\frac{\mathrm{kg}}{\mathrm{m}^3} \cdot (2750 - 2.40)\,\frac{\mathrm{kg}}{\mathrm{m}^3} = 17679.5$$
$$a = 1.1,$$
$$b = 3/7$$

and as boundary condition $Ar^* = Ar \cdot K_{Ar} \geq 2450$. With $Re_S = v_{F,mp} \cdot d_S \cdot \varrho_F / \eta_F$ and (4.37) follows from this:

$$\frac{v_{F,mp} \cdot d_S \cdot \varrho_F \cdot}{\eta_F} = \frac{v_{F,mp} \cdot 0.450 \cdot 10^{-3}\,\mathrm{m} \cdot 2.40\,\frac{\mathrm{kg}}{\mathrm{m}^3}}{18.26 \cdot 10^{-6}\,\mathrm{Pa\,s}} = 59.1457\,\frac{\mathrm{s}}{\mathrm{m}} \cdot v_{F,mp}$$
$$= 1.1 \cdot \left(17679.5 \cdot \left(0.9188 + 30 \cdot \left(\frac{1}{1 + 2.6459\,\frac{\mathrm{s}}{\mathrm{m}} \cdot v_{F,mp}}\right)^{0.35}\right)^{2.33}\right)^{3/7} \cdot 1.5947$$

$v_{F,mp}$ must be determined iteratively and results in:

$$v_{F,mp} = 17.11\,\frac{\mathrm{m}}{\mathrm{s}}$$

Verification of the validity range:

$$Ar^* = Ar \cdot K_{Ar} = 17679.5 \cdot \left(0.9188 + 30 \cdot \left(\frac{1}{1 + 2.6459\,\frac{\mathrm{s}}{\mathrm{m}} \cdot 17.11\,\frac{\mathrm{m}}{\mathrm{s}}}\right)^{0.35}\right)^{2.33}$$
$$= 2774937.2 \geq 2450.$$

- **Assessment:** The previously calculated gas velocities $v_{F,st} = 16.58\,\mathrm{m/s}$, $v_{F,sa} = 15.45\,\mathrm{m/s}$ and $v_{F,mp} = 17.11\,\mathrm{m/s}$ do show a deviation of $\Delta v_F \cong 1.66\,\mathrm{m/s}$ and a certain illogicality ($\rightarrow v_{F,st} > v_{F,sa}$) but show that they are greater than the intended initial conveying gas velocity of $v_{F,A} = 14.50\,\mathrm{m/s}$ and thus gas/solid separation = strand formation in the pipeline will occur. This was also observed in the experiments underlying Fig. 4.17. The loading μ of the planned design amounts to

$$\mu = \frac{\dot{m}_S}{A_R \cdot \varrho_{F,A} \cdot v_{F,A}} = \frac{5.556\,\frac{\mathrm{kg}}{\mathrm{s}}}{5.3456 \cdot 10^{-3}\,\mathrm{m}^2 \cdot 2.40\,\frac{\mathrm{kg}}{\mathrm{m}^3} \cdot 14.50\,\frac{\mathrm{m}}{\mathrm{s}}} = 29.87\,\frac{\mathrm{kg\,S}}{\mathrm{kg\,F}},$$

which, for a dilute phase pneumatic conveying of $\dot{m}_S = 20\,\mathrm{t/h}$ limestone chippings, the maximum possible loading is only

$$\mu_{sa} = \frac{\dot{m}_S}{A_R \cdot \varrho_{F,A} \cdot v_{F,sa}} = \frac{5.556\,\frac{\mathrm{kg}}{\mathrm{s}}}{5.3456 \cdot 10^{-3}\,\mathrm{m}^2 \cdot 2.40\,\frac{\mathrm{kg}}{\mathrm{m}^3} \cdot 16.4\,\frac{\mathrm{m}}{\mathrm{s}}} = 26.41\,\frac{\mathrm{kg\,S}}{\mathrm{kg\,F}}$$

($\rightarrow v_{F,sa} \cong 16.4\,\mathrm{m/s}$, average of the above limit velocities). Thus, the proportion

$$\frac{\dot{m}_{S,\mathrm{strand}}}{\dot{m}_S} = 1 - \frac{v_{F,A}}{v_{F,sa}} = 1 - \frac{\mu_{sa}}{\mu} = 1 - \frac{26.41\,\frac{\mathrm{kg\,S}}{\mathrm{kg\,F}}}{29.87\,\frac{\mathrm{kg\,S}}{\mathrm{kg\,F}}} = 0.116$$

of the total solid throughput, i.e. $\dot{m}_{S,\mathrm{strand}} = 0.116 \cdot 20\,\mathrm{t/h} = 2.32\,\mathrm{t/h}$, is transported as a strand at the bottom of the horizontal conveying pipe.
The limestone chippings to be conveyed here are a bulk material at the transition from Group 3 to Group 2, see Sect. 4.5 as well as Figs. 4.13 and 4.14. Due to its relatively wide grain size distribution ($\rightarrow \alpha_{\mathrm{RRSB}} < 60°$) and the resulting low gas permeability, it was expected that stable plug conveying would not be possible. Corresponding conveying tests confirmed this. The strand conveying that occurred during operation was relatively low in pulsation due to the considerable fine fraction.

- ***Pickup*-speed** $v_{F,pu}$:
  Since the boundary condition $d_{S,50} = 450\,\mu\mathrm{m} > 100\,\mu\mathrm{m}$ is fulfilled, (4.34) is used. From this follows:

$$v_{F,pu} = \left(0.0428 \cdot \left(\frac{d_{S,50} \cdot \varrho_{F,A}}{\eta_F}\right)^{0.175} \cdot \left(\frac{D_R}{d_{S,50}}\right)^{0.25} \cdot \left(\frac{\varrho_P}{\varrho_{F,A}}\right)^{0.75} \cdot \sqrt{g \cdot d_{S,50}}\right)^{1/0.825}$$

Validity range:

$$8 \le \left(\frac{D_R}{d_S} = \frac{82.5\,\mathrm{mm}}{0.450\,\mathrm{mm}} = 183.3\right) < 1340$$

$$700 \le \left(\frac{\varrho_P}{\varrho_{F,A}} = \frac{2750\,\frac{\mathrm{kg}}{\mathrm{m}^3}}{2.40\,\frac{\mathrm{kg}}{\mathrm{m}^3}} = 1145.8\right) < 4240$$

$Re_{S,pu}$-control only possible after calculation of $v_{F,pu}$.

$$v_{F,pu} = \left(0.0428 \cdot \left(\frac{0.450 \cdot 10^{-3}\,\mathrm{m} \cdot 2.40\,\frac{\mathrm{kg}}{\mathrm{m}^3}}{18.26 \cdot 10^{-6}\,\mathrm{Pa\,s}}\right)^{0.175} \right.$$
$$\left. \cdot\, 183.3^{0.25} \cdot 1145.8^{0.75} \cdot \sqrt{9.81\,\frac{\mathrm{m}}{\mathrm{s}^2} \cdot 0.450 \cdot 10^{-3}\,\mathrm{m}}\right)^{1/0.825}$$

$$v_{F,pu} = 5.71\,\frac{\mathrm{m}}{\mathrm{s}}$$

$$25 \le \left(Re_{S,pu} = \frac{v_{F,pu} \cdot d_{S,50} \cdot \varrho_{F,A}}{\eta_F} = \frac{5.71\,\frac{\mathrm{m}}{\mathrm{s}} \cdot 0.450 \cdot 10^{-3}\,\mathrm{m} \cdot 2.40\,\frac{\mathrm{kg}}{\mathrm{m}^3}}{18.26 \cdot 10^{-6}\,\mathrm{Pa\,s}} = 337.7\right)$$
$$< 5000$$

Since limestone chippings are a bulk material with a wide grain distribution and the coarse fraction should also be picked up by the gas, $v_{F,pu}$ is alternatively calculated with the maximum particle diameter $d_{S,\max} = 1.70\,\text{mm}$. This leads to

$$v_{F,pu,\max} = 11.32\,\frac{\text{m}}{\text{s}}$$

- **Assessment**: It should be taken into account that the $v_{F,pu}$ velocities only describe the beginning of the removal of a stationary bulk material body and provide no information about the removal mass flow or duration. In comparison with the *saltation* velocity $v_{F,sa}$, the calculated values are very low. The assumption expressed in Sect. 4.6.4 that for loadings of $\mu \gtrsim 10\, v_{F,sa} \cong v_{F,pu}$ will occur is obviously wrong. On the other hand, it should be considered that the present $v_{F,pu}$ measurements, in order to capture the influence of grain size, were generally carried out on bulk materials with a very narrow particle size distribution (→ Monograin). The limestone grit in question does not correspond to this. Since during the blowing out the considered conveying line, the gas velocity at the beginning of the line increases with increasing blowing time, i.e., continuously decreasing conveying pressure p_A, from $v_{F,A} = 14.50\,\text{m/s}$ in the direction of $v_{F,E}$, a relatively fast clearing of the line is ensured here. This is consistent with the experimental experience.

b) Vertical conveying line

For comparisons, the loosening speed $v_{F,L}$, the single-particle terminal velocity w_T and the terminal velocity $w_{T,\varepsilon}$ of a particle cloud are determined. All calculations are carried out with the assumed spherical median diameter $d_{S,50} = 0.450\,\text{mm}$ without shape correction and for the operating condition at the beginning of the conveying line.

- **Loosening speed** $v_{F,L}$:
  With $\varepsilon_L \cong (1 - \varrho_{SS}/\varrho_P) = (1 - 1600\,\text{kg/m}^3/2750\,\text{kg/m}^3) = 0.4182,\ Ar = 17679.5$ (→ already calculated above) and the simplifications $d_{S,SD} = d_{S,50}$ and $\psi = 1$ follows from (3.37):

$$v_{F,L} = 42.9 \cdot (1 - \varepsilon_L) \cdot \frac{\eta_F}{d_{S,50} \cdot \varrho_{F,A}} \cdot \left(\sqrt{1 + \frac{Ar}{3214} \cdot \frac{\varepsilon_L^3 \cdot}{(1 - \varepsilon_L)^2}} - 1 \right)$$

$$= 42.9 \cdot \frac{(1 - 0.4182)}{0.450 \cdot 10^{-3}\,\text{m}} \cdot \frac{18.26 \cdot 10^{-6}\,\text{Pa s}}{2.40\,\frac{\text{kg}}{\text{m}^3}}$$

$$\cdot \left(\sqrt{1 + \frac{17679.5}{3214} \cdot \frac{0.4182^3}{(1 - 0.4182)^2}} - 1 \right)$$

$$v_{F,L} = 0.20\,\frac{\text{m}}{\text{s}}$$

- **Single particle settling velocity** w_T:
  It is calculated using the equation valid for the entire *Re* range $Re_{S,T} > 3 \cdot 10^5$ (3.17):

$$w_T = 18 \cdot \frac{\eta_F}{d_{S,50} \cdot \varrho_{F,A}} \cdot \left(\sqrt{1 + \frac{1}{9} \cdot \sqrt{Ar}} - 1\right)^2$$

$$= 18 \cdot \frac{18.26 \cdot 10^{-6}\,\text{Pa s}}{0.450 \cdot 10^{-3}\,\text{m} \cdot 2.40\,\frac{\text{kg}}{\text{m}^3}} \cdot \left(\sqrt{1 + \frac{1}{9} \cdot \sqrt{17679.5}} - 1\right)^2$$

$$w_T = 2.69\,\frac{\text{m}}{\text{s}}$$

- **Settling velocity** $w_{T,\varepsilon}$ **of a particle cloud**:
  For this purpose, the equation recommended in [54] is used:

$$\frac{w_{T,\varepsilon}}{w_T} = 1 + (0.25 + k) \cdot \mu^{0.25} \quad \rightarrow \quad 50 \geq Re_{S,T,\varepsilon} \leq 1000 \tag{4.38}$$

with:

$k = 1$ Stokes range;
$k = 0.5$ Transition range;
$k = 0$ Newton range → see Sect. 3.1.1,

With $k = 0.5$, $w_T = 2.69\,\text{m/s}$ and $\mu = 29.87\,\text{kg}\,S/\text{kg}\,F$ (→ already calculated above) follows:

$$w_{T,\varepsilon} = \left(1 + (0.25 + 0.5) \cdot \left(29.87\,\frac{\text{kg S}}{\text{kg F}}\right)^{0.25}\right) \cdot 2.69\,\frac{\text{m}}{\text{s}}$$

$$w_{T,\varepsilon} = 7.41\,\frac{\text{m}}{\text{s}}$$

Validity range:

$$50 \leq \left(Re_{S,T,\varepsilon} = \frac{7.41\,\frac{\text{m}}{\text{s}} \cdot 0.450 \cdot 10^{-3}\,\text{m} \cdot 2.40\,\frac{\text{kg}}{\text{m}^3}}{18.26 \cdot 10^{-6}\,\text{Pa s}} = 438.3\right) \leq 1000$$

$Re_{S,T,\varepsilon}$ is in the transition range → $k = 0.5$. The approach according to (4.38) assumes a completely separated gas/solid flow.

- **Boundary transport velocity** $v_{F,tr}$: This is calculated with (3.40):

$$v_{F,tr} = 2.28 \cdot \frac{\eta_F}{d_{S,50} \cdot \varrho_F} \cdot Ar^{0.419} = 2.28 \cdot \frac{18.26 \cdot 10^{-6}\,\text{Pa s}}{0.450 \cdot 10^{-3}\,\text{m} \cdot 2.40\,\frac{\text{kg}}{\text{m}^3}} \cdot 17679.5^{0.419}$$

$$v_{F,tr} = 2.32\,\frac{\text{m}}{\text{s}}$$

The relation between $v_{F,tr}$ and w_T for the given Ar-number is consistent with that in Fig. 3.15.

- **Boundary velocity $v_{F,dp}$ at solid throughput $\dot{m}_S$:**
From the relevant equation here (4.23), it follows:

$$v_{F,dp} = \left(39.8 \cdot \sqrt{g \cdot d_S} \cdot \left(\frac{w_T \cdot d_{S,50} \cdot \varrho_{F,A}}{\eta_F}\right)^{-0.078}\right)^{0.763} \cdot \left(\frac{\dot{m}_S}{A_R \cdot \varrho_{F,A}}\right)^{0.237}$$

$$= \left(39.8 \cdot \sqrt{9.81\,\frac{\mathrm{m}}{\mathrm{s}^2} \cdot 0.450 \cdot 10^{-3}\,\mathrm{m}} \cdot \left(\frac{2.69\,\frac{\mathrm{m}}{\mathrm{s}} \cdot 0.450 \cdot 10^{-3}\,\mathrm{m} \cdot 2.40\,\frac{\mathrm{kg}}{\mathrm{m}^3}}{18.26 \cdot 10^{-6}\,\mathrm{Pa\,s}}\right)^{-0.078}\right)^{0.763}$$

$$\cdot \left(\frac{5.556\,\frac{\mathrm{kg}}{\mathrm{s}}}{5.3456 \cdot 10^{-3}\,\mathrm{m}^2 \cdot 2.40\,\frac{\mathrm{kg}}{\mathrm{m}^3}}\right)^{0.237}$$

$$v_{F,dp} = 7.51\,\frac{\mathrm{m}}{\mathrm{s}}$$

Gas velocities $v_F \leq v_{F,dp}$ cause the conveyance to collapse into a "dense" fluidized bed, see Fig. 4.21.

- **Type A-choking = Clogging limit $v_{F,pf}$ at the solid throughput $\dot{m}_S$:**
(4.19) results in:

$$v_{F,pf} = (21.6 \cdot \sqrt{g \cdot d_{S,50}} \cdot Ar^{0.105})^{0.6485} \cdot \left(\frac{\dot{m}_S}{A_R \cdot \varrho_F}\right)^{0.3515}$$

$$= \left(21.6 \cdot \sqrt{9.81\,\frac{\mathrm{m}}{\mathrm{s}^2} \cdot 0.450 \cdot 10^{-3}\,\mathrm{m}} \cdot 17679.5^{0.105}\right)^{0.6485}$$

$$\cdot \left(\frac{5.556\,\frac{\mathrm{kg}}{\mathrm{s}}}{5.3456 \cdot 10^{-3}\,\mathrm{m}^2 \cdot 2.40\,\frac{\mathrm{kg}}{\mathrm{m}^3}}\right)^{0.3515}$$

$$v_{F,pf} = 20.78\,\frac{\mathrm{m}}{\mathrm{s}}$$

Gas velocities $v_F \leq v_{F,pf}$ lead to solid backflows in the conveying pipe.

- **Verification of Type C-choking**:
From (4.20) it follows:

$$\frac{w_T}{\sqrt{g \cdot D_R}} = \frac{2.69\,\frac{\mathrm{m}}{\mathrm{s}}}{\sqrt{9.81\,\frac{\mathrm{m}}{\mathrm{s}^2} \cdot 0.0825\,\mathrm{m}}} = 2.99 > 0.35$$

Type C-choking, i.e., conveying with pronounced pressure pulsations, is to be expected at gas velocities $v_F < v_{F,pf}$. Equations for calculating/estimating the velocity $v_{F,ch,C} < v_{F,ch,A} = v_{F,pf}$ include, among others, [31, 32].

- **Assessment**: Due to the more uniform flow of bulk material particles across the cross-section of vertical conveying pipes, it is expected that these generally require significantly lower gas velocities for stable conveying operation than horizontal pipes. Conveying routes with horizontal and vertical sections are therefore often only checked/dimensioned using the velocity approaches for horizontal pipes. The model calculations carried out above confirm this expectation: Comparable, for example, are the limiting velocities $v_{F,A,\min} = 11.0\,\mathrm{m/s}$ (horizontal) $> v_{F,dp} = 7.51\,\mathrm{m/s}$ (vertical). On the other hand, the velocity $v_{F,pf}$ is obviously too large. This was confirmed in the context of the conveying tests carried out with limestone chippings - the test route contained a vertical section of approx. $6\,m$ height. Solid backflows and moderate pressure pulsations were only observed at gas velocities $v_F \lesssim 14\,\mathrm{m/s}$ at the foot of the vertical section. Possible causes: (4.19) was essentially developed based on measurement results with bulk materials of Geldart Group A, while limestone chippings are a Group B material, and the loadings realized in the considered example are significantly larger than those in the referenced literature [30, 31].

In general, when using a calculation approach from the literature, its permissible application range, the underlying boundary conditions/simplifications, and the data/measurement basis used must be carefully checked. Surprises are otherwise possible, as seen in the above results! The two conveyances discussed here were carried out with an initial conveying gas velocity of $v_{F,A} = 14.50\,\mathrm{m/s}$ is realized and works satisfactorily.

4.7 Pressure Loss in Pneumatic Conveying Lines

A typical conveying route consists of horizontal and vertical pipe sections as well as deflections, switches, etc. Each of these components causes a pressure loss Δp_i, which must be determined using an equation adapted to the specific pipe section. Thus, the following applies:

$$\begin{aligned}\Delta p_R = \sum_{i=1}^{n} \Delta p_i = {} & \Delta p(\text{Solid friction horizontal}) \\ & + \Delta p(\text{Solid friction vertical}) \\ & + \Delta p(\text{Solid lift}) \\ & + \Delta p(\text{Solid initial acceleration}) \\ & + \Delta p(\text{Solid deflection}) \\ & + \Delta p(\text{Conveying gas})\end{aligned} \tag{4.39}$$

(4.39) illustrates that a conveying route must not be calculated as a whole, but rather element by element. Depending on the chosen operating conditions, various flow states occur in the considered line, for the description of which, even for the same component, different Δp_i-calculation approaches may be required, see e.g. Fig. 4.5. Since the pressure loss of a conveying pipe element increases with increasing differential velocity $(u_F - u_S)$ between gas and solid, the motion equation of the flowing bulk material must also be known in addition to the pressure loss equation. This in turn depends on the flow pattern that develops. The indicated problem generally leads to iterative calculations to solve a given task. By suitable simplifications, the complexity of such designs can be reduced.

In the following, approaches for calculating the pressure loss of various pipe elements, taking into account the influence of the flow pattern, are presented and discussed. The permissible gas velocities can be found in Sect. 4.6, and the resulting velocity ratios $C = u_S/u_F$ are discussed in Sect. 4.8.

4.7.1 Solid Friction Horizontal Straight Pipe

In pneumatic conveying systems, the pressure loss caused by horizontal solid friction generally accounts for the majority of the total pressure loss. For its calculation, especially in the flight and strand conveying range, a so-called "standard method" referred to as λ_S-method has become established in the literature/at users [40, 54, 55], which will be presented first here. Special approaches, e.g. for plug or strand conveying, are discussed in Chap. 5.

The momentum balance describing the present problem, (2.56), can be further simplified for incompressible pneumatic conveying in a steady state to:

$$-A_R \cdot dp_{R,h} = F_{S,f} + F_{F,f} = -(A_R \cdot dp_{S,R,h} + A_R \cdot dp_{F,R,h}) \tag{4.40}$$

For the friction force $F_{F,f}$ caused by the conveying gas, the known approach applies, see Sect. 2.5.1, (2.76):

$$F_{F,f} = \varepsilon_F \cdot A_R \cdot \lambda_F \cdot \frac{dL}{D_R} \cdot \frac{\varrho_F}{2} \cdot u_F^2 \quad \rightarrow \quad -dp_{F,R,h} = \varepsilon_F \cdot \lambda_F \cdot \frac{dL}{D_R} \cdot \frac{\varrho_F}{2} \cdot u_F^2 \tag{4.41}$$

The resistance $F_{S,f}$ of the solid is predominantly caused by losses in particle/pipe wall collisions at very high conveying gas velocities and, to a lesser extent, particle/particle collisions. It can be described by the frictional approach analogous to (4.41)

$$F_{S,f} = (1 - \varepsilon_F) \cdot A_R \cdot \lambda_S^* \cdot \frac{dL}{D_R} \cdot \frac{\varrho_P}{2} \cdot u_S^2 = dm_S \cdot \lambda_S^* \cdot \frac{u_S^2}{2 \cdot D_R} \tag{4.42}$$

With decreasing gas velocity, more and more solid material is supported on the pipe bottom. In the limiting case of a completely demixed flow at very low gas velocities, the solid resistance is determined solely by the sliding friction of the bulk weight. For this, the following applies:

$$F_{S,f} = dm_S \cdot g \cdot \beta_R = (1 - \varepsilon_F) \cdot A_R \cdot dL \cdot \varrho_P \cdot g \cdot \beta_R \tag{4.43}$$

In the normal case, both resistance mechanisms overlap and can hardly be separated from each other by measurement. In order to arrive at a uniform design procedure that takes both effects into account, the following approach is taken: The current pressure loss of the solid is determined by considering $dm_S = \dot{m}_S \cdot d\tau_{\Delta L} = \dot{m}_S \cdot dL/u_S$, $d\tau_{\Delta L} =$ bulk material residence time in the pipe element dL, from the superposition of (4.42) and (4.43) to:

$$-dp_{S,R,h} = \frac{\dot{m}_S}{A_R} \cdot \frac{dL}{u_S} \cdot \left(\lambda_S^* \cdot \frac{u_S^2}{2 \cdot D_R} + g \cdot \beta_R\right)$$

If the pipe cross-sectional area is eliminated by $A_R = \dot{m}_F/(\varepsilon_F \cdot \varrho_F \cdot u_F)$ and for $\dot{m}_S/\dot{m}_F = \mu$, $u_S/u_F = C_h$, $Fr_R = u_F^2/(g \cdot D_R)$ set, after some elementary transformations, the result follows as

$$-dp_{S,R,h} = \varepsilon_F \cdot \mu \cdot \frac{dL}{D_R} \cdot \frac{\varrho_F}{2} \cdot u_F^2 \cdot \left(C_h \cdot \lambda_S^* + \frac{2 \cdot \beta_R}{C_h \cdot Fr_R}\right) = \varepsilon_F \cdot \mu \cdot \lambda_S \cdot \frac{dL}{D_R} \cdot \frac{\varrho_F}{2} \cdot u_F^2 \tag{4.44}$$

with the total resistance coefficient:

$$\lambda_S = C_h \cdot \lambda_S^* + \frac{2 \cdot \beta_R}{C_h \cdot Fr_R} \tag{4.45}$$

β_R can take the following values:

- in the flight conveying range: $\beta_R = w_T/v_F$, resulting from the lifting power of the lateral forces,
- in the strand conveying range: $\beta_R = \tan \varphi_W$, wall friction of the strand,
- in the plug conveying range: $\beta_R = \beta_R^* = \kappa \cdot \tan \varphi_W$, $\kappa > 1$, increased wall friction generated by bulk material wedging, see Sect. 4.4 ($\rightarrow \lambda_S^* = 0$).

(4.44) has the same structure as (4.41) and can be easily combined with it, see (4.48). Integration over not too large pipe lengths ΔL_h ($\rightarrow$ assuming incompressible flow!) yields with $\varepsilon_F = 1$ and $v_F = u_F$ the usage equation for calculating the additional pressure loss caused by the solid in horizontal pipe elements:

$$-\Delta p_{S,R,h} = \mu \cdot \lambda_S \cdot \frac{\Delta L_h}{D_R} \cdot \frac{\varrho_F}{2} \cdot v_F^2 \tag{4.46}$$

Measured λ_S-values are plotted according to (4.45) as a function of the Fr_R-number: λ_S turns out to be large Fr_R numbers approach an approximately constant value, while in the direction of smaller Fr_R numbers, it increases proportionally to Fr_R ($\rightarrow$ it applies: $C_h, \lambda_S^*, \beta_R \cong$ konst.). In Fig. 4.22, the compensation curves through measurement results with coarse-grained wheat, $d_{S,50} \cong 4.0\,\text{mm}$, through pipes of different diameters [55], in Fig. 4.23 the results of conveyances with titanium ore, $d_{S,50} \cong 20\,\mu\text{m}$, through a pipeline with $D_R = 82.5\,\text{mm}$ [56] are shown. The FLUIDCON results also plotted there will be discussed in Sect. 6.3.

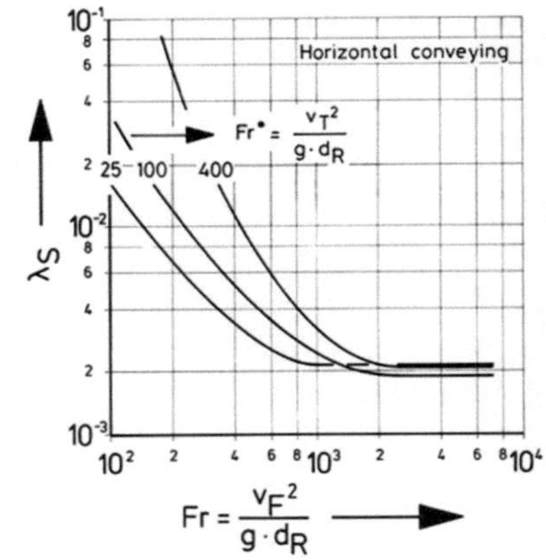

Fig. 4.22 λ_S-resistance coefficients of wheat, $D_R = (40–295)$ mm, $\mu \leq 10$, $w_T \cong 8.0\,\text{m/s}$

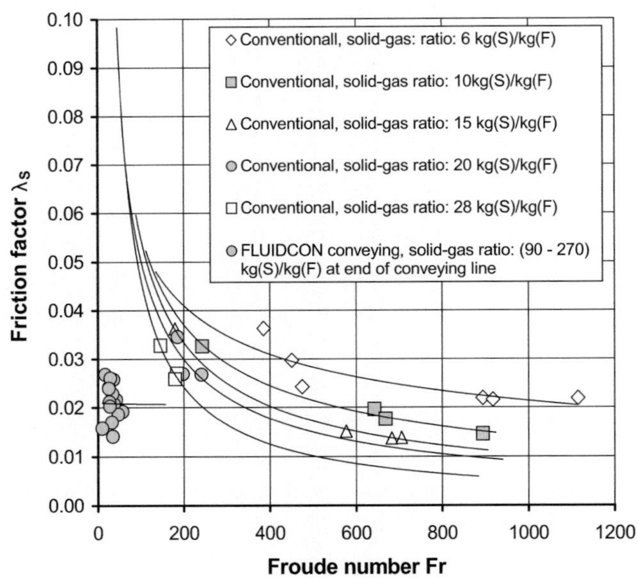

Fig. 4.23 λ_S-resistance coefficients of ground titanium ore

If pronounced dilute or dense phase conveying is present, then the following applies:

Dilute phase conveying: $(C_h \cdot \lambda_S^*) \gg (2 \cdot \beta_R/(C_h \cdot Fr_R))$, from this it follows: $\lambda_S \cong$ konst., $\Delta p_{S,R,h} \propto$ Massenträgheitskraft,

Dense phase conveying: $(2 \cdot \beta_R/(C_h \cdot Fr_R)) \gg (C_h \cdot \lambda_S^*)$, from this it follows: $\lambda_S \cong 1/Fr_R$, $\Delta p_{S,R,h} \propto$ Gewichtskraft

When considering the matter impartially, the λ_S-value is not a resistance coefficient, but rather an adaptation/correction factor, with which the various resistance mechanisms are transformed into a relationship proportional to the dynamic pressure/the mass inertia force. This approach has evolved historically: Initial scientific investigations were carried out on flight conveyors with only low loads, whose pressure losses could be adequately described by $\Delta p_{S,R,h} \propto \lambda_S^* \cdot (\varrho_P/2) \cdot u_S^2$. This was further developed. Since current pneumatic conveying systems predominantly operate in the strand/dense flow conveying range, i.e. $\Delta p_{S,R,h} \propto \varrho_S \cdot g \cdot \Delta L_h$, this leads to an unnecessarily complicated description of the conveying behavior. Example: As a correlation equation for the summary of the λ_S-values of fine-grained bulk materials, $d_{S,50} \lesssim 100\,\mu\text{m}$, in the strand and dense flow range, for example

$$\lambda_S = 2.1 \cdot \frac{Fr_{S,T}^{0.25}}{Fr_R} \cdot \frac{(\frac{D_R}{d_{S,50}})^{0.1}}{\mu^{0.3}} \tag{4.47}$$

with:

$Fr_{S:T}(= w_T^2/(g \cdot d_{S,50}))$, Froude number formed with the particle diameter $d_{S,50}$ and the single particle terminal velocity $w_T(d_{S,50})$

is proven [57]. This, for a given design case in (4.46) inserted, leads to a velocity-independent, weight-proportional pressure loss $\Delta p_{S,R,h} \propto \varrho_P \cdot g \cdot \Delta L_h$.

An additional – considerably more fundamental – problem of the λ_S-approach is its obviously limited scale-up capability. In [58] it was found that in $\lambda_S(Fr_R)$-representations of horizontally conveyed coarse-grained bulk materials (→ wheat, plastic granulate) the curves of different pipe diameters, $D_R = (50–400)$ mm, overlap. This would not be possible with a dimensionally correct description of the physical relationships, i.e., the approach neglects influencing variables of the relevance list or does not correctly represent the relationships in all cases due to the manipulations described above.

In practice, the splitting of the λ_S-value is circumvented and it is measured directly at appropriate test facilities. For this purpose, known = measured gas and solid mass flows $(\dot{m}_F, \dot{m}_S)$ are supplied to a horizontal measuring section of diameter D_R and length ΔL_h, to which sufficiently long inlet and outlet sections are connected upstream and downstream, respectively, to establish a steady-state conveying condition. The resulting pressures p_{in} at the inlet and p_{out} at the outlet of the measuring section is recorded. With the known measurement and system data, λ_S is calculated from the equation of the total pressure loss of gas and solid, (4.41) and (4.44),:

$$-\Delta p_{R,h} = (p_{\text{in}} - p_{\text{out}}) = (\mu \cdot \lambda_S + \lambda_F) \cdot \frac{\Delta L_h}{D_R} \cdot \frac{\varrho_F}{2} \cdot v_F^2 \tag{4.48}$$

Gas velocities v_F and gas densities ϱ_F are determined for the respective average pressure $p = (p_{\text{in}} + p_{\text{out}})/2$ in the measuring section. By systematically varying the operating variables, $\lambda_S(Fr_R)$-dependencies can be determined. These fan out, as (4.48) as well as Figs. 4.22 and 4.23 show, according to the involved dimensionless parameters into families of curves. For example, λ_S decrease at constant Fr_R with increasing load μ.

Subsequently, the λ_S-values of horizontal conveying sections are denoted by $\lambda_{S,h}$, those of vertical sections by $\lambda_{S,v}$.

4.7.2 Solid Friction Vertical Straight Pipe

A λ_S-approach analogous to (4.46) is used:

$$-\Delta p_{S,R,v} = \mu \cdot \lambda_{S,v} \cdot \frac{\Delta L_v}{D_R} \cdot \frac{\varrho_F}{2} \cdot v_F^2 \tag{4.49}$$

The frictional pressure loss $\Delta p_{S,R,v}$ of the solid in vertical sections is mainly caused by particle/pipe wall and particle/particle collisions. The sliding friction is negligible, i.e. $\beta_R \to 0$. This results in the λ_S-model:

$$\lambda_{S,v} \cong C_v \cdot \lambda_S^* \tag{4.50}$$

Compared to the resistance coefficient $\lambda_{S,h}$ of an identical horizontal conveyance and $C = C_v \cong C_h$ applies:

$$\frac{\lambda_{S,v}}{\lambda_{S,h}} = \frac{1}{1 + \frac{2 \cdot \beta_R}{C^2 \cdot \lambda_S^* \cdot Fr_R}} = f(Fr_R) \quad \to \quad \lambda_{S,v} < \lambda_{S,h} \tag{4.51}$$

Example: With $\beta_R = 0.5$, $C = 0.7$, $\lambda_S^* = 0.01$, $D_R = 0.1\,\text{m}$, $v_F = 15\,\text{m/s}$ delivers (4.51) the ratio$\lambda_{S,v}/\lambda_{S,h} = 0.529$, with $v_F = 20\,\text{m/s}$ follows $\lambda_{S,v}/\lambda_{S,h} = 0.666$. With increasing Fr_R-number, $\lambda_{S,v}/\lambda_{S,h} \to 1$.

The systematic evaluation of available measurement results from the dense phase conveying area, e.g. from [59], leads to $(\lambda_{S,v}/\lambda_{S,h})$-values in the relatively narrow limits of:

$$\frac{\lambda_{S,v}}{\lambda_{S,h}} = (0.45\text{-}0.50) \quad \to \quad \text{Dense phase conveying} \tag{4.52}$$

Here, the frictional pressure loss of a vertical conveying section is calculated like that of a horizontal section with the same operating setting, but with approximately halved resistance coefficient, i.e. $\Delta p_{S,R,v} \cong \Delta p_{S,R,h}/2$.

$\lambda_{S,v}$-values can be estimated by the limit values of the horizontal inflow $\lambda_{S,h}(Fr_R)$-curves, see Fig. 4.22. In the case of direct measurement, see Sect. 4.7.1, the measured pressure loss components due to gas friction and solid lifting must be eliminated.

4.7.3 Solid Lifting

It is necessary to consider the weight $(\Delta m_S \cdot g)$ of the bulk material mass currently located in the vertical conveying pipe. Δm_S is supported by the pressure difference force $(-\Delta p_{S,H} \cdot A_R)$. Using $\Delta m_S = \dot{m}_S \cdot \Delta L_v/\overline{u}_S$, $\overline{C}_v = \overline{u}_S/\overline{u}_F$ and $A_R = \dot{m}_F/(\overline{\varepsilon}_F \cdot \overline{\varrho}_F \cdot \overline{u}_F)$

$$-\Delta p_{S,H} = \frac{g}{\overline{C}_v} \cdot \frac{\dot{m}_S \cdot \Delta L_v}{A_R \cdot \overline{u}_F} = \overline{\varepsilon}_F \cdot \frac{g}{\overline{C}_v} \cdot \mu \cdot \overline{\varrho}_F \cdot \Delta L_v \tag{4.53}$$

with:

$\overline{u}_S$, $\overline{u}_F$ average solid, average gas velocity in the vertical section,
$\overline{\varrho}_F$ average gas density in the vertical section,
$\overline{\varepsilon}_F$ ($\overline{\varepsilon}_F \rightarrow 1$), average relative void volume in the vertical section.

Usually, $\overline{\varepsilon} = 1$ and thus $\overline{u}_F = \overline{v}_F$ is set. (4.53) illustrates a peculiarity regarding the position of a vertical pipe element along a conveying section, which is particularly important for dense phase conveying: The lifting pressure loss $\Delta p_{S,H}$ changes inversely proportional to the average gas velocity $\overline{v}_F$, that is, a lifting distance at the beginning of a conveying line causes a higher pressure loss $\Delta p_{S,H}$ than one at the end of the line. Cause: The lower gas/solid velocity at the beginning of the line results in a longer residence time $\Delta\tau_{\Delta L} = \Delta L_v/\overline{u}_S$ of the solid in the vertical pipe section and from this an enrichment of Δm_S, which must be compensated by a correspondingly larger $\Delta p_{S,H}$. Example: A given solid mass flow can be lifted with the same gas flow in the same conveying line either at the beginning of the line with $\overline{v}_{F,A} = 6\,\mathrm{m/s}$ or at the end of the line with $\overline{v}_{F,E} = 20\,\mathrm{m/s}$ to lift the same height. From this it follows:$\Delta p_{S,H}(\textit{start of the line})/\Delta p_{S,H}(\textit{end of the line}) \cong 20\,\mathrm{m/s}/6\,\mathrm{m/s} = 3.33$. In practice, this can lead to pressure losses on the order of $\Delta p_{S,H} \gtrsim 1.0\,\mathrm{bar}$ for greater lifting heights near the conveying pipe inlet.

The exact calculation of the pressure loss of a vertical section is further complicated by the fact that it is either preceded by a 90° deflection from the horizontal to the vertical or a material feed at the beginning of the vertical. Behind both, the bulk material must be accelerated to its steady-state velocity, see the following sections, i.e., a reduced solid velocity compared to the steady-state is established along the acceleration section, which, among other things, leads to a greater pressure loss. The issue of interaction between pipeline elements is discussed in Sect. 4.9. For initial estimates, $\overline{C}_v \cong 0.7$ can be set.

4.7.4 Solid Initial Acceleration

The solid material fed at the beginning of the line through a lock must be accelerated from $u_S = 0$ to the steady-state velocity $u_S = u_{S,A} = C_A \cdot u_{F,A} \cong C_A \cdot v_{F,A}$. The pressure difference required for this is $\Delta p_{S,B}$ and is calculated using the impulse theorem from

$$-\Delta p_{S,B} \cdot A_R = \Delta m_S \cdot \frac{\Delta u_S}{\Delta\tau} = \dot{m}_S \cdot \Delta u_S \tag{4.54}$$

With $\Delta u_S = u_{S,A} - 0 = u_{S,A} = C_A \cdot u_{F,A}$ follows:

$$-\Delta p_{S,B} = C_A \cdot \frac{\dot{m}_S}{A_R} \cdot u_{F,A} = C_A \cdot \mu \cdot \varrho_{F,A} \cdot u_{F,A}^2 \tag{4.55}$$

$v_{F,A} \cong u_{F,A}$ and $C_A \cong 0.7$ provides useful estimates.

4.7.5 Solid Deflection

In a pipe deflection, e.g., a 90° bend, gas and solid almost completely separate due to particle inertia and the acting centrifugal forces. The solid is pressed against the outer wall, slowed down by the increased wall friction and must be accelerated back to the steady-state velocity after the deflector. The pressure loss $\Delta p_{S,U}$ of a diversion is thus identical to the pressure difference required for the re-acceleration of the solid behind the diversion.

Application of (4.54) taking into account

$$\Delta u_S = (u_{S,\text{in}} - u_{S,\text{out}}) = \frac{\Delta u_S}{u_{S,\text{in}}} \cdot u_{S,\text{in}} = K_U \cdot C_U \cdot u_{F,U} \to K_U = \frac{\Delta u_S}{u_{S,\text{in}}}, C_U = \frac{u_{S,\text{in}}}{u_{F,U}} \tag{4.56}$$

results in the calculation equation

$$-\Delta p_{S,U} = K_U \cdot C_U \cdot \frac{\dot{m}_S}{A_R} \cdot u_{F,U} = K_U \cdot C_U \cdot \mu \cdot \varrho_{F,U} \cdot u_{F,U}^2 \tag{4.57}$$

with:

- $u_{S,\text{in}}$ Solid velocity at the deflector inlet = solid velocity that must be re-accelerated behind the deflector,
- $u_{S,\text{out}}$ Solid velocity at the deflector outlet,
- $u_{F,U}$ ($\cong v_{F,U}$), gas velocity before and after deflector,
- $\varrho_{F,U}$ Gas density behind deflector,
- K_U ($= 1 - u_{S,\text{out}}/u_{S,\text{in}}$), relative solid deceleration in the deflector, (4.56).

In (4.57), it is implicitly assumed that the solid velocity after the re-acceleration of the conveyed material behind the deflector has the same value as before the deflector. This is not always achievable, because, for example, due to a subsequent deflector, there is not a sufficiently long re-acceleration distance available for the first one. In this case, the actual re-acceleration velocity difference must be used. Δu_S can be calculated after the deflector.

Determining the current size of the pressure loss $\Delta p_{S,U}$ are the specific solid throughput ($\dot{m}_S/A_R$) or the loading μ. (4.57) illustrates that deflections at the pipe end cause a larger $\Delta p_{S,U}$ than at the pipe beginning ($\to v_{F,U} = v_{F,E} > v_{F,A}$), i.e., each deflection must be calculated individually. The design of the deflection, e.g., pipe bend, deflection head, T-bend, the deflection angle, and the spatial arrangement of the deflector are included in the value of K_U. The various design variants of deflections are discussed in Sect. 8.2.1. The determination of K_U and the resulting pressure loss $\Delta p_{S,U}$ is exemplified using pipe bends.

Figure 4.24 shows the force balance on the solid of an element $d\alpha_U$ of the bulk material strand forming on the outer wall of the bend. This is pressed against the wall by the centrifugal force F_C and decelerated along the deflection path due to the friction force F_f opposing the movement and the driving inertial force F_T. In the case of the deflec-

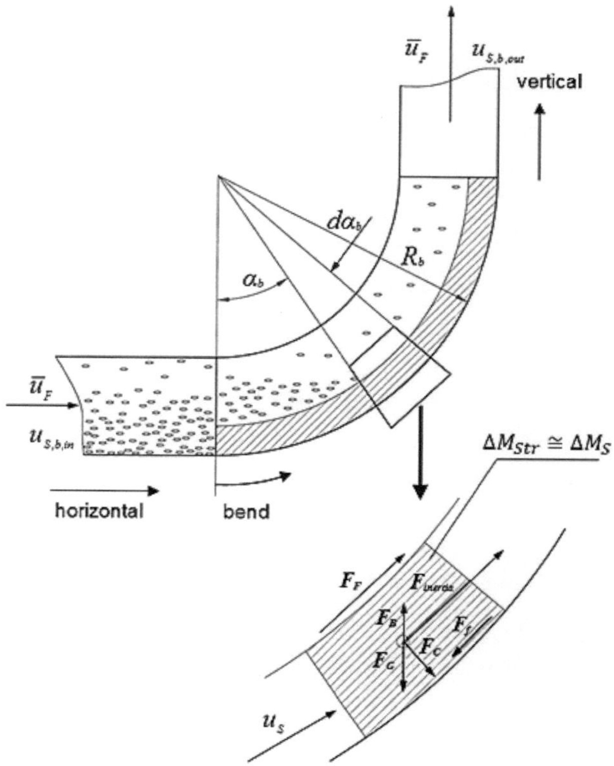

Fig. 4.24 Forces on the bulk material in a 90° bend from the horizontal to the vertical

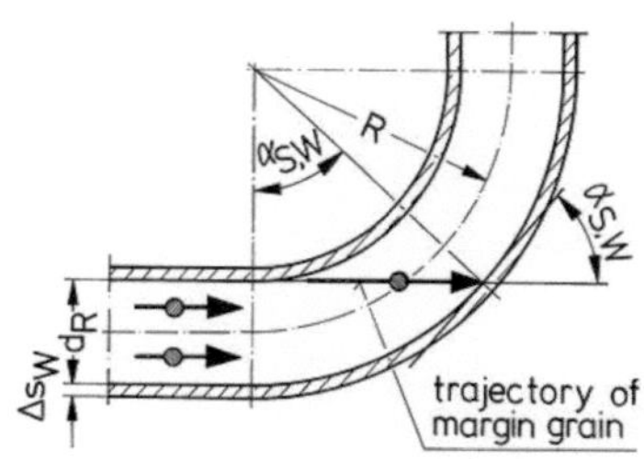

Fig. 4.25 Latest point of maximum separation

tion bend shown here from the horizontal to the vertical, this deceleration is additionally reinforced by the component of the solid's weight force F_G acting against the direction of movement. Generally neglected are the drive by the conveying gas flowing over the strand F_F and the buoyancy F_B. For the following considerations, a complete gas/solid separation along the entire deflection path is assumed. This assumption is largely fulfilled for dense phase conveying, especially of fine-grained bulk materials. If not, calculations should be made with complete separation from the point on the straight line extended flight path of the solid entering the bend on the inner side of the pipe, see angle $\alpha_{S,W}$ in Fig. 4.25. This is also the point where the greatest bend wear occurs.

a) Deflection in the horizontal plane

The derivation of the bulk material deceleration for this bend arrangement is shown in Box 4.4. The result, (4.59), combined with the definition equation for K_U, (4.56), results in:

$$K_U = 1 - \exp\left(-\frac{\alpha_U}{180^\circ} \cdot \pi \cdot \beta_R\right) \tag{4.58}$$

with:

α_U Deflection angle in [°].

Surprisingly, the bend radius R does not appear in (4.58) and thus has no influence on the pressure loss $\Delta p_{S,U}$, (4.57). A complete deceleration of the solid to $u_{S,\text{out}} = 0$ is also not possible within the scope of practical operating conditions according to (4.59). With the usual wall friction angles for steel pipes $\varphi_W \cong (15\text{-}30)^\circ \rightarrow \beta_R \cong (0.27\text{-}0.60)$ lie the K_U-values for 90°-bends ($\rightarrow \alpha_U = 90^\circ$) at $K_U \cong (0.35\text{-}0.61)$. Simplifying, the average value $K_U = 0.50$ is often used. For fine-grained bulk materials, a value for 90°-bends can be set to $(K_U \cdot C_U)_{90^\circ} \cong 0.35$. (4.58) in connection with the pressure loss equation (4.57) is often used in practice for arbitrary spatial bend arrangements.

Box 4.4: Bulk material deceleration in a deflection in the horizontal plane

Along the deflection circumference, with the simplifications mentioned in the text:

$$F_T = F_f$$

$\rightarrow$ Gravitational force F_G in conveying direction without influence!

$$F_T = -\Delta m_S \cdot \frac{du_S}{d\tau} = -\Delta m_S \cdot \frac{du_S}{dL_U} \cdot \frac{dL_U}{d\tau} = -\Delta m_S \cdot u_S \cdot \frac{du_S}{dL_U}$$

$$= -\Delta m_S \cdot u_S \cdot \frac{360^\circ}{2 \cdot R_U \cdot \pi} \cdot \frac{du_S}{d\alpha_U}$$

$$F_f = F_C \cdot \beta_R = \Delta m_S \cdot \frac{u_S^2}{R_U} \cdot \beta_R.$$

with:

$\beta_R = \tan \varphi_W$ Wall friction coefficient strand/pipe wall.

From this, the differential equation follows:

$$\frac{du_S}{u_S} = -\frac{\pi \cdot \beta_R}{180^\circ} \cdot d\alpha_U$$

The integration of $u_{S,\text{in}}$ to $u_{S,\text{out}}$ and $\alpha_U = 0^\circ$ to $\alpha_U = \alpha_U$ yields:

$$\frac{u_{S,\text{out}}}{u_{S,\text{in}}} = \exp\left(-\frac{\alpha_U}{180^\circ} \cdot \pi \cdot \beta_R\right) \tag{4.59}$$

b) Deflection from the horizontal to the vertical

The force balance on the strand element of Fig. 4.24 leads, with the mentioned simplifications, to the differential equation [60]

$$\frac{du_S}{d\alpha_U} = -\frac{R_U}{u_S} \cdot \left[\beta_R \cdot \left(\frac{(u_S)^2}{R_U} + g \cdot \cos \alpha_U\right) + g \cdot \sin \alpha_U\right] \tag{4.60}$$

with the solution

$$u_{S,\text{out}} = e^{-\beta_R \cdot \alpha_U} \cdot \sqrt{(u_{S,\text{in}})^2 + \frac{2 \cdot R_U \cdot g}{4 \cdot \beta_R^2 + 1} \cdot \left\{(2 \cdot \beta_R^2 - 1) - e^{2 \cdot \beta_R \cdot \alpha_U} \cdot [(2 \cdot \beta_R^2 - 1) \cdot \cos \alpha_U + 3 \cdot \beta_R \cdot \sin \alpha_U]\right\}}. \tag{4.61}$$

This requires the input of the deflection angle in radians ($\rightarrow 57.296° \equiv 1$ rad). For the 90° bend horizontal/vertical, this results in $\alpha_U = \pi/2 (\equiv 90°)$:

$$u_{S,\text{out}} = e^{-\beta_R \cdot \frac{\pi}{2}} \cdot \sqrt{(u_{S,\text{in}})^2 + \frac{2 \cdot R_U \cdot g}{4 \cdot \beta_R^2 + 1} \cdot \left\{(2 \cdot \beta_R^2 - 1) - 3 \cdot \beta_R \cdot e^{\beta_R \cdot \pi}\right\}} \tag{4.62}$$

K_U is calculated analogously (4.58), the pressure loss $\Delta p_{S,U}$ is calculated with (4.57). With an increasing deflection radius $R \cong R_U$ ($\rightarrow R_U =$ radius at average strand thickness) $u_{S,\text{out}}$ becomes smaller and $\Delta p_{S,U}$ thus larger. The evaluation of (4.61) and (4.62) shows that exit velocities $u_{S,\text{out}} \leq 0$, i.e., a complete solid deceleration, are possible. For the exit velocity $u_{S,\text{out}} = 0$ the limiting entry velocity $u_{S,\text{in}}$ can be calculated, above which no complete bulk material deceleration or no backflow of the bulk material occurs. In the case of the 90° bend, the following applies:

$$u_{S,\text{in}} \geq \sqrt{\frac{2 \cdot R_U \cdot g}{4 \cdot \beta_R^2 + 1} \cdot \left[1 + 3 \cdot \beta_R \cdot \left(e^{\pi \cdot \beta_R} - \frac{2}{3} \cdot \beta_R\right)\right]} \quad \rightarrow \quad 90°\text{-Bogen}, \quad u_{S,\text{out}} = 0 \tag{4.63}$$

In Fig. 4.26 for $\beta_R = 0.30$ and 0.50 ($\rightarrow$ wall friction angle: $\varphi_W \cong 17°$ or $27°$) the calculations carried out with (4.63) for a 90° pipe deflection are shown. $u_{S,\text{in}}(R, u_{S,\text{out}} = 0)$ is represented. For bulk material entry velocities $u_{S,\text{in}}$ below the individual β_R-curves, complete solid deceleration/backflow of the solid occurs, i.e., conditions that should be avoided for safe plant operation. With increasing bend radius $R =$ longer deceleration distance, the required entry velocity $u_{S,\text{in}}$ increases. The same applies to increasing friction coefficients β_R.

$u_{S,\text{out}} = 0$ does not necessarily mean that a pipe blockage must occur. Due to the accumulation of bulk material in the bend and the resulting gas flow, a plug-like transport with stronger pipe impacts occurs when sufficient pressure reserves are available. The resulting plugs generally do not dissolve along the further conveying path. Bends horizontal/vertical are the critical deflection elements of a pneumatic conveying path

Fig. 4.26 Minimum bulk material entry velocity into a 90° bend horizontal/vertical, $u_{S,\text{out}} = 0$

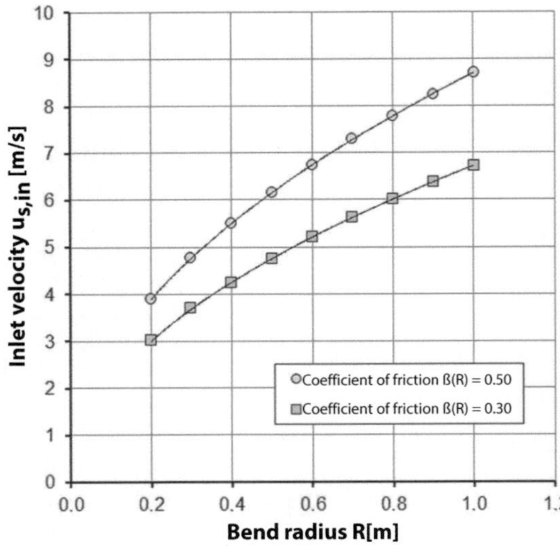

and must, especially for dense phase conveying, be checked for a solid exit velocity $u_{S,\text{out}} > 0$. With decreasing solid exit velocity $u_{S,\text{out}} \to 0$, $K_U \to 1$.

c) Further deflections

Each additional spatial bend arrangement requires its specific calculation approach and thus leads to other K_U-dependencies. Details on this can be found, among others, in [2, 60]. The pressure losses of these pipe elements can usually be estimated relatively reliably with (4.57) and (4.58) for a deflection lying in the horizontal plane.

The pressure losses $\Delta p_{S,U}$ of special designs of deflectors can also be determined with (4.57) and correspondingly adjusted K_U values. Example: The so-called Vortex-Elbow shown in Fig. 4.27 is used for wear reduction with very abrasive bulk materials. It avoids direct impact of the solid on the pipe wall by building up a bulk material cushion (→ autogenous wear protection, see Sect. 8.2.1). Systematic measurements [61] on a 90°/ DN 60-Elbow with fine-grained fly ash and loadings in the range of $\mu = (15\text{–}129)$ led to an average value $(K_U \cdot C_U)_{\text{Vortex}} = 0.434$. This is greater than that of a comparably flowed 90° pipe bend ($\to (K_U \cdot C_U)_{90°} = 0.35$). If no measurement results are available, then the maximum value $K_U = 1$ provides safe (possibly too safe!) results.

Elbows that are vertically flowed through from top to bottom and deflected into the horizontal plane must be equipped with larger bending radii to avoid tail back of the bulk material flow. This is especially true for longer upstream falling distances.

From the above, it follows that it is incorrect to install a pressure sensor in front of and behind the deflection for measuring the pressure loss of a deflection and to declare the differential pressure as elbow loss. It is necessary to determine the pressure gradients in the steady state by a series of pressure sensors along sufficiently long straight pipe

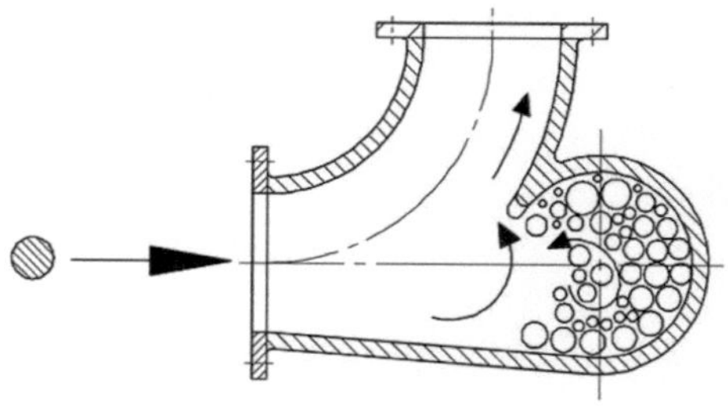
Fig. 4.27 Schematic of a vortex elbow

sections before and after the deflector. Their extrapolation to the central position of the unwound elbow circumference provides the pressure difference $\Delta p_{S,U}$.

The method presented above for calculating $\Delta p_{S,U}$ has proven successful in practice, see e.g. [62]. There are other approaches that assume an empirical resistance coefficient and a proportionality to $\varrho_{F,U} \cdot u_{F,U}^2/2$ [63]. In contrast, Schuchart [64] sets the solid pressure loss of a pipe elbow

$$\Delta p_{S,U} = 210 \cdot \left(\frac{D_R}{2 \cdot R}\right)^{1.15} \cdot \Delta p_{S,R,h} \tag{4.64}$$

proportional to the solid pressure loss $\Delta p_{S,R,h}$ of an equally long horizontal pipe element (e.g.: → 90°-bend: $\Delta L_h = 2 \cdot R \cdot \pi/4$). The proportionality factor contains the curvature radius R and pipe diameter D_R. The investigations in [64] were carried out with relatively coarse bulk material, $d_{S,50} \cong (1\text{–}2)$ mm, and loadings $\mu \lesssim 58$. While the approach according to (4.57) contains the limited in their value range and thus transferable to deviating situations parameters (K_U, C_U), such a transfer with (4.64) is difficult. An introduction to the problems of using deflections in pneumatic conveying systems can be found in [63].

4.7.6 Pressure Loss of Conveying Gas

The conveying gas itself rubs against the pipe wall, must be accelerated and lifted, and flows through installations. This leads to the inherent pressure loss

$$-\Delta p_F \cong \left(\lambda_F \cdot \frac{\Delta L}{D_R} + \sum_{i=1}^{n} \xi_i\right) \cdot \frac{\varrho_F}{2} \cdot u_F^2 + \left|\Delta p_{F,H}\right| \tag{4.65}$$

with:

ξ_i various resistance coefficients, see Sect. 2.5.

It is usually assumed that the conveying gas behaves in the present two-phase flow as in a single-phase flow. This is only approximately the case, e.g., its flow profile changes. For dense phase conveying, in general $(-\Delta p_F) \ll \sum(-\Delta p_{S,i})$, i.e., Δp_F can be neglected here.

4.7.7 Compressible Flow

The above equations for pressure loss calculation assume incompressible flow. In practice, however, the conveying gas expands due to the pressure gradient in the conveying direction. This leads to increasing velocities of gas and entrained solids, resulting in increased friction and an additional pressure loss Δp_{pac} for the permanent acceleration of the mass flows of both phases, see also Sect. 2.5.1. This is calculated according to (2.56) as:

$$-\Delta p_{\text{pac}} = \frac{1}{A_R} \cdot (\dot{m}_S \cdot \Delta u_S + \dot{m}_F \cdot \Delta u_F) = \frac{\dot{m}_F}{A_R} \cdot \Delta u_F \cdot (\mu \cdot C + 1) \tag{4.66}$$

with:

$\Delta u_F, \Delta u_S$ Velocity change of the gas or solid due to gas expansion in the calculation section,

C Velocity ratio of solid/gas in the calculation section.

It is common practice to calculate conveying sections incompressibly from the pipeline end, starting with the initial values (p_E, v_E). The gas expansion is corrected by a pressure correction of the operating/material data ($\rightarrow$ a. o. $u_F, \varrho_F, \lambda_S$) at the end of the considered = beginning of the subsequent calculation section as well as the incorporation of the pressure loss Δp_{pac} considered. Since for the Δu_F in (4.66) only after determining the friction pressure loss of the section, a first value is available, Δp_{pac} must be introduced subsequently. This results in a recursive iteration, which can also be used to increase the accuracy of the calculation of the friction pressure loss. The iteration generally requires only a few cycles to meet a meaningful termination criterion. The continuous course of state changes along the line is approximated by a step-shaped curve. Δp_{pac} often remains unconsidered in such design calculations and is often also negligibly small.

The pressure loss of long straight horizontal pipe sections can be calculated after integrating the (differential) equation (4.48) with:

$$-\Delta p = (p_{\text{in}} - p_{\text{out}}) = p_{\text{out}} \cdot \left(\sqrt{1 + (\mu \cdot \overline{\lambda}_S + \lambda_F) \cdot \frac{\Delta L_h}{D_R} \cdot \left(\frac{\dot{m}_F}{A_R}\right)^2 \cdot \frac{R_F \cdot T_F}{p_{\text{out}}^2}} - 1 \right) \tag{4.67}$$

The derivation of (4.67) corresponds to that of the single-phase equation (2.79): $(\mu \cdot \overline{\lambda}_S + \lambda_F)$ replaces λ_F, $\overline{\lambda}_S = \text{const.}$ is the average solid friction coefficient along the distance ΔL_h. Through (4.67), only the frictional pressure loss, but not Δp_{pac} is captured. A comparable integration for straight vertical sections does not lead to a closed solution.

In Box 4.5, the problem/approach for solving the basic differential equations describing the task is exemplified for a straight horizontal conveying section. As can be seen from Box 4.5, the system of equations (4.78, 4.79) can only be solved numerically, e.g., using a difference method. This requires an extensive calculation program. Ultimately, such a difference method corresponds to the practitioner's approach: section-wise pipeline calculation. In addition, task-specific and bulk material-adapted c_W- and λ_S-dependencies as well as suitable initial values, among others (u_F, u_S, p), must be provided. In both (4.69) and (4.72), various simplifications are possible, e.g., the gas acceleration can be neglected compared to the solid acceleration, and the resistance coefficient λ_F of the gas can be neglected compared to that $(\mu \cdot \lambda_S)$ of the solid in general.

Box 4.5: System of differential equations for calculating a straight horizontal conveying section

Elimination of $F_{\text{drag}} = F_{S,\text{drag}} = -F_{F,\text{drag}}$ from (2.51) and (2.55) results in the pressure loss equation

$$-A_R \cdot dp = F_{S,f} + F_{F,f} + \dot{m}_S \cdot du_S + \dot{m}_F \cdot du_F \tag{4.68}$$

which, with (4.41) and (4.44) as well as the continuity equations for $(\dot{m}_S, \dot{m}_F)$ can be transformed to:

$$-\frac{dp}{dL} = \varepsilon \cdot \varrho_F \cdot \left((\mu \cdot \lambda_S + \lambda_F) \cdot \frac{u_F^2}{2 \cdot D_R} + \frac{1-\varepsilon_F}{\varepsilon_F} \cdot \frac{\varrho_P}{\varrho_F} \cdot u_S \cdot \frac{du_S}{dL} + u_F \cdot \frac{du_F}{dL} \right) \tag{4.69}$$

$(A_R \cdot dp)$ from (2.51) and (2.56) eliminated, yields the motion equation:

$$F_{\text{drag}} = \varepsilon_F \cdot (F_{S,f} + \dot{m}_S \cdot du_S) - (1-\varepsilon_F) \cdot (F_{F,f} + \dot{m}_F \cdot du_F) \tag{4.70}$$

As an approach for the drag force F_{drag} of the solid particles present in the considered pipe element dL is simplified here

$$F_{\text{drag}} = \sum F_{W,i} = c_W \cdot \frac{\pi}{4} \cdot d_S^2 \cdot \frac{\varrho_F}{2} \cdot (u_F - u_S)^2 \cdot \frac{(1-\varepsilon_F) \cdot A_R \cdot \varrho_S \cdot dL}{\frac{\pi}{6} \cdot d_S^3 \cdot \varrho_P} \tag{4.71}$$

see (3.3). Taking into account (4.41), (4.44) and (4.71) and the continuity equations for $(\dot{m}_S, \dot{m}_F)$ follows from (4.70):

$$\frac{3}{4} \cdot c_W \cdot \frac{(u_F - u_S)^2}{d_S} = \varepsilon_F \cdot \left(\left(\frac{\varepsilon_F}{1-\varepsilon_F} \cdot \mu \cdot \lambda_S - \lambda_F \right) \cdot \frac{u_F^2}{2 \cdot D_R} + \frac{\varrho_P}{\varrho_F} \cdot u_S \cdot \frac{du_S}{dL} - u_F \cdot \frac{du_F}{dL} \right) \tag{4.72}$$

The gas velocity change du_F occurring in (4.69) and (4.72) is coupled to the pressure gradient dp and can be determined as follows: From the differentiated continuity equations of gas and solid

$$\varepsilon_F \cdot \varrho_F \cdot u_F = \frac{\dot{m}_F}{A_R} \rightarrow \frac{d\varepsilon_F}{\varepsilon_F} + \frac{d\varrho_F}{\varrho_F} + \frac{du_F}{u_F} = 0 \tag{4.73}$$

$$(1 - \varepsilon_F) \cdot u_S = \frac{\dot{m}_S}{A_R \cdot \varrho_P} \rightarrow -\frac{d\varepsilon_F}{1 - \varepsilon_F} + \frac{du_S}{u_S} = 0 \tag{4.74}$$

and the differentiated isothermal gas equation

$$\frac{p}{\varrho_F} = R_F \cdot T_F \quad \rightarrow \quad \frac{dp}{p} - \frac{d\varrho_F}{\varrho_F} = 0 \tag{4.75}$$

as well as

$$\mu = \frac{\dot{m}_S}{\dot{m}_F} = \frac{1 - \varepsilon_F}{\varepsilon_F} \cdot \frac{\varrho_P}{\varrho_F} \cdot \frac{u_S}{u_F} \quad \rightarrow \quad \frac{1 - \varepsilon_F}{\varepsilon_F} = \mu \cdot \frac{\varrho_F}{\varrho_P} \cdot \frac{u_F}{u_S} \tag{4.76}$$

follows:

$$du_F = -u_F \cdot (\mu \cdot \frac{\varrho_F}{\varrho_P} \cdot \frac{u_F}{u_S} \cdot \frac{du_S}{u_S} + \frac{dp}{p}) \tag{4.77}$$

(4.76) and (4.77) inserted into (4.69) yields:

$$\begin{aligned} -\frac{dp}{dL} = \varepsilon_F \cdot \varrho_F \cdot \Bigg(& (\mu \cdot \lambda_S + \lambda_F) \cdot \frac{u_F^2}{2 \cdot D_R} \\ & + \left(1 - \frac{\varrho_F}{\varrho_P} \cdot \left(\frac{u_F}{u_S}\right)^2\right) \cdot \mu \cdot u_F \cdot \frac{du_S}{dL} - \frac{{u_F}^2}{p} \cdot \frac{dp}{dL} \Bigg) \end{aligned} \tag{4.78}$$

(4.76) and (4.77) introduced into (4.72) and rearranged for du_S/dL results in:

$$\frac{du_S}{dL} = \frac{\frac{3}{4} \cdot \frac{c_W}{\varepsilon_F \cdot d_S} \cdot (u_F - u_S)^2 - (\frac{\varrho_P}{\varrho_F} \cdot \frac{u_S}{u_F} \cdot \lambda_S - \lambda_F) \cdot \frac{u_F^2}{2 \cdot D_R} - \frac{{u_F}^2}{p} \cdot \frac{dp}{dL}}{\frac{\varrho_P}{\varrho_F} \cdot u_S + \mu \cdot \frac{\varrho_F}{\varrho_P} \cdot u_F \cdot (\frac{u_F}{u_S})^2} \tag{4.79}$$

(4.79) can be inserted into (4.78) and this can be solved for dp/dL.

The equation of motion (4.72) can only be used in the presented form for a relatively uniform particle concentration distribution across the pipe cross-section without mutual influence on the resistance behavior of individual particles. In (4.70), the additional forces occurring and to be considered in a strongly demixed gas/solid flow are not included: In the case of strand conveying, for example, a "dense" gas/solid phase forms at the bottom of the pipe and a "dilute" flight phase above it. Gas and solid of the flight phase exert forces on the dense phase, which can be modeled as shear stresses on its surface. The distribution of the conveying gas across the pipe cross-section is extremely uneven, see Sect. 5.1.1.

When calculating conveying distances vertically upwards, the weight force of the bulk material must also be taken into account.

4.8 Velocity Ratio Solid/Gas

The solid/gas velocity ratio $C_h = u_S/u_F$ along a horizontal conveying distance can be determined with an approach corresponding to (4.79). After (dp/dL) is known by inserting (4.79) into (4.78), the change of u_S is calculated. If a section-wise incompressible gas flow is assumed, i.e. $du_F/dL = 0$, (4.72) and (4.79) simplify with $\varepsilon_F = 1$ and $v_F = u_F$ to:

$$\frac{dC_h}{dL} = \frac{\frac{3}{4} \cdot \frac{c_W}{d_S} \cdot (1 - C_h)^2 - (\frac{\varrho_P}{\varrho_F} \cdot C_h \cdot \lambda_S - \lambda_F) \cdot \frac{1}{2 \cdot D_R}}{\frac{\varrho_P}{\varrho_F} \cdot C_h} \tag{4.80}$$

With (4.80), the velocity course of the solid during acceleration, e.g., behind a deflection, as well as the required distance can be easily calculated.

If in (4.80) $dC_h/dL = 0$ or in (4.79) $du_S/dL = 0$ is set, the solid/gas velocity ratio in the steady state follows (→ here $C^{\#}$). Example 1: For a pneumatic conveying with $\lambda_S = C_h \cdot \lambda_S^* = \text{const.}$, $(\varrho_P/\varrho_F) \cdot C_h \cdot \lambda_S \gg \lambda_F$ and $dC_h/dL = 0$ follows from (4.80):

$$C_h^{\#} = \frac{1}{1 + w_T \cdot \sqrt{\frac{\lambda_S^*}{2 \cdot g \cdot D_R}}} \tag{4.81}$$

with:

w_T Terminal velocity of the characteristic individual particles, (3.11).

Already the consideration of λ_F in Example 1 leads to a quadratic equation for $C_h^{\#}$. Example 2: (4.80) provides with $\lambda_S = 2 \cdot \beta_R/(C_h \cdot u_F^2/(g \cdot D_R))$, $(\varrho_P/\varrho_F) \cdot C_h \cdot \lambda_S \gg \lambda_F$ and $dC_h/dL = 0$ formally for a completely demixed gas/solid flow:

$$C_h^{\#} = 1 - \frac{w_T}{u_F} \cdot \sqrt{\beta_R} \tag{4.82}$$

(4.82) does not correctly describe reality. See the comment at the end of Sect. 4.7.7. To determine the C_v-values of vertical conveyors, the corresponding approaches must be extended by the weight force of the bulk material.

For practical design calculations, often only the $C^{\#}$-values in the steady state are required. It is recommended to use (4.18) for horizontal straight conveying sections:

$$C_h^{\#} = 1 - \frac{v_{F,A,\min}}{v_F} \tag{4.83}$$

Vertical upward conveyance can be approximated by

$$C_v^{\#} = \frac{u_S}{u_F} = \frac{u_F - w_T}{u_F} = 1 - \frac{w_T}{u_F} \tag{4.84}$$

as described. Fig. 3.15 and Sect. 3.3.3, *fast fluidization*, show that (4.84) also correctly represents the real behavior of particle collectives as long as $Ar \gtrsim 10^4$ remains. For smaller values of Ar, i.e., finer-grained bulk material, higher discharge velocities $w_{T,\varepsilon} > w_T$ are required due to segregation effects. If $w_{T,\varepsilon}$ is set equal with the transport velocity $v_{F,tr}$, (3.40), this can be replaced over the entire Ar-range (4.84) by:

$$C_v^{\#} = 1 - \frac{v_{F,tr}}{v_F} \quad \rightarrow \quad v_{F,tr} = 2.28 \cdot \frac{\eta_F}{d_S \cdot \varrho_F \cdot} \cdot Ar^{0.419} = f(\varrho_F) \tag{4.85}$$

4.9 Calculation Example 8: Characteristic Curve of a Pneumatic Vertical Conveying

The conveying characteristic curve $\Delta p_R(v_F = v_{F,E}, \dot{m}_S = \text{const.})$ of a given bulk material is to be calculated through a vertically upward flowing conveying section. The solid feed is horizontal at the base of the line with the velocity $u_{S,A} = 0\,\text{m/s}$ in the conveying direction. The calculation should initially be incompressible. The admissibility of this assumption is to be checked and, if necessary, corrected.

Relevant design data:

- System data:
  Solid throughput: $\dot{m}_S = 108\,\text{t/h} = 30.0\,\text{kg/s}$,
  Conveying height: $\Delta L_v = 25\,\text{m}$,
  Conveying pipe diameter: $D_R = 206.5\,\text{mm}$, Cross-sectional area: $A_R = 3.3491 \cdot 10^{-2}\,\text{m}^2$,
  Conveying pressure at the beginning of the pipeline: $p_A = \text{variable}$,
  Pressure at the end of the pipeline: $p_E = 1.0\,\text{bar}$,
  Design/mixing temperature $\cong$ Solid temperature: $T_M = 293\,\text{K}$.
- Bulk material data:
  Average particle diameter: $d_{S,50} = 20\,\mu\text{m}$,
  Particle = solid density: $\varrho_P = 3200\,\text{kg/m}^3$,
  Resistance coefficient: $\lambda_{S,v} = C_v \cdot \lambda_S^* = 0.01 = \text{const.}$

- Gas data:
  Gas type: air,
  Gas density at the end of the pipeline: $\varrho_{F,E} = 1.20\,\text{kg/m}^3$,
  Dynamic viscosity: $\eta_F = 18.26 \cdot 10^{-6}\,\text{Pa s}$.
  Resistance coefficient: $\lambda_F = 0.02 = \text{const.}$

Calculation approach:
The pressure losses to be considered are, see Sect. 4.7:

$$\Delta p_R = \Delta p_{S,R,v} + \Delta p_{S,H} + \Delta p_{S,B} + \Delta p_F$$

With (4.48), (4.49), (4.53) and (4.55), $\varepsilon_F = 1$, $v_F = u_F$ as well as the loading $\mu = \dot{m}_S/\dot{m}_F = \dot{m}_S/(A_R \cdot \rho_{F,E} \cdot v_F)$ follows from this:

$$-\Delta p_R = \frac{\dot{m}_S}{A_R} \cdot \Delta L_v \cdot v_F \cdot \left(\frac{1}{2} \cdot \frac{\lambda_{S,v}}{D_R} + \frac{g}{C_v \cdot v_F^2} + \frac{C_A}{\Delta L_v} + \frac{1}{2} \cdot \frac{\lambda_F}{D_R} \cdot \frac{1}{\mu}\right)$$

In the case of pressure loss Δp_F of the gas, only the friction is considered. The loading μ changes with the given $\dot{m}_S$ with the gas velocity v_F. The velocity ratio $C_v = u_S/u_F = (1 - v_{F,tr}/v_F)$ is determined with (4.85).

Intermediate calculations:
Velocity $v_{F,tr}$, (4.85) or (3.40):

$$v_{F,tr} = 2.28 \cdot \frac{\eta_F}{d_{S,50} \cdot \varrho_{F,E}} \cdot \left(\frac{d_{S,50}^3 \cdot g}{\eta_F^2} \cdot \varrho_{F,E} \cdot (\varrho_P - \varrho_{F,E})\right)^{0.419}$$

$$= 2.28 \cdot \frac{18.26 \cdot 10^{-6}\,\text{Pa s}}{20 \cdot 10^{-6}\,\text{m} \cdot 1.20\,\frac{\text{kg}}{\text{m}^3}}$$

$$\cdot \left(\frac{(20 \cdot 10^{-6}\,\text{m})^3 \cdot 9.81\,\frac{\text{m}}{\text{s}^2}}{(18.26 \cdot 10^{-6}\,\text{Pa s})^2} \cdot 1.20\,\frac{\text{kg}}{\text{m}^3} \cdot (3200 - 1.2)\,\frac{\text{kg}}{\text{m}^3}\right)^{0.419}$$

$$= 1.662\,\frac{\text{m}}{\text{s}}$$

Velocity ratio C:

$$C_v = 1 - \frac{1.662\,\frac{\text{m}}{\text{s}}}{v_F} = C_v^\# \tag{4.86}$$

Loading μ:

$$\mu = \frac{\dot{m}_S}{\dot{m}_F} = \frac{\dot{m}_S}{A_R \cdot \varrho_{F,E} \cdot v_F} = \frac{30.0\,\frac{\text{kg}}{\text{s}}}{3.3491 \cdot 10^{-2}\,\text{m}^2 \cdot 1.20\,\frac{\text{kg}}{\text{m}^3} \cdot v_F} = \frac{746.469\,\frac{\text{m}}{\text{s}}}{v_F}$$

Pressure loss:

$$-\Delta p_R = \frac{30.0\,\frac{\mathrm{kg}}{\mathrm{s}}}{3.3491\cdot 10^{-2}\,\mathrm{m}^2}\cdot 25\,\mathrm{m}\cdot v_F\cdot\left(\frac{1}{2}\cdot\frac{0.01}{0.20\,\mathrm{m}}+\frac{9.81\,\frac{\mathrm{m}}{\mathrm{s}^2}}{\overline{C}_v\cdot v_F^2}+\frac{C_A}{25\,\mathrm{m}}+\frac{1}{2}\cdot\frac{0.02}{0.2\,\mathrm{m}}\cdot\frac{1}{\mu}\right)$$

$$= 22394.08\,\frac{\mathrm{kg}}{\mathrm{m}\cdot\mathrm{s}}\cdot v_F\cdot\left(0.025\,\frac{1}{\mathrm{m}}+\frac{9.81\,\frac{\mathrm{m}}{\mathrm{s}^2}}{\overline{C}_v\cdot v_F^2}+\frac{C_A}{25\,\mathrm{m}}+\frac{0.05\,\frac{1}{\mathrm{m}}}{\mu}\right)$$

Incompressible Flow Results:
The total and individual pressure losses Δp_i determined with the above equations are summarized in the following Table 4.1. $|\Delta p_R|$ is shown in Fig. 4.28 as a function of the gas velocity v_F. Since the gas velocity v_F used here is the final velocity $v_{F,E}$, the following also applies: $\overline{C}_v = C_A = C_{v,E}$.

Table 4.1 shows that the incompressible calculation can only provide a rough picture of the real conditions, as the relative pressure changes assume values of $(p_E + \Delta p_R)/p_E \gtrsim 1.2$.

Results compressible flow:
For a more accurate calculation, the following procedure can be used: The incompressible pressure loss Δp_R is considered as the first estimate for the real pressure loss. With it, the associated initial gas velocity follows from $v_{F,A} = v_{F,E}\cdot p_E/(p_E + \Delta p_R)$ and from $\overline{v}_F = (v_{F,A} + v_{F,E})/2$ the average gas velocity in the conveying pipe. The pressure losses Δp_i are now recalculated with $\overline{v}_F$. The result Δp_R is used again to calculate a new average gas velocity $\overline{v}_F$ and so on. This recursive iteration is repeated with the newly adjusting Δp_R until a predetermined accuracy limit is exceeded. Table 4.2 shows exemplary calculation values for $v_{F,E} = 25\,\mathrm{m/s}$ and $v_{F,E} = 10\,\mathrm{m/s}$ in comparison. The results of the compressible calculation are compared with those of the incompressible case in Fig. 4.28. Figure 4.28 shows that differences in the pressure differences Δp_R occur mainly in the area of dense phase conveying.

Table 4.1 Results of the incompressible calculation

$v_{F,E}$	[m/s]	30	25	20	15	10	5	3	2
$C_{v,E}$	[1]	0.9446	0.9335	0.9169	0.8892	0.8338	0.6676	0.4460	0.1690
μ	[1]	24.88	29.86	37.32	49.76	74.65	149.29	248.82	373.23
$\|\Delta p_{S,R,v}\|$	[mbar]	167.96	139.96	111.97	83.98	55.99	27.99	16.80	11.20
$\|\Delta p_{S,H}\|$	[mbar]	77.52	94.13	119.80	164.71	263.48	658.14	1641.90	6499.58
$\|\Delta p_{S,B}\|$	[mbar]	253.84	209.05	164.27	119.48	74.69	29.90	11.99	3.03
$\|\Delta p_F\|$	[mbar]	13.50	9.38	6.00	3.38	1.50	0.38	0.14	0.06
$\|\Delta p_R\|$	[mbar]	512.82	452.52	402.03	371.54	395.65	716.41	1670.81	6513.87

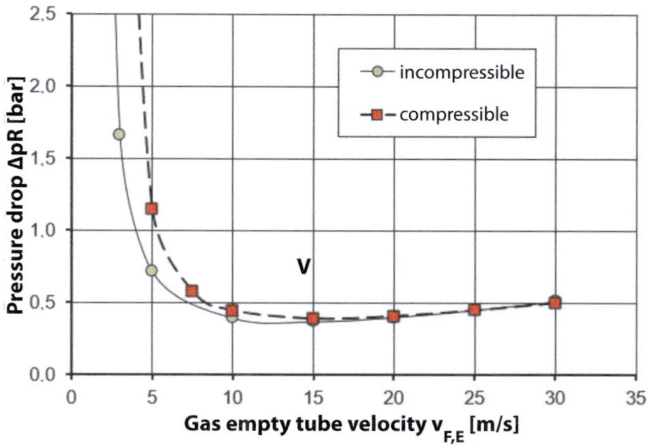

Fig. 4.28 Characteristic conveying curves $\Delta p_R(v_{F,E}, \dot{m}_S = \text{konst.})$ incompressible/compressible

The incompressible as well as the compressible calculation are based on some simplifications in the considered example: $C_v = C_v^{\#}$ is determined with (4.86), although the limit velocity $v_{F,tr}$ for the beginning of the line, i.e. with $\varrho_{F,A}$, would have had to be calculated, see (4.85). Since $v_{F,tr}$ with increasing conveying pressure $p_A =$ gas density $\varrho_{F,A}$ becomes smaller, the calculation is on the "safe" side. The lift pressure loss $\Delta p_{S,H}$ in Table 4.2 was calculated with $\overline{C}_v = \overline{C}_v^{\#}$, i.e. the respective solid/gas velocity ratio in the steady state. This assumes that the solid starting with $u_{S,A} = 0\,\text{m/s}$ reaches this steady state on a very short conveying distance. This will be discussed in more detail in the following Sect. 4.10. The pressure loss $\Delta p_{S,B}^* = \Delta p_{S,B} + \Delta p_{\text{pac}}$ contains both the initial and the permanent acceleration of the solid.

The recursive iteration procedure described above can generally be used for the sectional calculation of a pneumatic conveying line. More effective algorithms are available for this purpose than the one used above.

Table 4.2 Results of the compressible calculation

$v_{F,E}$	[m/s]	**25**	**25**	**25**	*10*	*10*	*10*	*10*	*10*
$v_{F,A}$	[m/s]	**25**	**17.21**	**17.26**	*10*	*7.62*	*6.94*	*6.92*	*6.91*
$\overline{v}_F$	[m/s]	**25**	**21.11**	**21.13**	*10*	*8.58*	*8.47*	*8.46*	*8.46*
$\overline{C}_v$	[1]	**0.9335**	**0.9213**	**0.9213**	*0.8338*	*0.8064*	*0.8037*	*0.8035*	*0.8035*
$C_{v,E}$	[1]	**0.9335**	**0.9335**	**0.9335**	*0.8338*	*0.8338*	*0.8338*	*0.8338*	*0.8338*
μ	[1]	**29.86**	**29.86**	**29.86**	*74.65*	*74.65*	*74.65*	*74.65*	*74.65*
$\lvert\Delta p_{S,R,v}\rvert$	[mbar]	**139.96**	**118.16**	**118.83**	*55.99*	*48.05*	*47.41*	*47.35*	*47.35*
$\lvert\Delta p_{S,H}\rvert$	[mbar]	**94.13**	**112.94**	**112.85**	*263.48*	*317.63*	*322.64*	*323.10*	*323.30*
$\lvert\Delta p_{S,B}^*\rvert$	[mbar]	**209.05**	**209.05**	**209.05**	*74.69*	*74.69*	*74.69*	*74.69*	*74.69*
$\lvert\Delta p_F\rvert$	[mbar]	**9.38**	**7.92**	**7.92**	*1.50*	*1.29*	*1.27*	*1.27*	*1.27*
$\lvert\Delta p_R\rvert$	[mbar]	**452.52**	**448.07**	**448.12**	*395.65*	*441.66*	*446.02*	*446.41*	*446.60*

4.10 Interaction of Pipeline Elements

For the sake of clarity and comprehensibility, the following considerations are presented using incompressible gas/solid flows.

In the previous calculation example 8, it was already indicated that the determination of the lifting pressure loss $\Delta p_{S,H}$, which occurred there for gas and solid in the steady state, is only correct if the proportion of the conveying distance ΔH_{crit} for solid acceleration from $u_S = u_{S,A} = 0\,\text{m/s}$ to $u_S = u_S^\#$, $u_S^\# =$ steady state velocity of the solid, applied to the entire vertical distance ΔL_v is negligibly small. This is approximately the case only for fine-grained solids that are conveyed with very low loading and evenly distributed over the pipe cross-section ($\rightarrow$ undisturbed single-particle flow). In practice, the solid is first accelerated over a finite distance ΔH_{crit} and then conveyed in a steady state over the distance $(\Delta L_v - \Delta H_{\text{crit}})$. Thus, its average integral velocity $\overline{u}_S$ and the associated velocity ratio $\overline{C}_v$ are lower than the corresponding values $(\overline{u}_S^\#, \overline{C}_v^\#)$ in the steady state. This has the consequence that the average solid residence time $\Delta\tau_{\Delta L} = \Delta L_v/\overline{u}_S$ in the vertical pipe, and thus the solid mass currently in this pipe Δm_S and as a result, the lifting pressure loss $\Delta p_{S,H}$ increases compared to those in the steady state. Strictly speaking, the friction pressure loss $\Delta p_{S,R,v}$ is also affected, since with constant gas velocity along the acceleration path ΔH_{crit} the value of C_v continuously changes, see Sects. 4.7.1 and 4.7.2.

The bulk material injection itself, as well as the type of solid feeder system, influence the pressure loss in the subsequent riser in the considered example. For example, a streaky bulk material feed would lead to different pressure losses than a uniformly distributed one across the pipe cross-section. Interactions of this kind occur in pneumatic conveying systems at various points. The following is a more detailed analysis of the practically occurring case of a 90° bend from the horizontal to the vertical with subsequent different vertical lengths in every pneumatic conveying system, often multiple times along the transport path [65].

4.10.1 Vertical Section with Upstream 90° Deflection

In Fig. 4.29, the profiles of the gas velocities $u_F \cong v_F$ and solid u_S as well as their ratio $C = u_S/u_F$ along the end area of the upstream horizontal section, the 90° foot bend and the subsequent vertical pipe are shown schematically. The solid flow leaving the horizontal section with $u_S = u_{S,\text{in}}$ and $C_h = C_{U,\text{in}}$ and entering the 90° horizontal-to-vertical deflection is decelerated in this to $u_{S,\text{out}}$ and $C_{U,\text{out}}$. $(u_{S,\text{in}}, C_{U,\text{in}})$ can be the steady-state values of the horizontal pipe, but may also assume smaller values depending on the pipe routing. The deceleration in the 90° bend can be calculated with (4.61) or estimated approximately with (4.59). Behind the bend, the solid is accelerated again. The required distance for this is ΔH_{crit}. The acceleration can, with sufficient length ΔL_v of the vertical section, reach the values $(u_{S,v}^\#, C_v^\#)$ of the steady-state condition, which then over the

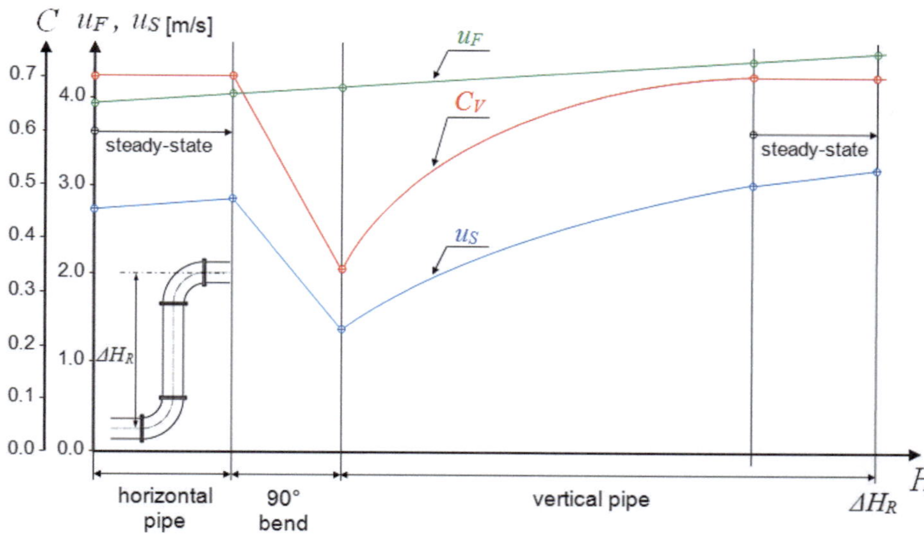

Fig. 4.29 Velocity profiles along the 90° foot point bend and the subsequent vertical section

remaining distance $(\Delta L_v - \Delta H_{\mathrm{crit}})$ remain constant, or, in the case of too short a vertical pipe, lead to end values $(u_{S,v}, C_v) < (u^{\#}_{S,v}, C^{\#}_v)$.

All individual pressure losses to be considered are largely identical to those of calculation example 8 and can be calculated with (4.48), (4.49), (4.53) and (4.87). The following equation (4.87) can be used instead of (4.57) for the bend pressure loss, see Sect. 4.7.5:

$$-\Delta p_{S,U} = \frac{\dot{m}_S}{A_R} \cdot (u^{\#}_{S,v} - u_{S,\mathrm{out}}) = \frac{\dot{m}_S}{A_R} \cdot u_F \cdot (C^{\#}_v - C_{U,\mathrm{out}}) \tag{4.87}$$

$(u^{\#}_{S,v}, C^{\#}_v)$ must be replaced by $(u_{S,v}, C_{v,R})$ in case of a too short pipe. The $\overline{C}_v$-value in (4.53) for calculating the lifting pressure loss $\Delta p_{S,H}$ is to be determined as a suitable average value of the area under the curve $C_v(\Delta L_v)$ (see below). For this purpose, the length ΔH_{crit} of the solids re-acceleration distance after the foot point bend must be known. Since gas and solid separate along this deflection and the solid leaves it as a more or less developed strand at the outer radius, ΔH_{crit} cannot be calculated as an acceleration distance for individual particles. A treatment as a bulk material assembly, whose spatial distribution and flow conditions change along the subsequent lifting distance, is thus necessary. This is currently not feasible with reasonable effort. Therefore, an empirical approach is used.

Critical lifting height ΔH_{crit}: In [54] a formula is proposed, with which the length of the conveying path L_{strand} can be calculated/estimated, along which strands, which can be caused by various structural conditions, are accelerated until complete dissolution. It generally applies:

$$L_{\text{strand}} = L_0 \cdot \left[1 + \Gamma_{\text{strand}} \cdot \left(\frac{\mu^{0.25} \cdot g \cdot D_R}{v_F^k \cdot w_T^{2-k}}\right)\right] \tag{4.88}$$

with:

L_0 Acceleration distance of a single solid particle,

Γ_{strand} Extension factor for strand dissolution,

k Exponent with the values: $k = 1$ – Stokes range; $k = 0.5$ – Transition range; $k = 0$ – Newton range.

The extension factors Γ_{strand} recommended in [54] for various situations are summarized in Table 4.3. In the case discussed here, the following applies: $\Gamma_{\text{strand}} = 15$.

Working hypothesis: The required acceleration distance ΔH_{crit} of the solid decelerated in the foot bend until the steady state is equal to the distance L_{strand}, which is required for the acceleration and resolution of the bulk material strand forming on the outside of the vertical pipe. Thus: $\Delta H_{\text{crit}} = L_{\text{strand}}$. This approach will be applied to *fine-grained solids* from the range of Stokes's resistance law $c_W(Re_{S,\text{rel}}) = 24/Re_{S,\text{rel}}$ as an example.

The exponent$k$ in (4.88) has the value $k = 1$. The acceleration distance L_0 is calculated for the individual particle with the diameter $d_S = d_{S,50}$ characterizing the current bulk material. The acceleration must be from $(u_{S,\text{out}}, C_{U,\text{out}})$ to $(u^{\#}_{S,v}, C^{\#}_v)$. L_0 can be determined analogously to the procedure in calculation example 4. Simplifications used are: $v_F = u_F$, $(\varrho_P - \varrho_F) = \varrho_P$. $(v_F - w_T) = v_F$. From this, it follows for

$$\Delta H_{\text{crit}} = v_F \cdot \Delta\tau_{\text{Stokes}} \cdot \left(\ln\left(\frac{1 - C_{U,\text{out}}}{1 - C^{\#}_v}\right) - (C^{\#}_v - C_{U,\text{out}})\right) \cdot \left[1 + 15 \cdot \left(\frac{\mu^{0.25} \cdot g \cdot D_R}{v_F \cdot w_T}\right)\right] \tag{4.89}$$

Table 4.3 Extension factors for determining the resolution distance of strands

Extension factor	Γ_{strand}
Behind a material feed point (into a horizontal line)	1 to 3
Free-flying strand	5
Bend strands:	
– on the upper pipe wall	10
– on the vertical pipe wall	15
– on the lower pipe wall	30

with:

$$\Delta H_{\text{crit}} = L_{\text{strand}}, \qquad C_{U,\text{out}} = \frac{u_{S,\text{out}}}{v_F}, \qquad w_{T,\varepsilon} = v_{F,tr}$$

$$\Delta \tau_{\text{Stokes}} = \frac{d_S^2 \cdot \varrho_P}{18 \cdot \eta_F}, \qquad C_v^{\#} = \frac{u_{S,v}^{\#}}{v_F} = 1 - \frac{w_{T,\varepsilon}}{v_F},$$

The terms before the square bracket describe the length L_0. Instead of $w_{T,\varepsilon} = v_{F,tr}$ other approaches for determining $w_{T,\varepsilon}$ are usable, see e.g. [36]. Furthermore, it applies: $C_v^{\#} < (1 - w_T/v_F)$. (4.89) provides results consistent with practical experience for the class of fine-grained bulk materials considered here only if the single-particle terminal velocity is used for w_T in the square bracket. In principle, $w_{T,\varepsilon}$ would also be possible, [54] is not clear here.

(4.89) can be converted into the following practical form by introducing $w_T = \Delta\tau_{\text{Stokes}} \cdot g$ as well as neglecting the "1" in the square bracket:

$$\Delta H_{\text{crit}} = L_{\text{strand}} = 15 \cdot \left[\ln\left(\frac{1 - C_{U,\text{out}}}{1 - C_v^{\#}} \right) - (C_v^{\#} - C_{U,\text{out}}) \right] \cdot \mu^{0.25} \cdot D_R \quad (4.90)$$

Fig. 4.30 shows an example of the evaluation of this equation for the cases ($C_{U,\text{out}} = 0.40, C_v^{\#} = 0.80$) and ($C_{U,\text{out}} = 0.35, C_v^{\#} = 0.70$) in dimensionless representation. The influence of the gas velocity v_F is carried out via ($C_{U,\text{out}}, C_v^{\#}$).

In [54] no references are given for the approach underlying (4.88). To verify this, own results and available literature, e.g. the measurements in [66], were used. A satisfactory agreement was obtained with these.

Average velocity ratio $\overline{C}_v$: This is required for the calculation of the lifting pressure loss $\Delta p_{S,H}$, (4.53), of the considered vertical distance. To determine it, the cases $\Delta L_v \geq \Delta H_{\text{crit}}$ and $\Delta L_v < \Delta H_{\text{crit}}$ are considered and described by different equations, see Fig. 4.31. The case $\Delta L_v = \Delta H_{\text{crit}}$ must be included as a boundary case in both approaches. In the curved course of the C-curves between $C_{U,\text{out}}$ at the foot of the verti-

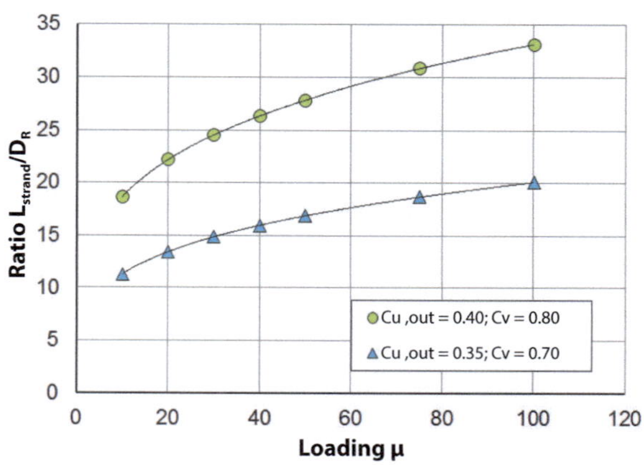

Fig. 4.30 Dimensionless distance for reacceleration/resolution of a strand behind a 90° bend horizontal-vertical, fine-grained bulk material

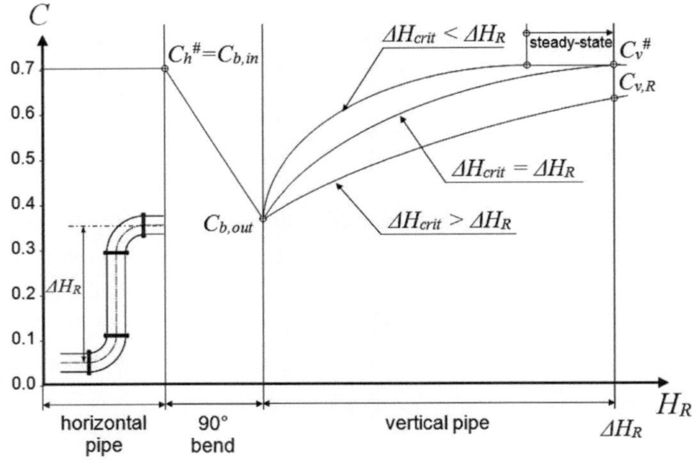

Fig. 4.31 Possible operating cases in a vertical section

cal distance and $C_v^{\#}$ at the critical lifting height ΔH_{crit}, or $C_{v,R}$ after the height ΔL_v ($\rightarrow$ operating case $\Delta L_v < \Delta H_{\text{crit}}$), it is obviously a concave function: large gas/solid velocity difference at the entrance of the vertical pipe = strong solid acceleration, followed by a steady decrease in these quantities. Since an analytical determination of the C-curve with the model described above is not possible, this curve branch must be characterized by a suitable average of its two limit values. This is done as follows: The *ln*-terms of the C-function in the square bracket of (4.90) are developed as logarithmic series, which are truncated after the 2nd term. This results in:

$$\ln\left(\frac{1 - C_{U,\text{out}}}{1 - C_v^{\#}}\right) - (C_v^{\#} - C_{U,\text{out}}) = \frac{1}{2} \cdot ((C_v^{\#})^2 - C_{U,\text{out}}^2) \tag{4.91}$$

In the modified (4.91) by (4.90), the velocity ratio $C_v^{\#}$ at the position ΔH_{crit} is replaced by C_x at the variable position $\Delta L_x \leq \Delta H_{\text{crit}}$ and this equation is divided by (4.90). This results in some rearrangements

$$C_x = \sqrt{\frac{\Delta L_x}{\Delta H_{\text{crit}}} \cdot (C_v^{\#})^2 + \left(1 - \frac{\Delta L_x}{\Delta H_{\text{crit}}}\right) \cdot C_{U,\text{out}}^2} \tag{4.92}$$

and leads with $(\Delta L_x / \Delta H_{\text{crit}}) = 1/2$ to the quadratic mean:

$$\overline{C}_x = \sqrt{\frac{1}{2} \cdot ((C_v^{\#})^2 + C_{U,\text{out}}^2)} \tag{4.93}$$

This is only an approximation of the actual conditions. Above ΔH_{crit} the $C_v^{\#} = \text{const.}$ is established. Thus, for the operating case:

$$\Delta L_v = \Delta H_{\text{crit}}: \quad \overline{C}_v = \sqrt{\frac{1}{2} \cdot ((C_v^{\#})^2 + C_{U,\text{out}}^2)} \tag{4.94}$$

If the current vertical distance ΔL_v is longer than the re-acceleration distance ΔH_{crit} of the solid strand, then the C-curve consists of a curved branch "x" from the base point to the critical height ΔH_{crit} with the mean value corresponding to (4.93) and a steady state branch "#" from ΔH_{crit} to ΔL_v with $C_v^{\#} = \text{konst.}$, see Fig. 4.31. For the lifting pressure loss $\Delta p_{S,H}$ the following applies:

$$-\Delta p_{S,H} = g \cdot \frac{\dot{m}_S}{A_R} \cdot \frac{\Delta L_v}{\overline{C}_v \cdot v_F} = g \cdot \frac{\dot{m}_S}{A_R \cdot v_F} \cdot \left(\frac{\Delta H_{\text{crit}}}{\overline{C}_x} + \frac{\Delta L_v - \Delta H_{\text{crit}}}{C_v^{\#}} \right) \tag{4.95}$$

Solving for $\overline{C}_v$ considering the approach for $\overline{C}_x$ yields the determination equation for the operating case:

$$\Delta L_v \geq \Delta H_{\textit{crit}}: \quad \overline{C}_v = \frac{1}{\frac{\Delta H_{\text{crit}}}{\Delta L_v} \cdot \frac{1}{\sqrt{\frac{1}{2} \cdot ((C_v^{\#})^2 + C_{U,\text{out}}^2)}} + (1 - \frac{\Delta H_{\text{crit}}}{\Delta L_v}) \cdot \frac{1}{C_v^{\#}}} \tag{4.96}$$

For $(\Delta H_{\text{crit}}/\Delta L_v) = 1$ goes (4.96) in (4.94) over.

If the re-acceleration distance ΔH_{crit} is longer than the vertical section ΔL_v, then the stationary velocity ratio $C_v^{\#}$ is not reached; the smaller value $C_{v,R}$ occurs, see Fig. 4.31. $C_{v,R}$ can be calculated using (4.92) if in this ΔL_x is replaced by ΔL_v:

$$C_{v,R} = \sqrt{\frac{\Delta L_v}{\Delta H_{\text{crit}}} \cdot (C_v^{\#})^2 + (1 - \frac{\Delta L_v}{\Delta H_{\text{crit}}}) \cdot C_{U,\text{out}}^2} \tag{4.97}$$

The average velocity ratio $\overline{C}_v$ for this operating case is thus calculated as:

$$\Delta L_v \leq \Delta H_{\text{crit}}: \quad \overline{C}_v = \sqrt{\frac{1}{2} \cdot (C_{v,R}^2 + C_{U,\text{out}}^2)} \quad \rightarrow \quad C_{v,R} \text{ from (4.96)} \tag{4.98}$$

For $(\Delta L_v/\Delta H_{\text{crit}}) = 1$ becomes $C_{v,R} = C_v^{\#}$ and (4.98) transitions to (4.94). Alternatively, the values for $(C_x, \overline{C}_x, C_{v,R})$ can be calculated iteratively from (4.90). For this, the respective ratio $(\Delta L_x/\Delta H_{\text{crit}})$ must be formed.

Relevance: In Table 4.4, lift pressure losses $\Delta p_{S,H}$ are given as examples, which were calculated taking into account the re-acceleration effect described above, are compared with the corresponding values $\Delta p_{S,H}^{\#}$ without this consideration (→ the bulk material is in a steady state over the entire conveying height). From (4.53) it follows:

$$\frac{\Delta p_{S,H}}{\Delta p_{S,H}^{\#}} = \frac{C_v^{\#}}{\overline{C}_v} \tag{4.99}$$

Table 4.4 Effects of the strand acceleration distance on the pressure loss

$\Delta H_{crit}/\Delta L_v$	$\Delta p_{S,H}/\Delta p^{\#}_{S,H}$	$\Delta p_{S,H}/\Delta p^{\#}_{S,H}$	$\frac{\Delta p_{S,H}-\Delta p^{\#}_{S,H}}{\Delta p^{\#}_{S,H}} \cdot 100\,\%$	$\frac{\Delta p_{S,H}-\Delta p^{\#}_{S,H}}{\Delta p^{\#}_{S,H}} \cdot 100\,\%$
0	1	**1**	0	**0**
0.25	1.073	**1.066**	7.3	**6.6**
0.50	1.146	**1.132**	14.6	**13.2**
0.75	1.219	**1.198**	21.9	**19.8**
1.00	1.292	**1.265**	29.2	**26.5**

$\overline{C}_v$ is determined with (4.96). Table 4.4 shows the ratio $\Delta p_{S,H}/\Delta p^{\#}_{S,H} = f(\Delta H_{crit}/\Delta L_v)$ for $C^{\#}_v = 0.90$ (**0,80**) and$C_{U,out} = 0.40$ (**0.40**).

It can be seen from Table 4.4 that with the increase of the relative acceleration length $\Delta H_{crit}/\Delta L_v$ of the strand, the relative pressure loss $\Delta p_{S,H}/\Delta p^{\#}_{S,H}$ also increases. In the examples shown, this increase is up to about 29% (**27%**) compared to the simplified calculated values $\Delta p^{\#}_{S,H}$.

If a fine-grained bulk material, such as assumed in calculation example 8, is horizontally fed into the vertical section, the above approach can also be applied. It is only necessary to set $(u_{S,A}, C_A) = (u_{S,out}, C_{U,out}) = 0$. Depending on the lock system = bulk material distribution over the cross-section of the conveying pipe, the extension factor Γ_{strand} may need to be adjusted. It is recommended not to fall below the value $\Gamma_{strand} = 5$. Application example: For $\Gamma_{strand} = 15$ ($\rightarrow$ one-sided solid feed = strand at the pipe wall), $\Delta H_{crit}/\Delta L_v = 1$, $C^{\#}_v = 0.80$ follows $\Delta p_{S,H}/\Delta p^{\#}_{S,H} = 1.414$. This corresponds to an increase in the lifting pressure loss of 41.4% compared to the steady state calculated case.

The calculation approach shown above provides particularly accurate and more operationally reliable pressure losses for dense phase conveying ($\rightarrow$ here, the total pressure loss of a vertical section is largely identical to its lifting pressure loss, see (3.42)) than the conventional method. For coarser bulk materials than the fine-grained solids analyzed here, i.e., for $k < 1$, the calculation equations for L_{strand} and L_0 must be modified accordingly, see (4.88) ff.

4.10.2 Potential Problem Sources in Conveying Routes

Below, some possible causes/triggers for problems in pneumatic conveying routes are qualitatively analyzed.

- **Consecutive Deflections**: Behind deflections, the bulk material decelerated in the bend must be accelerated back to its staedy state speed. With (4.88) in connection with the extension factors Γ_{strand} of Table 4.3, the necessary distances $L_{strand} = \Delta L_{crit}$

can be estimated. If the distances between two deflections specified by ΔL_{crit} are not maintained, the bulk material enters the subsequent bend with a speed lower than in the associated steady state and is further decelerated there. This can reduce the solid velocity to such an extent that deposits, dune/conglobation or plug formation with pronounced pressure pulsations or even clogging of the conveying line may occur. It is therefore essential to check the solid velocities using the equations given in Sect. 4.7.5. On the other hand, the total pressure loss $\sum \Delta p_{S,U}$ of a deflector combination with distances $\Delta L_U < \Delta L_{crit}$ is smaller than that of one with $\Delta L_U \geq \Delta L_{crit}$. For example, in the case of two consecutive 90° bends, in the case of $\Delta L_U \geq \Delta L_{crit}$ the solid must be re-accelerated to the full inertial velocity twice, while in the case of $\Delta L_U < \Delta L_{crit}$ a smaller velocity difference must be accelerated overall, see Sect. 4.7.5 for comparison.

Practical example: In a coal-fired power plant, electrostatic precipitator fly ash was to be transported from an existing collection bunker via a pneumatic dense phase conveying system over several 100 meters across the plant to a newly constructed loading silo. The solid material was fed in using a single pressure vessel. After approximately 40 meters, the conveying line opened into an existing underground channel via three immediately consecutive, spatially differently arranged 90° bends. The fly ash was a Group 1 bulk material from the left hatched area of the extended Geldart diagram, see Fig. 4.13 and Sect. 4.5: average particle diameter $d_{S,50} \cong (20\text{–}30)\,\mu\text{m}$, very broad particle size distribution with a slope angle in the RRSB diagram of $\alpha_{RRSB} \lesssim 40°$, low particle size fractions up to $d_S \cong 1000\,\mu\text{m}$, conveying gas initial velocity $v_{F,A} \cong 6\,\text{m/s}$. As expected, a strand conveying occurred. During operation of the plant, approximately every thirtieth pressure vessel charge led to a sudden pipe blockage, which could not be prevented by the installed non-clogging control circuit. The cause was identified as a deposit of coarse particles in the conveying pipe behind the last of the three 90° bends leading into the underground channel in the transport direction, which increased in thickness with the increasing number of charges. After about the thirtieth charge, more than half of the pipe cross-section was covered by the stationary coarse fraction of the solid material, so that it was then abruptly compressed into a plug due to the locally larger pressure gradient. The deceleration of the coarse fraction of the fly ash, which generally moved at a lower speed than the fine fraction, occurred in the present deflector combination until it came to a standstill. The problem was solved by installing a venturi nozzle-like device for remixing the bulk material flowing in the strand with the free gas flow above the strand directly before entering the three 90° bends.

- **Distance between bulk material feed and deflection**: It is common for the bulk material to be introduced into a horizontal conveying pipe. To maintain ground clearance, accessibility, and working possibilities in the installation space, the conveying line is generally led vertically upwards shortly after the solid feed to a higher pipe route. Here too, with (4.88) and the associated extension factor Γ_{strand} from Table 4.3, the distance between the lock and the lifting section required for solid acceleration up to the steady state ΔL_{crit} should be determined and checked whether the entry velocity $u_{S,\text{in}}$ of the solid in the 90° bend horizontal-vertical is greater than the minimum velocity $u_{S,\text{in,min}}$ to be calculated with (4.63). Distances $\Delta L_B < \Delta L_{\text{crit}}$ are possible as long as $u_{S,\text{in}} > u_{S,\text{in,min}}$ remains, see also Sect. 4.7.5b. If this condition is not met, the conveying gas velocity $v_{F,A}$ must be increased. Other (less sensitive) deflection arrangements should be checked analogously.

- **Long straight horizontal conveying sections**: At the end of very long straight and exclusively horizontal conveying pipeline sections, pronounced strong pressure pulsations are often observed, which among other things lead to the striking of subsequent pipeline elements, e.g., the vertical pipe at a receiving silo, and thus to considerable mechanical stresses on the support structure of the pipeline. At the same time, these pressure peaks can trigger a shutdown/response of the safety valve of the pressure generator or, to avoid this, require a reduction of the solid throughput $\dot{m}_S$. The observed pressure peaks are the result of bulk material dunes, conglobations, or plugs forming in such horizontal sections. Such effects have been observed so far, particularly with fine-grained bulk materials, including lignite and coal dusts as well as cements, all of which were conveyed in the strand mode. Along the horizontal section, the pulsations continuously increase in the conveying direction. At lengths $L_{P,\text{crit}} \gtrsim 100\,m$ they have such a strong impact on the overall behavior of the conveying system that countermeasures must be taken.

If the bulk material passes through the conveying pipe as a strand with a density of $\varrho_{\text{Str,x}} \leq \varrho_{SS}$, then its local cross-sectional coverage (→ after the conveying section "x") is calculated as follows:

$$\varphi_{\text{Str,x}} = \frac{A_{\text{Str,x}}}{A_R} = \frac{\dot{m}_S}{A_R \cdot \varrho_{\text{Str,x}} \cdot u_{S,x}} \tag{4.100}$$

Low area coverage = low bed height in the horizontal pipe leads to the initially strongly loosened/fluidized and liquid-like flowing strand degassing relatively quickly into the free space above the bed. As a result, internal bulk friction and wall friction are increasingly mobilized. The gas velocity $v_{F,O,x}$ in the free space above the strand increases only slightly to $v_{F,O,x} = v_{F,x}/(1 - \varphi_{\text{Str,x}})$. Assumption: The entire gas flow flows through the free space. From the free space gas flow and the particles carried along by it, shear forces are exerted on the strand surface, through which, at least in the front area of the conveying pipe, dunes running over the bed surface are generated. The lower bed area adjacent to the pipe wall is simultaneously decelerated by friction. Due to the increasing mobilization of wall friction and internal friction with the conveying path, the strand may come to a local standstill after a certain conveying distance. Subsequent bulk material accumulates and locally reduces the free pipe cross-section. After sufficient narrowing, the conveying gas = free space gas will abruptly set this deposit back in motion and transport it as a conglobation/plug. These processes repeat periodically. The more cohesive the bulk material and the shorter its venting time $\Delta\tau_E$, the more pronounced the described processes/pressure peaks will be.

Significantly responsible for the described behavior is that in the affected long straight horizontal pipe sections, there is no back-mixing of conveying gas and strand solids. More frequent periodic gas/solid back-mixing would repeatedly create a new highly loosened, more flowable strand state. Practical experience shows that the described effects do not occur or only occur strongly dampened when there are deflections/bends in such pipeline areas. By installing forcibly acting mixing elements in the pipeline, targeted influence can be taken. Example: Installation of several venturi nozzle-like pipe constrictions in the affected pipe section. It is crucial that this influence takes place before the first formation of the pulsation-causing conglobations/plugs: Once formed, such compressions can hardly be dissolved again. In Sect. 4.4, the presented problem is discussed in more depth.

- **Upward inclined sections**: Behind the transition of a horizontal section into an upward inclined pipe section, gas and solid separate into a solid-rich strand and a solid-poor gas flow above the strand, see Fig. 4.32. The forces driving the strand – pressure difference force $F_{\Delta p}$ and shear force F_τ due to gas flow and particle bombardment on the strand surface – must compensate the opposing forces – wall friction F_R of gas and solid as well as the downhill downforce F_H of the solid material. The Coulomb's solid friction mobilized here is generally significantly greater than that caused by particle wall impacts, while the shear force F_τ acting on the strand surface is lower than when driven by a gas flow around the particles. A solid velocity profile forms in the strand with high velocity at its surface and much lower velocity at the contact surface to the pipe wall, where the velocity at the pipe wall can also become zero or negative. The latter then leads to backflow of part of the bulk material flow against the conveying direction. Consequence: Strong pulsations, increased pressure loss, and possibly pipe blockages. The effects depend on the inclination angle α_{NS} of the incline against the horizontal and the length of the inclined section ΔL_{NS}. Inclined conveying routes with $\alpha_{NS} \cong (15\text{-}85)^\circ$ and $\Delta L_{NS} \gtrsim 5\,m$ should be avoided and designed as combinations of horizontal and vertical sections with 90° bends. If

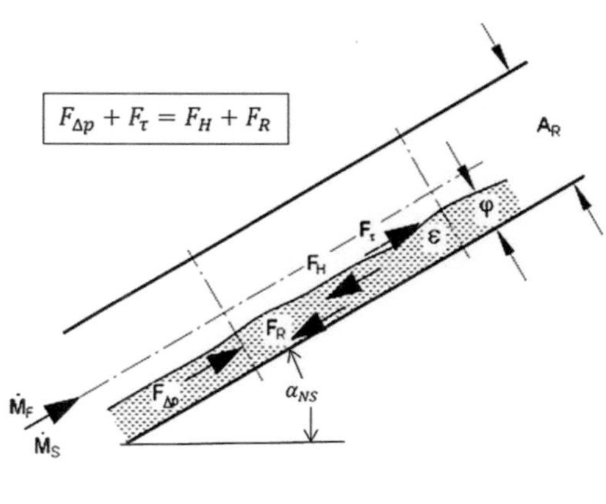

Fig. 4.32 Forces on the strand element of an inclined upward conveying

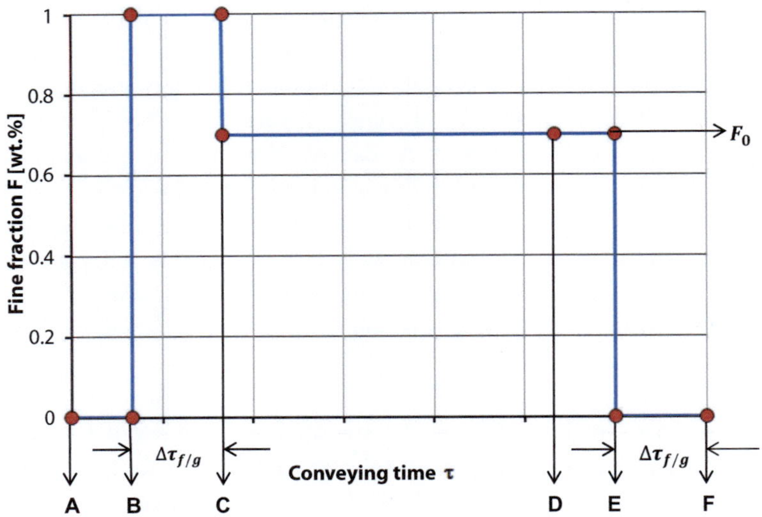

Fig. 4.33 Segregation during operation of a pneumatic conveying system. A – Start solid mass flow $\dot{m}_S$, B – first fine material "*f*" exits at the pipe end, C – first coarse material "*g*" exits at the pipe end, D – stop of solid mass flow $\dot{m}_S$, E – last fine material "*f*" exits at the pipe end, F – last coarse material "*g*" exits at the pipe end, $F_0 = \dot{m}_{S,f}/\dot{m}_S$ – target mass fraction of fine material "*f*" in the product

this is not possible, a significant increase in the conveying gas velocity is required to enhance the driving forces $(F_{\Delta p} + F_\tau)$.

- **Conveying vertically downward**: When the conveying pipe section is bent back into the horizontal at the end of such a section, this should be done using a bend with a large radius R to avoid a backlog of bulk material. Recommendation: $(R/D_R) \gtrsim 7$.
- **Bulk material separation**: Starting from a deposition-free pneumatic conveying, the following is described for a solid material, exemplarily consisting of two monograin fractions with the diameters $d_{S,f}$ and $d_{S,g}$ as well as $d_{S,f} \ll d_{S,g}$, estimated how its composition changes as a result of transport through the conveyor pipe. Figure 4.33 shows the idealized concentration profile of the fine fraction $F = \dot{m}_{S,f}/\dot{m}_S$, $\dot{m}_S$ = outgoing mass flow, in the product depending on the conveying time during start-up, continuous conveying, and shutdown of a conveying system. The mass fraction of the fine fraction in the un-conveyed bulk material is F_0. Both during start-up and shutdown, due to the different transport speeds of the two grain fractions $(\rightarrow \overline{u}_{S,f} \gg \overline{u}_{S,g})$ and the resulting different transit times through the conveying line $(\rightarrow \tau_g \gg \tau_f)$ it comes to segregation. This do not occur in continuous operation, i.e., after both the fine and coarse fractions have reached the end of the line during start-up until the stop of the supplied solid flow. In the case of the equation

$$\Delta\tau_{f/g} = \tau_g - \tau_f = \frac{L_R}{\overline{u}_{S,g}} - \frac{L_R}{\overline{u}_{S,f}} \cong \frac{L_R}{\overline{v}_F} \cdot \left(\frac{1}{\overline{C}_g} - \frac{1}{\overline{C}_f}\right) \tag{4.101}$$

with:

τ_f, τ_g	Transit times of fine and coarse grains through the conveying line,
L_R	Total length of the conveying line,
$\overline{u}_{S,f}, \overline{u}_{S,g}$	Average velocities of fine and coarse grains in the pipeline
$\overline{C}_f, \overline{C}_g$	$(= \overline{u}_{S,f/g}/\overline{u}_F \cong \overline{u}_{S,f/g}/v_F)$, velocity ratio of solid/gas,

calculated and in Fig. 4.33 by "$\Delta\tau_{f/g}$" marked periods, only fine material leaves during startup, $F = 1$, while only coarse material leaves during shutdown, $F = 0$, the conveying pipe, in between, the target composition F_0 is established due to mass conservation. The respective reject/deficiency quantities $F \neq F_0$ can be calculated with

$$\text{Startup, } F = 1: \Delta m_S = \Delta m_{S,f} = \dot{m}_{S,f} \cdot \Delta\tau_{f/g}$$

$$\cong \dot{m}_S \cdot F_0 \cdot \frac{L_R}{\overline{v}_F} \cdot \left(\frac{1}{\overline{C}_g} - \frac{1}{\overline{C}_f}\right) \tag{4.102}$$

$$\text{Departure, } F = 0: \Delta m_S = \Delta m_{S,g} = \dot{m}_{S,g} \cdot \Delta\tau_{f/g}$$

$$\cong \dot{m}_S \cdot (1 - F_0) \cdot \frac{L_R}{\overline{v}_F} \cdot \left(\frac{1}{\overline{C}_g} - \frac{1}{\overline{C}_f}\right) \tag{4.103}$$

Analogous calculations of the segregation of bulk materials with a broad particle size distribution through conveying lines with horizontal and vertical pipe elements as well as deflections are practically hardly feasible, since the velocities $(\overline{u}_{S,f}, \overline{u}_{S,g})$ cannot be calculated or can only be estimated very inaccurately due to the interactions between the grain fractions. There is at least the possibility to estimate the approximate segregation periods from the transit time difference $\Delta\tau_{f/g}$ of the finest and coarsest fraction of the currently present bulk material.

As can be seen from the above considerations and Fig. 4.33, a low-segregation pneumatic bulk material conveying requires both a deposit-free and a largely continuous transport. Batch-wise solid feeding into the conveying line, e.g., through pressure vessels, as well as frequent product changes, switching, stops, etc., should be avoided.

Further problem-causing factors will be discussed at the appropriate point.

4.11 Staggering of Conveying Lines

Pneumatic conveying lines are graduated in diameter to counteract an excessive increase in velocity in the conveying direction resulting from the expansion of the conveying gas. In particular, with larger conveying pressure differences, gas expansion in unstaggered conveying pipes = lines with a consistently constant diameter leads to unnecessary or undesirably high gas and thus solid velocities. This causes

- an additional increase in pressure loss and thus the power requirement of the conveying,
- increased conveying line wear,
- greater stress on the conveyed material, i.e., more grain breakage and grain abrasion.

Experience from practice shows that it makes sense to stagger also dense phase conveying, especially for fine-grained solids, in terms of reducing the drive power requirement [67], although their velocity level would not actually require staggering. The energetically advantageous strand conveying that occurs with this material class can be maintained by pipe expansion along the entire conveying path, see Fig. 4.36. Since there is a drop in gas velocity behind each staggering point and certain minimum limit velocities must not be undercut, the design of such staggerings, i.e., the determination of the starting position and the size of the expansion diameter, requires special care.

4.11.1 Calculation Model

First, staggerings in horizontal conveying pipe sections are considered. A staggered conveying line is simplified as a series connection of independent, i.e., with correct design, non-interactive, individual pipes. At each staging pipe beginning "i", the minimum horizontal conveying gas velocity $v_{F,min,i}$ defined by (4.13), $v_{F,min,i} = K_v \cdot D_{R,i}^k / \varrho_{F,i}^l$, must be exceeded. With $(R_F \cdot T_A) =$ const. along the transport path, instead of (4.13), (4.12), $v_{F,\mathrm{min},i} = K_v^* \cdot D_{R,i}^k / p_i^l$, can be used. The operating variables indexed with "i" are those at the beginning of each i-th stage tube. To the current conveying limit $v_{F,\mathrm{min},i}$ is a distance $\Delta v_{F,i} = (v_{F,i} - v_{F,\mathrm{min},i})$ to be maintained, depending, among other things, on the type of lock system. It can be shown that energetically optimal conveyances = deliveries with minimal energy input and simultaneously stable operating states are achieved when the condition $\Delta v_{F,i} = \Delta v_F =$ const. is maintained at all staggering points [67]. To adjust Δv_F, because the gas velocity $v_{F,i}$ is increased again in the conveying line behind each staggering point due to the further pressure loss, after a conveying distance $\Delta L_{\mathrm{opt,i}}$ the gas velocity can be reduced to $v_{F,i+1} = (v_{F,\mathrm{min},i+1} + \Delta v_F)$ again by increasing the pipe diameter, see Fig. 4.34. This implies, among other things, that to maintain the condition $\Delta v_{F,i} = \Delta v_F =$ const. only a certain number of stepped pipes N_{opt} can be installed in the considered pipeline, which depends on the respective operating conditions. For $N_i < N_{\mathrm{opt}}$ it follows that at one or more of the stepped pipe beginnings $\Delta v_{F,i} > \Delta v_F$, for $N_i > N_{\mathrm{opt}}$ accordingly $\Delta v_{F,i} < \Delta v_F$. The second case is the more critical one in terms of operational technology. The determination of the optimal staggering is described below.

Starting from a conveying line initially calculated without staggering for the given solid mass flow rate $\dot{m}_S$, this conveying line is staggered, if technically reasonable. For

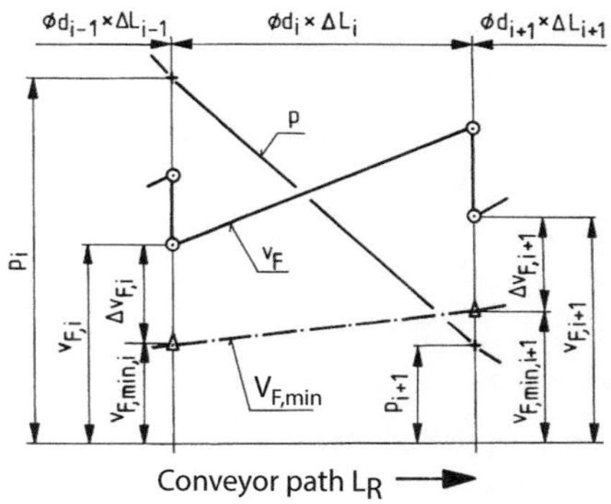

Fig. 4.34 Staggering of a pneumatic conveying line

this purpose, the calculated pipe diameter D_R^* can be expanded using an existing standard pipe series in the conveying direction or, alternatively, reduced at the beginning of the line and expanded in the rear pipe area. In the first case, the gas mass flow remains approximately constant, $\dot{m}_F \cong \dot{m}_F^*$, while the line pressure loss decreases, $\Delta p_R < \Delta p_R^*$. For the second solution, the following applies: $\dot{m}_F < \dot{m}_F^*$, $\Delta p_R \cong \Delta p_R^*$.

Starting with the values for the beginning of the conveying line, $i = 1$, selected operating conditions ($p_1 = \Delta p_R + p_E, D_1,\ v_{F,1}, \Delta v_{F,1} = \Delta v_F$) can, see Fig. 4.34, the optimal staggered pipe lengths $\Delta L_{\mathrm{opt,i}}$ along the conveying route be calculated as follows: continuity equation, ideal gas law, and (4.12) provide the dependency for the beginning of the $(i+1)$-staggered pipe:

$$v_{F,i+1} = v_{F,i} \cdot \frac{p_i}{p_{i+1}} \cdot \left(\frac{D_i}{D_{i+1}}\right)^2 = (v_{F,\min,i+1} + \Delta v_F) = \left(K_v^* \cdot \frac{D_{i+1}^k}{p_{i+1}^l} + \Delta v_F\right) \tag{4.104}$$

The still unknown pressure p_{i+1} is determined iteratively from (4.104). With its knowledge, the pressure difference $(p_i - p_{i+1})$ is defined and the length $\Delta L_{\mathrm{opt,i}}$ of the i-th staggered tube can be calculated. A first estimate for $\Delta L_{\mathrm{opt,i}}$ provides the integrated form of (4.44)

$$\Delta L_{\mathrm{opt,i}} = \frac{\pi^2}{16 \cdot R_F \cdot T_M} \cdot \frac{1}{\lambda_S} \cdot \frac{D_i^5}{\dot{m}_S \cdot \dot{m}_F} \cdot (p_i^2 - p_{i+1}^2) \tag{4.105}$$

in which $\lambda_S \cdot \mu \gg \lambda_F$ and $\lambda_S = \mathrm{const.}$ are considered, as well as the following sizes were replaced as follows: $\mu = \dot{m}_S/\dot{m}_F$, $v_F^2 \cdot \varrho_F = ((\dot{m}_F/D_i^2 \cdot \pi/4)^2/\varrho_F)$, $\varrho_F = p/(R_F \cdot T_M)$. Since (4.105) only considers straight horizontal pipe sections, it must be checked whether the determined length contains deflections and/or height sections. Their additional pressure loss shortens $\Delta L_{\mathrm{opt,i}}$. This must be calculated with an appropriately extended equation (4.105), possibly iteratively.

The described procedure is successively processed in the conveying direction from the beginning to the end of the pipeline and provides the optimal stepped pipe number

N_{opt}. It should be taken into account, among other things, that pipe expansions must be avoided directly before and after deflections and may only be installed after the solid matter has been re-accelerated behind a bend, see (4.88) and Table 4.3. The same applies to staggerings in vertical sections.

Practical relevance: Fig. 4.35 shows the results of conveying experiments with a coarse lignite ($\rightarrow d_S \cong (20\text{–}6300)\,\mu\text{m}$, $d_{S,50} \cong 1.0\,\text{mm}$) at a given 3-times staggered conveyor line with variation of the conveying gas quantity at a constant conveying pressure difference of $\Delta p_R \cong 1.75$ bar on the line. Plotted are the idealized curves of the gas velocities v_F, the local pressures p and the associated minimum velocities $v_{F,min,i} = v_{F,A,\min}$ along the conveying path L_R, here a total of 472 m. The $v_{F,A,\min}$-dependence according to (4.12) was known for the present bulk material from systematic measurements on unstaggered conveying lines with different diameters [68]. The curves shown in Fig. 4.35 were evaluated and presented only after the completion of the test series based on the measurement results ($\rightarrow$ among other things, pressures along the line). During the tests, only the conditions at the beginning of the conveying line were observed and evaluated.

The initially conducted experiment "A" with a gas starting velocity of approx. $v_{F,A} = v_{F,1} \cong 12.0\,\text{m/s}$ and a distance of $\Delta v_{F,1} \cong 3.5\,\text{m/s}$ to the absolute conveying limit showed a stable conveying behavior. $v_{F,A}$ was therefore reduced in experiment "B" to $v_{F,1} \cong 11.0\,\text{m/s}$, thus $\Delta v_{F,1} \cong 2.5\,\text{m/s}$, resulting in significantly larger pressure pulsations. The further reduction of the gas starting velocity to $v_{F,A} \cong 10.0\,\text{m/s}$,

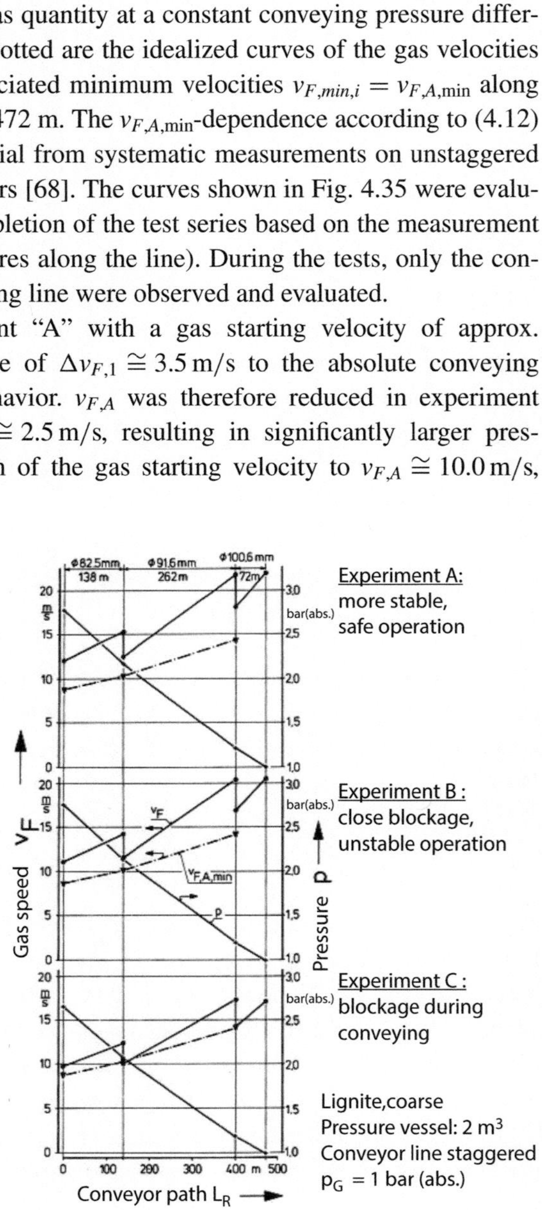

Fig. 4.35 Pressure and velocity profiles at different gas velocities in a staggered conveying section (coarse lignite)

i.e. $\Delta v_{F,1} \cong 1.5\,\mathrm{m/s}$, in another attempt "C" resulted in a sudden blockage of the conveyor line shortly after the start of the system. As already seen in Fig. 4.35, attempt "B", it is obviously not the gas velocity $v_{F,1}$ at the beginning of the line that is critical, but the one $v_{F,2}$ at the beginning of the second stage pipe. This falls below the local conveying limit in attempt "C", $v_{F,2} < v_{F,\min,2}$. By tapping the line, the plug could be located just behind the beginning of the second staggered pipe. The length ΔL_1 of the first staggered pipe is shorter in the considered example than the corresponding optimal staggering length $\Delta L_{\mathrm{opt},1}$, which at the same time is used to set a constant Δv_F, here: $\Delta v_{F,2} = \Delta v_{F,1}$, representing the minimum required pipe length. A reduction of the optimal stepped pipe lengths is possible if, by generally uneconomical raising of the overall gas velocity level at the critical step point, an appropriate distance Δv_F to the conveying limit is set, see Fig. 4.35, experiment "A".

Comparison with other calculation approaches: From the literature, two basic stepping criteria are known [69]. In [70] it is suggested to increase the pipe diameter for conveying fine-grained bulk materials with higher solid loadings when a critical dynamic pressure $p_{\mathrm{dyn,crit}}$, (4.106) in Box 4.6, is exceeded. In granulate conveying, according to [71], a pipe staggering at exceeding a critical Froude number Fr_{crit}, (4.108) in Box 4.6, leads to safe operating conditions. Calculation is done in both cases in the flow direction. Both criteria can be brought into an equivalent form of equations (4.12) or (4.13), i.e., the staging criterion used above, (4.107) and (4.109) in Box 4.6. A comparison of the diameter and pressure exponents of these equations with the measured exponents k and l from (4.12) shows that both provide essentially the same statements about the behavior of the respective bulk material class, see Sect. 4.6.2. Measured diameter exponents of fine-grained good flowing solids with high gas holding capacity in (4.12) tend to approach $k = 0$, while for coarse-grained solids with a narrow grain size distribution and thus high gas permeability ($\rightarrow$ granules) the pressure exponent tends towards $l = 0$. Alternating exponents can take values deviating from k or $l = 0.5$ [5]. The staggering criterion used here according to (4.12) thus includes the criteria mentioned above and clarifies that they can only be used in the area of the material classes assigned to them. (4.12) obviously represents a generalization of known stepping criteria.

Box 4.6 contains examples of further stepping criteria and their recommended applications. The approach according to [72], (4.111), fits well into the trends described above: coarse-grained solid, but with a broad grain size distribution; expected: pressure exponent $l \rightarrow 0$, recommended: $l = 0.15$. The criterion according to [73] for coal dust, (4.113), contradicts the previous explanations and own measurements: fine-grained and well-flowing solid; expected and previously measured: diameter exponent $k \rightarrow 0$, but recommended: $k = 0.5$. Consequence: with increasing conveying pipe diameter, steadily growing, oversized safety distances to the conveying limit, no optimal staggering point determination possible, etc.

Box 4.6: Staggering Criteria

Muschelknautz/Wojahn [70]:

→ fine-grained solids, high loadings

$$p_{\text{dyn,crit}} = \frac{\varrho_{F,\text{crit}}}{2} \cdot v_{F,\text{crit}}^2 = K_1 \tag{4.106}$$

$$v_{F,\text{crit}} = \sqrt{\frac{2 \cdot R_F \cdot T_M \cdot K_1}{p_{\text{crit}}}} = K_v^* \cdot \frac{D_R^0}{p_{\text{crit}}^{0.5}} \tag{4.107}$$

Bohnet [71]:

→ Granules

$$Fr_{\text{crit}} = \frac{v_{F,\text{crit}}^2}{g \cdot D_R} = K_2 \tag{4.108}$$

$$v_{F,\text{crit}} = \sqrt{K_2 \cdot g \cdot D_R} = K_v^* \cdot \frac{D_R^{0.5}}{p_{\text{crit}}^0} \tag{4.109}$$

Roski [72]:

→ Natural anhydrite (0–10) mm

$$SK = Fr_{\text{crit}} \cdot \left(\frac{\varrho_{F,\text{crit}}}{\varrho_{F,0}}\right)^{0.3} \tag{4.110}$$

$$v_{F,\text{crit}} = \sqrt{SK \cdot g \cdot D_R} \cdot \left(\frac{p_0}{p_{\text{crit}}}\right)^{0.15} = K_v^*(\mu) \cdot \frac{D_R^{0.5}}{p_{\text{crit}}^{0.15}} \tag{4.111}$$

with:

$$SK = f(\mu) = 114.4 + 159.2 \cdot \mu \quad \rightarrow \quad \text{Start of solid deposition} \tag{4.112}$$

Wypich/Reed [73]:

→ Coal dust

$$Fr_{\text{crit}} = 36 \tag{4.113}$$

leads to (4.109).

(4.12) takes into account the special properties of a conveyed material through the material-specific exponents k and l as well as the also bulk material-specific pre-factor K_v^*. These must be determined by experiment, see Sect. 4.6.2. Estimates/calculations without conveying tests are possible to a limited extent, according to the explanations in Sect.

4.6.2. If only a single limit value $v_{F,A,\min}$ is available from an operating plant for a bulk material, then an extrapolation to other operating and plant conditions should be carried out with at least $k = l = 0.5$.

4.11.2 Checks/Applications

In Fig. 4.36, the conveying characteristics of a limestone powder determined with different pressure differences Δp_R and gas velocities $v_{F,A}$ are compared for two equally long conveying lines. The measurement points underlying the displayed fitting curves are not plotted for the sake of clarity. The conveying line 3 is unstaggered, while the staggered conveying line 4 has the same inner pipe diameter, $D_1 = 82.5$ mm, as the ungraded line, for about three-quarters of its total length. After that, line 4 is increased to $D_2 = 91.6$ mm and $D_3 = 100.5$ mm. Significant advantages of the staggered route compared to the unstaggered one are only noticeable at relatively low conveying gas speeds $v_{F,A}$, i.e. in the slow/dense phase conveying range near the conveying limit $v_{F,A,\min}$. With increasing gas speed, the individual curves converge, and in some cases, lower solid throughputs $\dot{m}_S$ are realized in the staggered line than in the unstaggered one. This is surprising because the average diameter of the staggered conveying route, $\overline{D}_R \cong 86.6$ mm, calculated as volume-equivalent diameter, is larger than the diameter of the unstaggered line, $D_R = 82.5$ mm, and from this point of view, a throughput increase in the entire investigated speed range would have been expected. The geometric design of the staggering (→ position and size of the diameter jumps) is obviously satisfactory, as a retroactive effect on the minimum conveying gas speeds at the beginning of the line is not observed here.

In Fig. 4.37, the characteristic curves of a fly ash are plotted, which in all cases were conveyed with the same pressure difference of $\Delta p_R = 2.0$ bar on the conveying line over different distances, both through unstaggered lines with a diameter of $D_R = 82.5$ mm as well as through staggered routes with an initial diameter of $D_1 = 82.5$ mm which was

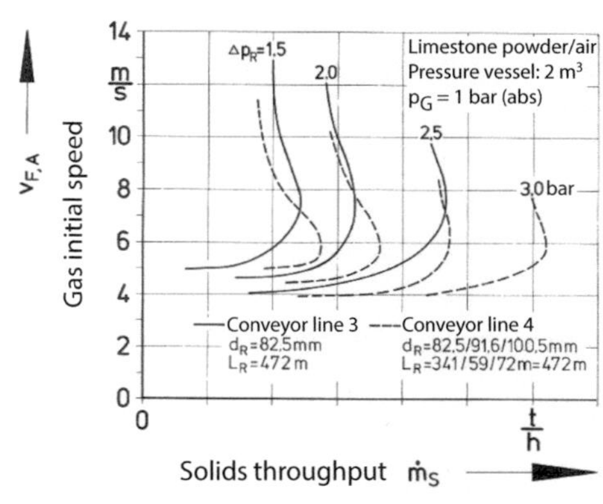

Fig. 4.36 Comparison of an unstepped conveyor line with a stepped conveyor line

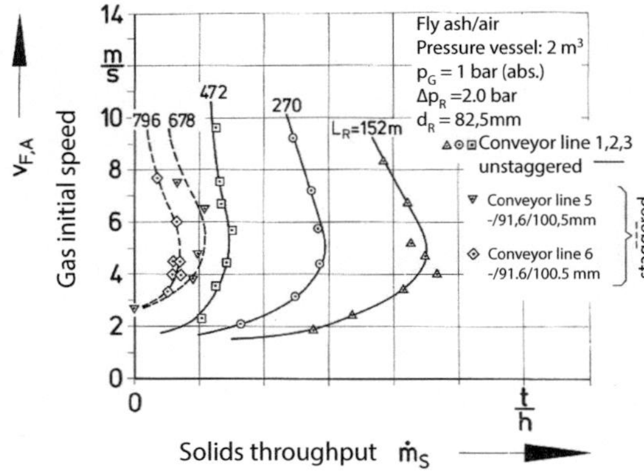

Fig. 4.37 Comparison of the conveying characteristics $\Delta p_R = 2.0$ bar of fly ash over various long staggered and non-staggered conveying routes, conveying routes available in the test center

extented over approximately half the conveying distance. Figure 4.37 first confirms (4.12), $v_{F,A,\min} = K_v^* \cdot D_R^k / p_R^l$, which states that in unstepped lines, the conveying limit speed $v_{F,A,\min}$ is constant for a given (D_R, p_R) regardless of the conveying distance. The fact that $v_{F,A,\min}$ is significantly increased in the two staggered conveying lines, is indicating an incorrect stepping for the considered application case.

Recalculations with the model described above show that in both cases the length of the second staggered tube is too short, i.e., less than $\Delta L_{\text{opt},2}$, see also Fig. 4.35. On the other hand, it can be seen from Fig. 4.37 that there is still an increase in the solid throughput in the velocity range $v_{F,A} \cong (4\text{-}7)\,\text{m/s}$ compared to a comparable non-staggered line. This advantage/gain is quickly lost at higher gas velocities.

Figure 4.38 shows the result of an analysis of an existing operating plant carried out on behalf of a customer. This conveyed the limestone grit already known from Fig. 4.17. Essential parameters: $d_{S,50} = 460\,\mu\text{m}$, $\varrho_S = 2750\,\text{kg/m}^3$, $\varrho_{SS} = 1320\,\text{kg/m}^3$. The problem with the system was that as it approached a critical solid throughput, which was significantly below the guaranteed throughput, the conveyance pulsed more and more, leading to the temporary formation of deposits, whose sudden re-acceleration, among other things, caused pipe supports to break off and regularly produced blockages when the critical solid throughput was reached. The latter operating state is shown in Fig. 4.38. The pipe blockages were detected in the initial area of the second staggered pipe $(D_2, \Delta L_2)$. Since the limestone to be conveyed was a sharp-edged, highly abrasive bulk material, the conveying route was staggered and executed using internally hardened steel pipes, which apparently were only available in limited diameters. At the same time, the system supplier had implemented the first staggered pipe as a short acceleration section with a relatively small diameter D_1 to reduce the conveying gas requirement. The area ratio at the transition of the staggered pipes 1 → 2 was only $(D_1/D_2)^2 = (125\,\text{mm}/150\,\text{mm})^2 = v_{F,2}/v_{F,1} \cong 0.69$, i.e., the local empty pipe gas velocity drops to about 69%. This, in combination with the too short length ΔL_1 of the first staggered pipe, led to the minimum conveying gas velocity being undercut. $v_{F,2} < v_{F,\min,2}$, and thus to the line blockages (→ the coefficients/exponents of the $v_{F,A,\min}$

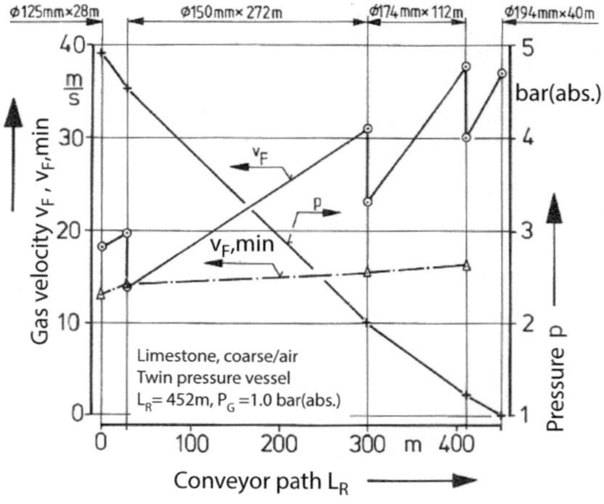

Fig. 4.38 Incorrect staggering of an operating plant, bulk material: coarse limestone

-equation (4.12) were known from older investigations!). It is evident that the problem can be solved by appropriately increasing the conveying gas flow $\dot{m}_F$, i.e., creating a gap $\Delta v_{F,2} > 0$ between gas velocity $v_{F,2}$ and corresponding limit velocity $v_{F,\min,2}$ or by extending the first staggered pipe ΔL_1. Since the gas supply of the plant was provided by a generously sized compressed air network, the conveying gas flow $\dot{m}_F$ was initially raised to a newly calculated value to maintain production, and only during a planned later operational shutdown was the length ΔL_1 of the first stepped pipe increased to approximately 2.5 times the existing value, and the conveying gas flow is reduced again. In both cases, the guaranteed solid throughput $\dot{m}_S$ was reliably achieved.

In Fig. 4.39, measured and calculated conveying characteristics of a fine-grained cement are compared using the design models described above. The investigations were carried out on two conveying lines: an unstaggered one with dimensions $D_R = 82.5\,\text{mm} \times L_R = 152\,\text{m}$ and an equally long, parallel, staggered line, whose diameter was expanded after $\Delta L_1 = 80\,\text{m}$ from $D_1 = 82.5\,\text{mm}$ to $D_2 = 100.0\,\text{mm}$. The pipe cross-section thus increases at the staggering point by approximately 47% and the gas velocity is reduced to approximately 68 % of its original value. This (too) large jump was deliberately chosen. All calculations were based on existing standard cement values and dependencies. Both for the unstaggered and the staggered line, there is a good agreement between calculated conveying characteristics and the corresponding measured values, see Fig. 4.39. In particular, the effects of the staggering of the conveying line are correctly reproduced by the calculation model from Sect. 4.11.1: raising the minimum permissible gas velocity at the beginning of the conveying line due to the too large diameter jump at the staggerng point, $\Delta v_{F,2} < \Delta v_{F,1}$, as well as the increase in solid throughput $\dot{m}_S$ in the entire investigated velocity range.

Based on the positive results with the calculation approach for the design of stepped conveying lines, a series of basic model calculations were carried out with this. Figure 4.40 shows related results with the measured characteristic data of a real fly ash,

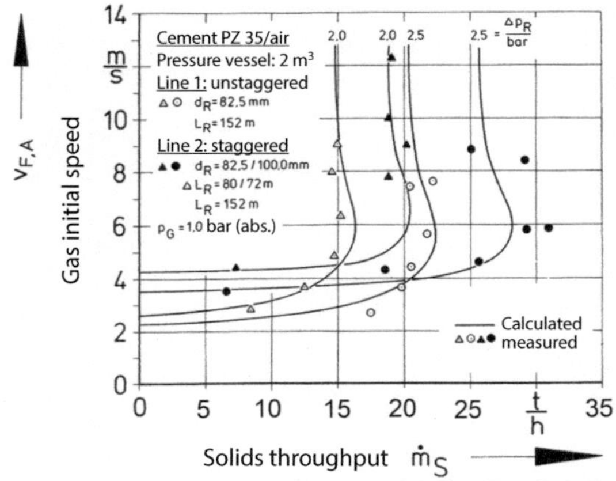

Fig. 4.39 Comparison between measured and calculated conveying characteristics, bulk material: cement

which served the analysis of different staggering designs of a $L_R = 152\,\text{m}$ long conveying distance. The conveying pipe initial diameter is in all cases $D_R = D_1 = 82.5\,\text{mm}$, the line pressure loss $\Delta p_R = 2.0\,\text{bar}$, the pressure at the end of the line $p_E = 1.0\,\text{bar}$. Line 1 is unstaggered, Line 2 is after $\Delta L_1 = 100\,\text{m}$ expanded to $D_2 = 89.0\,\text{mm}$ and used for all described operational states. The solids throughput $\dot{m}_s$ of the 2-times staggered conveying route is only in a gas velocity range near the conveying boundary greater than that in the unstaggered line 1. At higher velocities they are nearly identical. Reason: from the position of the staggering point in the back part of the pipe results $\Delta v_{F,2} > \Delta v_{F,1} = \Delta v_{F,A}$ and therefore $\Delta L_1 > \Delta L_{opt,1}$. Therefore due to the high conveying velocities at the end of the first and beginning of the second staggered pipe and in connection with the special pressure loss characteristic of the conveyed fly ash, the advantage of the expansion of the pipe diameter is compensated. In the range of lower gas velocities the staggering carried out is close to the optimal design, i. e. $\Delta L_1 \cong \Delta L_{opt,1}$.

If with the same diameter gradation, $D_1 = 82.5\,\text{mm}/D_2 = 89.0\,\text{mm}$, the position of the staggering point is chosen such that the condition $\Delta v_{F,2} = \Delta v_{F,1} = \Delta v_{F,A}$ is maintained at all operating speeds $v_{F,A}$ (→ "optimal" staggering), as the characteristic curve of line 3 in Fig. 4.40 shows, the solid throughput $\dot{m}_S$ can be increased in the entire speed range compared to both the unstaged line 1 and line 2. The staggering point is located at gas velocities $v_{F,A}$ near the conveying limit close to the end of the line and moves with increasing $v_{F,A}$ towards the beginning of the line, see also Fig. 4.41.

Line 4 in Fig. 4.40 represents the technical implementation of a "trumpet tube". In compliance with the boundary condition $\Delta v_{F,i} = \Delta v_{F,A} = \text{const.}$, the maximum possible number of staggered tubes with the shortest = optimal length was incorporated into the line. Pipe series used: DIN 2448. Figure 4.40 shows that this measure significantly increases the solid mass flow rate $\dot{m}_S$ throughout the entire velocity range. Compared to the unstaggered line 1, increases of about 40 % are recognizable. The number and length of the associated staggered tubes can be taken from Fig. 4.41 for a given gas velocity $v_{F,A}$. This illustrates that with increasing velocity $v_{F,A}$, i.e., increasing distance $\Delta v_{F,A}$ to

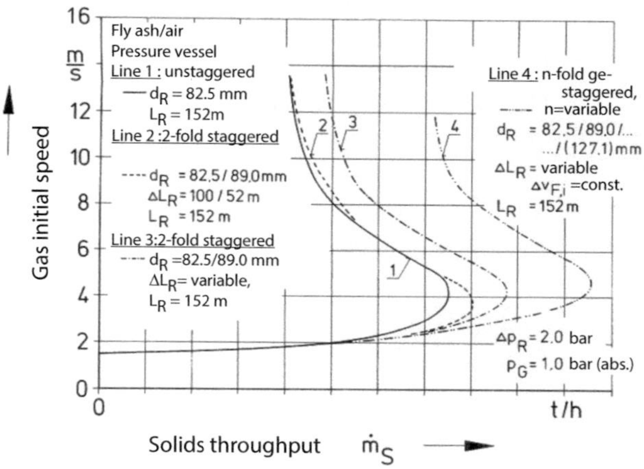

Fig. 4.40 Effects of different designs of a conveyor pipeline staggering, bulk material: fly ash

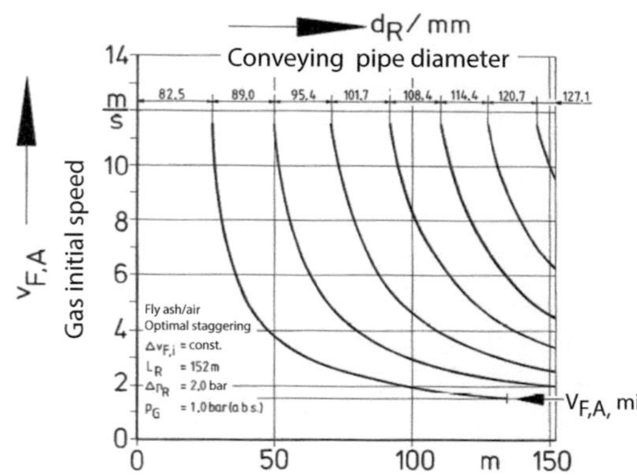

Fig. 4.41 Staggering of line 4 in Fig. 4.40

the limit velocity $v_{F,A,\min}$, which shifts the position of the individual staggering points towards the beginning of the conveying line and thereby increases the number of possible staggered pipes. Both can also be clearly seen from (4.104) and (4.105). The solid throughput $\dot{m}_S$ of the unstaggered line 1 = characteristic 1 in Fig. 4.40 can be increased to a maximum of characteristic 4 by means of conveying line staggering. The "trumpet pipe" represents the energetic staggering optimum from a conveying technology point of view. In practice, other factors must be taken into account, including the increasing costs for the pipeline and its support as the number of staggered pipes increases.

The above-considered staggerings were carried out in such a way that an ungraded conveying line was widened in diameter towards its end. This leads, given the pressure difference Δp_R to an increase in the solid throughput $\dot{m}_S$ or, given a predetermined throughput $\dot{m}_S$ to a reduced pressure loss Δp_R. In practice, the solid throughput $\dot{m}_S$ is usually determined by the task at hand. In this case, in general, it is more favorable to reduce the diameter of the unstaggered conveying line in the front part while maintaining the calculated pressure difference Δp_R and to widen it in the rear area, i.e., to treat the diameter of the unstaggered pipe as an average resistance-equivalent diameter. This generally leads to a reduction in the conveying gas demand $\dot{m}_F$ and, as a result, to a reduction in investment costs due to smaller pipe diameters, a smaller pressure generator, and a smaller receiving filter.

Note: Some of the pipeline pressure losses calculated in the preceding section were not determined using the calculation approach presented in Sect. 4.7, but with the scale-up model introduced only in Sect. 5.3. The reason is the simplified determination of resistance coefficients. This does not change the procedure described in Sect. 4.11.1 when applying the staggering model.

4.12 Energy Optimization

As already emphasized several times, the energy input required for pneumatic transport of bulk material is generally significantly greater than that of a comparable mechanical conveyance. This is exemplified in Fig. 4.42 [74]. Based on cement transports, the respective drive power requirements of a belt conveyor P_{Belt} are related to the power requirement figures P_{Pneu} of energetically optimized pipe conveyances (→ solid material feed with pressure vessel and screw feeder, conveyances in the dense phase range) as well as those of fluidized flow channel (= airslide) conveyances are shown. In the latter, the bulk material is transported in the form of a highly loosened fluidized bed over inclined sections in the conveying direction. Figure 4.42 illustrates that the energy requirement of a fluidized flow channel conveyance is approximately identical to that of a belt conveyor, while in pneumatic pipe transport, more than twenty times the energy input of the corresponding belt conveyor must be expended. This relation is maintained for other bulk materials [75] and increases considerably for lean phase pneumatic conveying and poorly designed systems. Table 4.5 shows a comparison in this regard, which also includes other mechanical transport systems [76].

The selection of an initially suitable, but also economical, conveying principle for a given task must be made taking into account the entire environment, i.e., based on the properties of the bulk material, the length and course of the conveying route, structural conditions, relevant safety and environmental protection regulations, as well as special operating conditions, e.g., conveying under protective gas. Alternative solutions must be evaluated and selected through a total cost analysis. Essential influencing factors are, on the one hand, the energy costs incurred during the entire life of the plant and, on the other hand, the one-time investment costs. As a general rule, of two conveying systems that fulfill the same task, the mechanical conveying system requires higher investment costs compared to a pneumatic system, while its energy consumption, as shown above, is generally significantly lower. The energy consumption of pneumatic conveying principles is also considerably more strongly influenced by the chosen/predetermined

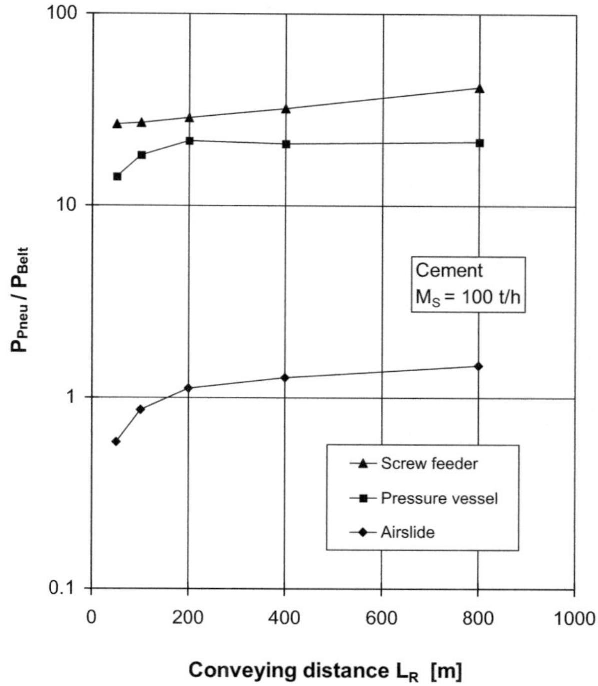

Fig. 4.42 Energetic comparison of conveying 100 t/h cement over different conveying distances, reference basis: belt conveyor

Table 4.5 Rough energetic comparison of pneumatic and mechanical conveying systems, horizontal conveying

Mechanical conveying system	P_{Pneu}/P_{Mech}
Belt conveyor, tubular belt conveyor	approx. 133
Trough chain conveyor, pocket wheel conveyor	approx. 5.7
Screw conveyor	approx. 1.3

operating and boundary conditions as well as the bulk material properties than that of mechanical conveyors. To keep pneumatic conveying systems competitive, their energy consumption required for safe operation must be reduced as a priority. The following describes the dependencies of the energy/power consumption of pneumatic conveyors.

4.12.1 Basic Relationships

The energy consumption P_{Pneu} of a pneumatic conveying system consists of the power consumption of the pressure generator P_V, the bulk solid lock P_{Sch} and the dedusting system P_{Est}:

$$P_{\text{Pneu}} = P_V + P_{\text{Sch}} + P_{\text{Est}} \tag{4.114}$$

P_{Est} is generally small compared to P_V, while P_{Sch} can take on a considerable size for some lock types, e.g., high-speed screw feeders, while for others, e.g., rotary valves and pressure vessels, only very low values are required or it approaches zero. The actual transport energy is supplied to the system with the conveying gas, i.e., with the portion of the pressure generator power P_V reduced by its efficiency η_V, which is supplied. It holds:

$$P_V = P_V(\Delta p_V, \dot{V}_V) = \frac{P_{th,V}}{\eta_V} \propto \dot{m}_S \cdot g \cdot L_R \cdot \beta_R \tag{4.115}$$

with:

$\Delta p_V, \dot{V}_V$ Pressure difference and suction volume flow of the pressure generator,
$P_{th,V}$ theoretical pressure generator power at $\eta_V = 1$.

In the summary resistance coefficient β_R the entire "physics" of the conveying process is included. $(\Delta p_V, \dot{V}_V)$ can deviate from the values required for the actual conveying Δp_R and $\dot{V}_F$ or $\dot{m}_F$. For example, in the case of rotary valve feeders, leakage gas losses can occur and due to $\Delta \dot{V}_{\text{Leck}}$ the volume flow $\dot{V}_V = (\dot{V}_F + \Delta \dot{V}_{\text{Leck}}) > \dot{V}_F$ is required, while for adjusting the gas distribution on pressure vessels, the additional pressure loss Δp_{Vert} is needed and thus must be compressed to $\Delta p_V = (\Delta p_R + \Delta p_{\text{Vert}}) > \Delta p_R$. By relating the total drive power P_{Pneu} to the actual transport/useful power $(\dot{m}_S \cdot g \cdot L_R)$ the performance coefficient

$$\psi = \frac{P_{\text{Pneu}}}{\dot{m}_S \cdot g \cdot L_R} \tag{4.116}$$

is defined. In industrial practice, however, the reference length of $L_R = 100\,\text{m}$ related specific power requirement

$$P_{sp} = \frac{P_{\text{Pneu}}}{\dot{m}_S \cdot L_R} \quad \left[\frac{\text{kW h}}{\text{t} \cdot 100\,\text{m}}\right] \tag{4.117}$$

with the units specified in (4.117) is enforced. This will be used in the following.

For each operating point $(\dot{m}_S, \Delta p_R, v_F)$ in the conveying diagram a drive power P_V is assigned, which results from the actual conveying data $(\dot{m}_F, \Delta p_R)$. The bulk material transport through the conveying pipe takes place largely isothermally, but the gas compression in the pressure generator is polytropic, approximately adiabatic. With $\eta_V = 1$ the theoretically specific compressor power, which is well suited for comparisons, is obtained:

$$P_{th,sp} = \frac{P_{th,V}}{\dot{m}_S \cdot L_R} = \frac{1}{\mu} \cdot \frac{p_E}{\varrho_{F,E} \cdot L_R} \cdot \frac{n}{n-1} \cdot \left[\left(\frac{p_A}{p_E}\right)^{\frac{n-1}{n}} - 1\right] \tag{4.118}$$

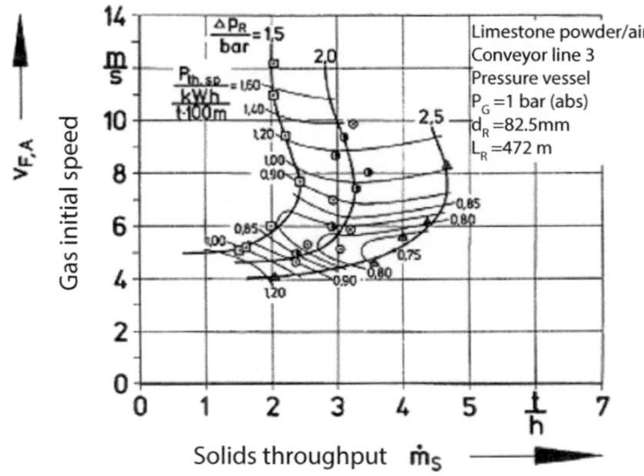

Fig. 4.43 Conveying diagram of a limestone powder conveying with $P_{th,sp}$-characteristics

As a polytropic exponent, we will further set $n = 1.3$. (4.118) illustrates that a minimal energy requirement is necessary when the loading μ becomes maximal. This operating condition occurs along a curve $\Delta p_R = (p_A - p_E) = \text{const.}$ in the modified delivery diagram, Fig. 4.2, at the contact point of the tangent from the coordinate origin to the respective considered Δp_R-curve $\rightarrow \mu \propto \dot{m}_S/v_{F,A}$, when $(\varrho_{F,A}, D_R) = \text{const.}$ In the normal state diagram, see Fig. 4.1, these operating points are to the left of the connecting line of the pressure minima of the $\dot{m}_S$-curves, i.e., at smaller gas velocities than these.

The possible energetic optimization potential for the conveying of fine-grained bulk materials is shown in Figs. 4.43 and 4.44 using concrete measurement results. Shown are limestone powder, $d_{S,50} \cong 23\,\mu\text{m}$, and fly ash, $d_{S,50} \cong 20\,\mu\text{m}$, at a counter pressure of $p_E = 1.0\,\text{bar}$ on a conveyor line with dimensions $D_R \times L_R = 82.5\,\text{mm} \times 472\,\text{m}$ determined state diagrams. In addition, lines of constant theoretical specific power requirements $P_{th,sp}$ are included. The $P_{th,sp}$-curves illustrate that the required energy input for transporting a given solid mass flow $\dot{m}_S$ can differ by more than a factor of 2 depending on the chosen conveying gas velocity $v_{F,A}$. Minimum energy requirement is established, as explained above, along the connecting line of the loading maxima of the individual Δp_R-curves. It increases towards larger and smaller gas velocities, i.e., with a sufficiently large measuring range $(\Delta p_R, v_{F,A})$, the lines $P_{th,sp} = \text{const.}$ form closed curves, comparable to the efficiency curves of a fan [27]. Since the size of $P_{th,sp}$ along the connecting line of the loading maxima of the individual Δp_R-curves change, it follows that an absolutely energy-minimal operating point $P^*_{th,sp}$ exists for the entire considered conveying diagram. In Fig. 4.43, conveyed material: limestone powder, this is obviously at high delivery pressures ($\rightarrow \Delta p^*_R > 2.5\,\text{bar}$), in the case of fly ash, Fig. 4.44, however, at significantly lower values ($\rightarrow \Delta p^*_R < 1.5\,\text{bar}$).

For a given pipeline and bulk material, the energy-minimal operating point for a conveying diagram is thus obtained as follows:

$$\begin{aligned} \text{Given: } L_R, D_R \quad &\Longrightarrow \quad \text{searched: } P^*_{th,sp} = P_{th,sp,\min} \\ &\Longrightarrow \quad \text{Operating point: } \Delta p^*_R, \dot{m}^*_S, v^*_{F,A} \end{aligned} \tag{4.119}$$

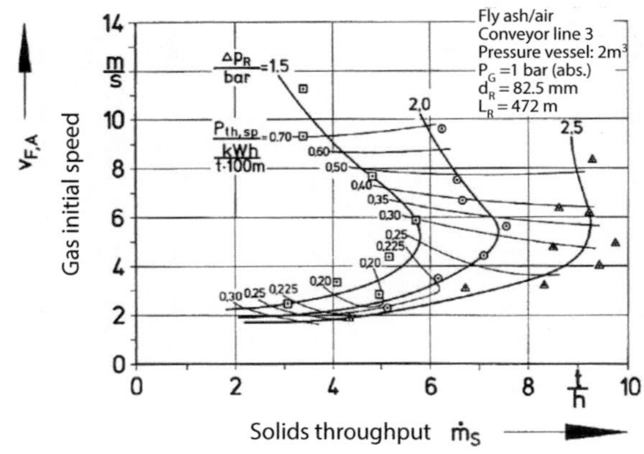

Fig. 4.44 Conveying diagram of a fly ash conveying with $P_{th,sp}$-characteristics

When creating a new conveying system, however, the following task is given:

$$\begin{aligned} \text{Given: } L_R, \dot{m}_S, \Delta v_{F,A} \quad &\Longrightarrow \quad \text{searched: } P_V^* = P_{V,\min} \\ &\Longrightarrow \quad \text{Operating point: } \Delta p_R^*, D_R^* \end{aligned} \tag{4.120}$$

The energy-minimal combination of conveying pipe pressure difference Δp_R^*, or p_V^* when considering the entire system, and the associated conveying pipe diameter D_R^* is searched. It can be shown that such overall optimal design points exist and can be realized. However, different tasks characterized by their products $(\dot{m}_S \cdot L_R)$ lead to completely different optimal solutions $(\Delta p_R^*, D_R^*)$ even for the same conveyed material. This means that a specific bulk material cannot be assigned a single characteristic energy-minimal conveying pressure Δp_R^*. In [77, 78] it is derived that Δp_R^* becomes larger with increasing product $(\dot{m}_S \cdot L_R)$ as well. Figure 4.45 shows this schematically. The position and slope of the $\Delta p_R^*(\dot{m}_S \cdot L_R)$-curves of the different bulk materials in Fig. 4.45 is significantly influenced by the dependence of the energy demand P_{th} of the conveying pipe diameter in the associated pressure loss equation ($\rightarrow \Delta p_R(D_R^y)$, includes the λ_S-dependency). The larger the exponent y, the lower the optimal pressure difference Δp_R^* for a given task. The model discussed in [77, 78] allows the inclusion of solid locks, pipe gradations, etc. in the optimization.

As shown in Sect. 4.12.2, energetic optimizations lead to significant energy-saving potentials, especially for fine-grained bulk materials, but to a lesser extent for coarse-grained products, e.g., plastic granules.

4.12.2 Applications

Figure 4.46a shows the conveying diagram of a fine-grained quartz powder, already presented in Fig. 4.16a, now extended by curves of constant theoretical specific power

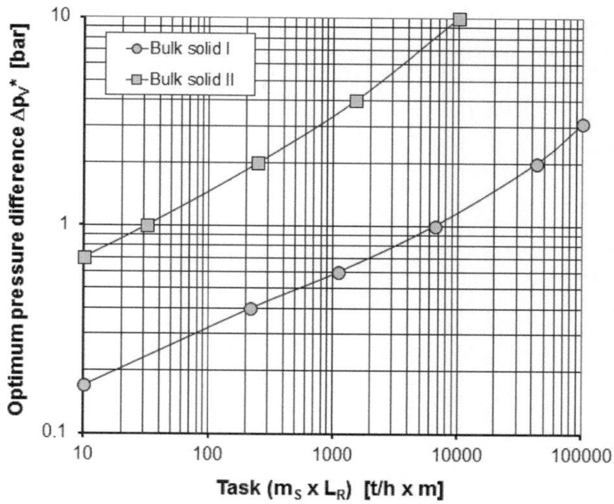

Fig. 4.45 Dependency of the energy-minimal conveying pressure p_V^* on the current task $(\dot{m}_S \cdot L_R)$, $\Delta v_{F,A} = \text{konst.}$, $p_E = 1.0\,\text{bar}$, schematic

demand $P_{th,sp}$. In Fig. 4.46b, the same diagram is shown with the loading μ instead of the solid mass flow $\dot{m}_S$ as abscissa and the curves $v_{F,A}(\mu_{\max}) =$ connecting line of the load maxima and $v_{F,A}(\dot{m}_{S,\max}) =$ connecting line of the throughput maxima. Along $v_{F,A}(\mu_{\max})$ the energy-minimal operating points of the various pressure loss curves $\Delta p_R = \text{const.}$ are arranged, which in this case are obviously too close to the conveying limit to be suitable for an operating system. In practice, energy-saving dense phase conveying is designed taking into account the respective boundary conditions and bulk material properties with gas velocities $v_{F,A}$ between $v_{F,A}(\mu_{\max})$ and $v_{F,A}(\dot{m}_{S,\max})$. The $v_{F,A}(\dot{m}_{S,\max})$-curves correspond to the connecting lines of the pressure loss minima in the normal state diagram, Fig. 4.1.

If the present quartz powder is to be conveyed, for example, with a throughput of $\dot{m}_S = 14.0\,\text{t/h}$ and $\Delta p_R = 2.0\,\text{bar}$ through the pipeline examined in Fig. 4.46, this can be done with an initial gas velocity of $v_{F,A} \cong 4.0\,\text{m/s}$, the corresponding final velocity of $v_{F,E} \cong (3\,\text{bar}/1\,\text{bar}) \cdot 4.0\,\text{m/s} = 12.0\,\text{m/s}$, a loading of $\mu \cong 140\,\text{kg}\,S/\text{kg}\,F$ as well as a specific power requirement of $P_{th,sp} \cong 0.20\,\text{kW h}/(\text{t} \cdot 100\,\text{m})$ or with $v_{F,A} \cong 11.5\,\text{m/s}$, $v_{F,E} \cong 34.5\,\text{m/s}$, $\mu \cong 48\,\text{kg}\,S/\text{kg}\,F$ and $P_{th,sp} \cong 0.60\,\text{kW h}/(\text{t} \cdot 100\,\text{m})$ occur. The first mentioned operating point represents a dense phase conveying, while the second one represents a lean phase pneumatic conveying. The energy demand of the lean phase conveying is three times greater than that of the dense phase conveying. Due to the very high gas velocities and the highly abrasive material being conveyed, significant pipeline wear will also occur in the lean phase pneumatic conveying system.

The following example demonstrates the relationships described above applied to an industrial plant. It consists of two identical single pressure vessel systems, designed to transport $\dot{m}_S = 15.0\,\text{t/h}$ of fly ash generated in the electrostatic precipitator of a coal-fired power plant over $L_R \cong 400\,\text{m}$ from a collection container to loading silos. Details

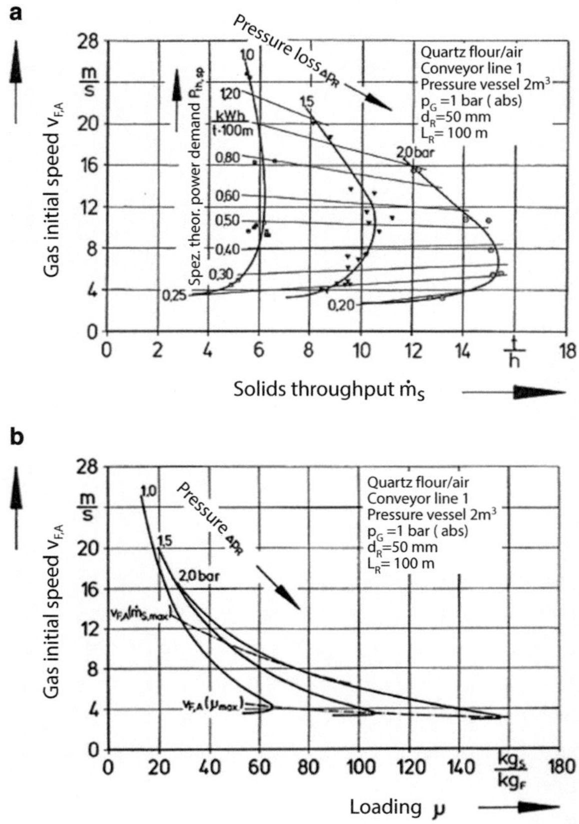

Fig. 4.46 Conveying diagrams, quartz powder/air, $d_{S,50} = 39\,\mu\text{m}$, $\varrho_S = 2610\,\text{kg/m}^3$, $\varrho_{SS} = 1210\text{kg/m}^3$, $D_R \times L_R = 50\,\text{mm} \times L_R = 100\,\text{m}$. **a** $v_{F,A} = f(\dot{m}_S, \Delta p_R, P_{th,sp})$, **b** $v_{F,A} = f(\mu, \Delta p_R)$

can be found in Table 4.6. Since this was a peak-load power plant with temporarily poor flowing/handling fly ash, pneumatic lean phase conveying systems were specified by the operator and designed by the supplier, see the column conveying gas volume flow $\dot{V}_{F,G} = 100\,\%$, corresponding to $v_{F,A} \cong 10.0\,\text{m/s}$, in Table 4.6. Based on initial operating experience and because one of the two conveyors only functioned as a standby system, the supplier suggested reducing the energy demand of one of the systems by decreasing the conveying gas throughput. A corresponding retrofit was carried out, see the column conveying gas volume flow $\dot{V}_{F,G} = 74\,\%$, corresponding to $v_{F,A} \cong 7.4\,\text{m/s}$, in Table 4.6.

Table 4.6 presents in the upper part design data, which were calculated assuming the most unfavorable fly ash and conveying behavior, and in the lower part measured operating results of both plants in comparison. The calculated power demand figures are, due to the assumptions underlying them, larger than the measured values. By reducing the conveying gas flow $\dot{V}_{F,G}$, a power saving of approximately 26.4 kW, corresponding to 31.2% based on the demand of the original pneumatic conveying system, was expected. The measured saving was, despite the lower demand level, 23.4 kW or 38.2%. The operator refrained from a further, in principle possible, reduction of the gas velocity level, as

Table 4.6 Design and operating data of an executed plant

Pressure vessel : size 4 ≙ 4 m^3 net volume
Conveying distance : 393m, of which 42m are vertical height, starting after 338m
4-times staggered pipeline, d_R =114.4/120.4/127.1/139.8 mm
Bends: 7x90°. 6x30°, 2x15°, R=1.5m
2 two-way valves
Conveying gas : Air
Solid : Fly ash

	Conveying gas -volume flow	$\dot{V}_{F,G}$		100%	74%
DESIGN	Solids throughput	$\dot{m}_e$	[t/h]	15	15
	Pressure loss in the delivery pipe	Δp_R	[bar]	2.1	2.0
	Power requirement	$\bar{P}$	[kW]	84.6	58.2
	Specific power demand	$\bar{P}/(\dot{m}_e \cdot L_R)$	[kWh/(t·100m)]	1.44	0.99
	Expected savings			26.4 kW ≙	31.2%
OPERATION	Solids throughput	$\dot{m}_e$	[t/h]	18.5	18.2
	Pressure loss in the delivery pipe	Δp_R	[bar]	1.7 – 1.9	1.5 – 1.7
	Power requirement	$\bar{P}$	[kW]	61.3	37.9
	Specific power demand	$\bar{P}/(\dot{m}_e \cdot L_R)$	[kWh/(t·100m)]	0.84	0.53
	Measured savings			23.4 kW ≙	38.2%

All performance figures at compressor coupling

concerns regarding operational reliability existed due to frequently changing coal types (→ world coal!) and occasionally observed strong hygroscopic ash behavior. The second line was also converted to the smaller conveying gas flow.

The approach to energetic optimization of a conveying system with the parameters specified by the task (bulk material, $L_R, \dot{m}_S, \Delta v_{F,A}$) is presented below: For the two bulk materials cement and fly ash, two independent pneumatic transport systems with the same solid throughput and the same pipeline layout are to be designed. Table 4.7 contains the essential operating and bulk material data. Since the smallest possible energy input is required, single pressure vessels and initial conveying gas velocities $v_{F,A}$ with a constant distance to the respective current minimum conveying gas velocity $v_{F,A,\min}$ of $\Delta v_{F,A} = (v_{F,A} - v_{F,A,\min}) = 2.0\,\mathrm{m/s}$ are chosen. By systematically specifying different conveying pressure differences or pipe diameters, the gas volume flows and pressure differences to be provided by the pressure generator can be determined, and from these, the power requirement of the pressure generator can be calculated or taken from the documents of corresponding manufacturers. Fig. 4.47 shows the result of such an analysis for the tasks described in Table 4.7.

Shown in Fig. 4.47 are the required coupling powers of possible pressure generators against their respective necessary final pressure p_V. For each p_V, a specific conveying pipe diameter D_R is assigned. Since the two bulk materials lead to different diameters at the same conveying pressure, the D_R-axis is dimensionless. In general, with increasing p_V, D_R becomes smaller. On the ordinate of Fig. 4.47, two power requirement values are plotted. Reason: Pressure vessel conveying is a batch operation that includes conveying phases and dead times, among other things, for filling and pressurising the transmitters (→ different depending on the circuit variant). As a result, a distinction must be made

Table 4.7 Characteristic data of the investigated conveying systems

Solids throughput	$\dot{m}_e$	[t/h]	50	
Total conveying length	L_R	[m]	200. unstaggered	
of which are: vertical height	H_R	[m]	20	
starting after	L	[m]	150	
Number of 90° bends	n_K		5	
Solid temperature	t_S	[°C]	20	
Max. Gas temperature	$t_{F,max}$	[°C]	120	
Altitude of plant	H_U	[m]	<500	
Counterpressure	p_G	[bar(abs)]	1	
Distance ($v_{F,A}$ -$V_{F,A,min}$)	$\Delta v_{F,A}$	[m/s]	2	
Conveying gas			Air	
Solid material			Cement	Fly ash
Bulk density	ρ_{SS}	[kg/m³]	1100	750
Particle density	ρ_S	[kg/m³]	3200	2200
Average particle Ø	$d_{S,50}$	[μm]	≈22	≈20

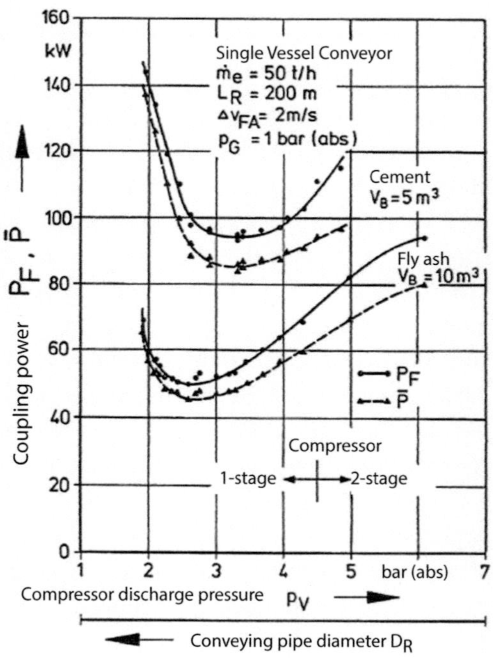

Fig. 4.47 Power requirement depending on compressor final pressure and bulk material

between the power P_F, which is necessary during the conveying phase, therefore represents a maximum value and ultimately determines the size of the pressure generator, and the average power $\overline{P}$, which provides the average over the total cycle or working time and thus determines the energy costs incurred for the operator. Details on this follow in Sect. 7.1.

The results of the calculations in Fig. 4.47 show that for both bulk materials, clear power minima exist at different pressures p_V^*, whose P_F^*- and $\overline{P}^*$-values are identical within the calculation accuracy. It should be taken into account that all onveyings included in Fig. 4.47 are dense phase transports with a distance of $\Delta v_{F,A} = 2.0\,\mathrm{m/s}$ to the current conveying limit, i.e., dense phase systems can also be optimized with regard to the energy-minimal combination (p_V^*, D_R^*). Corresponding investigations on lean phase pneumatic conveying systems, $\Delta v_{F,A} \gg 2.0\,\mathrm{m/s}$, would also lead to optimal combinations (p_V^*, D_R^*), but at considerably higher power demand values, see Fig. 4.46a. The above considerations had to take into account different pressure vessel net volumes, $V_B = 5\,\mathrm{m}^3$ for cement and $V_B = 10\,\mathrm{m}^3$ for fly ash, have been used as a basis, since the size of the "volume conveyor" pressure vessel for a given number of batches/time unit is determined by the bulk density of the respective conveyed material. Dry-running single and two-stage screw compressors were used as pressure generators, i.e., displacement machines whose current power requirements adapt, within limits, to changing counter-pressures = conveying pressures.

When using the method of pressure loss determination presented in Sect. 4.7, the above optimization calculations require the scale-up-capable knowledge of the resistance function λ_S of the present bulk material, i.e., its dependence on all relevant influencing factors, including the loading μ, the pipe diameter D_R, the gas velocity v_F, the characteristics of the bulk material, the conveying gas, possibly the pipe wall material, etc. The same applies to possible alternative calculation approaches. Furthermore, the minimum velocities limiting the conveying, e.g., $v_{F,A,\min}$ according to (4.13), should be known.

The results presented so far show that fine-grained bulk materials have a relatively large energy-saving potential. This is not the case for coarser-grained conveyed materials. This fact can be illustrated using the measurement results of a plastic granulate, $d_{S,50} \cong 3.9\,\mathrm{mm}$, plotted in Fig. 4.48. The solid throughput used as the abscissa value in Fig. 4.48a $\dot{m}_S$ ($\rightarrow$ maximum measured mass flow in the diagram $\dot{m}_{S,\max} \cong 20.5\,\mathrm{t/h}$) is replaced in Fig. 4.48b by the loading μ ($\rightarrow$ maximum measured loading $\mu \cong 26.0\,\mathrm{kg}\,S/\mathrm{kg}\,F$). Figure 4.48a shows that between the pneumatic flight conveying/incipient strand conveying area and the slow conveying area, there is a region of unstable conveying states that cannot be used for operating plants. Due to the relatively high conveying gas velocities in this area and the resulting large pressure differences across the forming conglobations/plugs, these are dissolved again from the end due to gas expansion, form anew, etc. After falling below this area, i.e., at sufficiently low gas velocities, a stable plug conveying is established for the coarse bulk material in question. For the dense phase tests shown here, conveying with controlled plug generation, cf. Sect. 6.1, was used, with the length of the plugs L_P being set equal in each case. The dense phase tests show an almost linear dependence $\dot{m}_S(v_{F,A})$, achieve a significantly lower solid throughput at the same conveying pressure as in the lean phase conveying range. The linear course $\dot{m}_S(v_{F,A})$ follows with the constant bulk density ϱ_b of the plug, the given pipe diameter D_R as well as $u_S \propto v_F$ directly from the continuity equation. The cause of the limited conveying capacity is the pipe cross-section, which is completely filled with bulk material in the plug area.

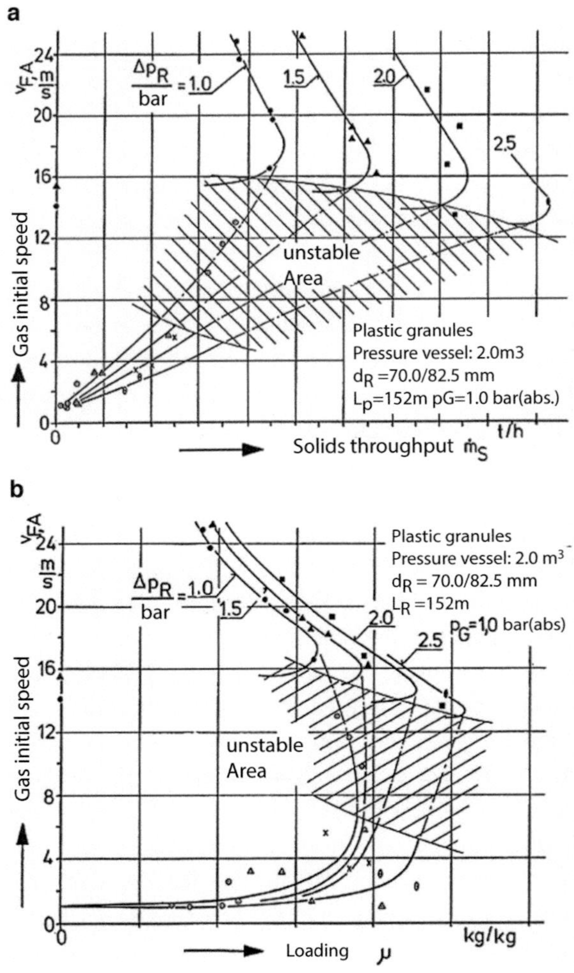

Fig. 4.48 Conveying diagrams, plastic granulate/air, lens-shaped monograin, $d_{S,50} \cong 3900\,\mu\text{m}$, $\varrho_S = 880\,\text{kg/m}^3$, $\varrho_{SS} = 490\,\text{kg/m}^3$, $D_R \times L_R = 70.0/82.5\,\text{mm} \times L_R = 152\,\text{m}$, staggered. **a** $v_{F,A} = f(\dot{m}_S, \Delta p_R)$, **b** $v_{F,A} = f(\mu, \Delta p_R)$

Fig. 4.48b illustrates that the maximum loadings achieved in dilute and dense phase μ are approximately equal, i.e., their theoretical specific power requirement $P_{th,sp}$ is identical. To achieve the solid mass flows of the dilute pneumatic conveying range in the dense phase range, correspondingly larger conveying pipe diameters are necessary. With the same operating setting (→ same plug and gas cushion lengths, same pressure difference, etc.), the power requirement $P_{th,sp}$ of the test conveying section is also established in the larger conveying pipe. In the considered example, no significant energy savings are possible with a dense phase = plug conveying system compared to an optimized dilute phase conveying. The benefit lies in a more product-friendly and less wear-intensive transport. However, it should be noted that the transition from an impact stress on the product and the pipe (→ dilute phase conveying) to a sliding stress (→ plug conveying) does not necessarily lead to reduced pipeline wear, grain abrasion or breakage. Only the stress mechanism is changed.

The above-described increase in solids throughput by increasing the pipe diameter D_R with constant plug and gas cushion length (L_P, L_F) leads to smaller length/diameter ratios (L_P/D_R) and thus less stable plugs. This can be counteracted for coarse-grained bulk materials ($\rightarrow$ linear dependency $\Delta p_P(L_P)$) by correspondingly longer plugs. The prerequisite is that in both operating cases the sum of the lengths of all plugs in the pipe, $\sum L_P$, remains the same, so it applies: $\Delta p_R = \sum \Delta p_P = \text{konst.}$

In the context of practical plant design, minimizations of power requirements are carried out, taking into account the advantages of pipe staggering as well as considering overall economic and operational aspects. Details on this can be found in, among others, [75, 77, 78].

4.12.3 Calculation Example 9: Energetic Optimization of a Fly Ash Conveying

A fly ash, which has been investigated/known for its handling and conveying behavior, $d_{S,50} \cong 15\,\mu\text{m}$, is to be pneumatically conveyed with the lowest possible energy input over a total conveying distance of $L_R = 800\,\text{m}$, including 30 m vertical height and various deflections, from the electrostatic precipitator of a coal-fired power plant to a group of loading silos: solid throughput $\dot{m}_S = 68\,\text{t/h}$. To reduce the wear of the conveying pipe, it is required that the solid velocity at the end of the route does not exceed the value $u_{S,E} = 18\,\text{m/s}$. The design temperature is $t_M = 80\,°\text{C}$, the altitude $H_{\text{ort}} \cong 100\,\text{m}$, the pressure at the end of the line $p_E \cong 1.0\,\text{bar}$. Conveying gas is ambient air. Pressure vessel senders in so-called multi-vessel circuit are used as solid locks, see Sect. 7.1.4. The energy-minimal combination $(\Delta p_V^*, D_R^*)$ is to be determined.

Note:
The required calculations can only be carried out with reasonable effort using sophisticated computer-aided calculation programs. One such program is used below. Due to space constraints, the intermediate results of the individual calculation steps cannot be shown. Only the procedure and the resulting outcome are described.

Procedure:

- Determination of the conveying gas velocity: Due to the long transport path and the possibly changing coal and thus fly ash qualities in a power plant, a constant distance of $\Delta v_{F,A} = 3.0\,\text{m/s}$ to the respective current minimum conveying velocity $v_{F,A,\min}$, (4.13), is set at the beginning of the conveying line for all designs. With this distance, even critical operating conditions can be controlled. In all cases, dense phase conveying is used.

- With a funding pressure appropriately estimated for the task p_V = output pressure of the pressure generator, a first ungraded pipe diameter D_R is calculated. In the considered case, for example, by choosing a single-stage screw compressor with $p_V = 4.0\,\text{bar}$. The calculation provides, among other things, the exact compressor final pressure p_V corresponding to the pipe diameter D_R, the pressure at the beginning of the conveying line p_A, the required gas volume flow $\dot{V}_V$ in the suction state and the gas velocities along the line. For the known values $(p_V, \dot{V}_V)$ the required coupling power P_V of the pressure generator is queried from the manufacturer or taken from its documents.
- Performing further calculations with systematic variation of the pipe diameter D_R to smaller and larger values. Evaluation as described above. Pipe diameters of a pipe series available at the customer are used. The result of these calculations is shown in Fig. 4.49. The power requirement P_V plotted on the ordinate corresponds to that during the respective conveying phase of the pressure vessel system. In the realized vessel interconnection, the difference to the average power $\overline{P}$ is small, and initially, only the position of the power minimum is of interest, which is usually identical in both cases. The energetically optimal design variant is found at $p_V^* = 3.2\,\text{bar}$ (= 2.2 bar (g.)) and $D_R^* = 260.4\,\text{mm}$. The associated gas flow is $\dot{V}_V^* = 2318\,\text{m}^3/\text{h}$ at (20 °C, 1 bar). On both sides of the energy minimum, the power requirement P_V increases steeply to significantly higher values.
- Operating points on the right curve branch in Fig. 4.49 lead to smaller pipe diameters D_R and lower gas volume flow rates $\dot{V}_V$ than the energy-optimal design. This indicates the possibility of an overall economic optimization (→ energy/operating costs + investment costs, etc.) of the plant. Further investigations on this subject can be found in [77].
- Verification of the final solid velocity $u_{S,E}$: The program provides the velocity $v_{F,E}$ of the conveying gas at the end of the pipeline. To determine $u_{S,E}$, the local solid/gas velocity ratio is required. It follows with (4.18) to:

$$C_{h,E} \cong \frac{u_{S,E}}{v_{F,E}} = 1 - \frac{v_{F,A,\min}}{v_{F,E}} \quad \rightarrow \quad v_{F,A,\min} \text{ from (4.13)}$$

The calculations show that the limit velocity $u_{S,E} = 18\,\text{m/s}$ is only exceeded at compressor pressures p_V significantly above the pressure p_V^* at the minimum performance in Fig. 4.49. Velocity-reducing pipeline staggerings are then required (→ Extension of the unstepped pipe in the conveying direction). Although these lead to a reduction in the respective power requirement P_V compared to the unstepped lines, but in the present example, they do not change the position and operating data of the energetic optimum determined above. In this consideration, it should be taken into account that the compressor pressure p_V is greater by an amount Δp_{ver} than the pressure p_A at the beginning of the conveying line. Δp_{ver} is required to cover the losses of the gas during the supply to the lock system, the gas distribution at this, and the bulk material discharge from the vessels: $v_{F,E} \propto (p_A/p_E)$.

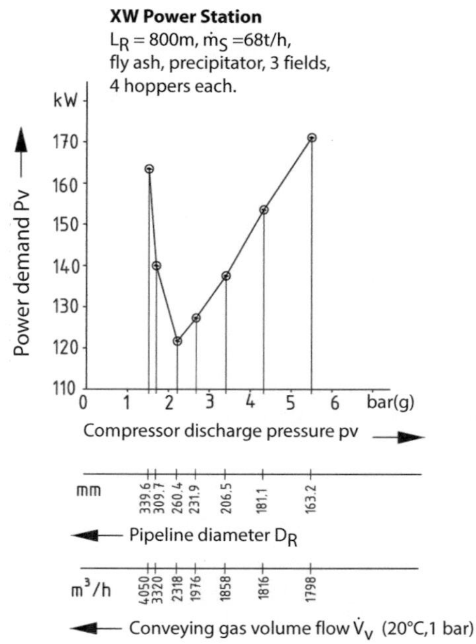

Fig. 4.49 Power requirement for conveying fly ash at various operating settings ($p_V, D_R, \dot{V}_V$), fly ash data: $d_{S,50} \cong 15\,\mu\text{m}$, $\varrho_S \cong 2200\,\text{kg/m}^3$

Result:

For the given task, an operationally reliable energy-minimal operating point ($p_V^*, D_R^*, \dot{V}_V^*$) is determined, which leads to a significantly lower power requirement compared to alternative (p_V, D_R) combinations: The transport is carried out by a single-stage compressor through a unstaggered pipeline, gas velocity at the beginning of the pipeline $v_{F,A} \cong 5.5\,\text{m/s}$, at the end of the pipeline $v_{F,E} \cong 14.8\,\text{m/s}$, loading during the conveying phase $\mu \cong 30\,\text{kg}\,S/\text{kg}\,F$.

4.13 Selected Measurement Results

To illustrate the behavior of conventional dilute and dense phase conveying, i.e., conveying without additional gas supply/injection along the transport route, as described in the preceding sections, the results of some supplementary measurements are presented and briefly explained below.

4.13.1 Length Dependency of the Conveying Diagram

For the fine-grained quartz powder already used in Fig. 4.16 and 4.46, Fig. 4.50 shows the relationship between solid throughput $\dot{m}_S$, conveying gas velocity $v_{F,A}$ and conveying distance L_R through a conveying pipe $D_R = 50\,\text{mm}$ at the constant pressure difference of

$\Delta p_R = 2.0$ bar on the line. Figure 4.50 shows that slow-/dense-phase conveying is considerably more sensitive to changes in operating conditions than dilute phase conveying: Exceeding the planned conveying distance L_R leads, for example, to significantly smaller solid throughputs $\dot{m}_S$. In the dilute pneumatic conveying range, this effect is significantly lower.

The maximum throughput increases with increasing pressure loss per unit length $(\Delta p_R/L_R)$ taking on an increasingly pronounced form, while the gas velocities $v_{F,A}(\dot{m}_{S,\max})$ and $v_{F,A}(\mu_{\max})$ quickly approach the limiting velocity $v_{F,A,\min}$ and also each other. As a result, the operating points of practical dense phase designs must be shifted towards higher gas velocities in order to maintain a minimum distance to $v_{F,A,\min}$. This moves further and further away from the energetically optimal operating state. In Fig. 4.50, $v_{F,A,\min} = $ const., since (p_A, D_R) are constants, cf. (4.13).

4.13.2 Solid Dependence of Conveyance

Figure 4.51 compares the measured conveying characteristics of five different bulk materials. These are fine-grained products from the Geldart classes A and C. Material 1 is the quartz powder already used several times above, $d_{S,50} = 39\,\mu$m. All were driven through the same conveying line with the same pressure difference Δp_R. The determined solid flow rates cover a range of $\dot{m}_S \cong$ (7-24) t/h, i.e. a range of approx. 1 : 3.5, while the minimum conveying gas initial velocities are between $v_{F,A,\min} \cong (1.5–3.5)$ m/s. At gas velocities in the dense flow range, all these bulk materials pass through the horizontal sections of the conveying route in the form of a more or less strongly fluidized strand

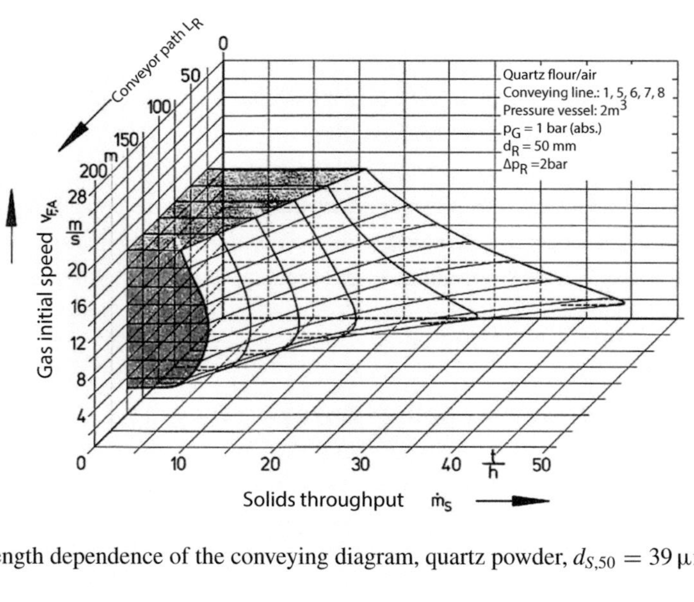

Fig. 4.50 Length dependence of the conveying diagram, quartz powder, $d_{S,50} = 39\,\mu$m

Fig. 4.51 Comparison of conveying characteristics of different bulk materials

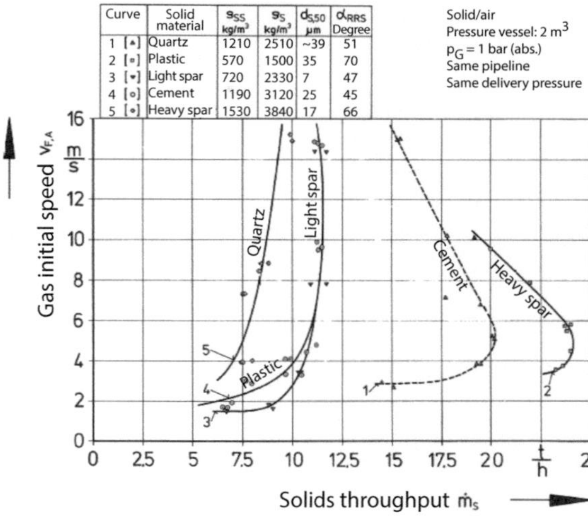

on the pipe bottom (→ observation through glass pipe elements). Surprisingly, this also applies to bulk material 3, light spar, formally a pronounced C-material. Due to its cohesiveness, the removal from the storage container and the introduction into the conveying route proved to be considerably more problematic for this material than the actual pipe transport.

The solid characteristics recorded in Fig. 4.51 do indeed show a clear trend in the influence, among other things, of $d_{S,50}$ and ϱ_S, on the relative position of the characteristic curves in the diagram, but are not sufficient for a complete characterization of the quantitative conveying behavior. For this purpose, further influencing variables with possibly varying weighting must be considered, e.g., width of the grain size distribution, grain shape description, specific surface area, gas holding capacity, and characteristic values that explicitly capture the impact and friction behavior between the solid and the pipe wall. Reliable models for predicting quantitative conveying results based on bulk material properties measured in the laboratory exist only rudimentarily, qualitative statements are possible, see Sect. 4.5. The specific conveying behavior of new, not yet pneumatically conveyed bulk materials is therefore generally determined by systematic conveying tests on a suitable pilot plant.

4.13.3 Influence of Grain Size Distribution

In Fig. 4.52, the grain size distributions of a coarse-grained limestone grit KS with $d_{S,50} = 460\,\mu m$, a limestone powder KSM produced from this by grinding with $d_{S,50} = 15\,\mu m$ and two different mixtures of these two products are shown: Mixture 1: 100 wt.-% KS + 20 wt.-% KSM, Mixture 2: 100 wt.-% KS + 40 wt.-% KSM. The KS

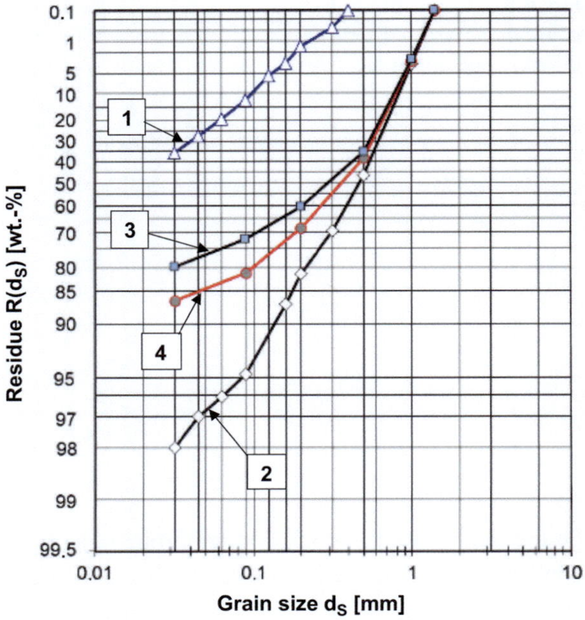

Fig. 4.52 Grain size distributions of the investigated limestones, (1) limestone powder = KSM, (2) limestone chippings = KS, (3) 100 M.-% KS + 20 M.-% KSM, (4) 100 M.-% KS + 40 M.-% KSM

corresponds to the coarse limestone already used in Fig. 4.17, $\varrho_S = 2750\,\text{kg/m}^3$. All four bulk materials were pneumatically transported through the same conveying line under identical operating conditions ($\rightarrow \Delta p_R \cong 2.0\,\text{bar}$, $D_R = 82.5\,\text{mm}$, $L_R = 152\,\text{m}$). The determined characteristic curves are plotted in Fig. 4.53. This illustrates how changes

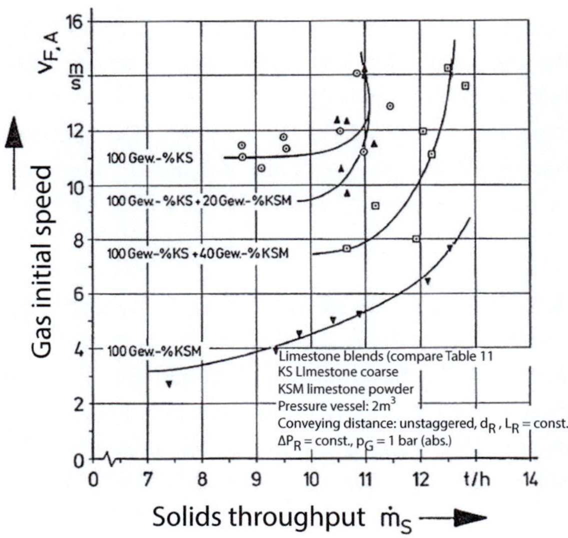

Fig. 4.53 Conveying characteristics of limestone mixtures

Fig. 4.54 Velocities of conveying gas as well as fine and coarse particles

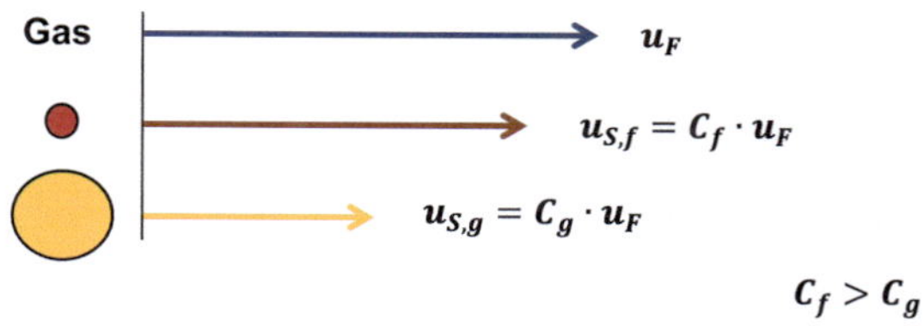

in the fine fraction of the conveyed product affect its minimum velocity $v_{F,A,\min}$ and the throughput $\dot{m}_S$: Increasing fine fraction in the coarse material initially leads to a lowering of the conveying limit and, for larger proportions, also to an increase in the solid mass flow. The reason for this, as schematically shown in Fig. 4.54, are the different velocities of the fine and coarse particles: The faster flying fine particles overtake the slower coarse particles, but cannot (always) avoid them due to their mass inertia, collide with them and thus drive them. This drive by impulse exchange acts in addition to that of the conveying gas and is also the cause for very coarse/large foreign bodies, which should not actually be conveyable under the respective operating conditions, still passing through the conveying line.

4.13.4 Unstable Operating Conditions of Specific Bulk Materials

As already described in earlier sections, e.g. Sects. 4.1.1 and 4.12.2, and illustrated in Fig. 4.48, coarser bulk materials ($d_{S,50} \gtrsim 0.3\,\text{mm}$) with narrow particle size distribution ($\rightarrow \alpha_{\text{RRSB}} \gtrsim 60°$) and compact particle shape ($\rightarrow$ approximately identical extension in all three spatial directions) pass through an unstable conveying zone, which cannot/must not be used for operating plants. To clarify the situation, two standard state diagrams with such unstable working areas are shown in Fig. 4.55 as an additional illustration. The upper gas velocities at which the instabilities characterized by extreme pressure pulsations, abrupt increase in pressure gradient, pipe impacts, etc., set in can be estimated by the clogging limit velocities $v_{F,St}$ of the respective bulk material, see Sect. 4.6.4.

This should be checked here with the equation

$$\frac{v_{F,St}}{\sqrt[5]{\frac{\dot{m}_S \cdot g^2}{\varrho_F}}} \cong 5 = K_{St}^* \tag{4.121}$$

(4.121) is a newer, the $v_{F,St}$-approaches of Sect. 4.6.4 supplementary, clogging limit equation for coarse-grained bulk materials [79, 80], which can be transformed by substitution of $\dot{m}_S = \mu_{st} \cdot A_R \cdot \varrho_F \cdot v_{F,st}$ into the form of (4.24), $\mu_{st} = K_{st} \cdot Fr_{R,st}^n$, with $n = 2$. Underlying bulk materials and operating conditions: different plastics, glass beads, quartz sands, $d_{S,50} \cong (0.54\text{-}3.67)\,\text{mm}$, $\mu = (0.2\text{-}50)\,\text{kg}\,S/\text{kg}\,F$, static pressures(1-20) bar,

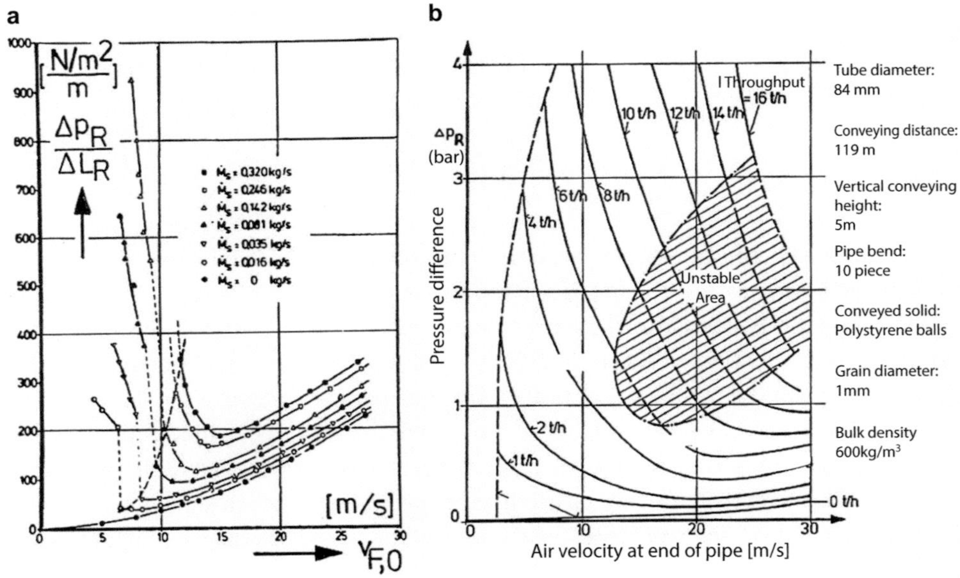

Fig. 4.55 Conveying characteristics of different polystyrene granulates. **a** $d_{S,50} = 2.3\,\text{mm}$, $D_R = 40\,\text{mm}$, $p_{\text{System}} = 1.2\,\text{bar}$ [4, 81], **b** $d_{S,50} = 1.0\,\text{mm}$, $D_R = 84\,mm, p_E = 1.0\,\text{bar}$ [82]

$D_R = (40, 80)\,\text{mm}$, horizontal pipe [80]. The application of (4.121) to the polystyrene granulate in Fig. 4.55a results in lower velocities $v_{F,St}$ than determined there. Example: $\dot{m}_S = 0.320\,\text{kg/s}$, calculated $v_{F,St} = 9.2\,\text{m/s}$, measured $v_{F,St} \cong 11.0\,\text{m/s}$. This illustrates that K^*_{St} should either be adapted material-specifically, as done in (4.27), or, as recommended in [80], be provided with a surcharge of (10-20) %. K^*_{St} should be chosen larger, the flatter the pressure generator characteristic curve is.

Starting from a given state $(p_E, v_{F,E})$ at the end of the conveying pipe, the current gas velocity $v_F \propto 1/\varrho_F \propto 1/p$ decreases towards the beginning of the pipeline, while the associated clogging velocity with $v_{F,st} \propto 1/\varrho_F^{0.2} \propto 1/p^{0.2}$ decreases significantly slower. This results in the operating condition that triggers instabilities ($v_F \leq v_{F,st}$) always occurs first at the beginning of a conveying pipe. Applied to the beginning of the instable state of the conveying characteristic curve $\dot{m}_S = 10\,\text{t/h}$ in Fig. 4.55b, for example, results in: $p_E = 1.0\,\text{bar}$, $\Delta p_R \cong 1.10\,\text{bar}$, $v_{F,E} \cong 24.0\,\text{m/s}$, from this: $p_A \cong 2.1\,\text{bar}$, $v_{F,A} \cong 11.4\,\text{m/s}$, from (4.121) it follows: $v_{F,st,E} = 14.74\,\text{m/s} < v_{F,E}$, $v_{F,st,A} = 12.7\,\text{m/s} > v_{F,A}$, i.e., the instabilities start from the beginning of the pipeline. The agreement between $v_{F,A}$ and $v_{F,st,A}$ is, although here measured on a conveyor line with horizontal and vertical sections as well as deflections, satisfactory due to the additional influencing factors not covered by (4.121).

The lower gas limit velocities of the instability range, which, when undercut, enable stable plug conveying, are in the range of $v_F \cong (6\text{–}8)\,\mathrm{m/s}$ and are currently generally determined by experiment. [83] contains extensive measured values and a semi-empirical calculation equation.

References

1 Hilgraf, P.: Pneumatische Dichtstromförderung: Grundlagen und Anwendungen. Lehrgangshandbuch zum Seminar „Pneumatische Förderanlagen für Dünn- und Dichtstrom". Technische Akademie Wuppertal, Wuppertal (2012)
2. Weber, M.: Strömungs-Fördertechnik. Krausskopf-Verlag, Mainz (1974)
3. Siegel, W.: Pneumatische Förderung: Grundlagen, Auslegung, Anlagenbau, Betrieb. Vogel Buchverlag, Würzburg (1991)
4. Wirth, K.-E.: Theoretische und experimentelle Bestimmungen von Zusatzdruckverlust und Stopfgrenze bei pneumatischer Strähnenförderung. Universität Erlangen-Nürnberg, Erlangen-Nürnberg (1980). Dissertation
5. Hilgraf, P.: Minimale Fördergasgeschwindigkeiten beim pneumatischen Feststofftransport. ZKG Int. **40**(12), 610–616 (1987)
6. Bohnet, M.: Experimentelle und theoretische Untersuchungen über das Absetzen, Aufwirbeln und den Transport feiner Staubteilchen in pneumatischen Förderleitungen. VDI-Forschungsheft 507. VDI-Verlag, Düsseldorf (1965)
7. Hilgraf, P.: Pneumatische Förderung – ein Überblick, Teil 1 und 2. ZKG Int. **46**(1), 25–29 (1993). Nr. 3, S. 141–148
8. Muschelknautz, E., Krambrock, W.: Vereinfachte Berechnung horizontaler pneumatischer Förderleitungen bei hohen Gutbeladungen mit feinkörnigen Produkten. Chem.-Ing.-Tech. **41**(21), 1164–1172 (1969)
9. Krambrock, W.: Dichtstromförderung. Chem.-Ing.-Tech. **54**(9), 793–803 (1982)
10. Konrad, K., Harrison, D., Nedderman, R.M., Davidson, J.F.: Prediction of the pressure drop for horizontal dense phase pneumatic conveying of particles. Proc. of Pneumatransport, paper E1. Organized by BHRA Fluid Engineering, Cranfield, Bedford, S. 225–244 (1980)
11. Rizk, F.: Pneumatische Förderung von Schüttgütern. Chem. Ind. **36**, 591–593 (1984)
12. Dixon, G.: How do different powders behave? Bulk-storage Mov. Control. **pp**, 81–88 (1979)
13. Mainwaring, N.J., Reed, A.R.: Permeability and air retention characteristics of bulk materials in relation to modes of dense phase pneumatic conveying. Bulk Solids Handl. **7**(3), 415–425 (1987)
14. Hilgraf, P.: Pneumatische Dichtstromförderung im Überblick, Teil 1 und 2. ZKG Int. **53**(12), 657–662 (2000). 54 (2001) Nr. 2, S. 94–105
15. Konrad, K.: Dense-phase pneumatic conveying: a review. Powder Technol. **49**, 1–35 (1986)
16. Lippert, A.: Die Staub-Luft-Förderung von Pulvern und Schüttgütern mit hohen Gutkonzentrationen im Gasstrom – Ein neuer Fördervorgang. Experimentelle und theoretische Untersuchungen. Technische Hochschule Karlsruhe, Karlsruhe (1965). Dissertation
17. Krambrock, W.: Möglichkeiten zum Verhindern der Stopfenbildung beim pneumatischen Transport. vt verfahrenstech. **12**(4), 190–202 (1978)

18. Legel, D.: Untersuchungen zur pneumatischen Förderung von Schüttgutpfropfen aus kohäsionslosem Material in horizontalen Rohren. TU Braunschweig, Braunschweig (1981). Dissertation
19. Tsuji, Y., Morikawa, Y., Honda, H.: A study of blowing off a stationary plug of coarse particles in a horizontal pipe. J. Powder Bulk Solids Technol. **3**(4), 30–35 (1979)
20. Jones, M.G., Mills, D.: Product classification for pneumatic conveying. Powder Handl. Process. **2**(2), 117–122 (1990)
21. Jones, M.G.: The influence of bulk particulate properties on pneumatic conveying performance. Thames Polytechnic, London (1988). Ph. D. Thesis
22. Geldart, D.: Types of gas fluidization. Powder Technol. **7**, 285–292 (1973)
23. Harder, J., Hilgraf, P., Zimmermann, W.: Optimization of dense phase pressure vessel conveying with respect to industrial application, part 1. Bulk Solids Handl. **8**(2), 205–209 (1988)
24. Mainwaring, N.J.: Characterisation of materials for pneumatic conveying. Powder Handl. Process. **6**(1), 23–27 (1994)
25. Jones, M.: Characterization for pneumatic conveyor design. In: McClinchey, D. (Hrsg.) Characterization of Bulk Solids. Blackwell, Oxford (2005), Kap. 5
26. Hilgraf, P.: Recent developments in efficient pneumatic conveying. In: Bhattacharya, J. (Hrsg.) Design and selection of bulk material handling equipment and systems, Bd. 1, S. 1–38. Wide Publishing, Kolkata (2011)
27. Hilgraf, P.: Untersuchungen zur pneumatischen Dichtstromförderung. Chem. Eng. Process. **20**(1), 33–41 (1986)
28. Hilgraf, P.: Untersuchungen zur pneumatischen Dichtstromförderung über große Entfernungen. Chem.-Ing.-Tech. **58**(3), 209–212 (1986)
29. Claudius Peters Projects test report: Systematic conveying tests with cold and hot Hismelt Iron Ore for Messrs. Lurgi Metallurgie GmbH/Oberursel. Buxtehude/Germany 2002, unveröffentlicht
30. Bi, H.T., Fan, L.-S.: Regime transitions in gas-solid circulating fluidized beds. AIChE Annual Meeting, Los Angeles. Paper # 101e., S. 17–22 (1991)
31. Bi, X.: Flow regime transitions in gas-solid fluidization and transport. University of British Columbia, Vancouver (1994). Ph.D. Thesis,
32. Bi, H.T., Grace, J.R., Zhu, J.X.: Types of choking in vertical pneumatic systems. Int. J. Multiph. **19**(6), 1077–1092 (1993)
33. Yang, W.C.: A mathematical definition of choking phenomenon and a mathematical model for predicting choking velocity and choking voidage. Aiche J. **21**, 1013–1021 (1975)
34. Fan, L.-S., Zhu, C.: Principles of gas-solid flows. Cambridge University Press, New York (1998)
35. Dhodapkar, S., Jacob, K., Hu, S.: Fluid-solid transport in ducts. In: Crowe, C.T. (Hrsg.) Multiphase Flow Handbook, S. 4.1–4.101. CRC Press, Boca Raton (2006)
36. Klinzing, G.E., Rizk, F., Marcus, R., Leung, L.S.: Pneumatic conveying of solids. Springer, Dordrecht (2010)
37. Cabrejos, F.J., Klinzing, G.E.: Pickup and saltation mechanisms of solid particles in horizontal pneumatic transport. Powder Technol. **79**, 173–186 (1994)
38. Plasynski, S., Dhodapkar, S., Klinzing, G.E., Cabrejos, F.J.: Comparison of saltation velocity and pickup velocity correlations for pneumatic conveying. AIChE Sym. Series „Advances in Fluidized Systems“, No. 281, Bd. 87., S. 78–90 (1991)
39. Welschof, G.: Pneumatische Förderung bei großen Fördergutkonzentrationen. VDI-Forschungsheft, Bd. 492. VDI-Verlag, Düsseldorf (1962)

40. Barth, W.: Strömungstechnische Probleme der Verfahrenstechnik. Chem.-Ing.-Tech **26**(1), 29–34 (1954)
41. Ochi, M., Ikemori, K.: Minimum transport velocity of granular materials at high concentration in a horizontal pipe. Bull. Jsme **21**, 1008–1014 (1978). Paper 156–10
42. Muschelknautz, E., Wojahn, H.: Auslegung pneumatischer Förderanlagen. Chem.-Ing.-Tech **46**(6), 223–235 (1974)
43. Matsumoto, S., Kikuta, M., Maeda, S.: Effect of particle size on the minimum transport velocity for horizontal pneumatic conveying of solids. J. Chem. Eng. Japan **10**(4), 273–279 (1977)
44. Cabrejos, F.J., Klinzing, G.E.: Minimum conveying velocity in horizontal pneumatic transport and the pickup and saltation mechanisms of solid particles. Bulk Solids Handl. **14**(3), 541–550 (1994)
45. Rizk, F.: Pneumatische Förderung von Kunststoffgranulaten in horizontalen Rohrleitungen unter Berücksichtigung des Gewichtseinflusses in Zusammenhang mit Gut- und Rohrwerkstoffeigenschaften, insbesondere im optimalen Förderbereich. Technische Hochschule Karlsruhe, Karlsruhe (1973). Dissertation
46. Cabrejos, F., Klinzing, G.: Incipient motion of solids particles in horizontal pneumatic conveying. Powder Technol. **72**, 51–61 (1992)
47. Rabinovich, E., Kalman, H.: Pickup, critical and wind threshold velocities of particles. Powder Technol. **176**, 9–17 (2007)
48. Rabinovich, E., Kalman, H.: Generalized master curve for threshold superficial velocities in particle-fluid systems. Powder Technol. **183**, 304–313 (2008)
49. Rabinovich, E., Kalman, H.: Boundary saltation und minimum pressure velocities in particle-gas systems. Powder Technol. **185**, 67–79 (2008)
50. Rabinovich, E., Kalman, H.: Incipient motion of individual particles in horizontal particle-fluid systems: A. Experimental analysis. Powder Technol. **192**, 318–325 (2009)
51. Rabinovich, E., Kalman, H.: Incipient motion of individual particles in horizontal particle-fluid systems: B. Theoretical analysis. Powder Technol. **192**, 326–338 (2009)
52. Rabinovich, E., Kalman, H.: Pickup velocity from particle deposits. Powder Technol. **194**, 51–57 (2009)
53. Rabinovich, E., Kalman, H.: Flow regime diagram for vertical pneumatic conveying and fluidized bed systems. Powder Technol. **207**, 119–133 (2011)
54. VDI-Gesellschaft Verfahrenstechnik und Chemieingenieurwesen (Hrsg.): VDI-Wärmeatlas, 10. Aufl. Abschnitt Laa1-Lac9. Springer, Berlin (2006)
55. Muschelknautz, E.: Theoretische und experimentelle Untersuchungen über die Druckverluste pneumatischer Förderleitungen. VDI-Forschungsheft, Bd. 476. VDI-Verlag, Düsseldorf (1959)
56. Claudius Peters Projects-Versuchsbericht: Untersuchung verschiedener Titanerzqualitäten im Hinblick auf ihr Verhalten beim Lagern und pneumatischen Fördern für Fa. Kerr-McGee Pigments GmbH/Krefeld. Buxtehude/Germany 2003, unveröffentlicht
57. Stegmaier, W.: Zur Berechnung der horizontalen pneumatischen Förderung feinkörniger Stoffe. F+h Fördern Heb. **28**(5/6), 363–366 (1978)
58. Siegel, W.: Experimentelle Untersuchungen zur pneumatischen Förderung körniger Stoffe in waagerechten Rohren und Überprüfung der Ähnlichkeitsgesetze. VDI-Forschungsheft, Bd. 538. VDI-Verlag, Düsseldorf (1970)
59. Werner, D.: Einfluss der Korngrößenverteilung bei der pneumatischen Dichtstromförderung in vertikalen und horizontalen Rohren. Technische Hochschule Karlsruhe, Karlsruhe (1982). Dissertation

60. Weidner, G.: Grundsätzliche Untersuchung über den pneumatischen Fördervorgang, insbesondere über die Verhältnisse bei Beschleunigung und Umlenkung. Forsch. Ing.-wesen **21**(5), 145–153 (1955). Dissertation, Technische Hochschule Karlsruhe
61. Claudius Peters Technologies research report: Examination of Vortex elbow. Buxtehude/Germany 2003, unveröffentlicht
62. Nickel, T.: Einfluss der Leitungsführung auf den Druckverlust bei der pneumatischen Dichtstromförderung in dünnen Leitungen. TUHH Technische Universität Hamburg-Harburg, Hamburg-Harburg (2000). Diplomarbeit
63. Dhodapkar, S., Solt, P., Klinzing, G.: Understanding bends in pneumatic conveying systems. Chem. Eng., 53–60 (April 2009). www.che.com
64. Schuchart, P.: Widerstandsgesetze für den pneumatischen Feststofftransport in geraden Rohren und Rohrkrümmern. TU Berlin, Berlin (1968). Dissertation
65. Hilgraf, P., Moka, M.A.: Pressure loss in vertical sections of pneumatic conveying lines. Cem. Int. **10**(6), 52–58 (2012)
66. Zipse, G.: Die Massenstromdichteverteilung bei der pneumatischen Staubförderung und ihre Beeinflussung durch Einbauten in die Förderleitung. Fortschr.-Ber. VDI-Z., Reihe 13, Nr. 3. VDI-Verlag, Düsseldorf (1966)
67. Hilgraf, P.: Zur Staffelung pneumatischer Förderleitungen unter besonderer Berücksichtigung der Dichtstromförderung. ZKG Int. **44**(4), 161–168 (1991)
68. Claudius Peters Technologies-Versuchsbericht: Untersuchungen zur pneumatischen Dichtstromförderung von Braunkohleprodukten der Fa. Rheinbraun AG/Köln. Buxtehude/Germany 1986, unveröffentlicht
69. Bohnet, M.: Fortschritte bei der Auslegung pneumatischer Förderanlagen. Chem.-Ing.-Tech. **55**(7), 524–539 (1983)
70. Muschelknautz, E., Wojahn, W.: Fördern. In: Ullmanns Enzyklopädie der technischen Chemie, Bd. 3, S. 131–184. Verlag Chemie GmbH, Weinheim (1973)
71. Bohnet, M.: Pneumatische Förderung von Schüttgütern. VDI-Bildungswerk BW 2470. VDI-Verlag, Düsseldorf (1974)
72. Roski, H.-J.: The influence of stepped pipelines in pneumatic long-distance transport of building materials. Pneumatech 3, Int. Conf. on Pneumatic Conveying Technology. Proc., S. 311–333 (1987)
73. Wypich, P.W.: The advantages of stepping pipelines for pneumatic transport of bulk solids. Powder Handl. Process. **2**(3), 217–221 (1990)
74. Hilgraf, P.: FLUIDCON – a new pneumatic conveying system for fine-grained bulk materials. Cem. Int. **2**(6), 74–87 (2004)
75. Hilgraf, P.: Der Energiebedarf pneumatischer Förderprinzipien im Vergleich mit mechanischen Förderungen. ZKG Int. **51**(12), 660–673 (1998)
76. Krause, F.: Anforderungen an die Schüttgutfördertechnik. Begleitband zur Fachtagung „Schüttgutfördertechnik“, Magdeburg, 10.10.1996. (1996)
77. Hilgraf, P.: Optimale Auslegung pneumatischer Dichtstrom-Förderanlagen unter energetischen und wirtschaftlichen Gesichtspunkten. ZKG Int. **39**(8), 439–446 (1986)
78. Hilgraf, P.: Einflussgrößen bei der energetischen Optimierung pneumatischer Dichtstrom-Förderanlagen. ZKG Int. **41**(8), 374–380 (1988)
79. Molerus, O.: Hydraulischer und pneumatischer Transport. In: Bohnet, M. (Hrsg.) Mechanische Verfahrenstechnik. Wiley-VCH, Weinheim (2004), Kap.~9
80. Heuke, U.: Horizontale pneumatische Förderung bei hohem Druck. Universität Erlangen-Nürnberg, Erlangen-Nürnberg (1998). Dissertation

81. Wirth, K.-E., Molerus, O.: Bestimmung des Druckverlustes bei der pneumatischen Strähnenförderung. Chem.-Ing.-Tech. **53**(4), 292–293 (1981). MS 898/81
82. Siegel, W.: Grenzen der pneumatischen Förderung. Chemie-Anlagen+Verfahren, 40–46 (Dezember 1981)
83. Pan, R., Mi, B., Wypich, P.W.: Pneumatic conveying characteristics of fine and granular bulk solids. KONA No. 12., S. 77–85 (1994)

5 Special Calculation Approaches, Scale-up

The model for calculating the pressure loss of conveying routes described in Sect. 4.7, regardless of the respective flow forms that develop, is the most frequently used design method in practice. The largest pressure loss component is generally provided by the friction loss of the solid in straight horizontal pipe sections, defined by the λ_S approach, (4.46). By means of the λ_S approach, the different forces acting on the system in the various flow forms are transformed into a representation proportional to the mass inertia, cf. Fig. 4.5 and the related comments in Sect. 4.7.1. In doing so, essential physical relationships/information for the conveying operation can be lost. Therefore, special models for describing the strand conveyance of fine-grained and the plug conveyance of coarse-grained bulk materials, as well as a proven approach for scaling up conveying systems, are presented below.

5.1 Strand Conveyance of Fine-Grained Bulk Materials

The dense phase conveying of fine-grained bulk materials preferably forms as strand conveying. In the following, only fine-grained solids, $d_{S,50} \lesssim 100\,\mu\text{m}$, bulk materials of group A and the adjacent area of group C according to Geldart, are considered. Deviating behavior of coarser products is pointed out. Visual observations on glass pipe systems show that for the slow conveying of the mentioned class of bulk materials, two different flow areas are established in the (horizontal) conveying pipe: Directly above the pipe bottom, a flowing bulk material strand is formed, loosened by the conveying gas. Above this, a gas stream with only a small amount of solid material flows through the upper conveying pipe cross-section. First, the gas mass flow distribution on both pipe cross-sections should be estimated.

P. Hilgraf, *Pneumatic Conveying*, https://doi.org/10.1007/978-3-662-67223-5_5

5.1.1 Gas Distribution Over the Conveying Pipe Cross-Section

The conveying pipe element shown in Fig. 5.1 ΔL_R is supplied with the conveying gas flow $\dot{m}_F$ and the solid flow $\dot{m}_S$. The gas mass flow $\dot{m}_F$ is divided into the shares $\dot{m}_{F,\mathrm{I}}$ and $\dot{m}_{F,\mathrm{II}}$, which represent cross-section I = strand cross-section and II = free space above the strand. From the assumption of a constant pressure p_i over the respective pipe cross-section A_R, a negligible pipe wall friction of the gas in strand cross-section I and a solid-free upper space II, i.e. $\dot{m}_{S,\mathrm{I}} = \dot{m}_S$, $\dot{m}_{S,\mathrm{II}} = 0$, follows the gas distribution derived in Box 5.1 ($\dot{m}_{F,\mathrm{II}}/\dot{m}_F$). The specific pressure loss $\Delta p_{\mathrm{II}}/\Delta L_R$ is calculated as pure gas flow, while the one $\Delta p_{\mathrm{I}}/\Delta L_R$ is calculated as the loss during the gas flow through a moving bed. Since fine-grained particles are present, the latter can be assumed as a flow in the gas viscosity-dominated Stokes region, see Sect. 3.2. A gas and/or solid exchange between the two flow areas is not considered.

The resistance coefficient $\lambda_{F,\mathrm{II}}$ of the gas flow through the upper space cross-section A_{II} consists of the friction components on the wetted pipe surface $=(U_{\mathrm{II}} - B_S) \cdot \Delta L_R$, U_{II} = circumference of the cross-section A_{II}, B_S = Strand width, and friction on the strand surface $= B_S \cdot \Delta L_R$. Visual observations show that with fine-grained bulk materials on the strand surface solid acceleration and dune formation occur. The drive required for this is provided by the gas flow $\dot{m}_{F,\mathrm{II}}$ and is taken into account by a pipe friction coefficient X_S increased by a factor λ_F. (5.14) in Box 5.1 shows the approach used. The bracket expression on the right side of the equation takes into account the different velocities of the upper space gas relative to the pipe wall and the strand surface. Due to the intensive mixing of solid and conveying gas at the beginning of the pipeline/after the injection, the (fine-grained) bulk material is transferred into a more or less loosened state, which is maintained over longer conveying distances due to the generally high gas holding capacity of these products.

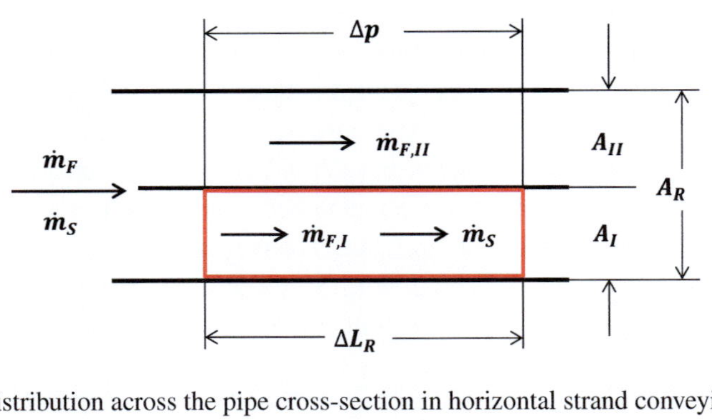

Fig. 5.1 Gas distribution across the pipe cross-section in horizontal strand conveying

Box 5.1: Calculation of gas distribution across the pipe cross-section in strand conveying, Stokes range, explanations in the text

$$\frac{\Delta p_{\mathrm{II}}}{\Delta L_R} = \frac{\Delta p_{\mathrm{I}}}{\Delta L_R} \tag{5.1}$$

$$\lambda_{F,\mathrm{II}} \cdot \frac{1}{D_{h,\mathrm{II}}} \cdot \frac{\varrho_F}{2} \cdot u_{F,\mathrm{II}}^2 = 150 \cdot \frac{(1-\varepsilon_{F,\mathrm{I}})^2}{\varepsilon_{F,\mathrm{I}}^3} \cdot \frac{\eta_F}{(d_{S,SD}^*)^2} \cdot (v_{F,\mathrm{I}} - \varepsilon_{F,\mathrm{I}} \cdot u_{S,\mathrm{I}}) \tag{5.2}$$

Inserting

$$\phi = \frac{A_{\mathrm{II}}}{A_R} \tag{5.3}$$

$$u_{F,\mathrm{II}} = \frac{\dot{m}_{F,\mathrm{II}}}{A_{\mathrm{II}} \cdot \varrho_F} = \frac{\dot{m}_{F,\mathrm{II}}}{A_R \cdot \varrho_F} \cdot \frac{1}{\phi}, \tag{5.4}$$

$$v_{F,\mathrm{I}} = \frac{\dot{m}_{F,\mathrm{I}}}{A_{\mathrm{I}} \cdot \varrho_F} = \frac{\dot{m}_{F,\mathrm{I}}}{A_R \cdot \varrho_F} \cdot \frac{1}{1-\phi} \tag{5.5}$$

$$u_{S,\mathrm{I}} = \frac{\dot{m}_S}{A_{\mathrm{I}} \cdot (1-\varepsilon_{F,\mathrm{I}}) \cdot \varrho_P} = \frac{\dot{m}_S}{A_R \cdot (1-\varepsilon_{F,\mathrm{I}}) \cdot \varrho_P} \cdot \frac{1}{1-\phi} \tag{5.6}$$

$$v_F = \frac{\dot{m}_F}{A_R \cdot \varrho_F}, \tag{5.7}$$

$$A_R = A_{\mathrm{I}} + A_{\mathrm{II}}, \tag{5.8}$$

$$\dot{m}_F = \dot{m}_{F,\mathrm{I}} + \dot{m}_{F,\mathrm{II}} \tag{5.9}$$

in (5.2) leads to a quadratic equation for $\dot{m}_{F,\mathrm{II}}$ with the solution:

$$\begin{aligned} \frac{\dot{m}_{F,\mathrm{II}}}{\dot{m}_F} &= \frac{k_1 \cdot k_2}{\frac{\dot{m}_F}{A_R}} \cdot \left(\sqrt{1 + \frac{\frac{\dot{m}_F}{A_R} - \frac{\varepsilon_{F,\mathrm{I}}}{1-\varepsilon_{F,\mathrm{I}}} \cdot \frac{\varrho_F}{\varrho_P} \cdot \frac{\dot{m}_S}{A_R}}{\frac{1}{2} \cdot k_1 \cdot k_2}} - 1 \right) \\ &= \frac{k_1 \cdot k_2}{\varrho_F \cdot v_F} \cdot \left(\sqrt{1 + \varrho_F \cdot \frac{v_F - \varepsilon_{F,\mathrm{I}} \cdot u_{S,\mathrm{I}} \cdot (1-\phi)}{\frac{1}{2} \cdot k_1 \cdot k_2}} - 1 \right) \end{aligned} \tag{5.10}$$

as well as:

$$k_1 = 150 \cdot \frac{(1-\varepsilon_{F,\mathrm{I}})^2}{\varepsilon_{F,\mathrm{I}}^3} \cdot \frac{\eta_F}{(d_{S,SD}^*)^2 \cdot \lambda_{F,\mathrm{II}}} \tag{5.11}$$

$$k_2 = \frac{D_{h,\mathrm{II}} \cdot \phi^2}{1 - \phi}, \tag{5.12}$$

$$D_{h,\mathrm{II}} \cong \frac{4 \cdot A_{\mathrm{II}}}{U_{\mathrm{II}}} = \frac{4 \cdot \phi}{U_{\mathrm{II}}} \cdot A_R \tag{5.13}$$

$$\lambda_{F,\mathrm{II}} = \frac{U_{\mathrm{II}} - B_S}{U_{\mathrm{II}}} \cdot \lambda_F + \frac{B_S}{U_{\mathrm{II}}} \cdot X_S \cdot \lambda_F \cdot \left(\frac{u_{F,\mathrm{II}} - u_{S,\mathrm{I}}}{u_{F,\mathrm{II}}}\right)^2 \tag{5.14}$$

This is despite simultaneous venting into the free upper space. In addition, there may be back-mixing of gas and solid along the conveying path, e.g., behind 90° deflection bends from the horizontal to the vertical or from the vertical to the horizontal. The reduction in bulk density of the strand caused by these effects is in the range $\varrho_{b,\mathrm{I}} \gtrsim 0.6 \cdot \varrho_{SS}$, with ϱ_{SS} = loosely poured bulk density [1]. For the following estimates, a general loosening of the strand density to $\varrho_{b,\mathrm{I}} \cong 0.75 \cdot \varrho_{SS}$ is assumed. The required average relative void volume $\varepsilon_{F,\mathrm{I}}$ of the strand can then be calculated from (2.21). The determination of the strand width B_S is carried out with an approximation equation from [2]:

$$B_S = D_R \cdot (4 \cdot \phi \cdot (1 - \phi))^{1/3} \tag{5.15}$$

U_{II} is easily determined from the ratio of the strand to the tube geometry.

The critical value of the upper space area fraction A_{II} of the total tube cross-section A_R is obviously at $\phi \leq A_{\mathrm{II}}/A_R \cong 0.50$. Due to the rapidly decreasing upper space cross-sectional area with increasing strand height H_{I} A_{II} can, for example, lead to sudden pipe blockages due to dune formation, as only very short plug lengths can be managed for the fine-grained bulk materials considered here. (5.10) allows, for given boundary conditions, the sizes $\varepsilon_{F,\mathrm{I}}, v_F, u_{S,\mathrm{I}}$ and ϕ to be determined independently of each other for the calculation of $(\dot{m}_{F,\mathrm{II}}/\dot{m}_F)$ to be given. This is only possible because the calculation approach shown in Box 5.1 does not contain any coupling to the driving/braking forces acting on the solid itself. This implies that not every selectable $(\varepsilon_{F,\mathrm{I}}, v_F, u_{S,\mathrm{I}}, \phi)$-combinations are also physically realizable.

The evaluation of (5.10) is to be demonstrated here using the example of the (real) transport of $\dot{m}_S = 50\,\mathrm{t/h} = 13.889\,\mathrm{kg/s}$ of a solid with the characteristic data $d^*_{S,SD} = 20\,\mu\mathrm{m}$, $\varrho_P = 3000\,\mathrm{kg/m^3}$, $\varrho_{SS} = 1000\,\mathrm{kg/m^3}$, $T_{S=F} = 293\,\mathrm{K}$ through a conveying pipe $D_R = 155.7\,\mathrm{mm}$ (→ pipe ∅ 168.3 mm × 6.3 mm). The local pressure at the considered pipeline position is $p = 2.00\,\mathrm{bar}$, the conveying gas mass flow $\dot{m}_F = 740\,\mathrm{kg/h} = 0.2056\,\mathrm{kg/s}$. The conveying gas is air with $\eta_F = 18.26 \cdot 10^{-6}\,\mathrm{Pa\,s}$ and $\varrho_F = 2.40\,\mathrm{kg/m^3}$. The gas friction coefficient is determined as $\lambda_F = 0.02$, the amplification factor is estimated as $X_S \cong 3$ and for the area ratio $\phi = A_{\mathrm{II}}/A_R$ whose critical value $\phi = 0.50$ is set. With these data, the following results from the equations mentioned above: $\varrho_{b,\mathrm{I}} = 750\,\mathrm{kg/m^3}$, $\varepsilon_{F,\mathrm{I}} = 0.75$, $B_S = D_R = 0.1557\,\mathrm{m}$, $U_{\mathrm{II}} = 0.400\,\mathrm{m}$. $\lambda_{F,\mathrm{II}} = 0.0356$ (→ it was set $(u_{F,\mathrm{II}} - u_S) \cong u_{F,\mathrm{II}}$), $k_1 = 28495630.5\,\mathrm{kg/(m^3\,s)}$, $D_{h,\mathrm{II}} = 0.0952\,\mathrm{m}$, $k_2 = 0.0476\,\mathrm{m}$,

$(\dot{m}_F/A_R) = 10.80\,\mathrm{kg/(m^2\,s)} \quad \rightarrow \quad v_F = 4.50\,\mathrm{m/s}, \quad (\dot{m}_S/A_R) = 729.46\,\mathrm{kg/(m^2\,s)} \quad \rightarrow$ $u_{S,\mathrm{I}} = 1.94\,\mathrm{m/s}$. Inserted into (5.10), this results in the gas mass flow ratio

$$\frac{\dot{m}_{F,\mathrm{II}}}{\dot{m}_F} = 0.8379 \quad \rightarrow \quad u_{F,\mathrm{II}} = 7.54\,\frac{\mathrm{m}}{\mathrm{s}},$$

meaning approx. 84 M.-% of the total conveying gas flow flows through the (almost) solid-free pipe cross-section above the strand. The gas velocities in the strand cross-section A_I are: $v_{F,\mathrm{I}} = 1.46\,\mathrm{m/s}$ or $u_{F,\mathrm{I}} = 1.95\,\mathrm{m/s}$, i.e. $u_{S,\mathrm{I}} \cong u_{F,\mathrm{I}}$. Increasing the strand density $\varrho_{b,\mathrm{I}}$, i.e. a reduction of the gap volume $\varepsilon_{F,\mathrm{I}}$, enlarges $(\dot{m}_{F,\mathrm{II}}/\dot{m}_F)$ and vice versa. Stationary strands ($\rightarrow \dot{m}_S, u_{S,\mathrm{I}} = 0$) reinforce the gas flow $\dot{m}_{F,\mathrm{II}}$ through the free upper space. Changes in ϕ along the conveying path cannot be calculated with (5.10): The quantities $(\dot{m}_F/A_R)$ and $(\dot{m}_S/A_R)$ are indeed constant and the gas density ϱ_F changes defined, $\varepsilon_{F,\mathrm{I}}$ can still be freely chosen, i.e., (5.10) lacks, as already indicated, the connection to the forces acting on the solid. The shear forces exerted by the upper space gas flow (and the solid material carried along by it) on the surface of the loosened strand lead to a large velocity gradient (du_S/dH_I) over the strand height H_I: very low solid velocities at the pipe bottom, solid velocities only slightly smaller than $u_{F,\mathrm{II}}$ at the strand surface. $u_{S,\mathrm{I}}$ and $u_{F,\mathrm{I}}$ are thus average velocities. Due to the reduced internal friction of the bulk material flow caused by its loosening, waves and dunes can form on its surface, driven by the applied shear stresses.

When considering coarser bulk materials, in (5.2), Box 5.1, the term on the right side must be replaced by the pressure loss approach of the corresponding product class. For the Newton region, for example, the equations shown in Box 5.2 apply (5.16) and (5.17), see Sect. 3.2. The evaluation of (5.17) leads to comparable $(\dot{m}_{F,\mathrm{II}}/\dot{m}_F)$-values as with the fine-grained bulk materials. Cause: The particle diameter $d^*_{S,SD}$ increases, but the relative void volume takes on significantly lower values, order of magnitude $\varepsilon_{F,\mathrm{I}} \cong 0.50$. There is little or no loosening of the bulk material observed.

Box 5.2: Calculation of gas distribution across the pipe cross-section in strand conveying, Newton region, explanations in the text

$$\lambda_{F,\mathrm{II}} \cdot \frac{1}{D_{h,\mathrm{II}}} \cdot \frac{\varrho_F}{2} \cdot u^2_{F,\mathrm{II}} = 1.75 \cdot \frac{1-\varepsilon_{F,\mathrm{I}}}{\varepsilon^3_{F,\mathrm{I}}} \cdot \frac{1}{d^*_{S,SD}} \cdot \varrho_F \cdot (v_{F,\mathrm{I}} - \varepsilon_{F,\mathrm{I}} \cdot u_S)^2 \tag{5.16}$$

with the definitions given in Box 5.1 follows:

$$\frac{\dot{m}_{F,\mathrm{II}}}{\dot{m}_F} = \frac{1 - \frac{\varepsilon_{F,\mathrm{I}}}{1-\varepsilon_{F,\mathrm{I}}} \cdot \frac{\varrho_F}{\varrho_P} \cdot \frac{\dot{m}_S}{\dot{m}_F}}{1 + \sqrt{k_3} \cdot \frac{1-\phi}{\phi}} \tag{5.17}$$

with:

$$k_3 = \frac{1}{3.50} \cdot \lambda_{F,\mathrm{II}} \cdot \frac{\varepsilon^3_{F,\mathrm{I}}}{1-\varepsilon_{F,\mathrm{I}}} \cdot \frac{d^*_{S,SD}}{D_{h,\mathrm{II}}} \tag{5.18}$$

The above approaches only approximate the real conveying behavior but provide ($\dot{m}_{F,\mathrm{II}}/\dot{m}_F$)-values, which are confirmed by measurements. The strand drive of fine and coarse-grained bulk materials is therefore carried out by the pressure difference force on the solid element, which is essentially determined by the gas pressure loss in the upper space, and by the shear stresses exerted on the strand surface by the upper space gas and the solid material carried along by it. The effects are different for fine and coarse-grained bulk materials: fine-grained solids form a significant velocity gradient over the strand height, while coarser products are rather displaced as compact blocks.

5.1.2 Calculation Approaches

a) Detailed Model

In Box 5.3, an exemplary equation system for the calculation of strand conveying along a straight horizontal conveying pipe is shown. Further boundary conditions are: isothermal flow, constant pressure over the respective pipe cross-section, consideration of a gas and solid exchange (→ momentum transfer) between the flow areas I and II along the conveying path, see Fig. 5.1, uniaxial consideration of areas I and II. A net transfer of gas and solid mass from the strand phase I to the flight phase II is expected/assumed, i.e., the strand height H_I decreases in the conveying direction. Cause: the gas velocity $u_{F,\mathrm{II}}$ in the upper space II can become greater than the *pickup* velocity $v_{F,pu}$ of the respective bulk material in strand I, thus solid and with it gas is taken up from the strand surface. The approach is based on the application of mass and momentum balances for the two flow areas, supplemented by suitable closure equations and builds, among other things, on the works [3, 4] as well as [2, 5].

Box 5.3: Approach for describing/calculating the strand conveying of fine-grained bulk materials

Volume fractions of gas ε_F and solid ε_S in phases I and II in the control volume:

$$\varepsilon_{F,\mathrm{I}} + \varepsilon_{S,\mathrm{I}} = 1, \tag{5.19}$$

$$\varepsilon_{F,\mathrm{II}} + \varepsilon_{S,\mathrm{II}} = 1 \tag{5.20}$$

Geometry equations:

$$\phi = \frac{A_\mathrm{II}}{A_R} \tag{5.21}$$

Strand width:

$$B_S = D_R \cdot (4 \cdot \phi \cdot (1 - \phi))^{1/3} = 2 \cdot \sqrt{H_\mathrm{I} \cdot D_R - H_\mathrm{I}^2} \tag{5.22}$$

Circumference of the pipe wall friction surface of strand phase I:

$$P_{\mathrm{I}} = D_R \cdot \cos^{-1}\left(1 - 2 \cdot \frac{H_{\mathrm{I}}}{D_R}\right) \tag{5.23}$$

Circumference of the pipe wall friction surface of flight phase II:

$$P_{\mathrm{II}} = D_R \cdot \pi - P_{\mathrm{I}} \tag{5.24}$$

Cross-sectional area of the strand phase I:

$$A_{\mathrm{I}} = (1 - \phi) \cdot A_R = \frac{D_R}{4} \cdot \left(P_{\mathrm{I}} - B_S \cdot \left(1 - 2 \cdot \frac{H_{\mathrm{I}}}{D_R}\right)\right) \tag{5.25}$$

Cross-sectional area of the flight phase II:

$$A_{\mathrm{II}} = \phi \cdot A_R = \frac{\pi}{4} \cdot D_R^2 - A_{\mathrm{I}} \tag{5.26}$$

Mass balances:

$$\dot{m}_S = \dot{m}_{S,\mathrm{I}} + \dot{m}_{S,\mathrm{II}}, \tag{5.27}$$

$$\dot{m}_F = \dot{m}_{F,\mathrm{I}} + \dot{m}_{F,\mathrm{II}} \tag{5.28}$$

From (2.39) it follows for a pipe element dL_R in differential form for a residue-free flow:

$$d(\dot{m}_i) = d(A_i \cdot \varepsilon_i \cdot \varrho_i \cdot u_i) = \dot{R}_{m,i} = \dot{r}_{m,i} \cdot dL_R \tag{5.29}$$

In connection with (2.41) the following applies for the case considered here:

$$\frac{d(A_{\mathrm{I}} \cdot \varepsilon_{S,\mathrm{I}} \cdot \varrho_P \cdot u_{S,\mathrm{I}})}{dL_R} = \dot{r}_{m,S}, \tag{5.30}$$

$$\frac{d(A_{\mathrm{I}} \cdot \varepsilon_{F,\mathrm{I}} \cdot \varrho_{F,\mathrm{I}} \cdot u_{F,\mathrm{I}})}{dL_R} = \dot{r}_{m,F} \tag{5.31}$$

$$\frac{d(A_{\mathrm{II}} \cdot \varepsilon_{S,\mathrm{II}} \cdot \varrho_P \cdot u_{S,\mathrm{II}})}{dL_R} = -\dot{r}_{m,S}, \tag{5.32}$$

$$\frac{d(A_{\mathrm{II}} \cdot \varepsilon_{F,\mathrm{II}} \cdot \varrho_{F,\mathrm{II}} \cdot u_{F,\mathrm{II}})}{dL_R} = -\dot{r}_{m,F} \tag{5.33}$$

Momentum balances: (2.44) leads in differential form for a pipe element dL_R and a residue-free flow to the differential equation:

$$d(\dot{\mathfrak{J}}_i) = d(\dot{m}_i \cdot u_i) = d(A_i \cdot \varepsilon_i \cdot \varrho_i \cdot u_i^2) = \sum_k F_{i,k} \tag{5.34}$$

For the case considered here, the following applies:

Solid in the strand area I:

$$\begin{aligned}\frac{d(A_{\mathrm{I}} \cdot \varepsilon_{S,\mathrm{I}} \cdot \varrho_P \cdot u_{S,\mathrm{I}}^2)}{dL_R} &= \dot{r}_{m,S} \cdot u_{S,\mathrm{I}} - \tau_{S,\mathrm{I}\to P,\mathrm{I}} \cdot \varepsilon_{S,\mathrm{I}} \cdot P_{\mathrm{I}} \\ &\quad + (\tau_{S,\mathrm{II}\to S,\mathrm{I}} \cdot \varepsilon_{S,\mathrm{II}} + \tau_{F,\mathrm{II}\to S,\mathrm{I}} \cdot \varepsilon_{F,\mathrm{II}}) \cdot \varepsilon_{S,\mathrm{I}} \cdot B_S \\ &\quad - \varepsilon_{S,\mathrm{I}} \cdot A_{\mathrm{I}} \cdot \frac{dp_{\mathrm{I}}}{dL_R} + F_{\mathrm{drag,I}}\end{aligned} \tag{5.35}$$

Gas in the strand area I:

$$\begin{aligned}\frac{d(A_{\mathrm{I}} \cdot \varepsilon_{F,\mathrm{I}} \cdot \varrho_{F,\mathrm{I}} \cdot u_{F,\mathrm{I}}^2)}{dL_R} &= \dot{r}_{m,F} \cdot u_{F,\mathrm{I}} - \tau_{F,\mathrm{I}\to P,\mathrm{I}} \cdot \varepsilon_{F,\mathrm{I}} \cdot P_{\mathrm{I}} \\ &\quad + (\tau_{S,\mathrm{II}\to F,\mathrm{I}} \cdot \varepsilon_{S,\mathrm{II}} + \tau_{F,\mathrm{II}\to F,\mathrm{I}} \cdot \varepsilon_{F,\mathrm{II}}) \cdot \varepsilon_{F,\mathrm{I}} \cdot B_S \\ &\quad - \varepsilon_{F,\mathrm{I}} \cdot A_{\mathrm{I}} \cdot \frac{dp_{\mathrm{I}}}{dL_R} - F_{\mathrm{drag,I}}\end{aligned} \tag{5.36}$$

Solid in the flight conveying area II:

$$\begin{aligned}\frac{d(A_{\mathrm{II}} \cdot \varepsilon_{S,\mathrm{II}} \cdot \varrho_P \cdot u_{S,\mathrm{II}}^2)}{dL_R} &= -\dot{r}_{m,S} \cdot u_{S,\mathrm{I}} - \tau_{S,\mathrm{II}\to P,\mathrm{II}} \cdot \varepsilon_{S,\mathrm{II}} \cdot P_{\mathrm{II}} \\ &\quad - (\tau_{S,\mathrm{II}\to S,\mathrm{I}} \cdot \varepsilon_{S,\mathrm{II}} + \tau_{F,\mathrm{II}\to S,\mathrm{I}} \cdot \varepsilon_{F,\mathrm{II}}) \cdot \varepsilon_{S,\mathrm{I}} \cdot B_S \\ &\quad - \varepsilon_{S,\mathrm{II}} \cdot A_{\mathrm{II}} \cdot \frac{dp_{\mathrm{II}}}{dL_R} + F_{\mathrm{drag,II}}\end{aligned} \tag{5.37}$$

Gas in in the flight conveying area II:

$$\begin{aligned}\frac{d(A_{\mathrm{II}} \cdot \varepsilon_{F,\mathrm{II}} \cdot \varrho_{F,\mathrm{II}} \cdot u_{F,\mathrm{II}}^2)}{dL_R} &= -\dot{r}_{m,F} \cdot u_{S,\mathrm{I}} - \tau_{F,\mathrm{II}\to P,\mathrm{II}} \cdot \varepsilon_{F,\mathrm{II}} \cdot P_{\mathrm{II}} \\ &\quad - (\tau_{S,\mathrm{II}\to F,\mathrm{I}} \cdot \varepsilon_{S,\mathrm{II}} + \tau_{F,\mathrm{II}\to F,\mathrm{I}} \cdot \varepsilon_{F,\mathrm{II}}) \cdot \varepsilon_{F,\mathrm{I}} \cdot B_S \\ &\quad - \varepsilon_{F,\mathrm{II}} \cdot A_{\mathrm{II}} \cdot \frac{dp_{\mathrm{II}}}{dL_R} - F_{\mathrm{drag,II}}\end{aligned} \tag{5.38}$$

The terms on the right sides of (5.35)–(5.38) describe, from left to right, per unit length dL_R: the momentum transfer from solid or gas, the friction forces of solid or gas with the respective contacted pipe wall, the shear forces exerted by solid and gas from region II on solid or gas at the strand surface, region I, the pressure difference forces on the pipe element, the resistance forces due to gas flow around particles or flow through the strand.

Closure equations:

Mass transfer:

This is calculated from the height change of the strand in the considered pipe element dL_R :

$$\dot{r}_{m,S} = \frac{1}{2} \cdot \varepsilon_{S,\mathrm{I}} \cdot \varrho_P \cdot u_{S,\mathrm{I}} \cdot B_S \cdot \frac{dH_\mathrm{I}}{dL_R}, \tag{5.39}$$

$$\dot{r}_{m,F} = \frac{1}{2} \cdot \varepsilon_{F,\mathrm{I}} \cdot \varrho_{F,\mathrm{I}} \cdot u_{F,\mathrm{I}} \cdot B_S \cdot \frac{dH_\mathrm{I}}{dL_R} \tag{5.40}$$

Other approaches are possible.

Friction forces [6]:

Shear stress between solid phase I and pipe wall I:

$$\tau_{S,\mathrm{I}\to P,\mathrm{I}} = \frac{\lambda_{S,\mathrm{I}\to P,\mathrm{I}}}{8} \cdot \varrho_P \cdot u_{S,\mathrm{I}}^2 \tag{5.41}$$

Shear stress between solid phase II and pipe wall II:

$$\tau_{S,\mathrm{II}\to P,\mathrm{II}} = \frac{\lambda_{S,\mathrm{II}\to P,\mathrm{II}}}{8} \cdot \varrho_P \cdot u_{S,\mathrm{II}}^2 \tag{5.42}$$

Shear stress between gas phase I and pipe wall I:

$$\tau_{F,\mathrm{I}\to P,\mathrm{I}} = \frac{\lambda_{F,\mathrm{I}\to P,\mathrm{I}}}{8} \cdot \varrho_{F,\mathrm{I}} \cdot u_{F,\mathrm{I}}^2 \tag{5.43}$$

Shear stress between gas phase II and pipe wall II:

$$\tau_{F,\mathrm{II}\to P,\mathrm{II}} = \frac{\lambda_{F,\mathrm{II}\to P,\mathrm{II}}}{8} \cdot \varrho_{F,\mathrm{II}} \cdot u_{F,\mathrm{II}}^2 \tag{5.44}$$

Shear stress between solid phase II and solid phase I:

$$\tau_{S,\mathrm{II}\to S,\mathrm{I}} = \frac{\lambda_{S,\mathrm{II}\to S,\mathrm{I}}}{8} \cdot \varrho_P \cdot (u_{S,\mathrm{II}} - u_{S,\mathrm{I}}) \cdot \left|u_{S,\mathrm{II}} - u_{S,\mathrm{I}}\right| \tag{5.45}$$

Shear stress between gas phase II and solid phase I:

$$\tau_{F,\mathrm{II}\to S,\mathrm{I}} = \frac{\lambda_{F,\mathrm{II}\to S,\mathrm{I}}}{8} \cdot \varrho_{F,\mathrm{II}} \cdot (u_{F,\mathrm{II}} - u_{S,\mathrm{I}}) \cdot \left|u_{F,\mathrm{II}} - u_{S,\mathrm{I}}\right| \tag{5.46}$$

Shear stress between gas phase II and gas phase I:

$$\tau_{F,\mathrm{II}\to F,\mathrm{I}} = \frac{\lambda_{F,\mathrm{II}\to F,\mathrm{I}}}{8} \cdot \varrho_{F,\mathrm{II}} \cdot (u_{F,\mathrm{II}} - u_{F,\mathrm{I}}) \cdot \left|u_{F,\mathrm{II}} - u_{F,\mathrm{I}}\right| \tag{5.47}$$

Interactions between gas and solid particles:

Drag force in flow range I, supplementary equation (3.25) with gas flow through the bulk solid-/strand in the Stokes range:

$$F_{\text{drag,I}} \cong 150 \cdot A_{\text{I}} \cdot \frac{\varepsilon_{S,\text{I}}^2}{\varepsilon_{F,\text{I}}^3} \cdot \frac{\eta_F}{(d_{S,SD}^*)^2} \cdot (v_{F,\text{I}} - \varepsilon_{F,\text{I}} \cdot u_{S,\text{I}}) \tag{5.48}$$

Drag force in flow region II, gas flow around a collective of individual particles in the Stokes region, (3.3) and (3.4):

$$F_{\text{drag,II}} \cong 18 \cdot A_{\text{II}} \cdot \varepsilon_{S,\text{II}} \cdot \frac{\eta_F}{(d_{S,SD}^*)^2} \cdot (u_{F,\text{II}} - u_{S,\text{II}}) \tag{5.49}$$

Extended approaches are possible.

The conveying gas is treated as an ideal gas, (2.2) and (2.3).

By combining the momentum balances for gas and solid while eliminating the respective drag forces or pressure differences, pressure loss or motion equations for flow regions I and II can be established, for example.

The calculation model shown in Box 5.3 is described by eight ordinary differential equations with the eight unknowns ($\varepsilon_{F,\text{I}}, \varepsilon_{F,\text{II}}, H_{\text{I}}, p, u_{F,\text{I}}, u_{S,\text{I}}, u_{F,\text{II}}, u_{S,\text{II}}$) and can therefore be solved using suitable numerical calculation methods. In [3, 4], an explicit 5th order Runge-Kutta method is used for this purpose, with the calculation running from the beginning of the line in the direction of flow. In addition to the necessity of providing suitable $\lambda_{a \to b}$ values, which is particularly problematic for the forces acting on the strand surface due to only a few available measurement results/studies in the literature, values for the flow parameters at the beginning of the line or calculation start must also be defined. Here, the strand height proves to be basically freely selectable. Model calculations in [3, 4] show that with different inlet heights of the strands (→ here: $H_{I,\text{in}} \cong 0.2 \cdot D_R$ as well as $H_{I,\text{in}} \cong 0.8 \cdot D_R$) and otherwise identical conveying conditions, after a certain startup distance, strands of approximately the same thickness form, see Fig. 5.2. The specification of a large initial strand height is considered "more robust" in [4]. Large strand heights at the entry of the conveying line also correspond more closely to the operating conditions at the entry into a practical conveying system: There, the lowest gas velocity and thus the highest solid concentration is established. Supplementary example: The bulk material discharge from a pressure vessel into the conveying line is carried out by means of a portion of the total conveying gas flow as a compact bulk material strand through an outlet diameter $D_{DG} \leq D_R$. Only after that is the remaining conveying gas flow supplied. The strand portion of the conveying pipe cross-section is thus on the order of: $(1 - \phi) \cong (D_{DG}/D_R)^2$. D_{DG} is generally only one nominal pipe size smaller than D_R.

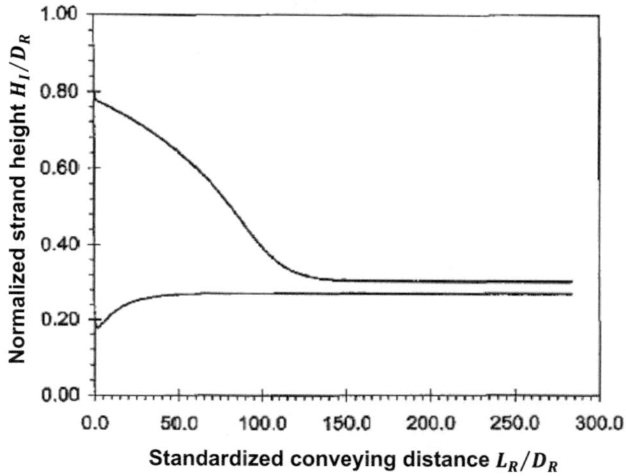

Fig. 5.2 Comparison of the effects of different strand heights at the beginning of the pipeline

In [3, 4] the model shown in Box 5.3 (→ here with: $\tau_{F,\mathrm{II}\rightarrow F,\mathrm{I}}$ and $\tau_{S,\mathrm{II}\rightarrow F,\mathrm{I}} = 0$) is used to simulate cement conveyance through a straight horizontal test section, $D_R \times L_R = 53\,\mathrm{mm} \times 20\,\mathrm{m}$, and corresponding measurement results from this system are compared. Characteristics of the cement: $d_{S,50} = 21\,\mu\mathrm{m}$, $\varrho_P = 3060\,\mathrm{kg/m^3}$, $\varrho_{SS} = 1070\,\mathrm{kg/m^3}$. Figure 5.3 shows the calculated courses of the strand thicknesses along the conveying route for the loadings $\mu = (50, 75, 100, 125)\,\mathrm{kg}\,S/\mathrm{kg}\,F$ at $H_{I,\mathrm{in}} \cong 0.8 \cdot D_R$. Control calculations with $H_{I,\mathrm{in}} \cong 0.2 \cdot D_R$ lead to largely identical strand heights after the respective run-up distance. Comparisons of the calculated strand heights with video recordings from the tests show a good "qualitative" agreement [3]. A comparison of the calculation and measurement results from the fully developed flow area (→ pressure losses, pressure gradients, etc.) leads to an overall satisfactory agreement. Details can be found in [3, 4]. Unfortunately, both works do not contain any information about the used $\lambda_{a\rightarrow b}$-values or their determination: A recalculation from the in [3, 4] used measurements is only partially possible, e.g. $\lambda_{S,\mathrm{II}\rightarrow S,\mathrm{I}}$ can only be determined indirectly with the described equipment. Alternatively, for $\lambda_{S,\mathrm{II}\rightarrow S,\mathrm{I}}$ a value optimized for the best possible agreement between measurement and calculation can also be found through systematic evaluation with the calculation model.

The above-discussed strand model does not take into account solid or gas velocity profiles in the respective flow cross-section, especially not the velocity distribution of the solid material over the strand height that can be expected for fine-grained products. So, there is still room for improvement in this regard. On the other hand, as can be seen from Sect. 5.1.1 and [3, 4], there is also potential for simplifications. With sufficient accuracy, for fine-grained bulk materials, e.g., $(\tau_{F,\mathrm{I}\rightarrow P,\mathrm{I}}, \tau_{F,\mathrm{II}\rightarrow F,\mathrm{I}}, \tau_{S,\mathrm{II}\rightarrow F,\mathrm{I}}) = 0$, $u_{S,\mathrm{I}} = u_{F,\mathrm{I}}$, $u_{S,\mathrm{II}} = u_{F,\mathrm{II}}$ can be set.

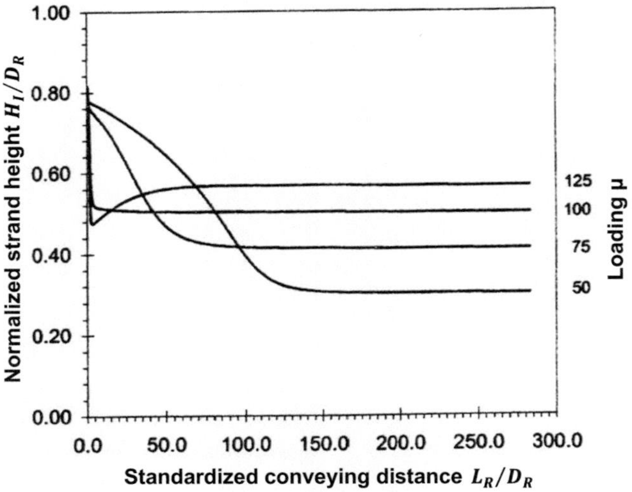

Fig. 5.3 Comparison of the strand heights and their profiles that occur at different loadings μ, $H_{I,\text{in}} \cong 0.8 \cdot D_R$

b) Simplified Model

That useful insights for practice can also be gained with an extremely simplified strand model is shown in [2]. There, the additional pressure loss caused by the bulk material $\Delta p_{S,R,h} = \Delta p$, of a horizontal straight strand conveying can be determined. For the calculation, the conveying gas is treated as sectionally incompressible, i.e., the momentum balances, (5.35–5.38), are reduced to simple force balances. Material exchange between the balance spaces I and II is not considered. The only relevant forces in the current calculation section ΔL_R are the pressure difference forces $F_{\Delta p,\text{I}}$ and $F_{\Delta p,\text{II}}$ at the balance spaces I and II, the strand driving force F_{B_S} induced by particle bombardment from the upper space II on the strand surface, and the friction force. $F_{S,f,W}$ of the solid strand at the tube bottom is considered, see Fig. 5.4. All other forces are neglected. The solid and gas velocities in the two flow areas are defined as follows:

$$u_{S,\text{I}} = u_{F,\text{I}} = u_{\text{I}}, \tag{5.50}$$

$$u_{S,\text{II}} = u_{F,\text{II}} = u_{\text{II}} \cong v_{\text{II}} \tag{5.51}$$

The respective velocity differences between gas and solid remain disregarded.

The pressure loss equation of balance area II is thus obtained as follows:

$$-\Delta p \cdot \phi \cdot A_R = F_{B_S} \tag{5.52}$$

and the one for the balance space I to:

$$-\Delta p \cdot (1 - \phi) \cdot A_R + F_{B_S} = F_{S,f,W} \tag{5.53}$$

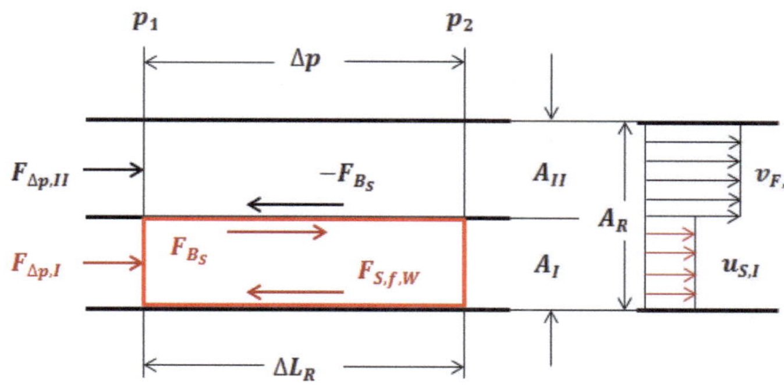

Fig. 5.4 Forces in the balance areas I and II

The friction force $F_{S,f,W}$ is calculated by multiplying the strand weight in the calculation section ΔL_R with the Coulomb's friction coefficient $\mu_W = \tan \varphi_W$, i.e., it holds:

$$F_{S,f,W} = \mu_W \cdot (\varrho_P - \varrho_F) \cdot (1 - \varepsilon_{F,\mathrm{I}}) \cdot (1 - \phi) \cdot A_R \cdot g \cdot \Delta L_R \tag{5.54}$$

From the combination of (5.52)–(5.54) the dimensionless pressure loss follows:

$$\frac{-\Delta p}{(\varrho_P - \varrho_F) \cdot (1 - \varepsilon_{F,\mathrm{I}}) \cdot \mu_W \cdot g \cdot \Delta L_R} = (1 - \phi) \tag{5.55}$$

Equating the pressure loss resolved equations (5.52) and (5.53) yields the motion equation of strand conveying:

$$F_{B_S} = \phi \cdot F_{S,f,W} \tag{5.56}$$

This must be further developed by definition of the term F_{B_S}. The strand driving force F_{B_S} is determined by the approach

$$F_{B_S} = \tau_{B_S} \cdot B_S \cdot \Delta L_R \tag{5.57}$$

with the shear stress acting on the strand surface

$$\tau_{B_S} = K \cdot \frac{\dot{m}_{S,\mathrm{II}}/(\phi \cdot A_R)}{m_P} \cdot m_P \cdot (v_{\mathrm{II}} - u_{\mathrm{I}}) = K \cdot \frac{\dot{m}_{S,\mathrm{II}}}{\phi \cdot A_R} \cdot v_{\mathrm{II}} \cdot \left(1 - \frac{u_{\mathrm{I}}}{v_{\mathrm{II}}}\right) \tag{5.58}$$

τ_{B_S} is here proportional to the number of particles with mass m_P crossing the upper space II per unit of time. The particles transfer part of their momentum to the strand upon impact. With the empirically determined coefficient K captures the actual portion of solid particles flowing through the pipe area II that impinges on the surface. The velocity v_{II} can be determined using the gas mass balance, (5.28),

$$v_{F,R} = \phi \cdot v_{\mathrm{II}} + (1 - \phi) \cdot u_{\mathrm{I}} \cdot \varepsilon_{F,\mathrm{I}} \tag{5.59}$$

with:

$v_{F,R}$ $(= \dot{m}_F/(A_R \cdot \varrho_F)$, empty pipe velocity of the total conveying gas flow $\dot{m}_F$ through the conveying pipe cross-section A_R,

taking into account the solid mass flow through the strand, $\dot{m}_{S,\mathrm{I}} = (1-\phi) \cdot A_R \cdot (1-\varepsilon_{F,\mathrm{I}}) \cdot \varrho_P \cdot u_\mathrm{I}$, to

$$v_{\mathrm{II}} = \frac{v_{F,R}}{\phi} \cdot \left(1 - \mu_\mathrm{I} \cdot \frac{\varepsilon_{F,\mathrm{I}} \cdot \varrho_F}{(1-\varepsilon_{F,\mathrm{I}}) \cdot \varrho_P}\right) \tag{5.60}$$

with:

μ_I $(= \dot{m}_{S,\mathrm{I}}/\dot{m}_F)$, loading of the strand

From the summary of (5.15) and (5.57)–(5.60), the calculation equation for the strand drive:

$$\begin{aligned} F_{B_S} &= \lambda_{B_S} \cdot \varrho_F \cdot v_{F,R}^2 \cdot D_R \cdot \Delta L_R \cdot \left(1 - \frac{u_\mathrm{I}}{v_\mathrm{II}}\right) \\ &\cdot \frac{(4 \cdot \phi \cdot (1-\phi))^{1/3}}{\phi^2} \cdot \left(1 - \mu_\mathrm{I} \cdot \frac{\varepsilon_{F,\mathrm{I}} \cdot \varrho_F}{(1-\varepsilon_{F,\mathrm{I}}) \cdot \varrho_P}\right) \end{aligned} \tag{5.61}$$

with:

λ_{B_S} $(= K \cdot \mu_\mathrm{II} = K \cdot \dot{m}_{S,\mathrm{II}}/\dot{m}_F)$, strand drive coefficient.

Inserting (5.54) and (5.61) into (5.56) yields the explicit motion equation of strand conveying:

$$\begin{aligned} &\frac{v_{F,R}^2}{g \cdot D_R \cdot \frac{\varrho_P - \varrho_F}{\varrho_F} \cdot (1-\varepsilon_{F,\mathrm{I}}) \cdot \mu_W} = Fri^2 \\ &= \frac{4}{\pi} \cdot \frac{1}{\lambda_{B_S}} \cdot \frac{\phi^3 \cdot (1-\phi)}{(4 \cdot \phi \cdot (1-\phi))^{1/3}} \cdot \frac{1}{1 - \frac{u_\mathrm{I}}{v_\mathrm{II}}} \cdot \frac{1}{1 - \mu_\mathrm{I} \cdot \frac{\varepsilon_{F,\mathrm{I}} \cdot \varrho_F}{(1-\varepsilon_{F,\mathrm{I}}) \cdot \varrho_P}} \end{aligned} \tag{5.62}$$

The term on the left side of (5.62) is to be interpreted as an extended Froude number and is represented as the square of the so-called friction number *Fri* [2, 7]. With (5.62), the relative upper space cross-section ϕ can be calculated from the usual known *Fri*-number. For this, however, the dependence of the velocity ratio $(u_\mathrm{I}/v_\mathrm{II})$ on ϕ must be known. This follows after some elementary transformations from the division of the gas mass balance, (5.59), by the solid mass flow $\dot{m}_{S,\mathrm{I}}$ of the strand to:

$$(1-\phi) = \frac{1}{1 + \left(\frac{(1-\varepsilon_{F,\mathrm{I}}) \cdot \varrho_P}{\varrho_F \cdot \mu_\mathrm{I}} - \varepsilon_{F,\mathrm{I}}\right) \cdot \frac{u_\mathrm{I}}{v_\mathrm{II}}} \tag{5.63}$$

Finally, if one introduces the approximations

$$\dot{m}_{S,\mathrm{I}} \cong \dot{m}_S, \tag{5.64}$$

$$\mu_\mathrm{I} \cong \mu \tag{5.65}$$

in (5.55), (5.62) and (5.63), then the additional pressure loss of the strand conveying can be calculated or with the six dimensionless characteristic numbers underlying this system

- Friction number: $Fri = v_{F,R}/\sqrt{g \cdot D_R \cdot (1-\varepsilon_{F,\mathrm{I}}) \cdot \mu_W \cdot (\varrho_P - \varrho_F)/\varrho_F} \rightarrow$ in general, given by the task,
- Volume flow ratio: $(\dot{V}_S/\dot{V}_F) = \mu \cdot \varrho_F/((1-\varepsilon_{F,\mathrm{I}}) \cdot \varrho_P) \rightarrow$ generally given by the task,
- relative gap volume of the strand: $\varepsilon_{F,\mathrm{I}} \rightarrow$ generally given by the task,
- dimensionless pressure loss: $(-\Delta p)/((\varrho_P - \varrho_F) \cdot (1-\varepsilon_{F,\mathrm{I}}) \cdot \mu_W \cdot g \cdot \Delta L_R) \rightarrow$ system response,
- relative pipe cross-sectional coverage of the strand: $(1-\phi) \rightarrow$ system response,
- Velocity ratio: $(u_\mathrm{I}/v_\mathrm{II}) \rightarrow$ system response,

which can be constructed in the state/pressure loss diagram of strand conveying shown in Fig. 5.5. The procedure to be followed is described in [2, 7].

The previous considerations assumed a strand sliding over the tube bottom. In the case that the force balance on the tube element ΔL_R, (5.52) and (5.53), leads to the inequality

$$-\Delta p \cdot A_R \leq F_{S,f,W} \tag{5.66}$$

the strand is coming to a standstill and only a flight conveying above the resting strand is established. Thus, it applies: $u_\mathrm{I} = 0$, $F_{B_S}/\phi \leq F_{S,f,W}$, $\mu_\mathrm{I} = 0$, $\mu = \mu_\mathrm{II}$ etc. Inserting these conditions into (5.55), (5.62) and (5.63) allows determining the pressure loss over the resting strand as well as the boundary curve between conveying with moving and stationary bulk material strands. To the left of the boundary curve "D" in Fig. 5.5, pneumatic transports over resting bulk material strands are established. Details can be found in [2, 7].

Since (5.55), (5.62) and (5.63) were created for equilibrium states, they do not allow statements about how an operating state reacts to an externally imposed disturbance. With constant friction number *Fri*, i.e., constant gas flow supplied by the pressure generator, the conveying remains stable if more solid material can be conveyed in the operating state created by the disturbance than is supplied to the system: the system then automatically returns to the initial state. The analysis of the operating points in the state diagram, Fig. 5.5, shows that this is the case when the characteristic curve of the volume flow ratio $(\dot{V}_S/\dot{V}_F) = \mu \cdot \varrho_F/((1-\varepsilon_{F,\mathrm{I}}) \cdot \varrho_P) = \text{const.}$ ($\rightarrow$ characteristic conveying curve) has a negative slope and at the same time the characteristic curve of the velocity ratio $(u_\mathrm{I}/v_\mathrm{II}) = \text{const.}$ ($\rightarrow$ characteristic curve of the flow state) has a positive slope. Of the three operating points A, B, C in Fig. 5.5, only point B meets these requirements, A

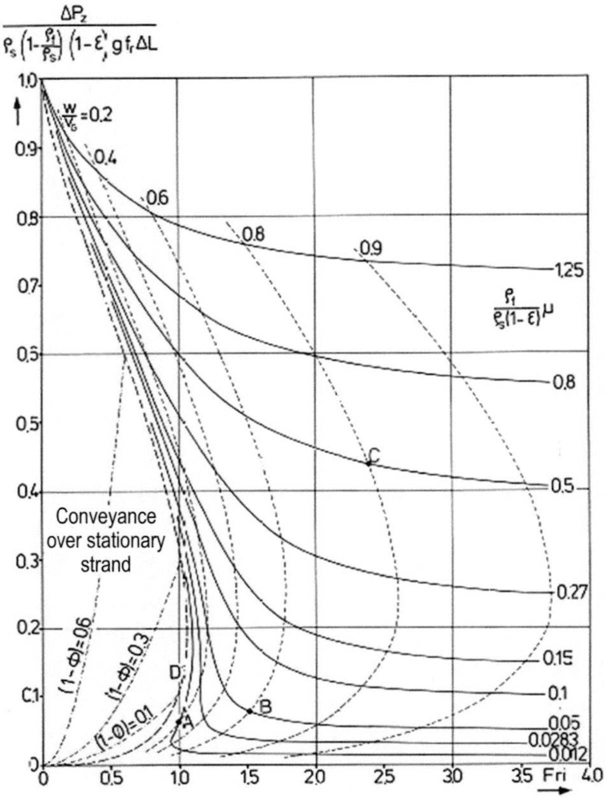

Fig. 5.5 Conveying diagram of a strand conveying, $\varepsilon_{F,\mathrm{I}} = 0.40$, $\lambda_{B_S} = 0.0826$

and C are unstable. Based on this result, boundary curves for the stable working range of a strand conveying can be calculated/defined using appropriate derivatives of the equations (5.55), (5.62) and (5.63). These are shown in Fig. 5.6. A stable strand conveyance is only possible to the right of curves "E" and "G" and below curve "F". Further details are discussed in [2, 7].

To create Figs. 5.5 and 5.6, the experimentally determined strand drive coefficient $\lambda_{B_S} = 0.0826$ and a strand porosity of $\varepsilon_{F,\mathrm{I}} = 0.40$ is used. The value $\varepsilon_{F,\mathrm{I}} = 0.40$ is suitable for describing a coarse-grained bulk material with low gas holding capacity, but not for fine-grained, fluidizable products. As already explained above, these form considerably larger relative gap volumes and do not move in the form of block flow assumed in the model. In order to arrive at somewhat realistic statements, the state diagram must therefore be recreated for each bulk material class. Example: Fig. 5.6 shows that with the coarse-grained bulk material shown there, a maximum relative pipe cross-sectional coverage by the strand (→ corresponds to the dimensionless pressure loss, (5.55)) of $(1 - \phi) \cong 0.23$ is possible. In the case of conveying the fine-grained cement

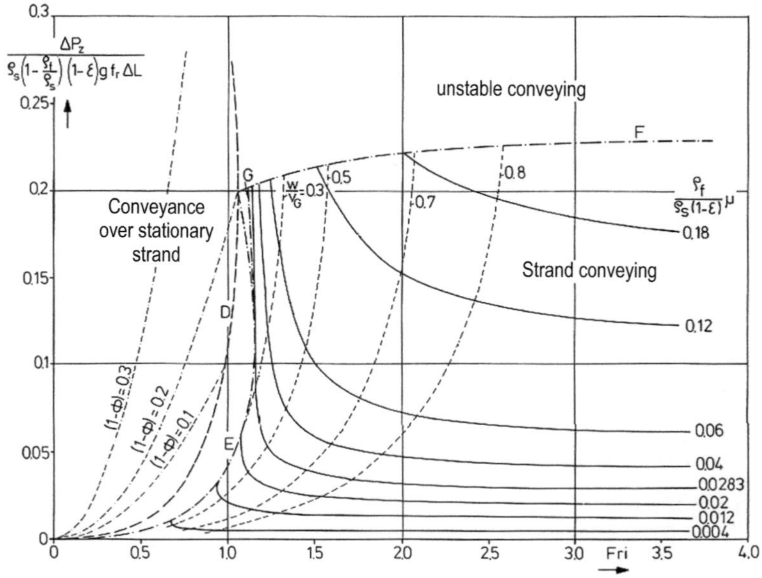

Fig. 5.6 Conveying diagram of a stable strand conveyance, $\varepsilon_{F,\mathrm{I}} = 0.40$, $\lambda_{B_s} = 0.0826$

from Fig. 5.3, loosely poured porosity $\varepsilon_F \cong 0.70$, however, normalized strand heights $(H_\mathrm{I}/D_R) > 0.50$, i.e. relative strand cross-sections $(1-\phi) > 0.50$, have been achieved. No instabilities are reported in [3, 4].

c) Remarks

The calculation models of strand conveying presented above require that the set/planned conveying state corresponds to that of a strand. This presupposes knowledge of the gas velocities corresponding to the respective bulk material, at which this flow form is established. For fine-grained, easily fluidizable solids, they are in the range of $v_{F;R} \cong (4-10)\,\mathrm{m/s}$. Since higher velocities generally occur in the rear part of conveying lines due to gas expansion, more and more solids enter the flight phase, i.e., the presented strand approaches can only be applied over limited areas of the transport route. Nevertheless, as attempted to show above, they provide well-founded indications for the design and operation of such a system. On the other hand, the reader should be shown how and with what means such problems can be approached.

Further information on pneumatic strand conveying can be found in [8–13], In [14], the model from [2], section b), is used with corresponding modifications for the calculation/description of vertically upward directed strand transports.

5.2 Plug Conveying of Coarse-Grained Bulk Materials

Bulk materials of the material group 2 described in Sect. 4.5 are coarse-grained, non-cohesive, have a relatively narrow particle size distribution, $\alpha_{RRSB} \gtrsim 60°$, and a very high gas permeability. Example: plastic granules in the mm range. They cannot be carried in flight by conveying gas streams with velocities $v_F \lesssim (10-15)$ m/s, fall out, form conglobations and plugs, which pass through the conveying route as compact units. The bulk material plugs are flowed through by the conveying gas.

Figure 5.7 shows the typical flow pattern that forms during the promotion of cohesionless bulk materials. A resting strand is deposited between the individual plugs. Each plug picks up bulk material from this strand at its front side and releases approximately the same amount at its rear side. Among other things, this results in a mechanical jamming force on the plug front. The conveying gas flows through the gap volume of the plug. From Fig. 5.7a to c, the real flow pattern is increasingly idealized with regard

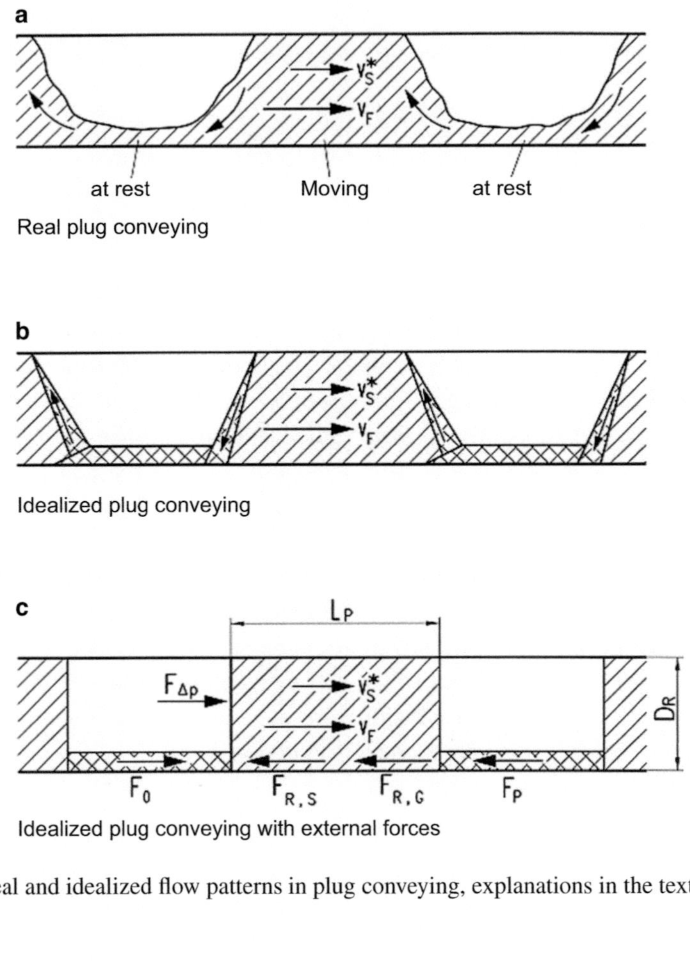

Fig. 5.7 Real and idealized flow patterns in plug conveying, explanations in the text

to a computational treatment. A set of equations suitable for calculating the conveying behavior of cohesionless individual plugs is exemplarily compiled in Box 5.4. It is based, among other things, on the idealizations of Fig. 5.7c and contains a number of simplifications and assumptions, e.g., the bulk material stresses varying over the height of the plug cross-section (→ decreasing towards the pipe apex) and flow velocities of the conveying gas (→ increasing towards the pipe apex) are not sufficiently considered. The length of the plug does not follow from the equations. It must be specified or determined from the respective operating conditions. Basic considerations for the calculation of plug conveying can be found, among others, in [8, 15–25]. The following will consider straight horizontal conveying sections.

In the bulk material of the plug, axial compressive stresses build up due to the acting mechanical forces. $\sigma_{S,\mathrm{ax}}$ in the conveying direction. These are partially deflected and result in radial compressive stresses $\sigma_{S,\mathrm{rad}}$ perpendicular to the pipe wall, see (2.15). The bulk material mechanics distinguish between active stress state (→ here: $\sigma_{S,\mathrm{ax}} > \sigma_{S,\mathrm{rad}}$) and passive stress state (→ here: $\sigma_{S,\mathrm{rad}} > \sigma_{S,\mathrm{ax}}$). Passive stress states in bulk material form when flowing through convergent channels, e.g., in a silo cone. For the calculation of plug conveying, knowledge of the stress transfer coefficient K is necessary, see (5.76) in Box 5.4. This requires a decision about the type of stress state in the plug. Due to the parallel conveying pipe walls, an active stress state is expected. Surprisingly, a series of investigation results, e.g., in [16, 21], can only be explained by assuming a passive stress state, i.e., the radial pressure stresses $\sigma_{S,\mathrm{rad}}$ are greater than the axial stresses $\sigma_{S,\mathrm{ax}}$. Figure 5.7b shows that this is possible if the plug front moves over a wedge-shaped strand sliding under the plug or if the conveying line contains deflections.

Box 5.4: Approach to describe/calculate the plug conveying of coarse-grained bulk materials

The boundary conditions of the following consideration are: conveying in the steady state, horizontal straight line, cohesionless bulk material, single homogeneous plug, i.e. $(\varepsilon_F, \varrho_S, K, \text{etc.}) = \text{const.}$ neglecting gas expansion along the plug and wall friction of the conveying gas.

From the force balance on the plug- = pipe element dL_x, i.e. on solid + conveying gas, without internal forces, see Fig. 5.8, follows:

$$dp \cdot A_R + d\sigma_{S,\mathrm{ax}} \cdot A_R + F_{Pf,W} = 0 \tag{5.67}$$

With $F_{Pf,W} = \overline{\tau}_W \cdot D_R \cdot \pi \cdot dL_x$, $\overline{\tau}_W =$ average wall shear stress, and $A_R = \pi/4 \cdot D_R^2$ goes (5.67) over to:

$$\frac{d\sigma_{S,\mathrm{ax}}}{dL_x} + \frac{dp_P}{dL_x} + \frac{4 \cdot \overline{\tau}_W}{D_R} = 0 \tag{5.68}$$

The friction force $F_{Pf,W}$ consists of two parts: on the one hand, the friction resulting from the uniformly distributed radial stress over the pipe circumference

$\sigma_{S,\mathrm{rad}} = K \cdot \sigma_{S,\mathrm{ax}}$, and on the other hand, the friction caused by the wall normal stress $\sigma_{S,G}$ of the bulk material weight in the pipe element dL_x. The latter, as shown in Fig. 5.9b, is unevenly distributed over the pipe circumference. The superposition of both components results in the normal stress distribution on the inner pipe wall shown in Fig. 5.9c. To determine $\sigma_{S,G}$, a simplified hydrostatic stress state of the bulk material over the height of the pipe diameter $D_R = 2 \cdot R$ is assumed, see Fig. 5.10 [16]. The wall normal stress $\sigma_{S,G}$ at any point on the pipe circumference is then calculated as:

$$\sigma_{S,G} = (1 + \cos\theta) \cdot \varrho_{b,P} \cdot g \cdot R \tag{5.69}$$

$\varrho_{b,P}$ is the bulk density of the plug. For the local total stress $\sigma_{S,\mathrm{tot}} = (\sigma_{S,\mathrm{rad}} + \sigma_{S,G})$ applies:

$$\sigma_{S,\mathrm{tot}} = \sigma_{S,\mathrm{rad}} + (1 + \cos\theta) \cdot \varrho_{b,P} \cdot g \cdot R \tag{5.70}$$

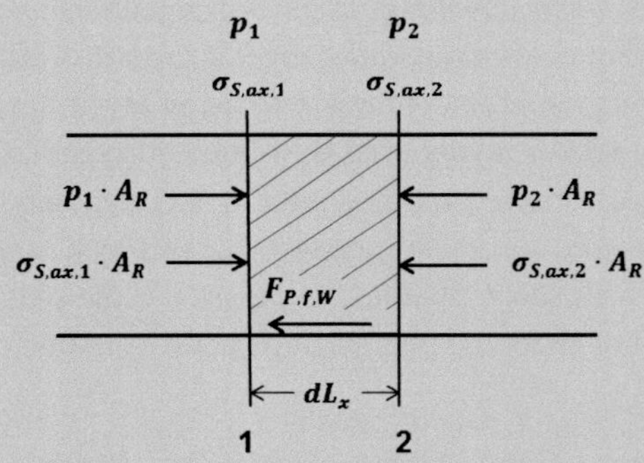

Fig. 5.8 Forces on the plug element

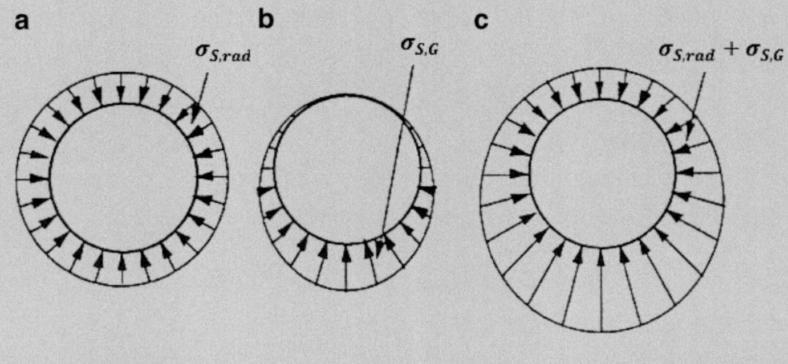

Fig. 5.9 Normal stresses on the pipe wall due to: **a** radial stresses in the bulk material, **b** bulk material weight, **c** resulting total stress

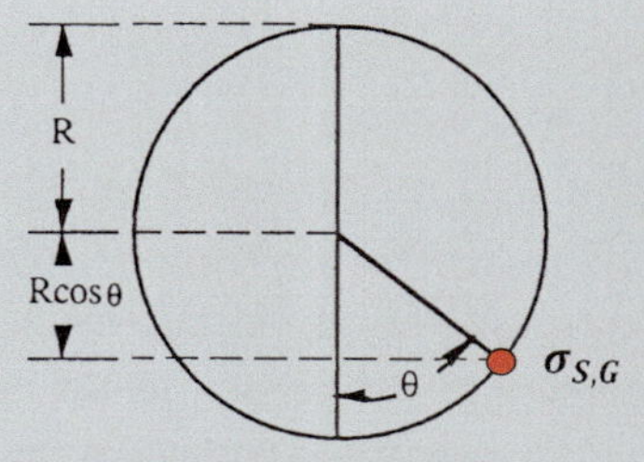

Fig. 5.10 Normal stress on pipe wall due to bulk weight

To determine the average wall shear stress $\overline{\tau}_W$, the local shear stresses $\tau_W = \mu_W \cdot \sigma_{S,\text{tot}}$, μ_W = sliding friction coefficient, integrated over the pipe circumference and normalized. It applies:

$$\begin{aligned}\overline{\tau}_W &= \frac{\int_0^{2\pi} \tau_W \cdot R \cdot d\theta \cdot dL_x}{2 \cdot R \cdot \pi \cdot dL_x} \\ &= \frac{\mu_W}{\pi} \cdot \int_0^{\pi} (\sigma_{S,\text{rad}} + (1 + \cos\theta) \cdot \varrho_{b,P} \cdot g \cdot R) \cdot d\theta \\ &= \mu_W \cdot K \cdot \sigma_{S,\text{ax}} + \frac{1}{2} \cdot \mu_W \cdot \varrho_{b,P} \cdot g \cdot D_R\end{aligned} \tag{5.71}$$

Taking into account (5.71) and the simplification

$$\frac{dp_P}{dL_P} = \frac{\Delta p_P}{\Delta L_P}$$

leads to the integration of (5.68) to the following dependence of the axial compressive stress $\sigma_{S,\text{ax}}$ on the length L_x:

$$\sigma_{S,\text{ax}} = C_{ax} \cdot \exp\left(-\frac{4 \cdot K \cdot \mu_W}{D_R} \cdot L_x\right) - \left(\frac{\Delta p_P}{\Delta L_P} + 2 \cdot \mu_W \cdot \varrho_{b,P} \cdot g\right) \cdot \frac{D_R}{4 \cdot K \cdot \mu_W} \tag{5.72}$$

C_{ax} is an integration constant with the dimension of a stress. Inserting the boundary conditions

$$L_x = L_P \rightarrow \sigma_{S,\text{ax}} = \sigma_{S,f}, \quad L_x = 0 \rightarrow \sigma_{S,\text{ax}} = \sigma_{S,b}$$

at the front and back sides of a plug in (5.72) and analysis of the two resulting equations shows that for practical plug lengths $\sigma_{S,f} \cong \sigma_{S,b}$ applies. This leads to the simple pressure loss equation [16]

$$-\frac{\Delta p_P}{L_P} = \frac{4 \cdot \mu_W \cdot K}{D_R} \cdot \sigma_{S,f} + 2 \cdot \mu_W \cdot \varrho_{b,P} \cdot g \tag{5.73}$$

with the jamming stress

$$\sigma_{S,f} = \frac{A_{\mathrm{Str}}}{A_R} \cdot \varrho_{b,\mathrm{Str}} \cdot u_{S,P}^2 = (1-\phi) \cdot \varrho_{b,\mathrm{Str}} \cdot u_{S,P}^2 \tag{5.74}$$

at the plug front. $u_{S,P}$ is the plug velocity in this case, $(1-\phi)$ and $\varrho_{b,\mathrm{Str}}$ are the cross-sectional proportion and the bulk density of the resting strand between the plugs. $A_{\mathrm{Str}}/A_R = (1-\phi)$ can be calculated with the equation

$$\frac{A_{\mathrm{Str}}}{A_R} = \frac{1}{1 + \frac{u_{S,P}}{0.542 \cdot \sqrt{g \cdot D_R}}} = \frac{1}{1 + \frac{\sqrt{Fr_P}}{0.542}} \tag{5.75}$$

according to [16]. Figure 5.11 shows an example comparison of (5.75) with measurements [26], further measurement results can be found in, for example, [18].

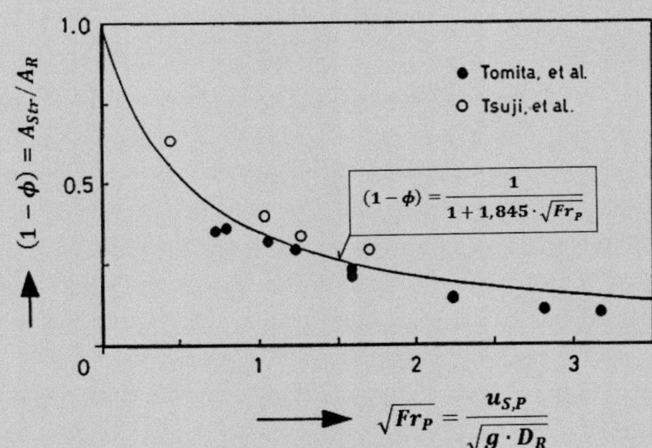

Fig. 5.11 Relative cross-section of the stationary strand between the plugs

The stress transfer coefficient K must be determined from a consideration of the Mohr's stress circle in the compressive stress/shear stress plane (σ_S, τ_S), see Fig. 5.12. Here, a distinction must be made between active and passive stress states. It applies:

$$K = \frac{\sigma_{S,\mathrm{rad}}}{\sigma_{S,\mathrm{ax}}} = \frac{1 \mp \sin\Phi \cdot \cos\omega \mp \varphi_W}{1 \pm \sin\Phi \cdot \cos\omega \mp \varphi_W} \rightarrow \text{Signum: } \frac{\text{active}}{\text{passive}} \tag{5.76}$$

with

$$\sin\omega = \frac{\sin\varphi_W}{\sin\Phi} \tag{5.77}$$

and the empirical formula for the angle Φ [18]:

$$\Phi = \frac{4}{3} \cdot \varphi_W \cdot \left(\frac{\varrho_{b,P}}{1000\,\mathrm{kg/m^3}} \right)^{1/3} \quad \rightarrow \quad \text{with: } \varphi_W \leq \Phi \leq \varphi_e \tag{5.78}$$

The meanings are: $\Phi =$ static internal friction angle of the bulk material, $\varphi_e =$ effective friction angle of the bulk material, $\omega =$ auxiliary angle in Fig. 5.12.

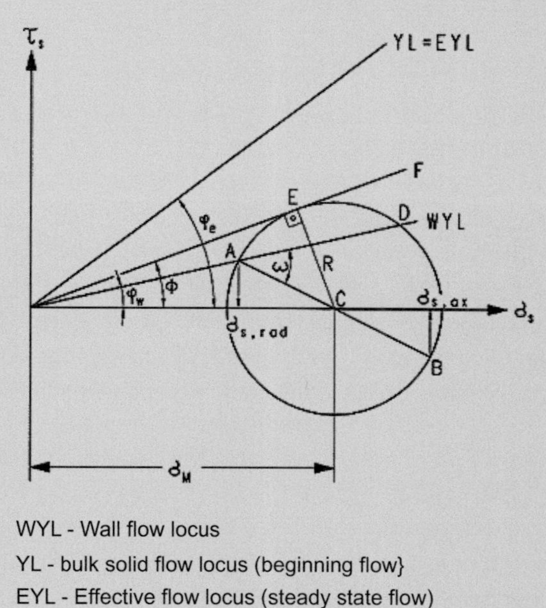

Fig. 5.12 Mohr circle representation of the stresses in the plug: The active stress state of a cohesionless bulk material is considered

The plug is flowed through by the conveying gas. The force balance on the conveying gas of the plug element results in the Ergun equation:

$$-\frac{\Delta p_P}{L_P} = f_1 \cdot \Delta v_{F,S} + f_2 \cdot \varrho_F \cdot \Delta v_{F,S}^2 \tag{5.79}$$

$$f_1 = 150 \cdot \frac{(1 - \varepsilon_{F,P})^2}{\varepsilon_{F,P}^3} \cdot \frac{\eta_F}{(d_{S,SD}^*)^2}, \tag{5.80}$$

$$f_2 = 1.75 \cdot \frac{(1 - \varepsilon_{F,P})}{\varepsilon_{F,P}^3} \cdot \frac{1}{d_{S,SD}^*} \tag{5.81}$$

The difference in velocity $\Delta v_{F,S}$ between the conveying gas and the solid/plug, related to the empty conveying pipe cross-section, is calculated using (3.31), Section 3.2.2:

$$\Delta v_{F,S} = v_F - \varepsilon_{F,P} \cdot u_{S,P} \tag{5.82}$$

To determine the still unknown solid/plug velocity $u_{S,P}$ equations (5.79) and (5.73–5.75) are equated:

$$f_1 \cdot \Delta v_{F,S} + f_2 \cdot \varrho_F \cdot \Delta v_{F,S}^2 = \frac{4 \cdot \mu_W \cdot K}{D_R} \cdot \frac{\varrho_{b,\mathrm{Str}} \cdot u_{S,P}^2}{1 + \frac{u_{S,P}}{0.542 \cdot \sqrt{g \cdot D_R}}} + 2 \cdot \mu_W \cdot \varrho_{b,P} \cdot g \tag{5.83}$$

The resolution for $u_{S,P}$ leads to the specification of the gas velocity v_F and the current boundary conditions lead to a cubic equation, from which $u_{S,P}$ can be determined using the known methods of algebra. It still needs to be determined which of the three solutions is relevant for the considered problem. It applies: $u_{S,P} < v_F/\varepsilon_{F,P}$. It is easier to calculate from (5.83) for given solid velocities $u_{S,P}$ the associated slip velocities $\Delta v_{F,S}$ and with (5.82) the gas velocities v_F required for the necessary drive. This leads to a quadratic equation with the solution

$$\Delta v_{F,S} = \frac{1}{2} \cdot \frac{f_1}{f_2 \cdot \varrho_F} \cdot \left(\sqrt{1 + 4 \cdot f_3 \cdot \frac{f_2 \cdot \varrho_F}{f_1^2}} - 1 \right) \tag{5.84}$$

with:

f_3 Term to the right of the equal sign in (5.83).

From the data determined in this way, the corresponding v_F value for the current speed $u_{S,P}$ can be determined. Alternatively, iterative methods are possible.

If the solid velocity used in (5.83), is $u_{S,P} = 0$ set, the f_3 reduces to $f_3 = 2 \cdot \mu_W \cdot \varrho_{b,P} \cdot g$ and the minimum gas velocity required for a plug drive $v_{F,\min}$ can be calculated. It holds:

$$v_{F,\min} = 42.9 \cdot \frac{1 - \varepsilon_{F,P}}{d_{S,SD}^*} \cdot \frac{\eta_F}{\varrho_F} \cdot \left(\sqrt{1 + \frac{1}{1607} \cdot \frac{\varepsilon_{F,P}^3}{(1 - \varepsilon_{F,P})^2} \cdot \frac{(d_{S,SD}^*)^3}{\eta_F^2} \cdot g \cdot \varrho_P \cdot \varrho_F \cdot \mu_W} - 1 \right) \tag{5.85}$$

with $\varrho_{b,P} = (1 - \varepsilon_{F,P}) \cdot \varrho_P$ and in dimensionless form with $d_{S,SD}^* = (\psi \cdot d_{S,SD})$:

$$Re_{\min} = 42.9 \cdot \frac{1 - \varepsilon_{F,P}}{\psi} \cdot \left(\sqrt{1 + \frac{\psi^3}{1607} \cdot \frac{\varepsilon_{F,P}^3}{(1 - \varepsilon_{F,P})^2} \cdot Ar \cdot \mu_W} - 1 \right) \tag{5.86}$$

$Re_{\min}$ and Ar are formed with $d_{S,SD}$.

With $\varepsilon_F = (1 - \varrho_{SS}/\varrho_P) \cong \varepsilon_{F,P} \cong \varepsilon_{F,\mathrm{Str}}$ and $\varrho_{SS} \cong \varrho_{b,P} \cong \varrho_{b,\mathrm{Str}}$ the above set of equations can be solved for $\Delta p_P/L_P$ and $u_{S,P}$. In general, it is necessary to iteratively adjust the pressure-dependent gas density ϱ_F.

When a stable plug moves through the conveying pipe, flowing/sliding occurs at the pipe wall, but no flow occurs within the bulk material itself. The Mohr's circle of stress in Fig. 5.12 must therefore intersect the wall flow locus WYL and may at most touch the material flow locus YL, which is identical to the effective (→ steady state) flow locus EYL for cohesionless bulk materials. The currently mobilized internal angle of friction $\varphi_i = \Phi$ is thus between the wall friction angle φ_W and the effective friction angle φ_e and must be determined experimentally. (5.78) in Box 5.4 provides reference values.

The evaluation of (5.85), which assumes cohesionless bulk materials, results in linear dependencies $u_{S,P}(v_F)$ or $u_{S,P}(u_F)$. Figure 5.13 shows exemplary measurement results in this regard from [18]. Corresponding curves in [17] can only be considered approximately linear. The intersection points of the curves in Fig. 5.13 with the abscissa provide the minimum empty tube gas velocities required for plug transport $v_{F,\min}$. Measured $v_{F,\min}$-values are by a factor $f_{\min} \cong (1.5-2.0)$ larger than those calculated with (5.85). The reason for this is that the considerations in Box 5.4 are based on a steady-state transport and thus the initial acceleration/inertia of the plug is not taken into account.

A comparison of (5.86) with that for determining the loosening velocity $v_{F,L}$ of a fluidized bed, (3.37), illustrates that $v_{F,\min} = v_{F,L}$ is obtained when the wall friction coefficient $\mu_W = 0.5$ amounts to. Even with other μ_W-values, the differences are small. According to the author's experience, a safety factor $f_{\min} \cong 1.75$ multiplied "measured" minimum fluidizing velocity $v_{F,L}$ gives a safe value for $v_{F,\min}$.

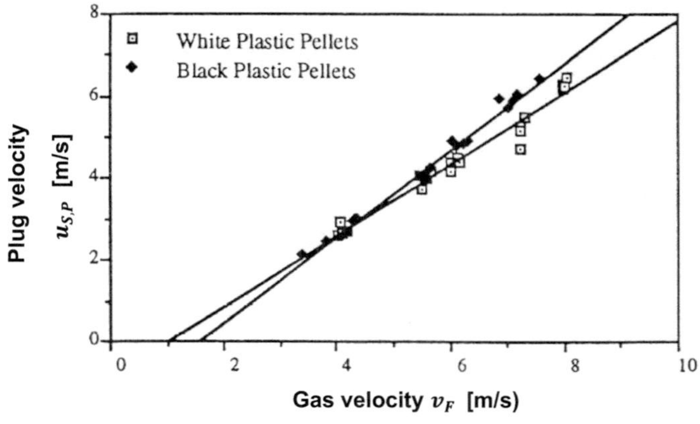

Fig. 5.13 Measured dependencies $u_{S,P}(v_F)$ of different plastic pellets: White/black pellets: $d_S = 3.12/3.76\,\mathrm{mm}$, $\varrho_P = 865/834\,\mathrm{kg/m^3}$, $\varepsilon_F = 0.431/0.451$, $\mu_W = 0.27/0.23$ against normal steel, $K = 0.756/0.806$, $D_R = 105\,\mathrm{mm}$, according to [18]

In order to finally be able to design a plug conveyance, the considerations on the individual plug must be extended to a sequence of plugs. As long as there is a linear dependency between gas pressure loss Δp_P and plug length L_P, see Sect. 4.4, for the cohesionless coarse-grained bulk materials studied so far, the individual plugs in the conveying line can be combined into a plug of the same total length. $L_{P,\text{tot}} = \sum L_{P,i}$ summed up and treated as a long single plug. For this, the bulk material filling degree Ψ of the current conveying line must be known. This follows from

$$\Psi = \frac{\Delta m_S}{V_R \cdot \varrho_{b,P}} = \frac{\dot{m}_S \cdot \Delta\tau_S}{A_R \cdot L_R \cdot \varrho_{b,P}} = \frac{\dot{m}_S \cdot L_R}{A_R \cdot L_R \cdot \varrho_{b,P} \cdot u_S} = \frac{\dot{m}_S}{A_R \cdot \varrho_{b,P} \cdot u_S} \tag{5.87}$$

with:

$\dot{m}_S$ solid mass flow rate of the considered system,
$\Delta\tau_S$ average residence time of the conveyed solid in the conveying line.

Thus, the following applies:

$$L_{P,\text{tot}} = \Psi \cdot L_R = \frac{\dot{m}_S}{A_R \cdot \varrho_{b,P} \cdot u_S} \cdot L_R \tag{5.88}$$

Using $L_{P,\text{tot}}$ instead of L_P in (5.73) and (5.79) allows calculating the total pressure loss of the conveying line.

This is no longer possible with a non-linear relationship $\Delta p_P = \Delta p_P(L_P)$, as found in fine-grained and cohesive bulk materials. For this, the interparticle cohesion and the particle/pipe wall adhesion must be taken into account in the above equations, see e.g. [16]. This leads to a complication of the calculation approaches and requires the individual consideration of the plugs along the conveying path. For fine-grained bulk materials in general and for the conveying of coarse-grained products at higher conveying pressures, the gas expansion along the plug can no longer be neglected.

The calculation approach shown in Box 5.4 is based on the very far-reaching idealizations of Fig. 5.7c, e.g. solid and plug velocities are equated. This simplifies, among other things, the description of the complicated processes taking place at the plug front when the resting strand is taken up into the plug. More detailed analyses can be found in e.g. [16, 25]. The approach in Box 5.4 is intended as a framework for more in-depth considerations. The required characteristic data, e.g. ε_F, ϱ_P, $d^*_{S,SD}$, $\varphi_W(\mu_W)$, φ_e, $v_{F,L}$, K, can be determined by means of laboratory experiments. Vertical plug conveyance is discussed, among others, in [16, 27, 28]. The practical implementation of plug conveyance is addressed in Sect. 6.1.

5.3 Scale-up of Conveying Systems

For the design of a pneumatic conveying system, at least two parameters must be known:

- The smallest conveying gas velocity, at which solid transport is just barely possible,
- the pressure loss of the conveying gas required for conveying.

Reputable suppliers of pneumatic conveying systems operate experimental conveying lines on a technical/industrial scale, from whose measurement results, with sufficient instrumentation with sensors, the process-characterizing parameters, e.g. $\lambda_S(v_F)$, can be calculated. For a transfer of the results to higher solid flow rates/other conveying pipe diameters, at least one more, preferably geometrically similar test line with a significantly different pipe diameter is required. The characteristic values resulting from the tests on these systems must be transformed into scale-up-appropriate dimensionless functions, with the help of which the target plant can be calculated. Since different flow forms/modes are generally passed through in technical conveying systems due to gas expansion, this requires both a step-by-step calculation of the test values and a corresponding pressure loss calculation of the new plant, and thus a considerable evaluation and calculation effort. For a new product, extensive conveying tests are therefore absolutely necessary for a safe and optimized design.

Due to the indicated problems, in practice, the approach is often taken to directly convert the results obtained from test plants using suitable scale-up methods to the plant to be designed. This simplifies the design considerably and allows direct access to test databases and the operating results of executed plants. Flexible, updatable model approaches are required for this, which can be adapted to the conveying behavior of a wide variety of bulk materials using appropriate measurement results. The scale-up system presented below is oriented towards the requirements of industrial tasks, has been used for practical plant designs for about 30 years, and has proven to be excellent [29].

The test results from a test plant with conveying distance L_R, the pipe diameter D_R as well as a characteristic conveying velocity $\Delta v_{F,A}$ obtained for solid mass flow rate $\dot{m}_S$ and conveying line pressure loss Δp_R are directly scaled up or converted to the operating data of the planned target system. This requires two steps:

1. Conversion of the test data to a conveying line geometrically similar to the test/reference system with the length, diameter, and solid throughput of the operating system.
2. Correction of the results from the reference geometry to the current conveying line geometry of the planned system.

Due to the large number of possible influencing factors and the generally only three relevant basic dimensions (L, M, T), see Sect. 2.4.2, the scale-up calculations for point 1 are not carried out using dimensionless numbers determined by dimensional analysis, but by means of simple power approaches/functions. Furthermore, and to increase the design accuracy, approaches are only created for individual specific bulk materials or bulk material classes. The following considerations are limited to conventional dilute and dense phase conveying through unstepped conveying lines. Application to stepped lines, pipes with internal installations, or secondary conveying gas addition is possible in a simple manner.

5.3.1 Minimum Velocity, Reference State

The concept of the required minimum velocity $v_{F,A,\min}$ at the entrance of horizontal conveying lines was already presented in Sect. 4.6.2. The following considerations, i.e., the experimental evaluation and the conversion to the current operating state, are carried out using (4.13). The solid constants (K_v, k, l) must be determined by the experiments.

Conveying is designed with a distance to the minimum velocity $v_{F,A,\min}$ appropriate for the task. The operating velocity $v_{F,A}$ thus follows from (4.16), $v_{F,A} = v_{F,A,\min} + \Delta v_{F,A}$. From the considerations of Sect. 4.6.2, it can be inferred that for $\Delta v_{F,A}$ the proportionality/approximation

$$\Delta v_{F,A} = v_{F,A} - v_{F,A,\min} \cong u_{S,A} \tag{5.89}$$

applies. Since the smallest gas velocity along the conveying line occurs at the beginning of the conveying line, this "critical" state is defined as the reference state for further scale-up considerations. As the characteristic gas velocity $\Delta v_{F,A}$, the distance of the current gas starting velocity $v_{F,A}$ to the corresponding current minimum speed $v_{F,A,\min}$ is used. The justification for this approach is based on Fig. 5.14: This shows that an experimental evaluation with gas speeds $v_{F,A}$ = const. leads to conveying speeds at different distances from the conveying limit and would therefore compare different conveying states with each other. Depending on the respective Δp_R- and/or D_R-values, gas speeds may also result that are not feasible. In the model presented here, states with the same distance $\Delta v_{F,A}$ to the conveying limit $v_{F,A,\min}$ are therefore evaluated, i.e., with $\Delta v_{F,A}$ = const. For each of the Δp_R-curves in Fig. 5.14, a new coordinate system is thus spanned along the $v_{F,A}$-axis with the origin $\Delta v_{F,A} = 0$ at $v_{F,A} = v_{F,A,\min}$. This takes into account the fact that below $v_{F,A,\min}$ no solid transport occurs. $\Delta v_{F,A}$ can therefore be interpreted as the characteristic solid velocity $u_{S,A}$.

5.3.2 Pipeline Pressure Loss

Since the majority of the total pressure loss of a conveying route generally consists of horizontal bulk material friction, a corresponding approach is chosen for the scale-up. This has the form:

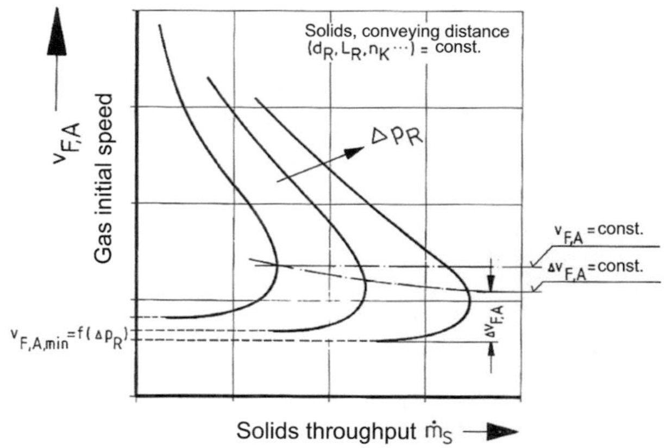

Fig. 5.14 Experimental evaluation with $v_{F,A}$ = konst. and $\Delta v_{F,A}$ = konst.

$$-\Delta p_R = K_R(\Delta v_{F,A}) \cdot \frac{\dot{m}_S^x \cdot L_R}{D_R^y} \tag{5.90}$$

$(K_R(\Delta v_{F,A}), x, y)$ must be determined experimentally. $K_R(\Delta v_{F,A})$ is the resistance coefficient or proportionality factor, which depends on $\Delta v_{F,A}$ and determines the pressure loss in the pipeline. Δp_R is linked with the other quantities. For the exponent x, $x \leq 1$ applies, since measurements show that the pressure loss coefficient $\lambda_{S,h}$ becomes smaller with increasing load μ, see, for example, Fig. 4.23. A linear dependency is assumed between Δp_R and L_R, which initially assumes an incompressible transport medium. The influence of the pipe diameter D_R leads to a dependency $y \gtrsim 2$. The simple application of the above approach requires velocity-independent exponents x and y.

The structure of (5.90) should be recognizable in the approaches proposed in the literature for calculating straight horizontal conveying sections. Therefore, some theoretical and empirical approaches from the literature for pressure loss calculation are analyzed in this sense. All examined approaches describe conveyances with incompressible flow. Their application to compressible flows requires the integration or sectional application of the corresponding equations. Both cases are considered one after the other.

The calculation model according to Barth/Muschelknautz [30, 31] described in Sect. 4.7.1 can be applied—at least formally—over the entire working range of pneumatic conveyances. Box 5.5 shows its transformation into a representation according to (5.90). The diameter exponent y is, as can be seen from (5.91)–(5.93), a velocity-dependent quantity with the limit values $y = 2$ in the dense phase/slow conveying range and $y = 3$ in the lean phase conveying area. The solids throughput exponent x results in $x = 1$. The characteristic conveying speed is the solid speed u_S. The right sides of (5.91–5.93) in Box 5.5 contain, besides u_S and D_R exclusively solid-specific characteristics, so that the coefficients of the particle velocity u_S can be considered constant for a given conveyed material. The function $(1/K_R) = f(u_S)$ according to (5.91) increases with increasing speed u_S starting from the coordinate origin up to a maximum value, and then decreases again.

Box 5.5: Transformation of the calculation approach according to Barth/ Muschelknautz

Barth/Muschelknautz [30, 31], **incompressible:**

$$\Delta p_R = \Delta p_F + \Delta p_S \quad \rightarrow -\Delta p_F \ll -\Delta p_S$$

$$-\Delta p_R = -\Delta p_S = \mu \cdot \lambda_S \cdot \frac{L_R}{D_R} \cdot \frac{\varrho_F}{2} \cdot v_F^2$$

with

$$\lambda_S = C \cdot \lambda_S^* + \frac{2 \cdot \beta_R}{C \cdot Fr_R}; \; C = \frac{u_S}{u_F} \cong \frac{u_S}{v_F}; \; Fr_R = \frac{v_F^2}{g \cdot D_R}; \; \mu = \frac{\dot{m}_S}{\dot{m}_F}$$

follows:

$$-\Delta p_R = \frac{\dot{m}_S \cdot L_R}{D_R^2} \cdot \left(\frac{2 \cdot \lambda_S^*}{\pi} \cdot \frac{1}{D_R} \cdot (C \cdot v_F) + \frac{4 \cdot g \cdot \beta_R}{\pi} \cdot \frac{1}{(C \cdot v_F)} \right)$$

as well as:

$$\left(\frac{1}{K_R}\right) = \left(\frac{\dot{m}_S \cdot L_R}{(-\Delta p_R) \cdot D_R^2}\right) = \frac{\frac{\pi}{4 \cdot g \cdot \beta_R} \cdot (C \cdot v_F)}{1 + \frac{\lambda_S^*}{2 \cdot g \cdot \beta_R} \cdot \frac{1}{D_R} \cdot (C \cdot v_F)^2} = \frac{a_0 \cdot u_S}{1 + \frac{b_0}{D_R} \cdot u_S^2} \quad (5.91)$$

Limiting cases:

a) Flight conveying/high conveying speeds:

$$\rightarrow (C \cdot \lambda_S^*) \gg \left(\frac{2 \cdot \beta_R}{C \cdot Fr_R}\right)$$

$$\left(\frac{1}{K_R}\right) = \left(\frac{\dot{m}_S \cdot L_R}{(-\Delta p_R) \cdot D_R^3}\right) = \frac{\pi}{2 \cdot \lambda_S^*} \cdot \frac{1}{(C \cdot v_F)} = \frac{b_1}{u_S} \quad (5.92)$$

b) Dense phase conveying/low conveying speeds:

$$\rightarrow (C \cdot \lambda_S^*) \ll \left(\frac{2 \cdot \beta_R}{C \cdot Fr_R}\right)$$

$$\left(\frac{1}{K_R}\right) = \left(\frac{\dot{m}_S \cdot L_R}{(-\Delta p_R) \cdot D_R^2}\right) = \frac{\pi}{4 \cdot g \cdot \beta_R} \cdot (C \cdot v_F) = a_1 \cdot u_S \quad (5.93)$$

In Box 5.6, three additional empirical approaches are exemplarily presented, which are particularly proposed for the calculation of dense phase conveying, and are transformed according to (5.90) and compared with each other. All other investigated approaches, e.g., [32–34], can be transformed in an analogous manner. Table 4.2 in [35] contains a compilation of "approximation equations" for calculating the pressure loss in conveying lines, which have the same structure as (5.90). The difference to this is that it is calculated with a constant, i.e., velocity-independent, coefficient K_R or this must be adapted to the respective operating conditions. Further calculation approaches, which were analyzed to secure the calculation process, can be found in [29]. All examined approaches have in common that they can generally be directly converted into a (5.90) analogous form. The right sides of these equations contain, in addition to the characteristic conveying speed, generally that of the solid u_S, almost exclusively solid properties, rarely those of the conveying gas (= gas density ϱ_F), in some cases the conveying pipe diameter D_R. The power product $(1/K_R)$ is thus, for a given conveyed material, essentially a function of the conveying speed u_S. The experimentally derived curves of $(1/K_R)$ consistently show increasing curves at low velocities and decreasing curves at high velocities. As a result, throughput exponents have values $x \leq 1$, diameter exponents y lie between $y = (0.5-3.0)$. Since only incompressible flows have been considered so far, the dependence of the pressure loss Δp_R on the delivery distance L_R is linear by definition.

Box 5.6: Calculation approaches for the dense flow region, incompressible

Wen/Simons [36]:

$$\left(\frac{1}{K_R}\right) = \left(\frac{\dot{m}_S \cdot L_R}{(-\Delta p_R) \cdot D_R^{2.25}}\right) = \frac{\pi}{4 \cdot g \cdot a_0} \cdot \frac{1}{d_{S,50}^{0.25}} \cdot u_S^{0.55} = a_1 \cdot u_S^{0.55} \quad (5.94)$$

with: $\frac{v_F}{\varepsilon_F} \cong 2 \cdot u_S$.

Stegmaier [37]:

$$\left(\frac{1}{K_R}\right) = \left(\frac{\dot{m}_S^{0.7} \cdot L_R}{(-\Delta p_R) \cdot D_R^{1.3}}\right) = \frac{a_0}{g} \cdot \frac{d_{S,50}^{0.1}}{Fr_{S,T}^{0.25}} \cdot \frac{v_F^{0.7}}{\varrho_F^{0.3}} = a_1 \cdot \frac{v_F^{0.7}}{\varrho_F^{0.3}} \quad (5.95)$$

with: $Fr_{S,T} = \frac{v_T^2}{g \cdot d_{S,50}}$.

Ostrovskii/Krivoi/Sokolov/Isakov [38]:

$$\left(\frac{1}{K_R}\right) = \left(\frac{\dot{m}_S \cdot L_R}{(-\Delta p_R) \cdot D_R^{2.22}}\right) = \frac{\pi}{4 \cdot a_0} \cdot \varrho_F^{0.5} = a_1 \cdot \varrho_F^{0.5} \quad (5.96)$$

The transition to conveyance with compressible flow requires the integration of the presented pressure loss approaches, taking into account the gas and solid accelerations along the conveying route. L_R. This is only possible approximately with strong simplifications, e.g., a constant velocity ratio $C = u_S/u_F \cong u_S/v_F$. Box 5.7 shows the integrated approaches from Box 5.5 and 5.6 with the corresponding simplifications. The reference state is the one at the beginning of the conveying line. As a consequence, it follows that in the integrated equations, the original pressure differences Δp_R must be replaced by pressure functions $f(p)$, the form of which is determined by the pressure-dependent quantities on the right-hand sides of the initial equations. The equation structure and exponents remain unchanged. The unsuitable pressure functions $f(p)$ can be replaced by simple power approaches over limited pressure ranges.

$$f(p) = a \cdot (-\Delta p_R)^z \tag{5.97}$$

For the operating case $p_E = 1.0$ bar these are also shown in Box 5.7.

Box 5.7: Calculation approaches for conveyance with compressible, isothermal flow

Barth/Muschelknautz [30, 31]**, Limit cases:**

a) Flight conveying/high conveying speeds:

$$\rightarrow (C \cdot \lambda_S^*) \gg \left(\frac{2 \cdot \beta_R}{C \cdot Fr_R}\right), \quad C, \lambda_S^* = \text{const.}$$

$$\left(\frac{1}{K_R}\right) = \left(\frac{\dot{m}_S \cdot L_R}{f(p) \cdot D_R^3}\right) = \frac{\pi}{4 \cdot \lambda_S^*} \cdot \frac{1}{(C \cdot v_{F,A})} = \frac{b_2}{u_{S,A}} \tag{5.98}$$

$$f(p) = \frac{p_A^2 - p_E^2}{p_A} = 1.50 \cdot (-\Delta p_R)^{0.85} \tag{5.99}$$

b) Dense phase conveying/low conveying speeds:

$$\rightarrow (C \cdot \lambda_S^*) \ll \left(\frac{2 \cdot \beta_R}{C \cdot Fr_R}\right), \quad C, \beta_R = \text{const.}$$

$$\left(\frac{1}{K_R}\right) = \left(\frac{\dot{m}_S \cdot L_R}{f(p) \cdot D_R^2}\right) = \frac{\pi}{4 \cdot g \cdot \beta_R} \cdot (C \cdot v_{F,A}) = a_1 \cdot u_{S,A} \tag{5.100}$$

$$f(p) = p_A \cdot \ln\left(\frac{p_A}{p_E}\right) = 1.41 \cdot (-\Delta p_R)^{1.25} \tag{5.101}$$

Wen/Simons [36]:

$$\rightarrow v_F/\varepsilon_F \cong 2 \cdot u_S = \text{const.}$$

$$\left(\frac{1}{K_R}\right) = \left(\frac{\dot{m}_S \cdot L_R}{f(p) \cdot D_R^{2.25}}\right) = \frac{\pi}{4 \cdot g \cdot a_2} \cdot \frac{1}{d_{S,50}^{0.25}} \cdot u_{S,A}^{0.55} = a_3 \cdot u_{S,A}^{0.55} \tag{5.102}$$

$$f(p) = p_A - p_A^{0.55} \cdot p_E^{0.45} = 0.54 \cdot (-\Delta p_R)^{1.14} \tag{5.103}$$

Stegmaier [37]:

$$\left(\frac{1}{K_R}\right) = \left(\frac{\dot{m}_S^{0.7} \cdot L_R}{f(p) \cdot D_R^{1.3}}\right) = \frac{a_0}{g} \cdot \frac{d_{S,50}^{0.1}}{Fr_{S,T}^{0.25}} \cdot \frac{v_{F,A}^{0.7}}{\varrho_{F,E}^{0.3}} = a_2 \cdot \frac{v_{F,A}^{0.7}}{\varrho_{F,E}^{0.3}} \tag{5.104}$$

$$f(p) = p_A^{0.70} \cdot p_E^{0.30} \cdot \ln\left(\frac{p_A}{p_E}\right) = 1.13 \cdot (-\Delta p_R)^{1.07} \tag{5.105}$$

Ostrovskii/Krivoi/Sokolov/Isakov [38]:

$$\left(\frac{1}{K_R}\right) = \left(\frac{\dot{m}_S \cdot L_R}{f(p) \cdot D_R^{2.22}}\right) = \frac{\pi}{6 \cdot a_0} \cdot \varrho_{F,E}^{0.5} = a_2 \cdot \varrho_{F,E}^{0.5} \tag{5.106}$$

$$f(p) = \frac{p_A^{1.5} - p_E^{1.5}}{p_E^{0.5}} = 1.86 \cdot (-\Delta p_R)^{1.23} \tag{5.107}$$

From the above, it follows that the approach according to (5.90) must be extended. Taking into account (5.97) leads to the generalized dependency:

$$\left(\frac{1}{K_R}\right) = \left(\frac{\dot{m}_S^x \cdot L_R}{(-\Delta p_R)^z \cdot D_R^y}\right) = f(\Delta v_{F,A}) \tag{5.108}$$

The exponent z describes, with otherwise constant boundary conditions, the length dependency $(-\Delta p_R) \propto L_R^{1/z}$ of the pressure loss in the pipeline. It is, as in e.g., (5.99), flight conveying, and (5.101), slow conveying, illustrated in Box 5.7, dependent on the velocity. This can be illustrated using Fig. 5.15: There, the expansion of a conveying gas in two differently long conveying lines from different starting points is considered in the state diagram. The following applies: $(\dot{m}_S, D_R) = \text{const}$. If the operating states at the beginning and end of the pipeline are on the left descending branch of the $\dot{m}_S$-characteristic, the pressure loss per unit length $(\Delta p_R/\Delta L_R)$ decreases continuously along the conveying path, see Fig. 5.15a. The average pressure loss per unit length $\overline{(\Delta p_R/\Delta L_R)}$ of a conveyor line decreases here with increasing conveying distance L_R. This leads in the relationship $(-\Delta p_R) \propto L_R^{1/z}$ to an exponent $(1/z) < 1$, i.e., to $z > 1$. Exactly opposite relationships result on the right ascending $\dot{m}_S$-characteristic branch, Fig. 5.15b. Here, $z < 1$. In the intermediate area, i.e., around the pressure loss minimum, Fig. 5.15c, $z \cong 1$ can be set. This area predominantly covers the working range of technical conveyances. With increasing back pressure = pressure at the end of the line p_E, the exponent z also approaches the value 1. The size and course of z are furthermore dependent on the chosen reference state, here: beginning of the delivery line.

Since the inclusion of the dependency $(-\Delta p_R) \propto L_R^{1/z}$ would drastically increase the required experimental effort to determine the exponents/coefficients, etc. in (5.108), see Sect. 5.3.3, we will first continue and evaluate with (5.90), i.e. $z = 1$. The resulting practical experiences confirm the admissibility of this approach.

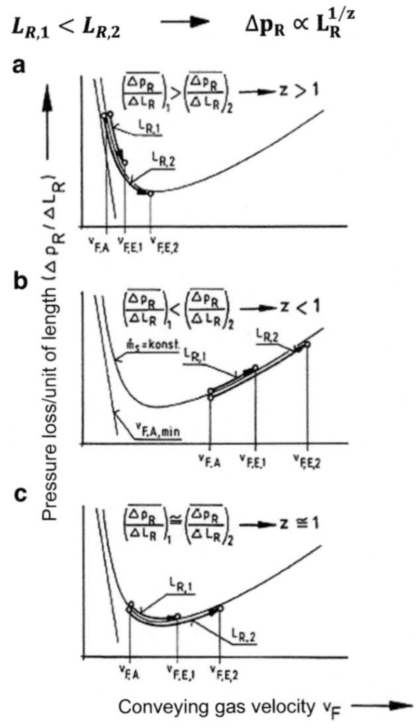

Fig. 5.15 Velocity dependence of the pressure loss exponent z

With (5.90), the operating data of the target system can be determined from the results of the experimental/reference system, see Sect. 5.3.5. However, these results are only correct for a pipeline layout of the operating system that is geometrically similar to that of the reference pipeline, i.e., it is assumed that, for example, the number of 90° deflections per unit length (N_{90°/L_R), the relative height changes ($\Delta L_{v,i}/L_R$), and their relative distances to the beginning of the pipeline ($\Delta L_{A\rightarrow v,i}/L_R$) etc. are identical in the test and target systems. Since this is generally not the case, the pressure loss of the planned system must be corrected accordingly. Practical experience with the presented model shows that the pressure loss differences due to differences between reference and actual target system geometry are usually very small. This allows the corrections to be kept simple. For example, a too large or too small 90° bend number ΔN_{90° can be calculated using (4.57) and $((-\Delta p_{S,U}) \cdot \Delta N_{90^\circ})$ and subtracted from or added to the total pressure loss of the target system determined with (5.90). When using (4.57), it is generally sufficiently accurate to use the pressure loss $(-\Delta p_{S,U})$ of all deflections. ΔN_{90° with the average gas/solid velocity along the conveying route to be determined. In a similar way, the other pressure loss components to be corrected can be dealt with. In all cases, the equations of Sect. 4.7 can be used. Further details can be found in [29].

5.3.3 Required Experimental Effort

The scale-up model described above specifies the experimental procedure that the solid to be conveyed must be subjected to *at least*. When evaluating according to (5.90), i.e., $z = 1$, the quantities K_v, k, l, x and y as well as the dependency $(1/K_R(\Delta v_{F,A}))$ must be determined specifically for the solid. Figure 5.16 shows the minimally required experimental program. At two equally long, preferably parallel, i.e., geometrically similar, conveying routes $L_{R,1} = L_{R,2}$ of different diameters $D_{R,1} \neq D_{R,2}$ two characteristic curves $(\Delta p_{R,1}, \Delta p_{R,2}) = \text{const.}$ on each line down to the conveying limit $v_{F,A,\min}$ must be measured. $D_{R,1}$ and $D_{R,2}$ as well as $\Delta p_{R,1}$ and $\Delta p_{R,2}$ should be as far apart as possible. The pressure differences $\Delta p_{R,1}$ and $\Delta p_{R,2}$ are the same in both conveying lines. Section 5.3.4 shows the practical evaluation. To also take into account the exponent z, systematic experiments on another conveying line with, for example, $D_{R,3} = D_{R,1}$ or $D_{R,2}$ as well as $L_{R,3} \neq L_{R,1}$ or $L_{R,2}$ can be performed. The difference between $L_{R,3}$ and $L_{R,1}$ or $L_{R,2}$ should be large, and the course of this line should follow that of the sections $L_{R,1}$ or $L_{R,2}$ be geometrically similar. If no geometrically similar test routes are available, a reference line profile must be defined and the measured results converted to this.

The author's investigations were usually carried out on the two reference conveying sections with the dimensions $D_R \times L_R = 52.4\,\text{mm} \times 152\,\text{m}$ and $82.5\,\text{mm} \times 152\,\text{m}$, which are still the basis of every scale-up. Each new bulk material is always measured/investigated in this way. For the reference line profile, the following applies: $(N_{90°}/L_R) = 5.25$ 90°-bends/100 m, a height section $(\Delta L_{v,i}/L_R) = 4\,\%$, $(\Delta L_{A\to v,i}/L_R) = 5\,\%$. Measurements on other lines are converted to the reference course.

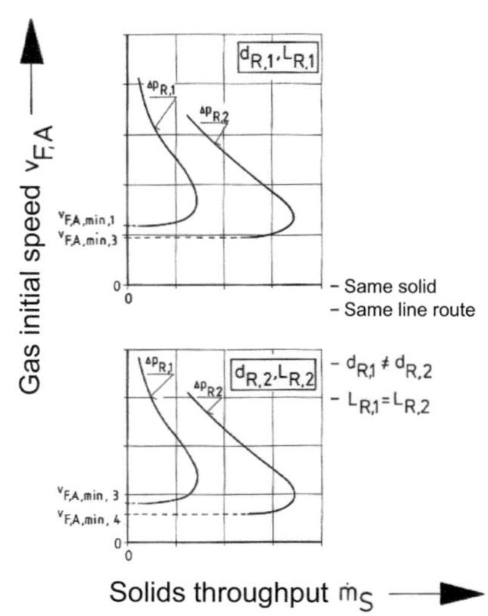

Fig. 5.16 Experiments to be carried out for (5.90)

5.3.4 Experimental Verification

With (5.90), extensive systematic conveying tests in operational scale, including [39–41], as well as the practical results of executed plants were evaluated. The examined solid material range includes both fine and coarse-grained bulk materials.

a) Exponent x

The exponent x in (5.90) describes the dependence of the conveying line pressure loss Δp_R on the solid throughput $\dot{m}_S$ with simultaneously constant values of D_R, L_R and $\Delta v_{F,A}$. A corresponding evaluation for the bulk material lignite dust with the data $d_{S,50} \cong 49\,\mu\text{m}$, $\varrho_P = 1400\,\text{kg/m}^3$ is shown in Fig. 5.17. The distance to the conveying limit is constantly $\Delta v_{F,A} = 2.0\,\text{m/s}$. In double logarithmic representation, length- and diameter-independent parallel lines result. The exponent x is identical to their slope and thus a constant for Fig. 5.17. The abscissa used $(\Delta p_R/\Delta L_R)$ clarifies that there is a linear dependency between Δp_R and L_R (→ see in particular the conveying lines with $D_R = 82.5\,\text{mm}$).

In Fig. 5.18, exponents x of the same lignite dust are plotted as a function of the distance $\Delta v_{F,A}$ to the respective conveying limit $v_{F,A,\min}$. In the investigated velocity range, x can

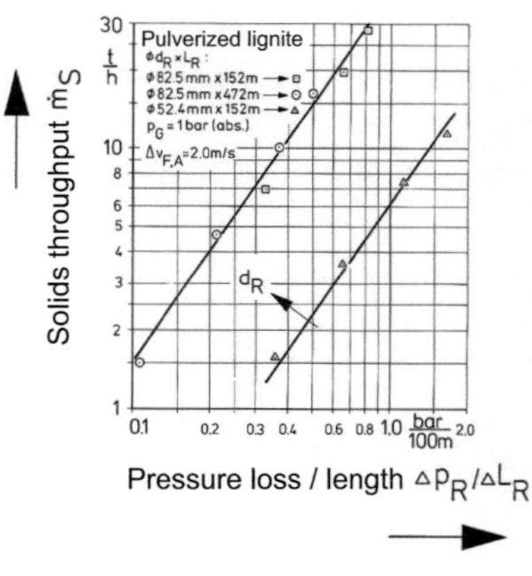

Fig. 5.17 Determination of the throughput exponent x

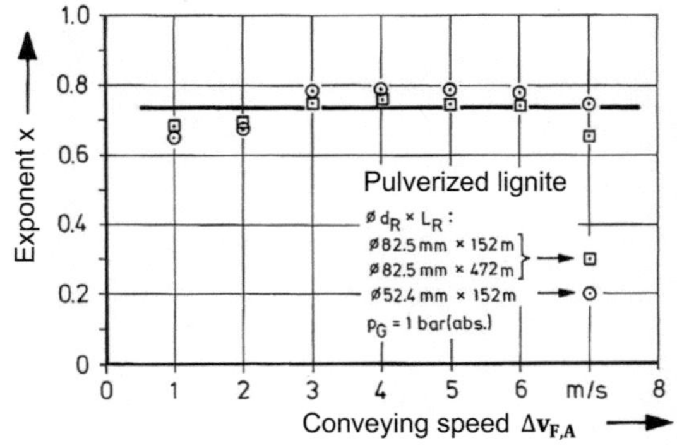

Fig. 5.18 Dependency of the throughput exponent x on the characteristic velocity $\Delta v_{F,A}$

be considered constant with sufficient accuracy. Comparable results were obtained for all bulk materials investigated so far. In this context, x proves to be only slightly dependent on the solid material. The range of measured values is between $x \cong (0.6-0.8)$. As an average value, $x = 0.70$ can be used.

b) Exponent y

The direct determination of the diameter exponent y in (5.90) is difficult in practice. It is easier to determine it indirectly via the quotient (y/x). This describes, for each constant value of Δp_R, L_R and $\Delta v_{F,A}$ the dependence of the solid throughput $\dot{m}_S$ on the conveying pipe diameter D_R. Figure 5.19 shows corresponding curves of lignite dust for $\Delta v_{F,A} = 2.0\,\text{m/s}$ in double logarithmic representation. The exponent (y/x) corresponds to the slope of the straight line, so it is a constant for Fig. 5.19. As can be seen from Fig. 5.20, this applies to the entire investigated range of the characteristic velocity $\Delta v_{F,A}$.

All previously analyzed conveyed materials yielded comparable results. In contrast to the exponent x, however, the exponent y, or the ratio (y/x), proves to be extremely dependent

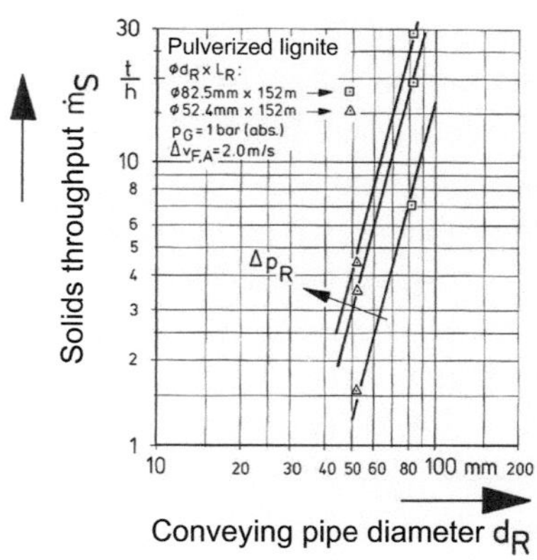

Fig. 5.19 Determination of the exponent (y/x)

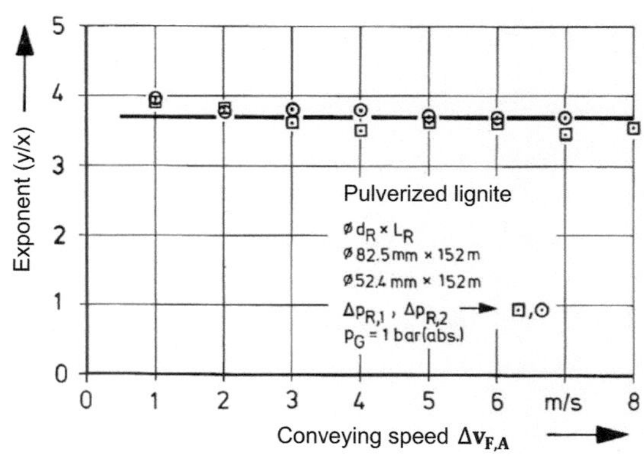

Fig. 5.20 Dependence of the exponent (y/x) on the characteristic velocity $\Delta v_{F,A}$

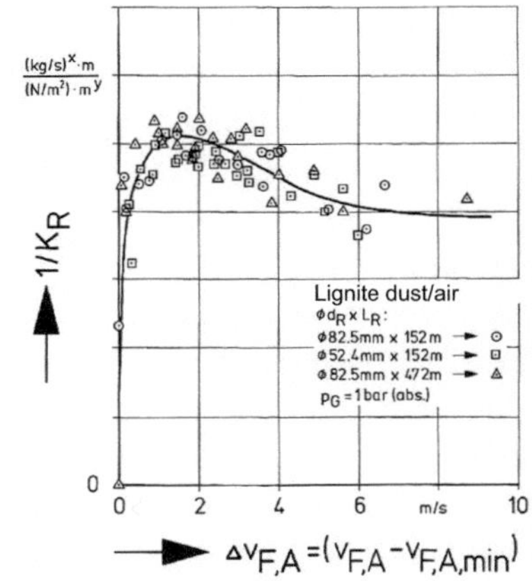

Fig. 5.21 Dependency $(1/K_R) = f(\Delta v_{F,A})$, Lignite dust

on the type of bulk material examined. Measured y-values lie between $y \cong (1.2{-}2.8)$. Thus, it applies:

$$\dot{m}_S \propto D_R^{y/x} \cong D_R^{(1.7-4.0)} \quad \rightarrow \quad \text{with: } \Delta p_R, L_R = \text{const.} \tag{5.109}$$

i.e., the solid mass flow $\dot{m}_S$ varies depending on the conveyed material, proportional to the first to the second power of the conveying pipe cross-sectional area ($\propto D_R^2$). This illustrates how crucial the precise knowledge of the diameter exponent y is for a plant design. There are close correlations between the size of y and the conveying behavior of the bulk material groups described in Sect. 4.5.

The results presented above justify the further application of (5.90).

c) Dependency $(1/K_R) = f(\Delta v_{F,A})$

In Fig. 5.21, the power product evaluated according to (5.90), or (5.108) with $z = 0$, is shown. $(1/K_R) = (\dot{m}_S^x \cdot L_R)/((-\Delta p_R) \cdot D_R^y)$ for the previously used lignite dust above the characteristic velocity $\Delta v_{F,A}$ is plotted. As in all previous and further diagrams, the required minimum velocities $v_{F,A,\min}$ were calculated with the equation adapted to the respective conveyed material and the current operating conditions (4.13). The displayed measurements can be balanced by a common curve without any noticeable systematic deviations of individual conveying lines. About 95% of all $(1/K_R)$-values lie within a range of $\pm 10\,\%$ around the compensation curve. Results of executed conveyor systems fit seamlessly. Fig. 5.22 shows the dependency $(1/K_R) = f(\Delta v_{F,A})$ of the conveyed material cement. The scatter of the measured values is comparable to that in Fig. 5.21. Systematic deviations are not recognizable here either. The power product $(1/K_R)$ is approximately constant over a wide speed range, i.e., independent of $\Delta v_{F,A}$. Such behavior is described, for example, by the approach according to [38], (5.96) in Box 5.6, or (5.106) in Box 5.7, and is consistent with practical experience.

The solid materials presented so far were fine-grained, $d_{S,50} < 50\,\mu m$. As an example of a coarser bulk material, the $(1/K_R)$-characteristic of a limestone grit with the average particle diameter $d_{S,50} \cong 460\,\mu m$ is plotted in Fig. 5.23. A comparison of the $(1/K_R)$-characteristics of selected bulk materials is shown in Fig. 5.24. To identify the bulk materials, their bulk densities ϱ_{SS} and average grain diameter $d_{S,50}$ are given. Currently, about 40 such characteristics are available. It is reminded that the $(1/K_R)$-values of the individual bulk materials are subject to different dimensions depending on the occurring exponents x and y, see calculation example 10, Sect. 5.3.6. For Fig. 5.24 the same basic dimensions were used. The compensation curves recorded in Fig. 5.24 as well as those of all other solid materials examined accordingly cover more than approx. 90% of the respective measured values from experimental and operational facilities in all cases examined with a deviation of $\pm 10\,\%$. These, in view of the measured objects of technical size, relatively moderate scatterings are largely identical with the inaccuracies/uncertainties of the measurements used.

Fig. 5.22 Dependency $(1/K_R) = f(\Delta v_{F,A})$, Cement

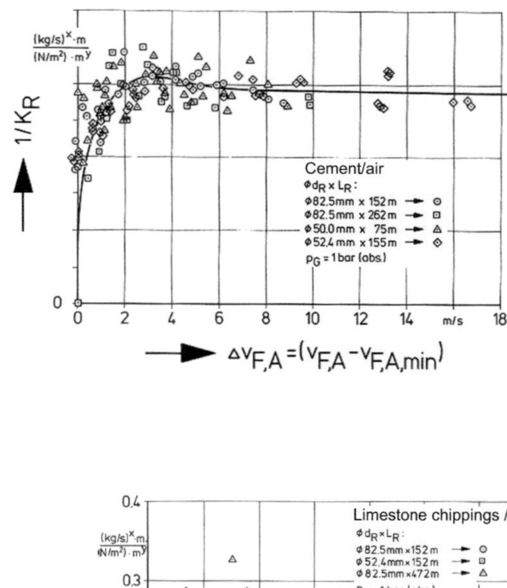

Fig. 5.23 Dependency $(1/K_R) = f(\Delta v_{F,A})$, limestone chippings

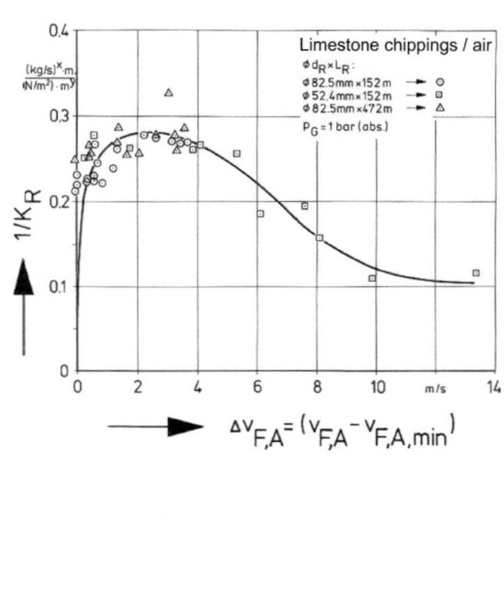

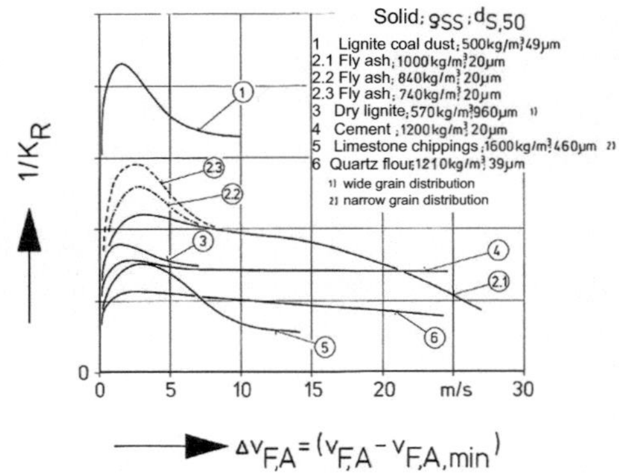

Fig. 5.24 Comparison of the $(1/K_R)$-characteristics of different bulk materials

The relationships discussed in the preceding sections are essentially confirmed by practice. For the bulk materials that have been measured from dense phase to the lean phase conveying, see Fig. 5.24, the entire working range can be uniformly represented without loss of accuracy. The pressure loss/conveying behavior relevant for plant design is thus compressed into a single solid-specific curve. When using, for example, the λ_S-model, the knowledge of several λ_S-curves ($\rightarrow f(D_R, \mu, \text{etc.})$) and thus a greater experimental effort would be required.

d) Applications

Given an appropriate reference dataset, the presented scale-up method allows, in addition to the design of operating plants, see Sect. 5.3.5, for example, the uncomplicated and fast creation of conveying diagrams of the planned plant. Characteristic fields of the conveying route calculated $D_R \times L_R = 52.4\,\text{mm} \times 152\,\text{m}$ from the characteristic data of the product lignite dust are exemplarily plotted in Figs. 5.25 and 5.26.

The energetic plant optimizations described in Sect. 4.12 can be carried out particularly efficiently with the described scale-up method. Calculation example 9, Sect. 4.12.3, was calculated in this way. For a task $(\dot{m}_S^x \cdot L_R)$ optimal operating conditions result $((-\Delta p_R) \cdot D_R^y)$ where the power product $(1/K_R)$ reaches a maximum value. Line staggerings can be easily incorporated ($\rightarrow$ assumption of a series connection of non-interactive individual pipes, see Sect. 4.11). The small number of solid-specific parameters, the constants K_v, k, l, x and y as well as a single $(1/K_R)$ -curve, allow/support bulk material comparisons and the formation of conveying material classes. Figure 5.24 shows, for example, that within a product class, here fly ash, the $(1/K_R)$ -dependencies with decreasing bulk density ϱ_{SS} increasingly pronounced and rising maxima towards larger values. This applies generally.

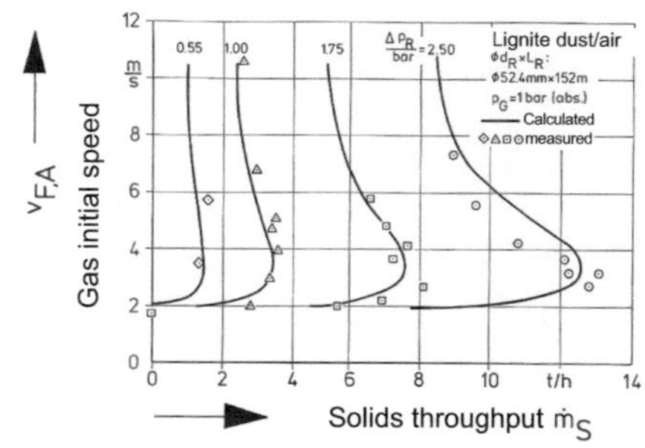

Fig. 5.25 Comparison of a measured and a calculated conveying diagram $v_{F,A}(\dot{m}_S, \Delta p_R)$, lignite dust, $D_R \times L_R = 52.4\,\text{mm} \times 152\,\text{m}$

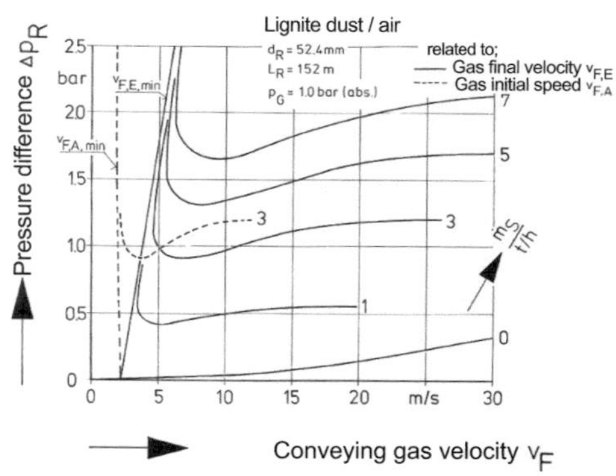

Fig. 5.26 Calculated conveying diagrams $\Delta p_R(v_{F,A}$ or $v_{F,E}, \dot{m}_S)$, lignite dust, $D_R \times L_R = 52.4\,\text{mm} \times 152\,\text{m}$

5.3.5 Procedure for Scale-up

In order to dimension a new conveying system based on an existing reference data set, the following steps are required:

- Determination of an appropriate distance $\Delta v_{F,A}$ to the conveying limit, based on the given task. This is done independently of the pipe diameter D_R and/or the pressure difference in the conveying line Δp_R.
- Determination of the associated K_R-value from the $(1/K_R)$-diagram of the present solid, see, for example, Fig. 5.21.
- According to the task, D_R or Δp_R can now be calculated from (5.90) ($\rightarrow$ degree of freedom of the design is, depending on the specification, either the pipe diameter D_R or the pressure loss Δp_R).
- Adjustment of the course of the reference line to that of the planned system, i.e., corresponding correction of the pressure loss.
- Since now D_R and Δp_R are known, i.e., for a given counter pressure p_E also p_A, the current minimum speed $v_{F,A,\min}$ can be calculated using (4.13).

Table 5.1 Dimension matrix (the upper three rows) and transformations of the core matrix into a unit matrix

	$\dot{m}_S$	D_R	$\Delta v_{F,A}$	$-\Delta p_R$	L_R	g	
M	1	0	0	1	0	0	
L	0	1	1	−1	1	1	
T	−1	0	−1	−2	0	−2	
Z1	1	0	0	1	0	0	= M
Z2	0	1	1	−1	1	1	= L
Z3	0	0	−1	−1	0	−2	= T + Z1
Z4	1	0	0	1	0	0	= Z1 = M
Z5	0	1	0	−2	1	−1	= Z2 + Z3
Z6	0	0	1	1	0	2	= −Z3

- $(v_{F,A,\min} + \Delta v_{F,A}) = v_{F,A}$ provides the current initial conveying gas velocity and by means of the continuity equation $\dot{m}_F = A_R \cdot \varrho_{F,A} \cdot v_{F,A}$, with: $\varrho_{F,A}$= gas density at the beginning of the pipeline, i.e., at the pressure p_A, the gas mass flow required for conveying.

By choosing the same distance in both test and operational facilities at the beginning of the conveying line $\Delta v_{F,A}$ to the current conveying limit, a K_R -value dependent only on this speed is defined, with the help of which the respective missing operating variable ($\rightarrow \Delta p_R$ or D_R) is determined. An iterative approach like with the λ_S-method is not necessary. $\Delta v_{F,A}$ can be considered as a characteristic (fictitious) solid velocity $u_{S,A}$ at the beginning of the conveying pipe.

5.3.6 Calculation Example 10: Dimensional Analysis Verification of the Scale-up Approach

A dimensionally "correct" representation of the solid-specific pressure loss approach presented in Sect. 5.3.2 is to be derived.

- **Relevance list:**
  The influencing variables used in Sect. 5.3.2 and extended by the gravitational acceleration are used ($\rightarrow$ gravitational acceleration: different conditions would apply on the moon):
  - Target size: $(-\Delta p_R)$,
  - geometric influencing variables: D_R, L_R,
  - material influencing variables: Gas and solid = const.,
  - process-related influencing variables: $\dot{m}_S$, $\Delta v_{F,A} = u_{S,A}$,
  - other influencing variables/dimensioned constants: $g = 9.81\,\text{m/s}^2$.
- **Dimension matrix**:
- In the upper part of Table 5.1, the resulting dimension matrix is shown, taking into account the arrangement notes of Sect. 2.4.2. In the parts of Table 5.1 follow-

ing the dimension matrix, the core matrix is transformed into a unit matrix. Here, ZX = Zeile X means.

- **Dimensionless numbers:**
  The core matrix shows that the rank of the dimension matrix $r = 3$ amounts to and is identical to the number of basic dimensions d. The $n = 6$ influencing process variables can thus be reduced to $m = (n - r) = (6 - 3) = 3$ dimensionless numbers. From the lower part of Table 5.1, lines Z4–Z6, the following applies to these according to the rules given in Sect. 2.4.2:

$$\Pi_1 = \frac{-\Delta p_R}{\dot{m}_S \cdot D_R^{-2} \cdot \Delta v_{F,A}} = \frac{(-\Delta p_R) \cdot D_R^2}{\dot{m}_S \cdot \Delta v_{F,A}}$$

$$\Pi_2 = \frac{L_R}{D_R}$$

$$\Pi_3 = \frac{g}{D_R^{-1} \cdot \Delta v_{F,A}^2} = \frac{g \cdot D_R}{\Delta v_{F,A}^2} = \frac{1}{Fr_{R,\Delta v}} \quad \rightarrow \quad \text{Froude number}$$

With the transformation

$$\Pi_1^* = \frac{\Pi_1}{\Pi_2 \cdot \Pi_3} = \frac{(-\Delta p_R) \cdot D_R^2 \cdot \Delta v_{F,A}}{\dot{m}_S \cdot g \cdot L_R}$$
$$\rightarrow \quad \propto \frac{\text{power input by gas}}{\text{power required for bulk material transport}}$$

results in the relationship $F(\Pi_1^*, \Pi_2, \Pi_3) = 0$ and from this:

$$\left(\frac{(-\Delta p_R) \cdot D_R^2 \cdot \Delta v_{F,A}}{\dot{m}_S \cdot g \cdot L_R}\right) = K_{\Delta p} \cdot \left(\frac{D_R}{L_R}\right)^a \cdot \left(\frac{\Delta v_{F,A}^2}{g \cdot D_R}\right)^b \tag{5.110}$$

with:

$K_{\Delta p}, a, b$ constants and exponents to be determined experimentally.

- **Comparison with the used scale-up approach:**
  Rearrangement of (5.110) according to the pressure loss Δp_R yields:

$$-\Delta p_R = (K_{\Delta p} \cdot g^{1-b}) \cdot \frac{\dot{m}_S \cdot L_R^{1-a}}{D_R^{2+b-a}} \cdot \Delta v_{F,A}^{2 \cdot b-1}$$

Written in the form of (5.108), it follows that:

$$\left(\frac{1}{K_R}\right) = \left(\frac{\dot{m}_S \cdot L_R^{1-a}}{(-\Delta p_R) \cdot D_R^{2+b-a}}\right) = \frac{1}{(K_{\Delta p} \cdot g^{1-b})} \cdot \frac{1}{\Delta v_{F,A}^{2 \cdot b-1}} \tag{5.111}$$

Or:

$$\left(\frac{1}{K_R}\right) = \left(\frac{\dot{m}_S \cdot L_R^{1/z}}{(-\Delta p_R) \cdot D_R^y}\right) = \frac{c}{\Delta v_{F,A}^w} \tag{5.112}$$

(5.112) corresponds to (5.108); with $z = 1$ or $a = 0$ results (5.90). Both, however, with an exponent of the solid mass flow of $x = 1$. Since practical measurements, yield values $x < 1$, this indicates an incomplete relevance list as the basis for the dimensional analysis. Consequence: further dimensionless characteristic numbers.

In principle, both the above dimensional analysis and practical experience confirm the chosen scale-up approach. This is also considerably easier to handle in terms of plant design.

5.4 Calculation with Equivalent Lengths

The method of so-called equivalent lengths, also known as apparent length method, which was formerly popular among practitioners for calculating the pressure loss of conveying lines, is now of little importance and is therefore only briefly described. However, in some applications, the method has advantages over alternative methods. One such application is presented in conclusion.

5.4.1 Calculation Approach

The approach is based on the fact that equivalent pipe lengths are assigned to the various pipeline elements along a conveying line, e.g., deflectors, vertical sections, fittings, in such a way that the pressure loss of the considered element becomes equal to that of a correspondingly long horizontal pipe section. The equivalent length $L_{äq,h}$ of a horizontal pipe element thus corresponds to its true length ΔL_h. After summing up all equivalent lengths of the respective conveying line, its pressure loss can be calculated as a continuous horizontal line with the total length $\sum L_{äq,i}$. For this, only the resistance coefficient of the horizontal gas/solid flow is required. Equivalent lengths are determined empirically by the offering specialist companies. They differ depending on the company and application or the boundary conditions of the task, sometimes considerably. Examples can be found in [42]. However, they can also be calculated. This will be demonstrated using the example of the equivalent length $L_{äq,U}$ of a pipe bend. From the equation of (4.48) and (4.57) follows with

$$\lambda_{\text{tot}} = \frac{\lambda_F}{\mu} + \lambda_S \quad \rightarrow \quad \lambda_{\text{tot}} \cdot \mu = \lambda_F + \mu \cdot \lambda_S \quad \text{as well as} \quad \Delta L_h = L_{äq,U} \tag{5.113}$$

Table 5.2 Calculation of equivalent lengths

Pressure loss due to	Basic equations	Equivalent lengths
Straight horizontal pipe	$-\Delta p_h = \lambda_{\text{tot}} \cdot \mu \cdot \frac{\Delta L_h}{D_R} \cdot \frac{\varrho_F}{2} \cdot u_F^2$	$L_{äq,h} = \Delta L_h$
Initial acceleration	$-\Delta p_{S,B} = C \cdot \mu \cdot \varrho_F \cdot u_F^2$	$L_{äq,B} = \frac{2 \cdot C \cdot D_R}{\lambda_{\text{tot}}}$
Deflection	$-\Delta p_U = K_U \cdot C \cdot \mu \cdot \varrho_F \cdot u_F^2$ $K_U = K_U(\alpha_U, \beta_R)$	$L_{äq,U} = K_U \cdot \frac{2 \cdot C \cdot D_R}{\lambda_{\text{tot}}}$
Straight vertical pipe, friction	$-\Delta p_{R,v} = \frac{\lambda_{\text{tot}}}{2} \cdot \mu \cdot \frac{\Delta L_v}{D_R} \cdot \frac{\varrho_F}{2} \cdot u_F^2$	$L_{äq,R,v} = \frac{1}{2} \cdot \Delta L_v$
Straight vertical pipe, lift	$-\Delta p_H = (1 + \frac{\mu}{C}) \cdot \varrho_F \cdot g \cdot \Delta L_v$	$L_{äq,H} \cong 2 \cdot \frac{g}{C} \cdot \frac{D_R \cdot \Delta L_v}{\lambda_{\text{tot}}} \cdot \frac{1}{u_F^2}$
Throttle valve, fitting	$-\Delta p_D = \xi(A_D) \cdot \frac{\varrho_{\text{tot},D}}{2} \cdot u_{\text{tot},D}^2$ $\rho_{\text{tot},D}, u_{\text{tot},D}$ Mixture density, -velocity A_D Opening cross-section $\xi(A_D)$ Resistance coefficient	$L_{äq,D} = \frac{\xi(A_D)}{\lambda_{\text{tot}}} \cdot (\frac{A_R}{A_D})^2 \cdot D_R \cdot (1 + \mu) \cdot (\frac{1}{\mu} + \frac{\varrho_F}{\varrho_S})$

$$\lambda_{\text{tot}} \cdot \mu \cdot \frac{L_{äq,U}}{D_R} \cdot \frac{\varrho_F}{2} \cdot u_F^2 = K_U \cdot C \cdot \mu \cdot \varrho_F \cdot u_F^2 \tag{5.114}$$

and from this the equivalent length of a pipe bend with an arbitrary deflection angle α_U:

$$L_{äq,U} = 2 \cdot \frac{K_U \cdot C \cdot D_R}{\lambda_{\text{tot}}} \tag{5.115}$$

In Table 5.2, equations for determining the equivalent lengths of different pipe elements are compiled. When applying these, it must be taken into account that all $L_{äq,i}$-lengths with the local values of u_F, C, ϱ_F and λ_{tot} were formed, see in this regard (5.114) and (5.115). As can be seen in Table 5.2, some of these local quantities are also part of the resulting $L_{äq,i}$-equations. Since the transport gas expands along the conveying route, their values are still unknown when determining the pipeline pressure loss and must be estimated. This leads, as in the design method presented in Sect. 4.7, to iterative calculations and offers no advantage in this regard. If certain design situations are given general $L_{äq,i}$-lengths assigned, however, simple orienting rough calculations are possible.

The $L_{äq,i}$-equations in Table 5.2 illustrate that, in addition to operating conditions, the type of conveyed solid also influences the current values of $L_{äq,i}$. Example: The gas/solid velocity ratio C when conveying coarse bulk material is on the order of $C_g \cong 0.5$, of finer-grained at $C_f \cong 0.9$. The ratio of the equivalent lengths of two pipe bends, cf. (5.115), is thus under identical operating conditions: $L_{äq,U,f}/L_{äq,U,g} = C_f/C_g = 0.9/0.5 = 1.8$.

5.4.2 Applications

Various industrial processes are supplied with solid material in parallel across multiple task points using pneumatic conveying. Typically, an equal distribution of the individual solid flows to the various injection points is required. The quality of this equal distribution must be guaranteed. A prominent example of such a process is the feeding of blast furnaces for pig iron production with coal dust as a reducing agent and heat carrier [43, 44]. Depending on the furnace throughput, this is continuously blown in pneumatically against the furnace pressure (generally several bar) via up to approx. 40 individual lances arranged along the furnace circumference/the wind form. Two possible systems for realizing such a task are schematically shown in Fig. 5.27: Fig. 5.27a shows a distributor that divides a pneumatically supplied gas/solid mixture $\dot{m}_{\text{tot}} = (\dot{m}_F + m_S)$ onto N outgoing parallel lines, while in Fig. 5.27b the distribution to the N lines takes place in a fluidized/suitably aerated pressure vessel. In principle, the individual supply of the lines with conveying gas and solid material would also be possible, but with an increasing number of

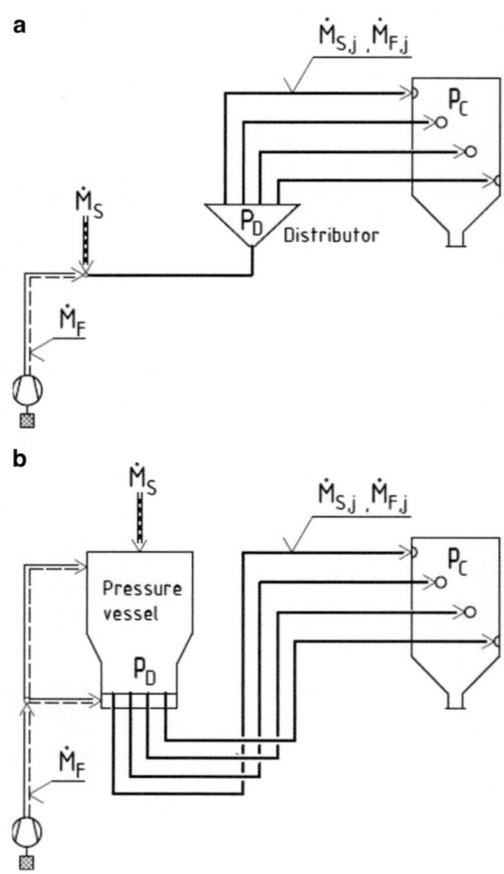

Fig. 5.27 Distribution systems [45]

lines, this quickly leads to considerable investment costs. In the following, it is assumed that the individual parallel lines have an identical inner diameter D_R [45].

From Fig. 5.27, it can be seen that the same pressure difference $\Delta p_{D,C} = (p_D - p_C)$, with: p_D = pressure in the distributor/sender, p_C = pressure in the reactor/receiver, is applied to all parallel individual lines. Since, in general, both the lengths and the spatial courses of these lines to their delivery points at the reactor are different, different mass flows $\dot{m}_{\text{tot},j}$ of the gas/solid mixture to be conveyed are established in the individual lines. Through the line with the greatest length, or with the greatest specific resistance $(\rightarrow \propto (\sum L_{\ddot{a}q,i}/\Delta p_{D,C}))$, the smallest mass flow occurs and vice versa. An equal distribution must be enforced by installing additional conveying pipe elements — straight pipe sections, bends, etc. — in the various lines, i.e., by adjusting the individual specific resistances to those of the longest/critical line. Fig. 5.28 shows the practical implementation of such a balancing station for charging a blast furnace.

In the injection systems considered here, a solid mass flow $\dot{m}_S$ is to be evenly distributed over N parallel individual lines. The resulting conveying gas distribution is not clear. However, it must be known for the conveying calculation and is also relevant in cases where the gas acts as a reaction partner at the same time, e.g., when charging coal dust burners with air. In the following, it is assumed that with an equal distribution of the solid, the conveying gas is also distributed equally. Thus, it applies: $\mu = (\dot{m}_S/\dot{m}_F) = \mu_j = (\dot{m}_{S,j}/\dot{m}_{F,j}) = \text{const.}$ and $\dot{m}_{\text{tot},j} = (\dot{m}_{\text{tot}}/N) = \text{const.}$ for the $j = 1 - N$ parallel lines. This approach seems plausible because the gas flows faster than the solid and thus quasi through a bed whose high resistance forces an equal distribution to all pipes.

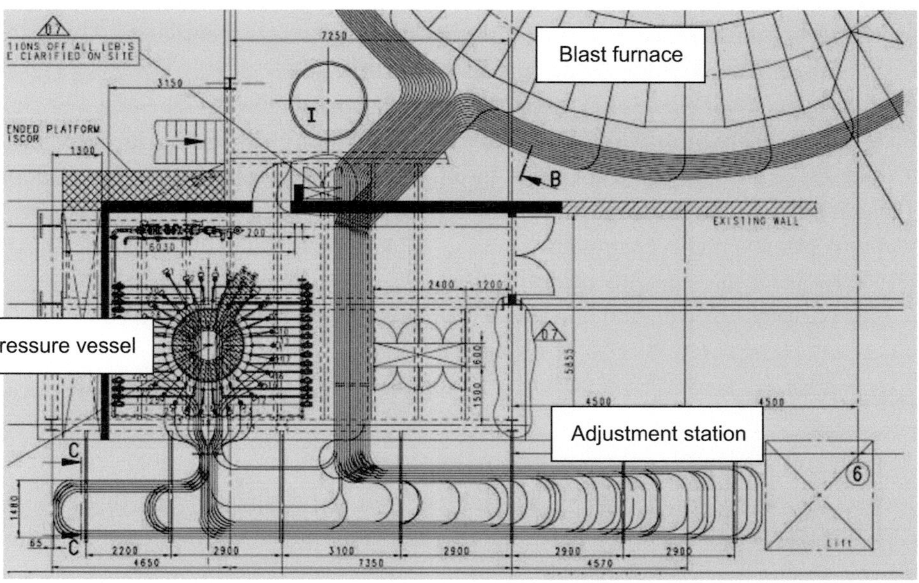

Fig. 5.28 Example of a balancing station [45]

Starting from the known pressure p_C in the reactor, the required solid mass flow $\dot{m}_{S,j} = \dot{m}_{S,\text{tot}}/N$ is divided by the $j = 1 - N$ individual lines and a chosen or predetermined pipe diameter D_R the pressure loss $\Delta p_{D,C}$ of the previously identified longest/critical line is calculated and thus also p_D is determined. The balancing of the distribution system can then, for example, be carried out using the calculation method described in Sect. 4.7. This requires the sequential input of all pipeline elements of the parallel conveying routes, each from the distributor to the reactor, i.e., lengths, positions, and spatial orientation of straight pipe sections, of deflections including their deflection angles and radii, possibly the opening cross-sections of necessary fittings, etc., must be specified. With the known pressures (p_D, p_C), the pipe diameter D_R and the courses of the N individual lines, the current solid flow rates $\dot{m}_{S,j}$ can be calculated through the various conveying pipes. Based on this data, by planning various additional pipeline elements into the individual lines, a comparison of the $\dot{m}_{S,j}$ values is carried out. Subsequent control calculations show the new $\dot{m}_{S,j}$ distribution, which can then be continuously improved in further input steps. In this process, the local boundary and implementation conditions of the plant must be taken into account, i.e., there is a mutual interaction between construction and pneumatic calculation. This approach, even if supported by suitable calculation programs [46, 47], is clearly time-consuming. Therefore, the method of equivalent lengths will be presented below as a simplified method of line balancing.To do this, the equivalent total lengths are first calculated. $(\sum L_{äq,i})_j = L_{äq,\text{ges},j}$ of the N planned parallel lines is determined. Since, given pressures (p_D, p_C) and the desired equal distribution, an identical average gas density and a (approximately) equal velocity profile are established in all parallel lines, the individual $L_{äq,i}$-values are determined with an average solid/gas velocity ratio $\overline{C}$ and an average gas density $\overline{\varrho}_F$ instead of the local values (C, ϱ_F). This simplifies the calculation and does not lead to a significant reduction in accuracy for length adjustment. The line with the largest equivalent length, for example $L_{äq,\text{ges},N}$, is flowed through by the smallest mass flow $\dot{m}_{\text{tot},N}$ or $\dot{m}_{S,N}$ and must therefore be dimensioned for the desired target flow rate. The flow rates of the equivalent shorter lines are larger, $\dot{m}_{\text{tot},j\neq N} > \dot{m}_{\text{tot},N}$, and are adjusted by aligning their equivalent lengths to that $L_{äq,\text{ges},N}$ of the longest line is throttled to the setpoint. Using the equations in Table 5.2, the difference lengths $\Delta L_{äq,j} = (L_{äq,\text{ges},N} - L_{äq,\text{ges},j\neq N})$ can be transformed into real pipe elements with corresponding $\Delta L_{äq,j}$ values. Their installation position along the respective line can be flexibly determined.

The pressure loss equation of a single line j is simplified by the described method to:

$$(\Delta p_{D,C})_j = \overline{\lambda}_{\text{tot}} \cdot \mu \cdot \frac{L_{äq,\text{ges},j}}{D_R} \cdot \frac{\overline{\varrho}_F}{2} \cdot \overline{u}_F^2 = \frac{1}{2} \cdot \overline{\lambda}_{\text{tot}} \cdot \frac{L_{äq,\text{ges},j}}{D_R \cdot A_R^2} \cdot \frac{\dot{m}_{S,j} \cdot \dot{m}_{F,j}}{\overline{\varrho}_F} \tag{5.116}$$

with: $\overline{u}_F \cong \overline{v}_F$, $L_{äq,\text{ges},j} = (\sum L_{äq,i})_j$.

With the equivalent length model and competing calculation approaches, pipeline adjustments carried out differ only slightly. The latter are somewhat more accurate, as the real positions of the various pipe elements along the conveying routes are taken into account. However, the effects on the pipeline adjustment are negligible, because larger $\dot{m}_{S,j}$-inequalities are caused by manufacturing inaccuracies, tolerances of pipe diameters, etc. Measured and calculated solid distribution accuracies differ by less than ±2%. Further details, e.g., on the behavior of the system during load changes, can be found in [45].

5.4.3 Calculation Example 11: Reactor Feeding with Parallel Conveying Lines

Fig. 5.29 shows a pressure vessel system with $N = 4$ outgoing conveyor lines, which are to continuously supply a reactor with a total of $\dot{m}_S = 8.0\,\text{t/h}$ coal dust. An equal distribution of the coal mass flows is required, i.e. $\dot{m}_{S,j} = 2.0\,\text{t/h}$. Nitrogen is used as the conveying gas; $\dot{m}_F = 126.0\,\text{kg/h}$. Pressure in the sender: $p_D = 2.25\,\text{bar(abs)}$; Pressure in the reactor: $p_C = 1.0\,\text{bar(abs)}$; thus: $\Delta p_{D,C} = 1.25\,\text{bar}$. Steel pipe lines, $(\varnothing 33.7 \times 2.6)\,\text{mm}$,

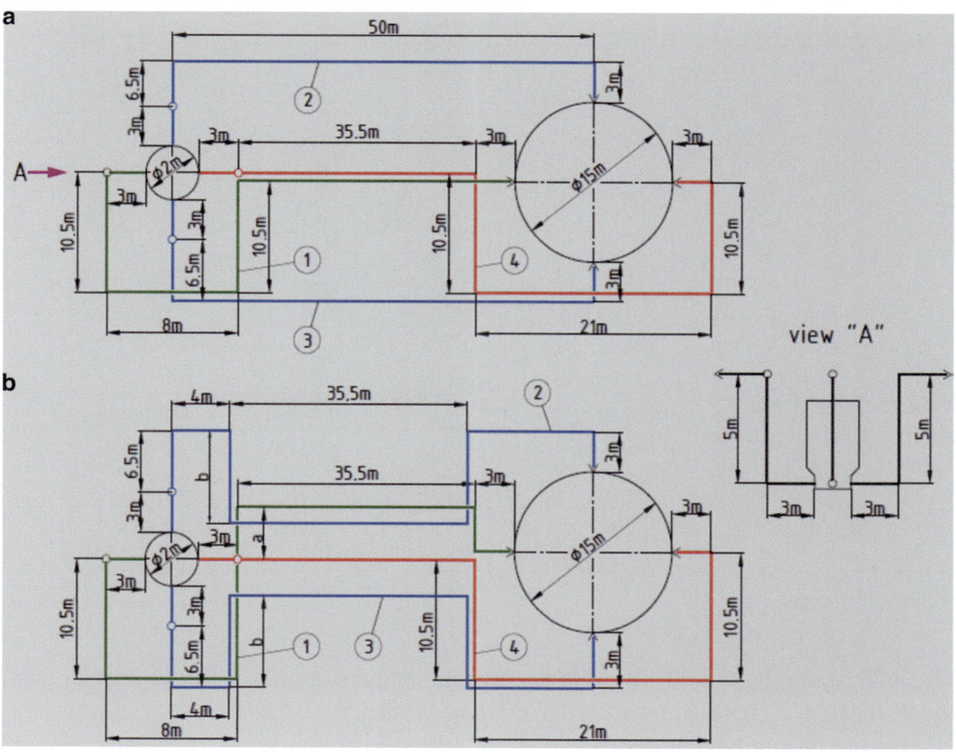

Fig. 5.29 Pipe routes in the calculation example. **a** not balanced, **b** balanced

Table 5.3 Unbalanced lines

Line number		1	2 + 3	4
Number of 90° bends n_U		5	4	6
$L_{äq,h} = \Delta L_h$	[m]	70.50	62.50	83.50
$L_{äq,R,v} = 0.5 \cdot \Delta L_v$	[m]	2.50	2.50	2.50
$L_{äq,B}$	[m]	5.13	5.13	5.13
$n_U \cdot L_{äq,U}$	[m]	12.83	10.26	15.39
$L_{äq,H}$	[m]	5.76	5.76	5.76
$L_{äq,\mathrm{ges},j}$	[m]	96.72	86.15	112.28
$\Delta L_{äq,j} = L_{äq,\mathrm{ges},4} - L_{äq,\mathrm{ges},j}$	[m]	15.56	26.13	0
$\dot{m}_{S,j}$	[t/h]	2.43	2.83	2.00
Deviation	[%]	+21.5	+41.5	±0

Table 5.4 Adjusted lines

Line number		1	2 + 3	4
Number of 90° bends n_U		7	8	6
$L_{äq,h} = \Delta L_h$	[m]	80.94	78.37	83.50
$L_{äq,R,v} = 0.5 \cdot \Delta L_v$	[m]	2.50	2.50	2.50
$L_{äq,B}$	[m]	5.13	5.13	5.13
$n_U \cdot L_{äq,U}$	[m]	17.95	20.52	15.39
$L_{äq,H}$	[m]	5.76	5.76	5.76
$L_{äq,\mathrm{ges},j}$	[m]	112.28	112.28	112.28
$\Delta L_{äq,j} = L_{äq,\mathrm{ges},4} - L_{äq,\mathrm{ges},j}$	[m]	0	0	0
$\dot{m}_{S,j}$	[t/h]	2.00	2.00	2.00
Deviation	[%]	±0	±0	±0

Table 5.5 Effects of alignment inaccuracies on Line 1

$\Delta L_{äq,1}$ [m]	$\dot{m}_{S,1}$ [t/h]	Deviation [%]
0	2,000	±0
1.0	2,025	+1.25
2.0	2,050	+2.50
15.56	2,430	+21.50

$D_R = 28.5\,\mathrm{mm}$, $A_R = 0.63794 \cdot 10^{-3}\,\mathrm{m}^2$. Design temperature: $T_M = 20\,°\mathrm{C}$. Further data: $\overline{\lambda}_{\mathrm{tot}} = 0.010$, $\overline{C} = 0.90$, $\overline{\varrho}_F = 1.868\,\mathrm{kg/m}^3$, $\beta_R = 0.44$.

Due to structural conditions, the pipe route shown in Fig. 5.29a is initially planned. 90° bends with a radius $R_b = 0.5\,\mathrm{m}$ are used. Table 5.3 contains the equivalent lengths calculated for this design as well as the solid mass flow rates $\dot{m}_{S,j}$ determined at the pressure difference $\Delta p_{D,C} = 1.25\,\text{bar}$ calculated with (5.116). The differences $\Delta L_{äq,j}$ in the equivalent lengths result in impermissibly large deviations of the solid mass flow rates $\dot{m}_{S,j}$ from the target value. $\dot{m}_{S,4} = 2.0\,\text{t/h}$. A line balance is required. In the present case, the $\Delta L_{äq,j\neq4}$-values are converted according to possible alternative line routes in 90° bends ($\rightarrow L_{äq,U} = 2.565\,\text{m}$) and straight horizontal pipe sections. Figure 5.29b shows the selected implementation variant and Table 5.4 the associated operating data for this (complete) balance.

To set $\Delta L_{äq,j} = 0$ to be adjusted, the distances (a, b) in Fig. 5.29b would have to be chosen as $a \cong 5.218\,\text{m}$ and $b \cong 7.935\,\text{m}$. In practice, such accuracies are not achievable, so that residual lengths $\Delta L_{äq,j} \neq 0$ will remain. Table 5.5 shows the effects of different $\Delta L_{äq,j}$-residual lengths on the distribution accuracy using the example of line 1.

References

1. Hilgraf, P.: Einfluß von Schüttgütern im Hinblick auf ihre Handhabung. VDI-Berichte 1918., S. 63–93 (2006).
2. Wirth, K.-E.: Theoretische und experimentelle Bestimmungen von Zusatzdruckverlust und Stopfgrenze bei pneumatischer Strähnenförderung. Universität Erlangen-Nürnberg, Erlangen-Nürnberg (1980). Dissertation.
3. Levy, A., Mason, D.J.: Two-layer model for non-suspension gas-solids flow in pipes. Powder Technol. **112**, 256–262 (2000).
4. Mason, A.D., Levy, A.: A model for non-suspension gas-solids flow of fine powders in pipes. Int. J. Multiph. Flow **27**, 415–435 (2001).
5. Hong, J., Shen, Y., Liu, S.: A model for gas-solid stratified flow in horizontal dense-phase pneumatic conveying. Powder Technol. **77**, 107–114 (1993).
6. Truckenbrodt, E.: Lehrbuch der angewandten Fluidmechanik. Springer, Berlin (1988).
7. Molerus, O.: Fluid-Feststoff-Strömungen. Springer, Berlin (1982).
8. Muschelknautz, E., Krambrock, W.: Vereinfachte Berechnung horizontaler pneumatischer Förderleitungen bei hohen Gutbeladungen mit feinkörnigen Produkten. Chem.-Ing.-Tech. **41**(21), 1164–1172 (1969).
9. Bohnet, M.: Experimentelle und theoretische Untersuchungen über das Absetzen, Aufwirbeln und den Transport feiner Staubteilchen in pneumatischen Förderleitungen. VDI-Forschungsheft, Bd. 507. VDI-Verlag, Düsseldorf (1965).
10. Muschelknautz, E., Wojahn, H.: Auslegung pneumatischer Förderanlagen. Chem.-Ing.-Tech. **46**(6), 223–235 (1974).
11. Bohnet, M.: Fortschritte bei der Auslegung pneumatischer Förderanlagen. Chem.-Ing.-Tech. **55**(7), 524–539 (1983).
12. Wirth, K.-E., Molerus, O.: Bestimmung des Druckverlustes bei der pneumatischen Strähnenförderung. Chem.-Ing.-Tech. **53**(4), 292–293 (1981). MS 898/81.
13. Wypich, P.W., Yi, J.: Minimum transport boundary for horizontal dense-phase pneumatic conveying of granular materials. Powder Technol. **129**, 111–121 (2003).
14. Wirth, K.-E.: Zirkulierende Wirbelschichten. Springer, Berlin (1990).
15. Weber, M.: Strömungs-Fördertechnik. Krausskopf-Verlag, Mainz (1974).

16. Konrad, K., Harrison, D., Nedderman, R.M., Davidson, J.F.: Prediction of the pressure drop for horizontal dense phase pneumatic conveying of particles. Proc. of Pneumatransport, paper E1. Organized by BHRA Fluid Engineering, Cranfield., S. 225–244 (1980).
17. Legel, D.: Untersuchungen zur pneumatischen Förderung von Schüttgutpfropfen aus kohäsionslosem Material in horizontalen Rohren. TU Braunschweig, Braunschweig (1981). Dissertation.
18. Mi, B., Wypich, P.W.: Pressure drop prediction in low-velocity pneumatic conveying. Powder Technol. **81**, 125–137 (1994).
19. Pan, P., Wypich, P.W.: Pressure drop and slug velocity in low-velocity pneumatic conveying of bulk solids. Powder Technol. **94**, 123–132 (1997).
20. Dhodapkar, S.V., Plasynski, S.I., Klinzing, G.E.: Plug flow movement of solids. Powder Technol. **81**, 3–7 (1994).
21. Pan, R., Wypich, P.W.: Pressure drop prediction in single-slug pneumatic conveying. Powder Handl. Process. **7**(1), 63–68 (1995).
22. Destoop, T.: Sizing of discontinuous dense phase conveying systems. Powder Handl. Process. **5**(2), 139–144 (1993).
23. Iltzsche, M.: Pfropfenmodell mit Druck-, Geschwindigkeits- und Axialpressungsverteilungen kohäsionslosen Schüttguts bei horizontaler pneumatischer Förderung, Teil 1 und 2. Hebezeuge Förderm. **29**(8), 232–235 (1989). Nr. 9, S. 266–269.
24. Niederreiter, G.: Untersuchung zur Pfropfenentstehung und Pfropfenstabilität bei der pneumatischen Dichtstromförderung. Technische Universität München, München (2005). Dissertation.
25. Lecreps, I.J.M.: Physical mechanisms involved in the transport of slugs during horizontal pneumatic conveying. Technische Universität München, München (2011). Dissertation.
26. Tsuji, Y.: Recent works of pneumatic conveying in Japan. Bulk Solids Handl. **3**(3), 589–595 (1983).
27. Konrad, K., Totah, T.S.: Vertical pneumatic conveying of particle plugs. Can. J. Chem. Eng. **67**(2), 245–252 (1989).
28. Strauß, M., McNamara, S., Herrmann, H.J., Niederreiter, G., Sommer, K.: Plug conveying in a vertical tube. Powder Technol. **162**, 16–26 (2006).
29. Hilgraf, P.: Auslegung pneumatischer Dichtstromförderungen auf der Grundlage von Förderversuchen—Untersuchungen zum Scale-up. ZKG Int. **42**(11), 558–566 (1989).
30. Barth, W.: Strömungstechnische Probleme der Verfahrenstechnik. Chem.-Ing.-Tech **26**(1), 29–34 (1954).
31. Muschelknautz, E.: Theoretische und experimentelle Untersuchungen über die Druckverluste pneumatischer Förderleitungen. VDI-Forschungsheft 476. VDI-Verlag, Düsseldorf (1959).
32. Keys, S., Chambers, A.J.: Scaling pneumatic conveying characteristics for pipeline pressure. Powder Handl. Process. **7**(1), 59–62 (1995).
33. Wypich, P.W.: Pneumatic conveying of bulk solids. University of Wollongong (1989). Thesis.
34. Mallick, S.S., Wypich, P.W.: An investigation into modeling of solids friction for dense phase pneumatic conveying of powders. Part. Sci. Technol. **28**, 51–65 (2010).
35. Siegel, W.: Pneumatische Förderung: Grundlagen, Auslegung, Anlagenbau, Betrieb. Vogel Buchverlag, Würzburg (1991).
36. Wen, C.-Y., Simons, H.P.: Flow characteristics in horizontal fluidised solids transport. A.i.ch.e. J. **5**(2), 263–267 (1959).
37. Stegmaier, W.: Zur Berechnung der horizontalen pneumatischen Förderung feinkörniger Stoffe. F+h Fördern Heb. **28**(5/6), 363–366 (1978).
38. Ostrovskii, G.M., Krivoi, V.T., Sokolov, V.N., Isakov, V.P.: english translation. J. Appl. Chem. Ussr **50**, 1117–1119 (1977).

39. Hilgraf, P.: Untersuchungen zur pneumatischen Dichtstromförderung. Chem. Eng. Process. **20**(1), 33–41 (1986).
40. Hilgraf, P.: Untersuchungen zur pneumatischen Dichtstromförderung über große Entfernungen. Chem.-Ing.-Tech **58**(3), 209–212 (1986).
41. Hilgraf, P.: Minimale Fördergasgeschwindigkeiten beim pneumatischen Feststofftransport. ZKG Int. **40**(12), 610–616 (1987).
42. Hesse, Th.: Pneumatische Förderung. Helen M. Brinkhaus, Rossdorf (1984)\Hack{\ enlargethispage{4mm}}.
43. Hilgraf, P., Nolde, H.-D.: New design methods for coal distribution systems for blast furnace coal injection. ICSTI'09, 5th International Congress on the Science and Technology of Ironmaking, Shanghai, 19–23 October 2009. Collected proceedings, paper B2-2.07, Bd. 2., S. 783–787, 2009.
44. Aue-Klett, C., Nickel, T., Hartge, E.-U., Werther, J., Hilgraf, P., Reppenhagen, J., Eichinger, F.: Split of pneumatic conveying line to multiple parallel lines—prediction of mass flow distribution and control. Chem.-Ing.-Tech. **73**(6), 705–708 (2001).
45. Hilgraf, P., Moka, M.A.: Simplified method of feed pipe balancing at pneumatic reactor feed. ZKG Int. **66**(11), 50–56 (2013). http://www.zkg.de/download/604561/ClaudiusPeters_Hilgraf_Moka_final_201213.pdf.
46. Nickel, T.: Einfluß der Leitungsführung auf den Druckverlust bei der pneumatischen Dichtstromförderung in dünnen Leitungen. Diplomarbeit, TUHH Technische Universität Hamburg-Harburg 2000, unveröffentlicht.
47. Klett, C.: Entwicklung einer Regelung zur gezielten Aufteilung einer pneumatischen Förderung auf mehrere Teilströme. Diplomarbeit, TUHH Technische Universität Hamburg-Harburg 2000, unveröffentlicht.

Modern Dense Phase Conveying Methods

6

Dense phase conveying (DPF) is slow conveying, which is operated close to the respective conveying limit and with which, in general, only a limited amount of bulk materials can be conveyed stably. To increase operational reliability and to expand the application range of conventional dense phase conveying, DPF methods have been developed that use various additional control and action principles to support conveying. Conventional DPF is understood here as conveying in the form of highly loaded strands or as a sequence of dunes, conglobations, and/or plugs through smooth pipes without special additional devices. The most suitable bulk materials for this are in the two hatched areas of the extended Geldart diagram, see Fig. 4.13 (→ Bulk materials of groups 1 + 2 according to Sect. 4.5). Plant designs must be very precisely adapted to the specific properties of the bulk material to be conveyed. In order to meet the requirements of plant operators for largely trouble-free conveying operation and trouble-free restarting after possible disturbances/blockages, and on the other hand to be able to transport bulk materials that are not conventionally conveyable in dense phase, various new pneumatic conveying methods have been developed and offered on the market. Some of these are described below based on their action principles, advantages and disadvantages, and recommended application areas. Subsequently, a relatively new pneumatic conveying method, the so-called FLUIDCON conveying, is presented in more detail.

6.1 Controlled Plug Generation

Coarse-grained bulk materials with a relatively narrow particle size distribution tend to form plugs in the conveying line naturally. To support this random process, it is obvious to generate the plugs in a controlled manner. This is done by alternately inserting bulk material plugs and solid-free gas cushions into the conveying line. The lengths and

P. Hilgraf, *Pneumatic Conveying*, https://doi.org/10.1007/978-3-662-67223-5_6

distances of the plugs must be matched to each other and to the respective bulk material in such a way that a wedging effect due to too long plugs is avoided and the convergence of plugs is reliably prevented. A plug dissolution must be counteracted by maintaining a stable minimum length. Figure 6.1 shows an implementation variant: During conveying, a small constant gas flow is supplied to the sender, while the majority of the conveying gas flow is introduced into the conveying line in a clocked manner via a special mixing nozzle arranged behind the sender. Valve V1 is opened and closed in a controlled manner for this purpose. The solid outlet of the sender remains constantly open. The arrangement of the mixing nozzle/the gas inlet point into the conveying line must be at such a distance from the lock outlet that the bulk material can collapse and form a stable slope/angle of repose in the upstream pipe element when gas is supplied via V1. Other circuit variants are possible, e.g., the alternating opening of valve V1 and that to the sender. Plug generation by periodically opening and closing the material outlet valve of the sender is absolutely unsuitable for mechanical and wear reasons, as a flowing bulk material column must be traversed by the closing body each time.

Highly suitable for the controlled plug formation process are coarse-grained solids with a narrow particle size distribution from the right hatched area of Fig. 4.13, i.e., bulk materials of material group 2. In principle, it can also be used for the transport of fine-grained solids. Due to their specific properties (→ very low gas permeability, high displacement pressure) only short plugs, separated by longer gas cushions, may be introduced into the conveying line. The merging of plugs should be avoided. Due to the generally good gas retention capacity and the easy fluidizability of fine-grained products, these plugs already disintegrate after a short conveying distance and pass through the line as a continuous strand. The achievable loadings are generally significantly smaller than those possible with a conventional DPF. An exception may be fine, highly cohesive products that are capable of forming very stable plugs.

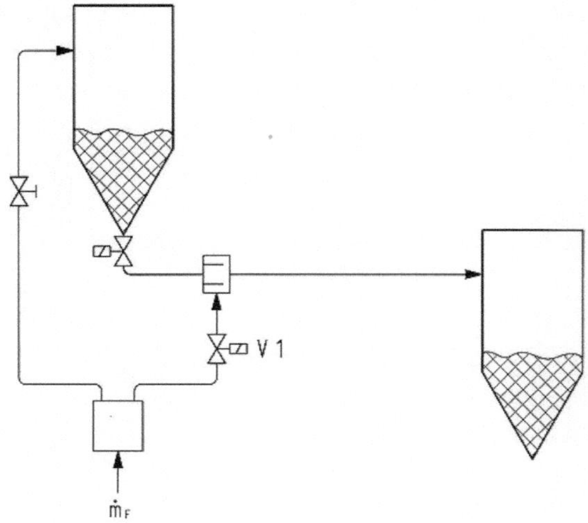

Fig. 6.1 Controlled plug generation

Own measurement results from conveyance with controlled plug production and bulk materials of group 2 are exemplarily shown in Figs. 4.48 and 6.2. The operating conditions determined in Fig. 4.48 below the unstable range were determined with different cycle time ratios “mixing nozzle open” $\Delta\tau_{auf}$ to “mixing nozzle closed” $\Delta\tau_{zu}$. Optimal conveying conditions were established at $\Delta\tau_{auf}/\Delta\tau_{zu} \cong 1$ and are in the material-specific order of magnitude of $\Delta\tau_{auf} = \Delta\tau_{zu} \cong (2-4)$ s. The extremely abrasive blast furnace coke shown in Fig. 6.2 was investigated to minimize the wear of the conveying line, both in an unstaggered and staggered conveying line. In both cases, the diameter at the line entry was $D_R = 82.5$ mm. The aim of the investigations was to convey both low-wear and abrasion- and grain-destruction-free.

An important insight provided by Fig. 6.2 is that the characteristic curves of the non-staggered and staggered lines merge/connect with each other. The solid throughput $\dot{m}_S$ is therefore significantly influenced by the initial conveying pipe diameter $D_{R,A}$. This suggests the transport mechanism schematically shown in Fig. 6.3: The individual bulk material plugs completely fill the respective conveying pipe cross-section along the pipeline. In the downstream stepped pipes, the solid matter falling out and re-accumulating leads to a reduction in the length/diameter ratio ($\Delta L_P/D_R$) of the plugs. As long as the $(\Delta L_P/D_R)_i$ ratio remains sufficiently large and thus no plug dissolution occurs, the bulk material is conveyed stably as described. The conveying limit velocity $\overline{v}_{F,A,\min}$ (→ calculated from the time-averaged conveying gas flow), which is $\overline{v}_{F,A,\min} \cong 1.40$ m/s for the present blast furnace coke, must not be undercut at any stepping point. Thus, for this

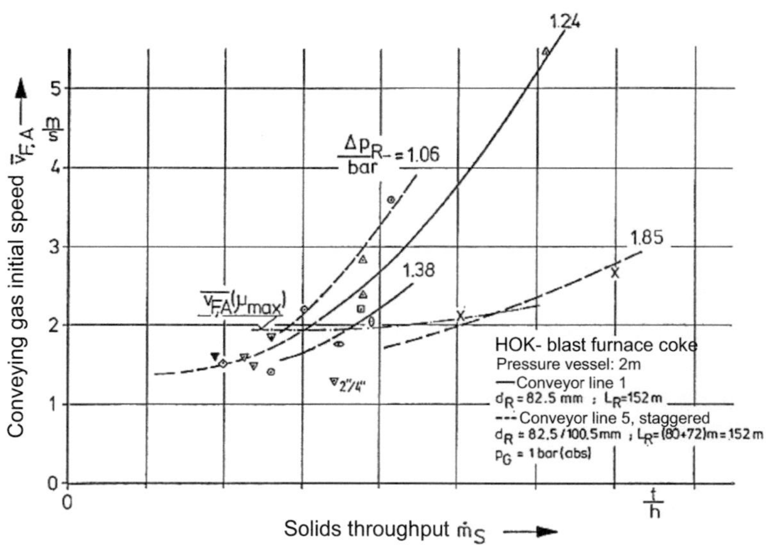

Fig. 6.2 Controlled plug production: conveying diagrams $\overline{v}_{F,A}(\dot{m}_S, \Delta p_R)$ of a blast furnace coke, $d_{S,50} \cong 2.0$ mm, $\varrho_P = 500\,\text{kg/m}^3$, through the two conveying lines $D_R \times L_R = 82.5\,\text{mm} \times 152\,\text{m}$, non-staggered, and $D_R \times L_R = 82.5/100.5\,\text{mm} \times 80/72\,\text{m}$, staggered

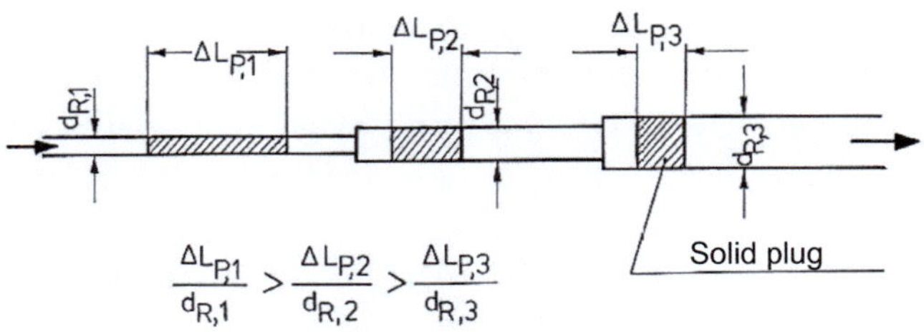

Fig. 6.3 Plug transport in a staggered conveyor line

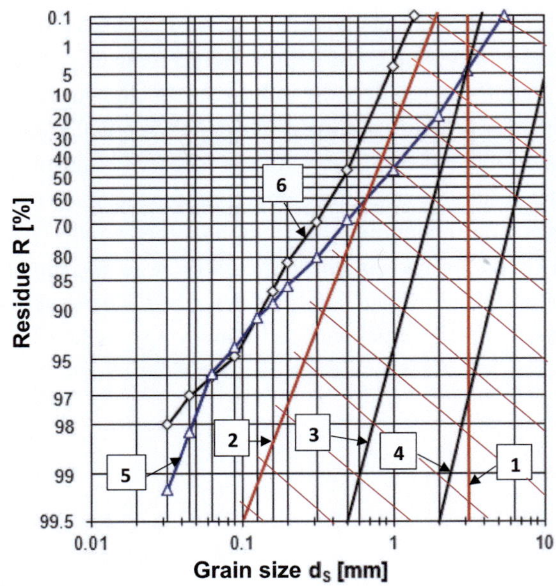

Fig. 6.4 Bulk materials suitable for plug conveying, explanations in the text

type of pneumatic conveying, a stepping criterion deviating from the method described in Sect. 4.11 must be applied. Overall, the conveying process reacts relatively insensitively to stepping errors. In conveying experiments through a multiply stepped pipeline (→ "trumpet pipe"), gas velocity profiles with $v_{F,E} < v_{F,A}$ is realized.

For plug conveying, especially with controlled plug generation, bulk materials are suitable whose particle size distributions lie within the hatched area of Fig. 6.4. Curve "1" shows the ideal, curve "2" a barely acceptable particle size characteristic. Required are: $\alpha_{RRSB} \gtrsim 60°$. The demarcation is based on systematic conveying tests and has proven itself in practice. Examples of suitable products are the particle size distributions of blast furnace coke (3), plastic granulate (1), and cylindrical activated carbon (4) in Fig. 6.4. Not suitable are, for example, the limestone chippings (6) already described in Fig. 4.17 and in calculation example 7, Sect. 4.6.5, and the coarse-grained lignite (5) from Fig. 4.35. While the limestone chippings immediately lead to pipe blockages, an indifferent state occurs with the lignite, whose particle size distribution partly lies in the permissible range: With continuous gas actuation, there is temporary conveying, then its standstill, renewed conveying, etc. There is obviously a "borderline material" for the described conveying method.

Details on the dimensioning of plug conveyors can be found in Sect. 5.2.

6.2 Targeted Plug Dissolution

Plugs/local densifications of the bulk material are localized here in the stage of their formation and dissolved by targeted injection of a partial flow of the conveying gas before exceeding a critical length. The partial gas flow is guided through an internal or external bypass line with overflow connections to the main line past the bulk material to the critical point. Figure 4.12, Sect. 4.4, illustrates the principle. The displacement pressure Δp_P required for the transport of the forming plug is greater than the available conveying pressure Δp_R. The conveying gas thus flows more strongly through the bypass line and re-enters the main line at a point where $\Delta p_R > \Delta p_P$ is. The plug part behind the re-entry is removed and/or dissolved by the entering gas flow. Long plugs are thus divided from the end, i.e., against the conveying direction, into shorter manageable elements. The plug detection is self-regulating through the pressure build-up in front of the plug. In the example shown in Fig. 4.12, the correct adjustment of the pressure loss of the bypass line to the expected conveying pressure loss of the bulk material is crucial: Under normal operating conditions, the bypass must only be flowed through by a very small part of the conveying gas (→ high flow resistance = small pipe diameter). The distances of the overflow openings must be adapted to the critical plug length of the bulk material manageable under the chosen operating conditions. The described method is commercially marketed with an internal bypass line, variant Fig. 6.5a, under the name FLUIDSTAT conveying. Variant Fig. 6.5a is, as a comparison with Fig. 6.5b illustrates, the preferred embodiment due to the lower manufacturing costs.

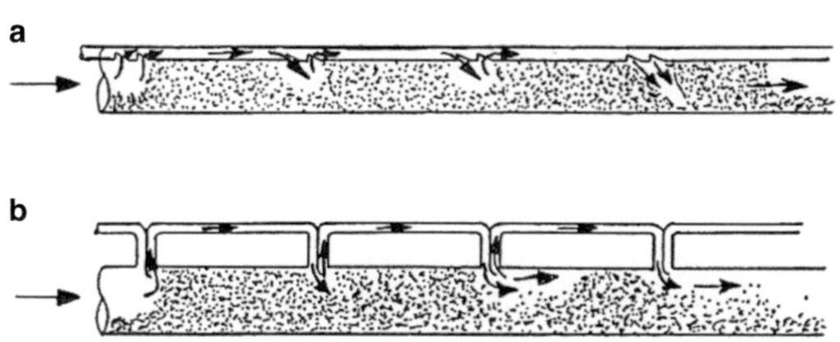

Fig. 6.5 Plug dissolution by bypass gas: FLUIDSTAT conveying

In alternative systems, the pressure build-up in front of a plug, i.e., a locally larger pressure gradient, causes the targeted opening of overflow valves in the plug area. Figure 6.6 shows an example of the structure and mode of operation of the so-called PNEUMOSPLIT conveying. The conveying line is accompanied along its entire length by a secondary line, which is connected to the main line at intervals of approximately (0.5–1.5) m through suitable check valves and is permanently flowed through by a partial gas flow. After each (2–4) overflow connections, a pressure sensor P is arranged in the main line. In undisturbed operation, the main and secondary lines (R, B) establish the (→ idealized) linear gas pressure distributions ($p_{R,0}$, $p_{B,0}$). An impending blockage leads to the disturbed distributions (p_R, p_B). The first sensor P in the conveying direction, which detects a higher gas pressure in the secondary line than in the conveying line, closes the secondary line (→ valve S2). Thus, bypass gas flows through the upstream overflow valves into the conveying pipe. The pressure profiles in Fig. 6.6 show that always the first pressure sensor behind the endangered pipe section blocks the secondary line and plugs are thus reliably detected.

The last described principle of monitoring the conveying state = pressure profile along the conveying route is implemented in the industry in various ways or realized by apparatus. Mechanically working as well as fully electronic solutions are available, e.g., so-called booster systems [1] (→ Note: The system evaluations in [1] differ from those of the present text). Bulk materials suitable for these methods are located throughout the Geldart diagram, particularly in the field between the hatched areas of Fig. 4.13, i.e., they can be used for the transport of bulk materials of material group 3 that cannot be conveyed with a conventional DPF.

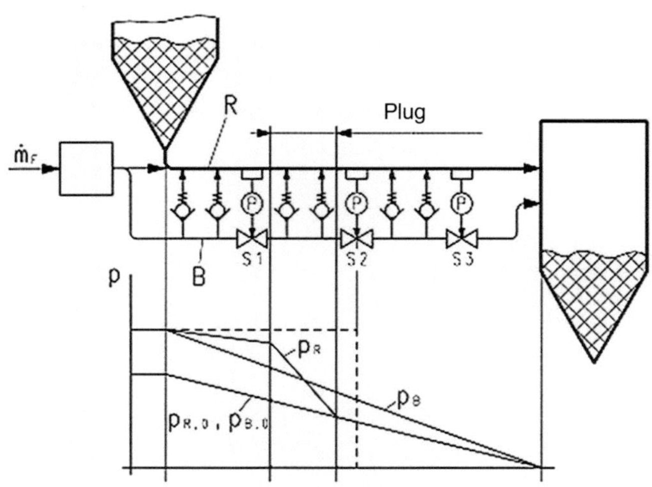

Fig. 6.6 Plug dissolution by targeted gas supply: PNEUMOSPLIT conveying

Advantages of such methods are:

- Stable operation near the conveying limit.
- Generally reliable prevention of pipeline blockages.
- Restarting against a filled pipeline, e.g., after a power failure, is generally possible.

Disadvantages:

- Higher investment costs; at least about 30% higher than with conventional DPF.
- Penetration of bulk material into and transport through the bypass line can lead to contamination, blockage and/or wear problems.
- Blowing out, especially internal bypass lines, is hardly possible and can cause problems with frequent product changes (→ contamination) and long downtimes (→ hardening bulk material).
- Overflow valves, boosters, sensors are sensitive to penetrating solids.

Overall, these systems require regular maintenance and functional checks.

6.3 Suspension Methods

These utilize and consistently support the generally long gas retention capacity of fine-grained/powdered bulk materials. Such products demix at low conveying speeds into a strand and a gas phase flowing over it. Due to the intensive gas/solid mixing during the bulk material feeding, the strand behind it is highly loosened, i.e., its "viscosity" and thus its flow resistance are low. However, as the conveying distance increases, the strand vents into the gas headspace and thus compresses. In suspension methods, transport gas and solid are therefore forcibly remixed at regular intervals along the conveying path to maintain the advantageous state of a highly loosened/fluidized strand. The mixing can be done by gas mixing from a permanently flowed through bypass line, Fig. 6.7a, or by suitable static pipe internals, Fig. 6.7b. The liquid-like state of the bulk material strand in the conveying pipe reduces its "viscosity" = internal friction and prevents the formation of plugs/wedges. With suspension methods, bulk materials of material group 1 and the border areas of the adjacent groups 3 and 4 can be stably conveyed, see Sect. 4.5.

The method shown in Fig. 6.7a is commercially offered under the name TURBU-FLOW conveying: A bypass line with specially designed outlet openings is inserted into the apex of the conveying pipe. Its diameter is relatively large, so its pressure loss remains low and it is permanently flowed through by a larger amount of gas even during normal conveying operation. The outlet areas of this secondary line are designed in such a way that part of the incoming bypass gas flow is directed with high turbulence into the bulk material strand running underneath and loosens or fluidizes it. Behind the outflow point, a corresponding gas flow is taken up again from the main line into the bypass line.

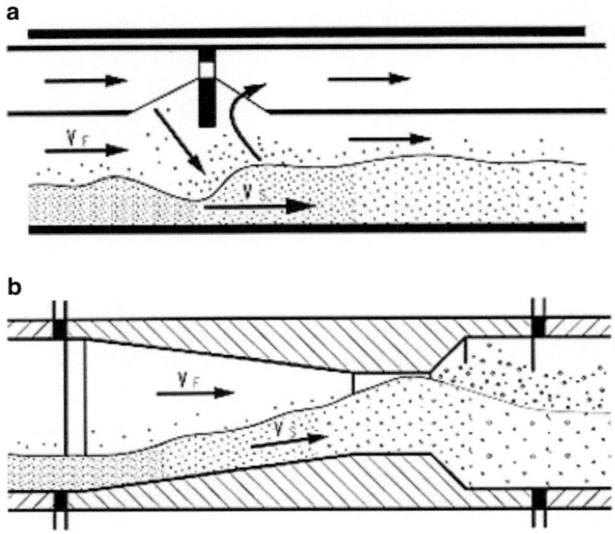

Fig. 6.7 Examples of suspension methods. **a** Gas mixing from bypass line, TURBUFLOW conveying, **b** static mixer installation

Despite the external similarity to the FLUIDSTAT method, see Figs. 6.5a and 6.7a, the TURBUFLOW method implements a completely different principle of action.

Figure 6.7b shows a static mixer installation, which brings together and mixes the conveying gas and solid strand using a specially designed pipe constriction/nozzle. In all suspension methods, the distance between the gas/solid mixing points is determined by the gas retention capacity of the respective bulk material: With decreasing venting time $\Delta\tau_E$ this distance approaches zero. This also indicates the limits of this method.

6.4 Processes with Uncontrolled Gas Addition

Two such processes are schematically shown in Figs. 6.8 and 6.9. In the GATTYS process, see Fig. 6.8, a perforated bypass line extending over the full conveying distance is inserted into the conveying line. Additional gas from a separate gas source is permanently supplied to the main line along the entire length through this hose. However, since the additional gas follows the path of least resistance, it avoids building plugs/solid compactions and flows around them. A targeted dissolution of plugs is therefore not possible,

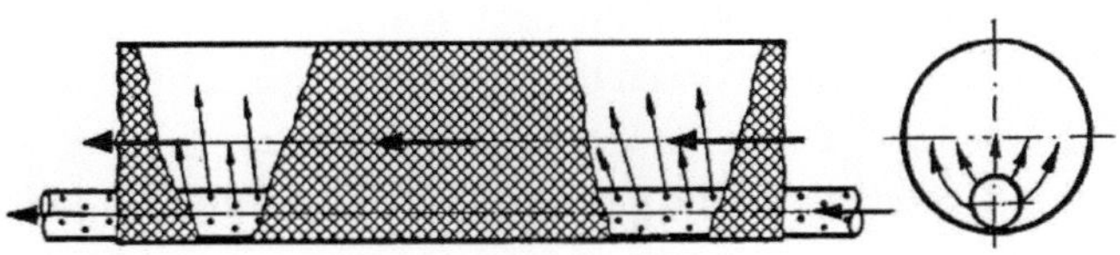

Fig. 6.8 Conveying with the GATTYS process

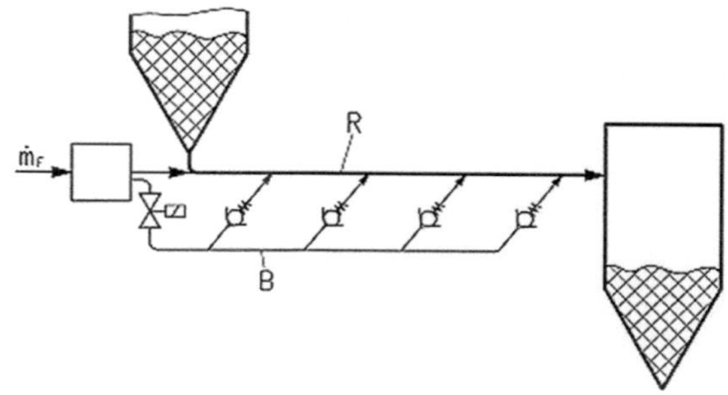

Fig. 6.9 Conveying with uncontrolled gas distribution, FLUIDSCHUB conveying

but a certain loosening and fluidization effect supporting the conveying process is achieved along the conveying pipe. A disadvantage is that the supply of bypass gas constantly increases the conveying gas flow in the transport direction. This leads to an additional velocity increase along the conveying path beyond the pure gas expansion, which may have to be reduced by pipe staggering. The GATTYS process is no longer commercially available.

In the conveying process shown in Fig. 6.9, the bypass line running parallel and outside the actual conveying pipe is connected to the main line via overflow/check valves. The gas supply to the bypass line can be switched on and off or regulated via a valve activated by a conveying pressure sensor. However, this does not change the fact that the distribution of the bypass gas along the actual conveying route is largely uncontrolled and the additional gas, as in the GATTYS process, will bypass local bulk material compactions with steeper pressure gradients via the bypass line, only to flow back into the conveying line behind the disturbance. The gas flow introduced into the incipient plug remains small.

The ability to specifically dissolve plugs and/or bulk material compactions is low for the conveying processes with uncontrolled gas addition for the reasons indicated. On the other hand, the permanent gas supply leads to loosening, reduction of the internal friction angle, and, especially for finer-grained solids, fluidization of the bulk material. This prevents plug formation to a certain extent. The bulk materials that can be safely conveyed with these processes are fine-grained and identical to those of the suspension processes.

6.5 Notes on Process Selection

The potential problem areas mentioned in Sect. 6.2 as disadvantages of processes with targeted plug dissolution apply in principle to all conveying processes with bypass gas supply to the conveying line: Their price is inevitably higher, and penetration of particularly fine-grained bulk materials into the auxiliary lines arranged inside or outside the conveying route and the associated fittings is not reliably preventable even by using check valves and requires appropriate monitoring or periodic inspections. This should be taken into account when selecting the process to be used. The following recommendations are provided:

- Bulk materials from the left hatched area of the extended Geldart diagram, Fig. 4.13, i.e. materials of substance group 1, can be easily conveyed using conventional DPF. Figure 6.10 compares, for example, the conveying characteristics of a lime meal from this Geldart area, which were determined both conventionally and with the PNEUMOSPLIT process on the same conveying route [2]. It can be seen that with PNEUMOSPLIT, lower conveying gas velocities $v_{F,E}$ are possible at the end of the line than with conventional DPF, but these are in a range that no longer offers any advantages for the examined lime meal: The same applies to the vast majority of solids to be conveyed. Possible exceptions would be the transport of extremely abrasive or abrasion-sensitive bulk materials. Figure 6.10 also shows no significant energetic advantages of PNEUMOSPLIT compared to conventional conveying. Dense phase conveying of bulk solids of group 1 should therefore be implemented as conventional DPF.
- Bulk materials from the right hatched area of the extended Geldart diagram, Fig. 4.13, i.e. materials of substance group 2, should preferably be transported using controlled plug formation methods. The relatively large pressure and power reserves of the pressure generator to be planned, which are necessary for a natural, i.e. uncontrolled, plug conveying due to randomly forming excessively long bulk material plugs and/or the merging of shorter plugs, are significantly reduced.

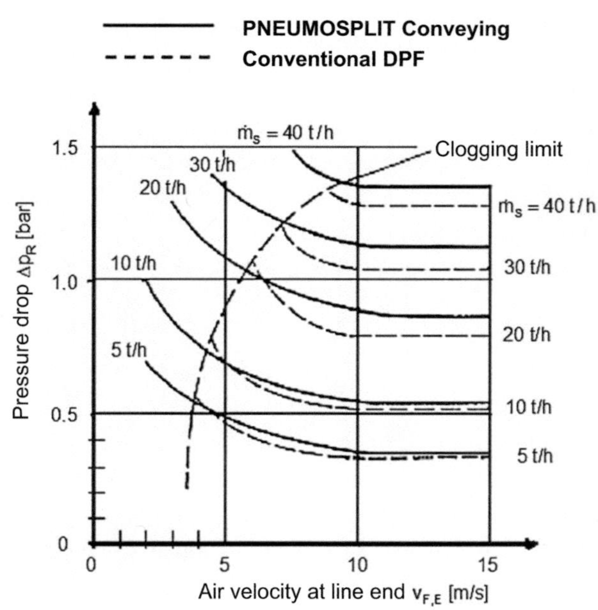

Fig. 6.10 State diagram of lime powder in conventional DPF and conveying with the PNEUMOSPLIT method, bulk density: $\varrho_{SS} = (1020\text{–}1430)\ \text{kg/m}^3$, conveying distance: $L_R = 61.5$ m including $\Delta L_v = 5$ m conveying height and $n_U = 6 \times 90°$-bends

- The conveying of bulk materials of substance group 3, i.e. products between the two hatched areas of the extended Geldart diagram, should preferably be realized with the methods of targeted plug dissolution, or is sometimes only possible with these methods.
- Suspension methods and those with uncontrolled gas addition should be used for conveying bulk materials from the edge areas of groups 3 and 4 adjacent to substance group 1.
- The suitability of a bulk material from groups 4 and 5, see Sect. 4.5, for dense phase conveying must be analyzed on a case-by-case basis and, if necessary, secured by conveying tests.

Basic rule: The conveying method used must support the natural bulk material behavior of the solid to be conveyed and not work against it. Example: Transporting fine-grained bulk materials that tend to form strands as plugs is nonsensical, as they will dissolve back into a strand after a short conveying distance anyway. In this case, a strand conveying method, possibly with intensive gas mixing into the strand, should be implemented from the outset.

6.6 FLUIDCON Conveying

From Figure 4.42, Sect. 4.12, it can be seen that the energy requirement of a fluidzed flow channel conveying for the transport of fine-grained cement is about 20 times lower than that of a well-designed conventional pneumatic pipe conveying. This relation is also maintained for other bulk materials [3]. The energetic advantage of pneumatic flow channel conveying, in which the bulk material is transported in the form of a highly fluidized bed, see Fig. 6.11, is offset by the disadvantages of the required channel inclination in the conveying direction for stable conveying operation and sufficient residual emptying, and the resulting restrictions on the routing, e.g., no vertical sections and complicated deflections. It is obvious to combine the advantages of fluidized flow channel conveying—low energy consumption—with those of pneumatic pipe conveying—unlimitedly flexible pipeline routing. In the following FLUIDCON process, this approach is consistently implemented: A fully or partially fluidizable conveying pipe (→ flow channel principle) is additionally axially flowed through by a "driving gas stream" (→ pipe

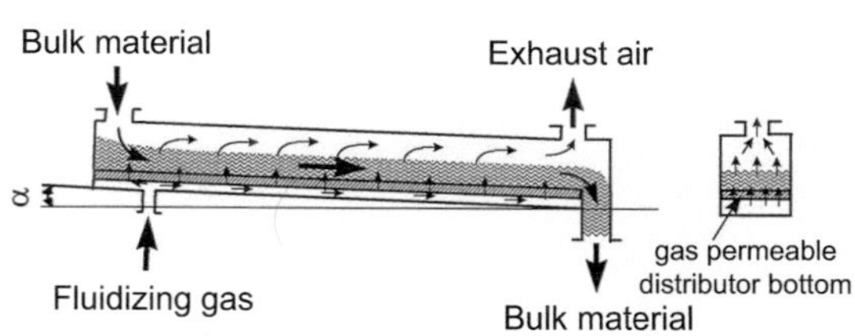

Fig. 6.11 Schematic structure of a fluidized flow channel

conveying principle). Its pressure loss replaces the flow channel inclination α_{WS}. Both suction and pressure conveying are possible.

6.6.1 Structure and Function

Fine-grained bulk materials, e.g., cement and fly ash, can be fluidized with very low gas velocities. They then behave similarly to liquids and can be transported on slightly inclined conveying channels under the influence of gravity while maintaining fluidization/gas flow [4, 5]. Figure 6.11 shows the schematic structure of the plant. The conveying channel consists of a lower box, through which the fluidizing gas is supplied, a generally gas-permeable fabric-made aeration floor, and an upper box for gas extraction. Specific fluidizing gas flows $\dot{q}_{WS}$ and channel inclination angles α_{WS} are, depending on the respective bulk material, in the order of magnitude of: $\dot{q}_{WS} \cong (1.0-3.0)\,\mathrm{m}^3/(\mathrm{m}^2\,\mathrm{min})$, $\alpha_{WS} \cong (3-8)°$. The axial transport speed is $u_S \cong 2.0\,\mathrm{m/s}$. In the FLUIDCON process, pipe conveying is supported by an integrated fluidizing system. The currently used aeration system is schematically shown in Fig. 6.12. For gas distribution, normal fluidized flow channel fabric is used, which can be covered by a perforated metal plate. The perforated plate then serves as additional wear protection. The use of suitable metal fabrics is also possible at high operating temperatures.

The schematic structure of a FLUIDCON conveying system can be seen in Fig. 6.13. A positive pressure conveying system with bulk material feed via a single lock is considered. Multipoint tasks are discussed later in Sect. 6.6.1. The total conveying gas flow $\dot{m}_F$ coming from the pressure generator is divided into a fluidizing gas flow $\dot{m}_{F,WS}$ with

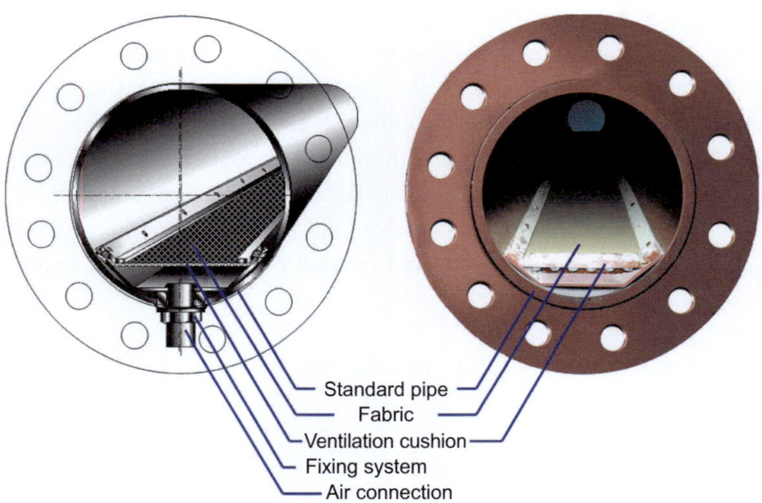

Fig. 6.12 Structure of a FLUIDCON pipe

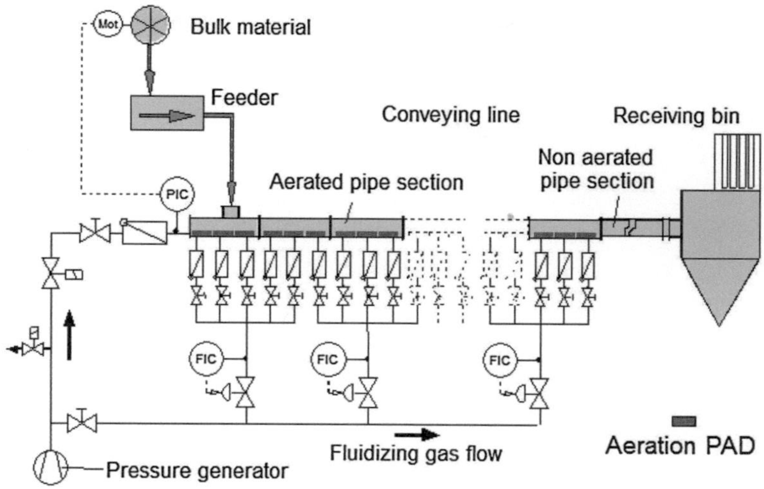

Fig. 6.13 Gas distribution scheme of a FLUIDCON conveying system

outlets for supplying the fluidization system and into an axial driving gas flow $\dot{m}_{F,\mathrm{treib}}$ fed to the beginning of the conveying pipe. Through fluidization, the bulk material is transferred into a liquid-like state with low internal friction and is lifted off the pipe floor and into the driving gas flow. These are optimal conveying conditions, with which the same transport speeds as on a fluidized flow channel can be realized.

The FLUIDCON fluidizing system is divided along the conveying route into individual gas-supplied aeration elements of length $\Delta L_{WS} \cong 3.0\,\mathrm{m}$. (2m– max. 5m) of these elements/pads are supplied with a defined gas flow by a regulator FIC, which operates without external energy, depending on the current pressure gradient and bulk material. This ensures that in the event of an impending blockage in the conveying pipe, i.e., a local pressure increase, the driving gas does not escape via the fluidizing gas line bypassing the plugging point. At the same time, the gas quantity limitation/regulation in the fluidizing gas line is a prerequisite for the use of all measures monitoring the conveying process, e.g., plugging circuits, conveying pressure-guided throughput controls, etc.

Describing the fluidization state of the bulk material in the conveying pipe by the ratio

$$\mathcal{F} = \frac{\dot{q}_{WS}}{v_{F,L}} \tag{6.1}$$

with:

$\mathcal{F} > 1$	Fluidization degree,
$\dot{q}_{WS}$	current local fluidization/operating speed,
$v_{F,L}$	current local loosening/minimum fluidization speed,

the bulk material is then fluidized/expanded equally along the transport path when $\mathcal{F} = \text{konst.}$ is set. With FLUIDCON, fine-grained bulk materials with particle diameters

$d_{S,50} \lesssim 200\,\mu\text{m}$ are conveyed. Their loosening speed $v_{F,L}$ during fluidization with a gas is determined by the equation limited to the Stokes range.

$$v_{F,L} = \frac{g}{150} \cdot \frac{\varepsilon_L^3}{1-\varepsilon_L} \cdot \frac{(d_{S,SD}^*)^2 \cdot \varrho_P}{\eta_F} \tag{6.2}$$

see (3.25) and (3.36). The dynamic viscosity η_F is independent of the operating pressure p up to pressures of approx. 10 bar, but changes with the operating temperature T_M, i.e. $v_{F,L} \neq f(p)$, $v_{F,L} = f(T_M)$. At a given temperature T_M $v_{F,L}$ thus remains constant along the conveying line, i.e., in order $\mathcal{F}$ to maintain constant, $\dot{q}_{WS}$ must also be kept constant and the distribution of the local specific gas mass flow $\Delta\dot{m}_{F,WS} = \dot{q}_{WS} \cdot \varrho_F$, $\varrho_F = \varrho_F(p)$, can be adjusted accordingly using the fluidizing gas control. This leads to specifically larger gas mass flows $\Delta\dot{m}_{F,WS}$ at the beginning of the pipeline than at the end of the pipeline. Such a ramp is usually calculated and set assuming a linear pressure gradient. Readjustment is possible in a simple way. The temperature dependence of the loosening speed $v_{F,L}$ can be represented using the so-called Sutherland equation [6]. Operating conditions of FLUIDCON systems: Typical gas velocities at the beginning of the conveying line are in the range of $v_{F,\text{treib}} \cong (1-3)\,\text{m/s}$, the specific fluidizing gas flows at $\dot{q}_{WS} \cong (0.3-1.0)\,\text{m/min}$. The gas supply can generally be provided by blowers, $\Delta p_V \leq 1.0\,\text{bar}$. This becomes immediately apparent when the slope driving force driving a normal fluidized chute conveyor is simplistically equated to the pressure force resulting from the pressure difference Δp_R for driving an equally long horizontal FLUIDCON conveyor. For the length-related pressure loss, this results in:

$$\left(-\frac{\Delta p_R}{L_h}\right) \cong \varrho_{WS} \cdot g \cdot \sin\alpha_{WS} \tag{6.3}$$

with:

- ϱ_{WS} bulk density in the fluidized state,
- α_{WS} flow channel inclination, which is required for the given bulk material and the selected specific fluidizing gas flow,
- L_h (horizontal) conveying distance.

With $\alpha_{WS} = 5°$, $\varrho_{WS} = (300-750)\,\text{kg/m}^3$ follows as an example: $(-\Delta p_R/L_h) \cong (255-645)\,\text{Pa/m}$, i.e., approximately $(0.26-0.65)\,\text{bar}/100\,\text{m}$ of conveying distance is required. In the FLUIDCON process, the pressure losses for overcoming lifting distances, deflections, etc. must also be taken into account. The relatively small pressure differences have the additional effect of reducing the increase in gas velocity caused by gas expansion. Conveying gas velocities at pipe end in the range $v_{F,E} \cong (7-13)\,\text{m/s}$ are aimed for.

By supplying fluidizing gas, the conveying gas flow along the conveying path increases from $\dot{m}_{F,\text{treib}}$ at the beginning of the conveying pipe by $\dot{m}_{F,x} = \int_0^{L_x} d\dot{m}_{F,WS}$ after the path L_x, i.e., the gas velocity increases in the conveying direction both due to gas expansion and the increase in gas mass flow. If the local gas velocity $v_{F,x}$ exceeds a critical value $v_{F,\mathrm{crit}}$, then the local conveying gas flow $\dot{m}_{F,x}$ is capable of conveying the bulk material stably without fluidization [7]. A fluidizing gas supply is no longer necessary, and the conveying line can be continued as a normal unaerated pipe. However, it is generally more advantageous to maintain the fluidized flow channel character of the conveying and the associated low transport speeds over the entire conveying distance. In this case, the FLUIDCON line is staggered, i.e., the pipe diameter is increased in sections.

Vertical sections and deflections/bends in a FLUIDCON conveying line are designed as unaerated pipes. Depending on the position of a vertical section along the conveying path, to adjust stable operating conditions, their pipe diameter $D_{R,v}$ is designed to be correspondingly smaller than the diameter D_R of the aerated horizontal conveying pipe. The locally increased gas velocity thus prevents gas/solid separation and associated pressure pulsations.

FLUIDCON allows conveying at inclination angles up to $\alpha_{FC} \cong 30°$ (→ tested so far) against the horizontal inclined uphill, which should be avoided in conventional conveying due to the risk of *backflows* (→ the friction force F_R in Fig. 4.32 runs against $F_R \rightarrow 0$ in FLUIDCON conveying). The arrangement of the inclined section along a pipeline is arbitrary. The only exception is the position directly at the beginning of the pipeline: The bulk material feed should always be into a horizontally arranged pipe. After a sufficiently long horizontal acceleration section, the transition to the inclined section may then be made.

For FLUIDCON suitable and tested bulk material locks are: rotary valve, pressure vessel, screw feeder, various flap locks.

The realization of multipoint tasks, i.e. the more or less simultaneous supply of bulk material via several parallel locks into a conveying line, is a requirement for pneumatic conveying systems, which occurs, for example, in the power plant industry. There, the fly ash accumulating in different filter funnels is fed into a common conveying line connecting a larger number of these funnels. The bulk material, which behaves like a liquid due to fluidization below the feeding points and along the conveying route, reacts with a quasi-instantaneous dissolution of material accumulations and compressions caused by the possibly irregular, portion-wise, often simultaneous solid discharge from the locks. Short moderate pressure peaks at the beginning of the line accompany this process. The high loadings μ achievable with FLUIDCON with simultaneously low conveying pressure p_R allow the use of simple and cost-effective lock systems, e.g. rotary valves, suitable double pendulum flaps, etc., which is of considerable importance for the large number of feeding points under a power plant filter.

Restarting a conveying process that has been interrupted, e.g. by a power failure, i.e. starting against a filled line, is absolutely unproblematic with FLUIDCON. For this purpose, the conveying gas is supplied to the conveying process in a time-staggered manner: After starting the fluidizing gas supply, the driving gas flow is switched on with a slight time delay of a few seconds. This captures the previously fluidized deposited bulk material and conveys it evenly and without pressure fluctuations. After significantly less than a minute, the original stationary conveying state has been re-established. The method has proven successful with all bulk materials tested so far and is implemented as a standard in FLUIDCON systems.

6.6.2 Suitable Bulk Materials

Particularly suitable for FLUIDCON are all bulk materials that can be fluidized with only low gas velocities. A long gas retention capacity is additionally advantageous. Bulk materials with these properties can be found in the hatched area of the Geldart diagram in Fig. 6.14 [8, 9]. The various bulk materials entered in Fig. 6.14 have already been successfully conveyed with FLUIDCON. They cover the entire recommended application range. The suitability of products outside the hatched area in Fig. 6.14 must be analyzed on a case-by-case basis and, if necessary, secured by fluidization tests. Usually, bulk materials to the left of the hatched area behave extremely cohesively and are therefore not or only conditionally fluidizable with mechanical support. The fluidizing gas flows through these products in the form of individual channels/rat holes, while the surrounding bulk material remains unfluidized and is supported on the pipe bottom. To the right of the hatched area, the gas velocities and thus the fluidizing gas flows accumulating over a given conveying distance generally become uneconomically large.

Through FLUIDCON, a relatively wide range of bulk materials is made accessible to a simply constructed dense phase/slow conveying system. In particular, it also covers conventionally only pneumatically conveyable bulk materials in the transition range of Geldart groups A/B with initial conveying gas velocities of $v_{F,A} \gtrsim 12\,\mathrm{m/s}$ or exclusively by complex bypass methods with lower velocities. These products include, for example, the extremely abrasive sandy alumina ($\rightarrow Al_2O_3$, $d_{S,50} \cong 80\,\mu\mathrm{m}$, $\varrho_S = 4400\,\mathrm{kg/m^3}$), which could be transported in a FLUIDCON system with gas velocities of $v_F \cong (3.0-16.0)\,\mathrm{m/s}$ over $L_R = 410\,\mathrm{m}$. The required velocities of conventional conveying systems and alternative bypass systems are significantly higher for this bulk material. Pipeline wear and grain abrasion/breakage are thus minimized [10, 11].

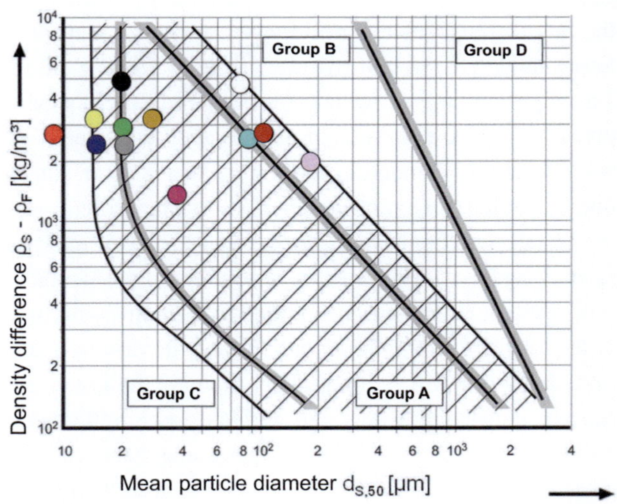

Fig. 6.14 Bulk materials suitable for FLUIDCON

6.6.3 Calculation Approach

In order to design a FLUIDCON system, a relationship between the variables conveying line pressure loss Δp_R, solid throughput $\dot{m}_S$, conveying distance L_R, conveying pipe diameter D_R, fluidizing gas velocity $\dot{q}_{WS}$ and driving gas velocity $v_{F,\text{treib}}$ as well as the line layout and the properties of the respective bulk material must be established. In principle, this can be done by fluidized flow channel models [4, 5], e.g., treating the fluidized bulk material as a non-Newtonian fluid, or by approaches from the field of pneumatic pipe conveying. In the former case, the influence of the driving gas flow must be incorporated, in the latter, that of fluidization. Since a FLUIDCON line also contains vertical sections, deflections, etc., the standard approach of pneumatic pipe conveying presented in Sect. 4.7 is applied below [8].

As an example, a horizontal pipe section of length ΔL_R is considered. Its pressure loss is calculated for conventional conveying from (4.48). With the permissible simplification $(\mu \cdot \lambda_S) \gg \lambda_F$ as well as $\mu = \dot{m}_S/\dot{m}_F = \dot{m}_S/(A_R \cdot \varrho_F \cdot v_F)$ this is transformed as follows:

$$-\Delta p_R = \frac{1}{2}\lambda_S \cdot \dot{m}_S \cdot \frac{\Delta L_R}{A_R \cdot D_R} \cdot v_F \quad \rightarrow \quad \text{with: } \lambda_S = \lambda_S\left(Fr_R = \frac{v_F^2}{g \cdot D_R}\right) \tag{6.4}$$

(6.4) can then be used to determine the pressure loss of straight horizontal FLUIDCON conveying sections when instead of

$A_R \rightarrow A_{FC}$ = the free axial flow cross-section of the FLUIDCON pipe above the fluidizing floor,

$D_R \rightarrow D_{FC}$ = the hydraulic diameter of the free axial flow cross-section,

$\lambda_S \rightarrow \lambda_S^{(FC)}$ = the bulk material resistance coefficient of horizontal FLUIDCON conveying

is set. The above transformation of (4.48) in (6.4) takes into account that due to the supply of fluidizing gas along the conveying path, the loading μ is no longer a constant, but decreases towards the end of the line: Only the solid mass flow $\dot{m}_S$ remains unchanged along the transport route.

Figure 6.15 shows the schematic curves of the resistance coefficients $\lambda_S(Fr_R)$ of a conventional conveying and those $\lambda_S^{(FC)}(Fr_R)$ of a FLUIDCON conveying. The right horizontal curve branch of the conventional conveying describes the impact losses in the pneumatic conveying area, the left curve branch with decreasing Froude number Fr_R the increasing sliding friction losses due to the intensifying gas/solid separation. The function $\lambda_S^{(FC)}(Fr_R)$ deviates from $\lambda_S(Fr_R)$ in two essential points:

a) the horizontal curve branch of the FLUIDCON conveying is significantly extended towards smaller Froude numbers,
b) this curve branch is slightly above that of conventional conveying.

Regarding (a): The extension of the "flight conveying" branch of the $\lambda_S^{(FC)}(Fr_R)$-resistance curve into the range of low Froude numbers, i.e., lower conveying speeds for a

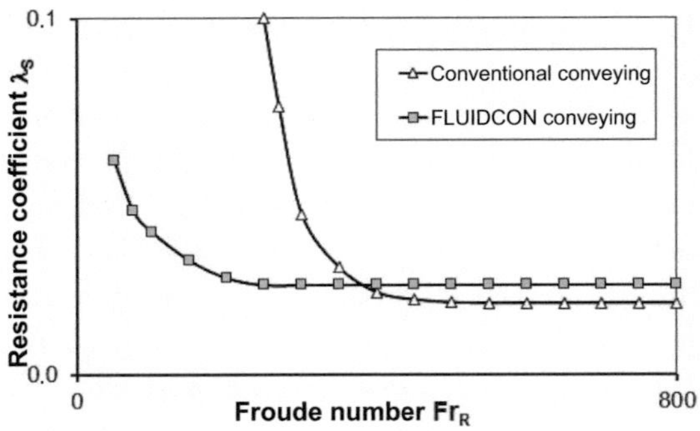

Fig. 6.15 Schematic course of the resistance coefficients λ_S and $\lambda_S^{(FC)}$

given conveying pipe, directly shows the positive effects of fluidizing the bulk material along the conveying path. By adding the fluidizing gas, the bulk material is lifted off the pipe floor and into the conveying gas stream. It does not support itself on the pipe wall. The pressure loss essentially results from the impact losses of the particles with the surrounding pipe wall area; sliding friction does not occur. This type of solid transport can be characterized as "dense" flight conveying. In [12], it is referred to as pneumatic fluidized flow conveying. The characteristic feature is the significantly smaller resistance coefficient at low Froude numbers, i.e., in the dense flow range, compared to conventional conveyances. Figure 4.23 shows an example of a comparison of measured λ_S-values of conventional conveyances of ground titanium ore ($\rightarrow d_{S,50} \cong 19\,\mu\text{m}$, $\varrho_S = 4700\,\text{kg/m}^3$) with $\lambda_S^{(FC)}$-values of corresponding FLUIDCON conveyances, which impressively confirm this.

Regarding (b): Assuming that the acceleration pressure loss of a conventional conveyance due to the permanent gas expansion is already included in the $\lambda_S(Fr_R)$, the following applies to a FLUIDCON conveyance:

$$\begin{aligned} -\Delta p_R &= \frac{1}{2} \cdot \lambda_S^{(FC)} \cdot \dot{m}_S \cdot \frac{\Delta L_R}{A_{FC} \cdot D_{FC}} \cdot v_F \\ &= \frac{1}{2} \lambda_S \cdot \dot{m}_S \cdot \frac{\Delta L_R}{A_{FC} \cdot D_{FC}} \cdot v_F + |\Delta p_{F,B}| + |\Delta p_{S,B}| \end{aligned} \tag{6.5}$$

Herein,

$$|\Delta p_{F,B}| \cong \frac{\Delta \dot{m}_{F,WS}}{A_{FC}} \cdot v_F \tag{6.6}$$

describes the additional acceleration of the fluidizing gas flow $\Delta \dot{m}_{F,WS}$ supplied to the pipe section ΔL_R from its axial inlet velocity of zero to v_F and

$$|\Delta p_{S,B}| \cong \frac{\dot{m}_S}{A_{FC}} \cdot C \cdot \Delta v_F = \left(\frac{\dot{m}_S}{A_{FC}}\right) \cdot \left(\frac{\Delta \dot{m}_{F,WS}}{A_{FC}}\right) \cdot \frac{C}{\varrho_F} \tag{6.7}$$

the additional acceleration of the bulk material mass flow $\dot{m}_S$ in the pipe section ΔL_R due to the gas mass flow increase by $\Delta \dot{m}_{F,WS}$ caused by the additional axial velocity increase Δv_F. In general, it holds: $\Delta p_{F,B} \ll \Delta p_{S,B}$.

The dependencies $\lambda_S^{(FC)}(Fr_R)$ of the bulk materials previously conveyed or investigated with FLUIDCON can be described with sufficient accuracy by the function

$$\lambda_S^{(FC)} = \frac{A(\dot{q}_{WS})}{Fr_R^B} \tag{6.8}$$

Fig. 6.16 illustrates the situation: In this figure, resistance curves $\lambda_S^{(FC)}(Fr_R)$ obtained with a cement, $d_{S,50} \cong 29\,\mu\text{m}$, $\varrho_S = 3110\,\text{kg/m}^3$, are shown. They show a behavior typical for FLUIDCON: With increasing specific fluidizing gas flow $\dot{q}_{WS}$ reduces $\lambda_S^{(FC)}$ and runs into a limit curve, which is not undercut even with further increase of $\dot{q}_{WS}$, i.e., $A(\dot{q}_{WS})$ thus takes on a constant final value specific to the bulk material A_{limit}. The exponent B of the Froude number Fr_R has been determined so far with values between $B = (0.6–0.7)$. It seems to be largely independent of the type of bulk material. With increasing pipe diameter = larger axial flow cross-section A_{FC}, $\lambda_S^{(FC)}$ decreases in a defined way: A scale-up is possible.

The existence of the limit value A_{limit} indicates the possibility of simple energetic optimization of a FLUIDCON conveying. For example, in the case of the cement shown in Fig. 6.16, an increase in the specific fluidizing gas flow above $\dot{q}_{WS} \cong 0.50\,\text{m/min}$ does not lead to any further reduction of $\lambda_S^{(FC)}$, but it increases the energetic and apparatus effort. Comparable results are also shown in Fig. 6.17, in which the dependence of the solid throughput $\dot{m}_S$ on the fluidizing gas flow $\dot{q}_{WS}$ for four bulk materials at constant boundary conditions ($\rightarrow \Delta p_R, v_{F,\text{treib}}$) is plotted. Plaster of Paris (Stucco) and cement enter a horizontal curve after exceeding a critical value $\dot{q}_{WS,\text{crit}}$. Greater specific aeration does not lead to any further increase in $\dot{m}_S$. Optimal operating points are thus at the tran-

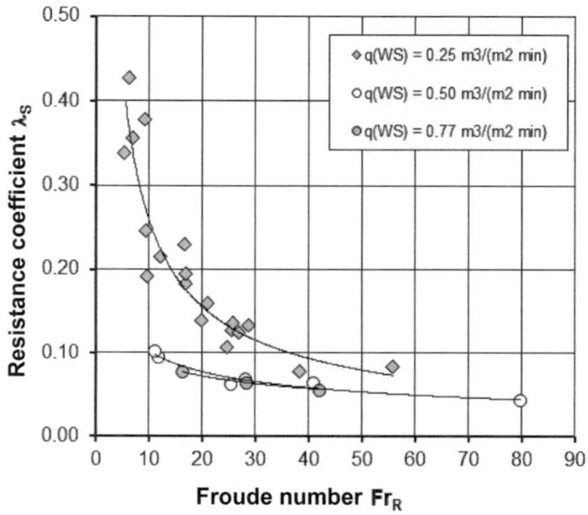

Fig. 6.16 Resistance coefficients $\lambda_S^{(FC)}$ of cement

sition to the horizontal branch of the respective characteristic curve, i.e., at $\dot{q}_{WS,\text{crit}}$. With $\dot{q}_{WS,\text{crit}}$ is also A_limit is defined. The sandy alumina and the petroleum coke in Fig. 6.17 are still above their critical fluidization. For petroleum coke, stable conveying operation was possible with only $\dot{q}_{WS} \cong 0.08\,\text{m/min}$ and $v_{F,\text{treib}} < 0.5\,\text{m/s}$.

The required specific fluidizing gas flows $\dot{q}_{WS}$ of a FLUIDCON conveyance are generally significantly lower than those of comparable fluidized flow channel conveyances. Example: Cement: $\dot{q}_{WS}$(FLUIDCON) $\cong 0.50\,\text{m/min} \rightarrow \dot{q}_{WS}$(Flow channel) $\cong 2.0\,\text{m/min}$. The cause is the supporting effect of the axial gas flow.

6.6.4 Application Examples

The basis for the design of FLUIDCON systems for new, not yet conveyed bulk materials are systematic conveying tests on test plants of operational size, the results of which are then incorporated into the above-mentioned calculation models. To secure the scale-up of the pipe diameter, measurements should be taken on at least two lines of different nominal diameters. Here: DN 100 × 150 m length and DN 150 × 55 m length, including approximately 6 m vertical section and installed glass pipe elements for visual observation of the conveying behavior. Both tracks can be combined with various bulk material locks. Based on such investigations and the developed calculation approaches, the following exemplary operating plants were dimensioned [9].

a) Alumina conveying/Russia

It conveys $\dot{m}_S = 135\,\text{t/h}$ so-called primary alumina over $L_R \cong 410\,\text{m}$. The plant supplies the storage silos in front of the electrolysis cells, where aluminum is produced from the alumina. Alumina is a bulk material at the transition of Geldart groups A/B with an average particle diameter of $d_{S,50} \cong 80\,\mu\text{m}$ and a solid density of $\varrho_S \cong 4400\,\text{kg/m}^3$, see

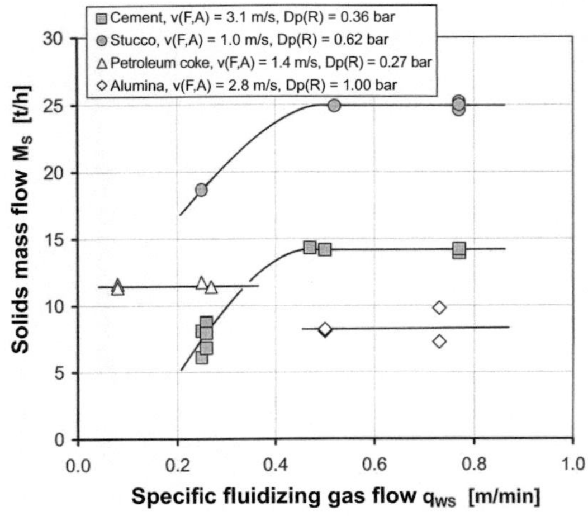

Fig. 6.17 Effect of the specific fluidizing gas flow $\dot{q}_{WS}$ on the solid throughput $\dot{m}_S$ of different bulk materials

Fig. 6.14. It can be fluidized with low gas velocities but degasses very quickly when the gas supply is interrupted [10]. Alumina = Al_2O_3 is an extremely abrasive product. Its wear effect can be characterized by its Vickers hardness, $HV(Al_2O_3) \cong 18\,kN/mm^2$, in relation to that of the surrounding wall material. The comparison with steel St 37, $HV(St37) \cong 1.3\,kN/mm^2$, illustrates that so-called high-level wear is to be expected: $HV(Al_2O_3)/HV(St37) \cong 14 \gg 1$ This also occurs with other metallic materials, $HV(Metalle) \lesssim 5\,kN/mm^2$, (→ see Chapt. 8 on the topic of wear).

The system has some special features: Despite the abrasiveness of the alumina, the bulk material is fed into the conveyor line using a screw feeder. This was necessary due to existing structural conditions. A pressure vessel feed would have been common for this product, but it could only have been realized with an additional intermediate transport. The screw feeder used is a high-speed rotating, $n_P \cong (500-1500)\,min^{-1}$, two-sided bearing press screw system, whose screw channel is completely filled with bulk material. The sealing of the conveying pressure Δp_R is achieved by the moving bulk material filling. When starting and stopping, a check valve closes the screw channel. Details on the operating principle can be found in Sect. 7.2 and [13]. To mitigate wear problems, the screw speed was set to $n_P = 730\,min^{-1}$ and the pressure difference in the line was limited to $\Delta p_R = 1.0$ bar and the screw itself was designed with wear-resistant armor.

Another special feature of the conveyor line is a 153-meter long straight pipe section inclined 10° upwards against the horizontal, see Fig. 6.18. The design data of the system are compiled in Table 6.1.

The commissioning of the plant proceeded as follows: Unexpectedly high wear occurred at the central nozzle supplying the conveying gas and at the end wing of the screw lock. Constructive changes, e.g., a roof above the nozzle, and an even higher quality wear protection eliminated this problem. The service life of the critical wear parts

Fig. 6.18 Inclined section in the alumina conveying system/Russia

Table 6.1 System data of the alumina conveying/Russia

Bulk solid		**Sandy alumina**
Conveying gas		**Air**
Type of conveying system		**FLUIDCON**
Type of solid feeder		**Screw feeder**
Solids mass flow	$\dot{m}_S$ [t/h]	135
Total conveying distance	L_R [m]	410
Including: total height	H_R [m]	35
No. of height steps along pipe	[1]	3; 2 vertical, 1 at 10° inclined above horizontal
No. of 90° bends	[1]	7
Pipe diameter	D_R [mm]	388.8 (∅406.4 × 8.8)
Total gas volume flow	$\dot{V}_F$ [m^3/h at 20 °C, 1 bar]	**7485**
Average spec. fluidization gas flow	$\dot{q}_{WS}$[m^3/(m^2 min)]	0.65
Gas velocity at pipe inlet	$v_{F,\text{treib}}$ [m/s]	3.0
Gas velocity at pipe outlet	$v_{F,E}$ [m/s]	16.2
Pipe pressure difference	$\|\Delta p_R\|$ [bar]	1.00
Total pressure difference	$\|\Delta p_{\text{vor}}\|$ [bar]	**1.30**
Power consumption of compressor	P_C [kW]	254
Type of screw feeder	[–]	**CP X-pump**
Screw diameter	D_P [mm]	300
Screw speed	n_P [min^{-1}]	730
Power consumption of feeder	P_P [kW]	68
Solid air ratio at pipe inlet	μ_A [(kg/h)$_S$/(kg/h)$_F$]	41.2
Solid air ratio at pipe outlet	μ_E [(kg/h)$_S$/(kg/h)$_F$]	15.4
Total specific power consumption	P_{spec} [kW h/(t · 100 m)]	**0.582**

is currently significantly more than 6000 operating hours. The fabric of the fluidization elements, protected by a perforated metal sheet, shows no signs of wear after several years of operation. Of the 90° pipe bends made of thick-walled steel pipe, the last one before the pipe end was replaced by a cast basalt bend.

Surprisingly, the conveying pressure was about 0.25 bar higher than the design. This was caused by the 153 m long inclined section and could be attributed to a stronger deceleration of the alumina velocity and thus a higher bulk material filling level of the conveying pipe in this section. The consequence is an increased lifting pressure loss. In executed FLUIDCON conveyances with shorter inclined sections, this effect had not been observed or recognized yet. Since the plant was designed with sufficient reserves, no changes resulted from this.

An increase in the product fines content, specifically the proportion $\leq 45\ \mu m$, which is problematic for the operation of the downstream electrolysis cells, was not observed due to the conveying process. Figure 6.19 shows exemplary measured particle size distributions of the alumina before and after a conveyance.

A direct energy comparison of the described alumina conveyance with competing processes is not possible here, as comparative data are not available. However, the current state of the art is that the otherwise used bypass conveying systems are operated with conveying gas initial velocities in the range of $v_{F,A} \cong (6\text{–}8)\ \mathrm{m/s}$ and significantly higher conveying pressures. A pressure vessel feed would save the power requirement of the screw feeder drive motor, although the non-continuous operation of pressure vessel systems partially offsets this advantage due to the required increased bulk material mass flow during the actual conveying phase [14]. The expected specific total power requirement, see Table 6.1, increased due to the increased pressure loss of the inclined section to $P_{\mathrm{spec}} = 0.631\ \mathrm{kW\,h/(t \cdot 100\,m)}$.

b) Cement Conveying/Russia

These are two largely identical systems that convey different cement qualities from various storage silos to a packing station. Typical cement characteristics are: bulk material at the transition of Geldart groups C/A, $d_{S,50} \cong 15\ \mu\mathrm{m}$, residue $R(d_S = 90\ \mu m) \cong 3$ M.-%, $\varrho_S \cong 3200\ \mathrm{kg/m^3}$, see Fig. 6.14. In Table 6.2, the operating data of one of these FLUIDCON systems is compared with the design data of an optimized conventional conveying system. In both systems, the bulk material is fed in using screw feeders, and both conveying pipe diameters D_R are identical. The use of screw feeders for cement was standard in the minerals industry at the time the systems were built.

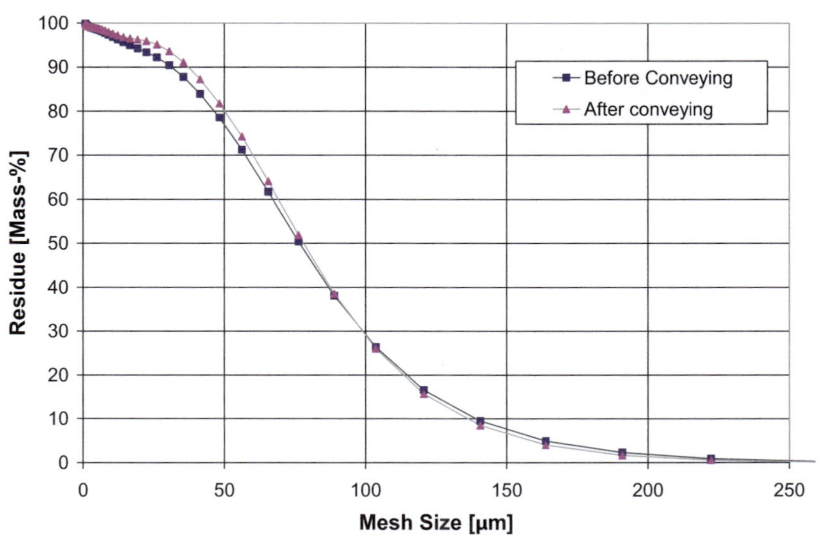

Fig. 6.19 Particle size distributions of the alumina before and after conveyance

Table 6.2 Comparison of a FLUIDCON cement conveying system with a conventionally designed pneumatic conveying system

Bulk solid		Cement	
Conveying gas		Air	
Type of solid feeder		X-Pump	
Type of conveying system		FLUIDCON	Conventional
Solids mass flow	[t/h]	135	135
Total conveying distance	[m]	153	153
Including: total height	[m]	9	9
Total gas volume flow	[m^3/h at 20 °C, 1 bar]	**2237**	**4364**
Gas velocity at pipe inlet	[m/s]	3.0	10.2
Gas velocity at pipe outlet	[m/s]	11.2	22.7
Solid/air ratio at pipe inlet	[kg_S/kg_F]	92	27.2
Pipe pressure difference	[bar]	1.10	1.30
Total pressure difference	[bar]	**1.40**	**1.60**
Power consumption of compressor	[kW]	86	173
Power consumption of X-Pump	[kW]	70	87
Total power consumption	[kW]	**156**	**260**
Total specific power consumption	[kWh/(t · 100 m)]	0.755	1.259
Power consumption relative to conventional conveying	[%]	60.0	100

From Table 6.2, it can be seen that the FLUIDCON conveying is operated with empty pipe gas velocities between $v_{F,\text{treib}} = 3.0\,\text{m/s}$ and $v_{F,E} = 11.2\,\text{m/s}$, while conventional conveying only starts at $v_{F,A} = 10.2\,\text{m/s}$. This results in a halving of the conveying gas flow and thus a significant size reduction in the required receiving filter. Overall, the total drive power requirement can be reduced by FLUIDCON to 60% of that of conventional conveying. A further reduction in the drive power of the two plants compared in Table 6.2 would be possible under the same operating conditions by using pressure vessels. As already noted under a) Alumina conveying/Russia, this does not achieve the full power of the screw feeder motor, as the operation of a pressure vessel system requires a higher drive power of the compressor.

Both FLUIDCON plants were started up at the push of a button and accepted by the customer after proof of all guarantees. They have been in operation for several years.

c) Blast furnace sand powder conveying/Germany

The plant conveys $\dot{m}_S = 100\,\text{t/h}$ blast furnace sand powder = ground blast furnace slag from a silo group over $L_R \cong 300\,\text{m}$ to a loading silo. The conveyor line contains two sections inclined upwards at an angle of approximately 14.5° to the horizontal. The longer of the inclines is about 30 m long. The blast furnace slag is a very abrasive bulk material with the following characteristics: $d_{S,50} \cong 10\,\mu\text{m}$, $R(d_S = 90\,\mu\text{m}) \cong 0.1\,\text{M.-\%}$, $\varrho_S \cong 2950\,\text{kg/m}^3$, $\text{HV(Schlacke)} \cong 7.5\,\text{kN/mm}^2$. The formal classification in the Geldart diagram, see Fig. 6.14, results in a cohesive Group C material. However, laboratory tests show good fluidizability, long gas retention, and Group A behavior. In Table 6.3, the operating data of the FLUIDCON system are compared with those of an optimized conventional conveying system. Both systems are designed with the same pipe diameter D_R and in both, the bulk material is fed to the conveying pipe by means of a screw feeder.

Compared to conventional conveying, the use of FLUIDCON halves the conveying gas volume flow and thus also the size of the required receiving filter area of this system. The gas velocities increase from $v_{F,\text{treib}} = 2.8\,\text{m/s}$ at the beginning of the conveying line to $v_{F,E} = 12.8\,\text{m/s}$ at the end of the line, thus being in a range where the lowest possible wear is expected: With FLUIDCON, the total drive power requirement of the system can be reduced to 41% of that of conventional conveying. The power requirement of the

Table 6.3 Comparison of FLUIDCON blast furnace slag conveying with a conventional pneumatic conveying system

Bulk solid		Ground blast furnace slag	
Conveying gas		Air	
Type of solid feeder		X-Pump	
Type of conveying system		FLUIDCON	Conventional
Solids mass flow	[t/h]	100	100
Total conveying distance	[m]	300	300
Including: total height	[m]	22	22
Total gas volume flow	[m^3/h at 20 °C, 1 bar]	**3662**	**7088**
Gas velocity at pipe inlet	[m/s]	2.8	11.3
Gas velocity at pipe outlet	[m/s]	12.8	24.2
Solid/air ratio at pipe inlet	[kg_S/kg_F]	56	12
Pipe pressure difference	[bar]	0.84	1.18
Total pressure difference	[bar]	**1.14**	**1.48**
Power consumption of compressor	[kW]	114	305
Power consumption of X-Pump	[kW]	44	77
Total power consumption	[kW]	**158**	**382**
Total specific power consumption	[kWh/(t · 100 m)]	0.527	1.273
Power consumption relative to conventional conveying	[%]	41	100

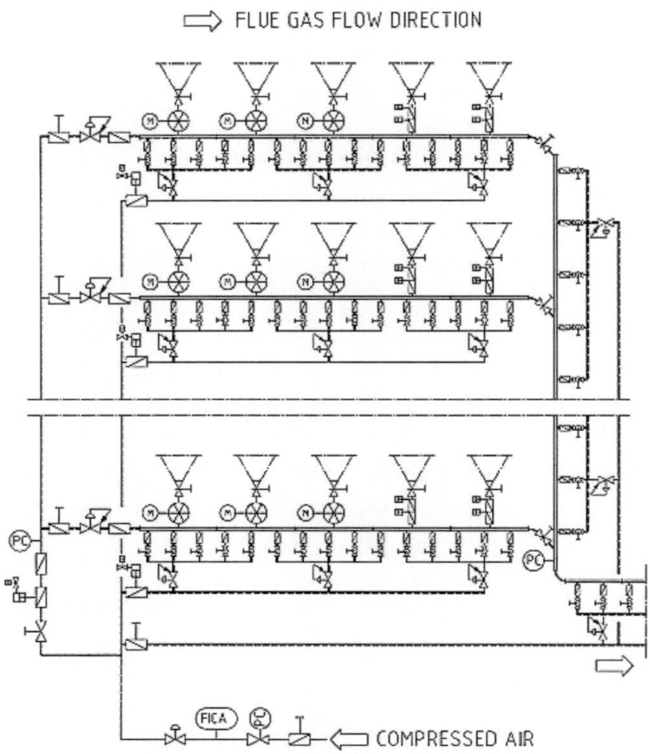

Fig. 6.20 Basic structure of a FLUIDCON fly ash removal system below an electrostatic precipitator

FLUIDCON system was guaranteed and met by the supplier. The customer subsequently ordered two more FLUIDCON conveying systems.

In addition, reference [15] is cited, in which a raw meal transport realized with FLUIDCON is analyzed and compared to conventional conveying. Various FLUIDCON systems with multi-point feeding from the power plant industry are in operation. Example: 55 t/h of coal fly ash from 30 parallel accumulation points are transported via a common conveying line. Figure 6.20 shows their basic structure [16]. From the experience gained with approximately 100 built FLUIDCON systems, it follows that the power requirement for a FLUIDCON conveying system is on average about 50% of that of an energetically optimized conventional pneumatic pipe conveying system.

Note: From the above illustrations, it follows that FLUIDCON conveyors should be used for predominantly horizontal conveying routes. Reason: The vertical sections are designed as normal conveyors without fluidization and therefore offer no advantages.

References

1. Klintworth, J., Markus, R.D.: A review of low-velocity pneumatic conveying systems. Powder Handl. Process. **5**(4), 747–753 (1985)
2. Siegel, W.: Grenzen der pneumatischen Förderung. Chemie-anlagen + Verfahr., 40–46 (Dezember 1981)
3. Hilgraf, P.: Der Energiebedarf pneumatischer Förderprinzipien im Vergleich mit mechanischen Förderungen. ZKG Int. **51**(12), 660–673 (1998)
4. Keuneke, K.: Fluidisierung und Fließbettförderung von Schüttgütern kleiner Teilchengröße. VDI-Forschungsheft 509. VDI-Verlag, Düsseldorf (1965)
5. Muschelknautz, E.: Die Berechnung der pneumatischen Fließförderung. transmatic 76, Teil II: Pneumatische und hydraulische Förderung. Krauskopf-Verlag, Mainz, S. 29–43 (1976). C1
6. Hirschfelder, J., Curtiss, C.F., Bird, R.B.: Molecular theory of gases and liquids. Wiley, New York (1954)
7. Hilgraf, P.: Minimale Fördergasgeschwindigkeiten beim pneumatischen Feststofftransport. ZKG Int. **40**(12), 610–616 (1987)
8. Hilgraf, P.: FLUIDCON—a new pneumatic conveying system for fine-grained bulk materials. Cem. Int. **2**(6), 74–87 (2004)
9. Hilgraf, P., Dikty, M.: Pneumatic conveying with FLUIDCON—operating experience and results. Cem. Int. **6**(6), 46–55 (2008)
10. Wolf, A., Hilgraf, P.: FLUIDCON—A new pneumatic conveying system for alumina. 2006 TMS Annual Meeting & Exhibition, San Antonio, March 12–16. Collected proceedings: light metals—alumina & Bauxit., S. 81–87 (2006)
11. Wolf, A., Hilgraf, P., Hilck, A., Marshalko, S.V.: Operational experience with a brown-field expansion project in Sayanogorsk, Russia. TMS Annual Meeting & Exhibition, New Orleans, 09–13. March. Collected proceedings: light metals, Bd. 2008., S. 51–56 (2008)
12. Weber, M.: Strömungs-Fördertechnik. Krausskopf-Verlag, Mainz (1974)
13. Hilgraf, P., Paepcke, J.: Der Eintrag von Schüttgut in pneumatische Förderstrecken mittels Schneckenschleusen. ZKG Int. **46**(7), 368–375 (1993)
14. Hilgraf, P.: Optimale Auslegung pneumatischer Dichtstrom-Förderanlagen unter energetischen und wirtschaftlichen Gesichtspunkten. ZKG Int. **39**(8), 439–446 (1986)
15. Dikty, M., Schwei, P.: Entscheidungsmatrix für den Schüttguttransport. ZKG Int. **60**(7), 56–66 (2007)
16. Hilgraf, P., Blumenberg, C., Hielscher, J.: Innovations in coal fired units: coal injection and fly ash handling. POWER-GEN Europa 2010, Amsterdam, 08–10 June 2010. Collected Proceedings: Track 4, Session 5, 2010

Bulk Material Locks

7

The task of the bulk material lock in a pressure conveying system is to introduce a predetermined solid mass flow against the conveying line overpressure into the conveying gas stream/transport line while sealing the system overpressure against the environment or the upstream system components, i.e., keeping the gas leakage across the feeder as low as possible. In suction systems, the lock must realize the discharge of the solid separated from the conveying gas at the end of the conveying distance from the system's underpressure to the ambient pressure while avoiding the penetration of leakage gas. Compare this to Fig. 1.2. In pressure conveying systems, the gas mass flow $\dot{m}_{F,V}$ supplied by the pressure generator is reduced by the gas leakage to the usable conveying gas flow for solid transport $\dot{m}_{F,F} = (\dot{m}_{F,V} - \dot{m}_{F,\text{leck}})$. In suction conveying systems, the downstream blower only sucks the gas flow $(\dot{m}_{F,V} - \dot{m}_{F,\text{leck}})$ through the actual conveying distance. In both cases, too great values of $\dot{m}_{F,\text{leck}}$ can lead to impairments of the conveyance or, to avoid this, the pressure generator must be designed for a more energetically unfavorable larger gas mass flow. Changing operating conditions, e.g., due to pressure fluctuations in the conveying line, and the wear of the lock due to runtime should not or only slightly influence the locking process.

The solid material inlet and outlet can be realized using different operating principles. The simplest is the sealing by a correspondingly high bulk material column. It must be taken into account that its height must fulfill the condition

$$H_{SS} > \frac{|\Delta p_{\text{Sch}}|}{g \cdot (1 - \varepsilon_F) \cdot \varrho_S} \tag{7.1}$$

see (3.35) in Sect. 3.3.1. This leads to very large and generally hardly feasible construction heights for larger pressure differences Δp_{Sch} at the lock. However, this principle is used for special applications.

P. Hilgraf, *Pneumatic Conveying*, https://doi.org/10.1007/978-3-662-67223-5_7

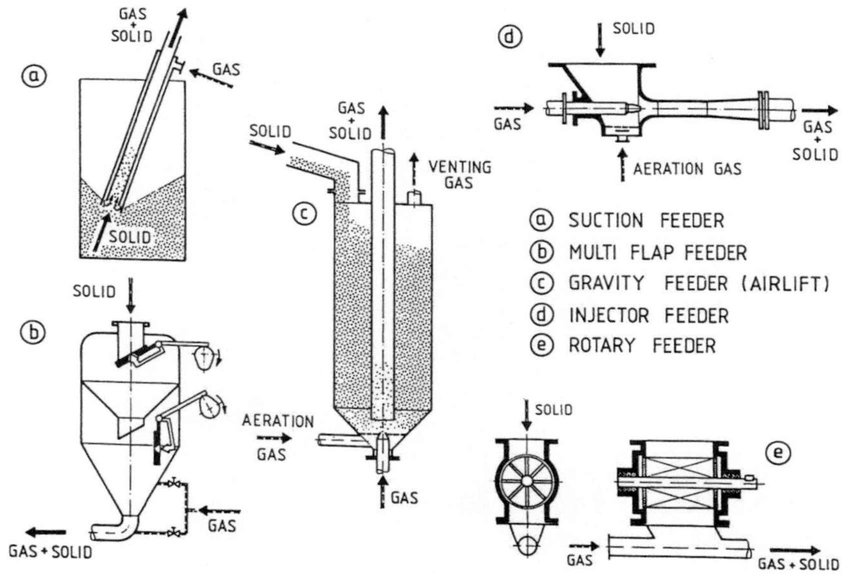

Fig. 7.1 Examples of bulk material feeder systems

Table 7.1 provides an overview of the application areas of various lock systems. Some of these systems are schematically illustrated in Fig. 7.1. These and other systems will be discussed in more detail below. Both in Table 7.1 and in Fig. 7.1, the term *suction feeder* does not refer to the discharge system, but to the bulk material pick-up device. The application areas mentioned in Table 7.1 are rough guidelines.

The following presents the basic process engineering design of various lock systems. Constructive requirements (→ e.g., the pressure vessel is a "pressure" container subject to special regulations!) are not discussed, wear-related aspects are covered in Sect. 8.2.

7.1 Pressure Vessel Locks

The most universal and common feed device for dense phase conveying is the pressure vessel. It exists in the two basic design variants shown in Fig. 7.2. The most common, especially for large-volume transmitters, is the gravity-assisted bottom-side solid discharge. The gas flow $\dot{m}_F$ provided by the pressure generator is divided into a portion $\dot{m}_{F,B}$ supplied to the pressure vessel and another partial flow directly introduced into the conveying line $\dot{m}_{F,M}$. The gas flow $\dot{m}_{F,B}$ to the vessel can, depending on the requirements of the bulk material present, be further divided into a portion to the vessel cone to support a trouble-free and uniform solid discharge from the container and a portion to the

Table 7.1 Application areas of different bulk material locks

Installation type	Feeder type	Solids mass flow	Conveying distance [m]	Pressure drop [bar]	Suitable bulk solids (i.a. Geldart group)	Operation mode	Conveying type
Suction conveyance	Suction nozzle / pipe -	≤400 t/h	≤150	≤0,5	C/A, A, B, D, flowing, non-caking	Continuous	Lean phase
Pos. pressure conveyance	Injector -	≤5 t/h	≤75	≤0,2	C, A, B, D, all pneumatically conveyable	Continuous	Lean phase
	Multi flap -	≤60 m³/h	≤150	≤0,5	C/A, A, B, D, flowing, non-caking, low abrasiv	Continuous	Lean phase
	Gravity (Airlift) -	≤500 t/h	≤125 vertical	≤0,6	C/A, A, B, flowing	Continuous	Lean phase
	Rotary valve -	≤150 m³/h	≤250	Normal: ≤1,0 Special: ≤3,0	C/A, A, B, D, flowing, non-caking, low/non-abrasiv	Continuous	Lean phase +dense phase
	Screw -	≤500 m³/h	≤900	≤2,5	C/A, A, A/B, wide particle size range necessary, non-caking, non-plastifying	Continuous	Lean phase +dense phase
	Pressure vessel- - Single p.v. - Two p.v. in parallel - Two p.v. in sequence	 ≤120 m³/h ≤250 m³/h ≤250 m³/h	 ≤1000 ≤1500 ≤2000	≤6,0	C, A, B, D, flowing, resp. flowing by auxiliary means (e.g. by gas injection, stirring etc.)	 Discontinuous Quasi-continuous Continuous	Lean phase +dense phase

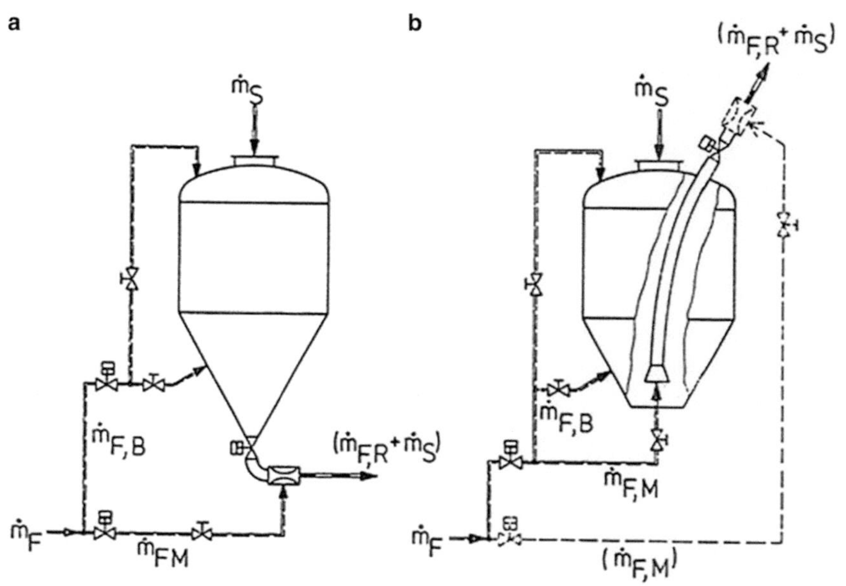

Fig. 7.2 Design variants of pressure vessels. **a** solid discharge downwards, **b** solid discharge upwards (→ *bottom/top discharge system*)

container head. For coarse-grained bulk materials, cone aeration is generally not necessary or not effective. By changing the (solid-specific) ratio $\dot{m}_{F,B}/\dot{m}_{F,M}$ the solid throughput $\dot{m}_S$ can be changed/adjusted within limits. The gas distribution is adjusted once using manual valves.

The pressure vessel lock is a discontinuous process, consisting of the steps: filling the container with bulk material, pressurizing to conveying pressure, conveying = emptying the container, and relaxing the vessel to filling pressure. To achieve a largely continuous conveying, various vessels must be connected in parallel or in series. This is illustrated in Sects. 7.1.2–7.1.4.

7.1.1 Single Pressure Vessel

Fig. 7.3 shows the structure of a single pressure vessel system, which is connected downstream of a continuously operating production step, e.g., bulk material grinding. To decouple the two continuous/discontinuous process stages, a correspondingly large buffer container must be connected upstream of the single vessel transmitter. In Fig. 7.4, an example of a possible design variant of a pressure vessel for fine-grained bulk materials is shown: The cone aeration is carried out here via a fluidizing cone that introduces the gas over a larger area with an inlet diameter of $D_K = 600\,\text{mm}$ = maximum bridging diameter within a hemispherical bottom. Due to the short residence times of the solid in the transmitter, this diameter, in conjunction with the aeration, is generally completely sufficient for safe bulk material discharge. Other design variants, e.g., aeration using nozzles or flat fluidizing units over the entire cone height, are also used. After removing the hemispherical bottom, this opening can be used as a manhole for inspections/checks, etc., of the vessel. The valve designations used in Fig. 7.4 are referred to in the following text.

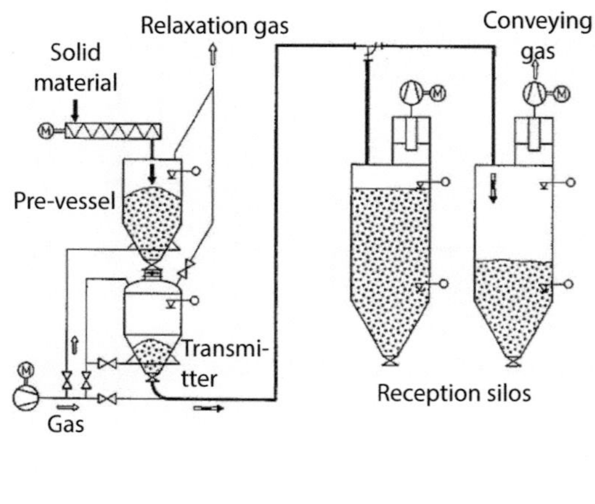

Fig. 7.3 Schematic structure of a pressure vessel system with a single vessel

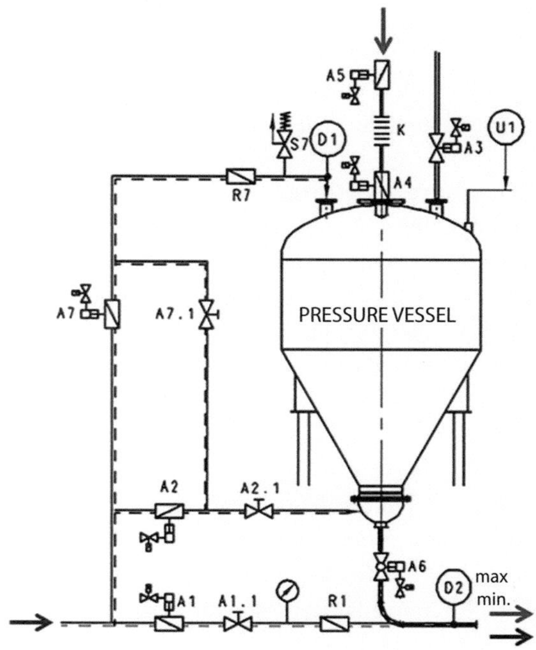

Fig. 7.4 Structure of a single pressure vessel

a) Surcharge factor, solid mass flow during the conveying phase:
Since the single vessel operates intermittently = in batches, a working/conveying cycle of duration $\Delta\tau_{Ch}$ consists of dead times $\Delta\tau_{\text{tot}}$, during which no conveying takes place, and the actual conveying time $\Delta\tau_{\text{för}}$:

$$\Delta\tau_{Ch} = \Delta\tau_{\text{för}} + \Delta\tau_{\text{tot}} \tag{7.2}$$

To transport a given solid mass Δm_S in a given time $\Delta\tau$, i.e., the nominal (constant) mass flow $\dot{m}_{S,n} = \Delta m_S/\Delta\tau$, from A to B, only the portion $\Delta\tau_{\text{för}}/\Delta\tau_{Ch} = 1 - \Delta\tau_{\text{tot}}/\Delta\tau_{Ch}$ of the batch time $\Delta\tau_{Ch}$ is available to the single pressure vessel. This requires an increased solid throughput $\dot{m}_S$ during the actual conveying phase $\Delta\tau_{\text{för}}$. It therefore applies: $\dot{m}_{S,n} \cdot \Delta\tau_{Ch} = \dot{m}_S \cdot \Delta\tau_{\text{för}}$. This results in $\dot{m}_S = \dot{m}_{S,n} \cdot \Delta\tau_{Ch}/\Delta\tau_{\text{för}}$. The ratio

$$f_Q = \frac{\dot{m}_S}{\dot{m}_{S,n}} = \frac{\Delta\tau_{Ch}}{\Delta\tau_{\text{för}}} = \frac{\Delta\tau_{Ch}}{\Delta\tau_{Ch} - \Delta\tau_{\text{tot}}} \tag{7.3}$$

is subsequently referred to as the surcharge factor. Its size is determined, among other things, by the type of pressure vessel connection (→ here single vessel; further arrangements are discussed in Sects. 7.1.2–7.1.5), the chosen number of batches/time unit, the conveying gas flow, as well as the conveying pressure and the respective vessel size. Since the surcharge factor f_Q determines the solid mass flow $\dot{m}_S = f_Q \cdot \dot{m}_{S,n}$ during the conveying phase and thus determines the conveying line and pressure generator dimensions, but is not explicitly predictable, f_Q or $\dot{m}_S$ must be determined iteratively. The f_Q-values of a single vessel are in the range of $f_Q \cong (1.2-1.5)$. Smaller values are assigned to dilute phase conveyance, larger values to dense phase conveyance.

b) Pressure vessel size

Through the conveying characteristics of a pressure vessel transmitter, the nominal solid mass flow rate $\dot{m}_{S,n}$ is divided into $\dot{N}_{Ch} = \Delta N_{Ch}/\Delta\tau$ individual portions/time unit of the size $(V_{B,\text{Netto}} \cdot \varrho_{b,B})$, which are transported one after the other. With a given temporal batch number $\dot{N}_{Ch}$, the net volume $V_{B,\text{Netto}}$ of the individual vessel = the bulk material filling volume of the container is thus calculated:

$$V_{B,\text{Netto}} = \frac{\dot{m}_{S,n}}{\varrho_{b,B} \cdot \dot{N}_{Ch}} \quad \rightarrow \quad \text{with: } \dot{N}_{Ch} \leq \dot{N}_{Ch,\max} \tag{7.4}$$

Suitable container sizes can be taken from the series of various suppliers, which generally cover volumes between $V_{B,\text{Netto}} \cong (0.2-20)\,\text{m}^3$. The corresponding gross volumes $V_{B,\text{Brutto}}$ (= the total volume of the respective sender) is roughly $V_{B,\text{Brutto}} \cong 1.1 \cdot V_{B,\text{Netto}}$. This difference is necessary for measurement and wear-related reasons (→ among others, level measurement, closing of valves outside the bulk material, etc.). The maximum number of batches per unit of time $\dot{N}_{Ch,\max}$ is a manufacturer-specific empirical value that depends, among other things, on the abrasiveness of the bulk material to be conveyed, the current conveying pressure, and the vessel size itself: $\dot{N}_{Ch,\max}$ values increase with decreasing vessel volume and decrease with increasing conveying pressure and more abrasive bulk material. Example: low abrasive bulk material, conveying pressure < 4 bar, $V_{B,\text{Netto}} = 0.2/2.0/20.0\,\text{m}^3$, $\dot{N}_{Ch,\max} = 30/15/6\,\text{Batches/h}$. By specifying $\dot{N}_{Ch,\max}$ the smallest permissible vessel size is selected.

The bulk density $\varrho_{b,B}$ of the solid material filled into the sender depends on the type of product to be conveyed and its venting time = filling time $\Delta\tau_1$ of the vessel, values equal to or less than its bulk density ϱ_{SS} are possible. For short filling times $\Delta\tau_1$ and slowly venting (fine-grained) bulk materials, $\varrho_{b,B} < \varrho_{SS}$.

Systematic measurements lead to the relationship:

$$\varrho_{b,B} \cong \varrho_{SS} - \frac{1}{4} \cdot (\varrho_{SS} - \varrho_{b,\text{fluid}}) \quad \rightarrow \quad \text{when: } \Delta\tau_1 < 2 \cdot \Delta\tau_E \tag{7.5a}$$

$$\varrho_{b,B} \cong \varrho_{SS} \quad \rightarrow \quad \text{when: } \Delta\tau_1 \geq 2 \cdot \Delta\tau_E \tag{7.5b}$$

$\varrho_{b,\text{fluid}}$ is in this case the one described in Sect. 3.3.2. Fig. 3.12 shows the standardized fluidized bed apparatus for measuring this fluidized density. $\Delta\tau_E$ is the associated venting time of the bulk material. The ratio of fluidized density to bulk density is in the order of $\varrho_{b,\text{fluid}}/\varrho_{SS} \cong 0.65$ for the fine-grained bulk materials of group 1, see Sect. 4.5, and increases to group 2 to $\varrho_{b,\text{fluid}}/\varrho_{SS} = 1$. Increased temperatures reduce the $(\varrho_{b,\text{fluid}}/\varrho_{SS})$ ratio of a gas/solid mixture.

A larger number of batches per unit of time $\dot{N}_{Ch}$ reduces the required transmitter size $V_{B,\text{Netto}}$, but leads to a higher stress, especially on the fittings working in the bulk material. At the same time, with increasing $\dot{N}_{Ch}$ the surcharge factor f_Q and with this the design mass flow $\dot{m}_S$ of the actual conveying phase increase. It can be seen that this results in an optimization problem with regard to minimizing the sum of investment, operating, and energy costs. Corresponding calculations were carried out in [1]. The results show that, given a specific task, the total costs per year decrease steeply with increasing $V_{B,\text{Netto}}$ and level off at an approximately constant limit value when a critical transmitter volume is exceeded, i.e., larger pressure vessel volumes have cost advantages in operation.

c) Calculation of the pressure vessel cycle:
To achieve a chosen temporal batch number $\dot{N}_{Ch}$, the sum $\sum_i \Delta\tau_i = \Delta\tau_{Ch}$ of the individual times $\Delta\tau_i$ of the conveying cycle (→ Filling, pressurizing, conveying, etc.) must fulfill the condition.

$$\xi_{Ch} \cdot \left(\frac{1}{\dot{N}_{Ch}}\right) = \sum_i \Delta\tau_i \tag{7.6}$$

In (7.6), the $(1/\dot{N}_{Ch})$ is the maximum time duration available for a single conveying cycle and $\xi_{Ch} \leq 1$ is a utilization factor that can cover, for example, temporal uncertainties. Example: $\xi_{Ch} = 0.95$ states that the currently available batch time, i.e., $(1/\dot{N}_{Ch})$, is only utilized to 95% and a time reserve of 5% should be planned. The value of $\sum_i \Delta\tau_i = \Delta\tau_{Ch}$ must now be adjusted by the plant design in such a way that $(\xi_{Ch}/\dot{N}_{Ch})$ is not exceeded. The following procedure is necessary for this:

- From the known nominal solid mass flow $\dot{m}_{S,n}$, an assumed batch number $\dot{N}_{Ch}$ and a density $\varrho_{b,B}$ determined from the available bulk material data, a pressure vessel size is calculated using (7.4) and adapted to the series available from the manufacturer. For the selected $V_{B,\text{Netto}}$, $\dot{N}_{Ch}$ is recalculated.
- Taking into account the chosen vessel size, the wear properties of the bulk material, and the planned operating conditions, an initial surcharge factor f_Q is selected and with this, the solid mass flow $\dot{m}_S = f_Q \cdot \dot{m}_{S,n}$ during the actual conveying phase is determined, see (7.3). Starting values for f_Q are generally tabulated by the providers.
- With $\dot{m}_S$ the unknown quantities conveying pipe diameter D_R, pressure difference Δp_R at the conveying line and required conveying gas flow $\dot{m}_F$ can be determined for the respective current boundary conditions. This is done using the calculation approaches described in earlier sections.
- The gas flow $\dot{m}_{F,V}$ to be provided by the pressure generator during the conveying phase consists of the actual conveying gas flow $\dot{m}_F$ and the so-called top gas flow $\dot{m}_{F,O}$ together:

$$\dot{m}_{F,V} = \dot{m}_F + \dot{m}_{F,O} \tag{7.7}$$

$\dot{m}_{F,O}$ replaces the solid volume flow rate discharged from the transmitter during conveyance ($\dot{m}_S/\varrho_S$) and thus maintains the container pressure p_B upright/constant. In particular, in dense phase conveying $\dot{m}_{F,O}$ takes on non-negligible large values. It applies:

$$\dot{m}_{F,O} = \left(\frac{\dot{m}_S}{\varrho_S}\right) \cdot \varrho_{F,B} \tag{7.8}$$

with:

$\varrho_{F,B}$ current gas density in the sender, $\varrho_{F,B}(p_B, T_B = T_S)$.
To discharge the bulk material from the sender, for gas distribution at the pressure vessel, and to overcome the resistance of the pipeline between the sender and the pressure generator, the additional pressure loss in total Δp_B is required. The pressure generator must therefore be designed for at least the operating pressure

$$p_V = |\Delta p_R| + \Delta p_B + p_E \tag{7.9}$$

Δp_B can be in the order of magnitude of $\Delta p_B \cong 0.5$ bar

- Using the now available operating data, the individual times $\Delta\tau_i$ of a conveying cycle are calculated (→ see below) and their sum $\sum_i \Delta\tau_i = \Delta\tau_{Ch}$ with the available total time $(\xi_{Ch}/\dot{N}_{Ch})$ compared, see (7.6). Required is: $\Delta\tau_{Ch} \leq (\xi_{Ch}/\dot{N}_{Ch})$, optimal: $\Delta\tau_{Ch} = (\xi_{Ch}/\dot{N}_{Ch})$. If these conditions are not met, the surcharge factor f_Q must be corrected accordingly, and the calculations described above must be repeated until there is sufficient agreement between $(\xi_{Ch}/\dot{N}_{Ch})$ and $\Delta\tau_{Ch}$ is given.

d) Steps of a conveying cycle:

The above analysis requires the determination of the individual time steps $\Delta\tau_i$ of the respective pressure vessel cycle. Fig. 7.5 shows its idealized course and the work steps to be considered. The following consideration is based on a transmitter design according to Fig. 7.4, a bulk material extraction from a filled pre-container, and a directly assigned

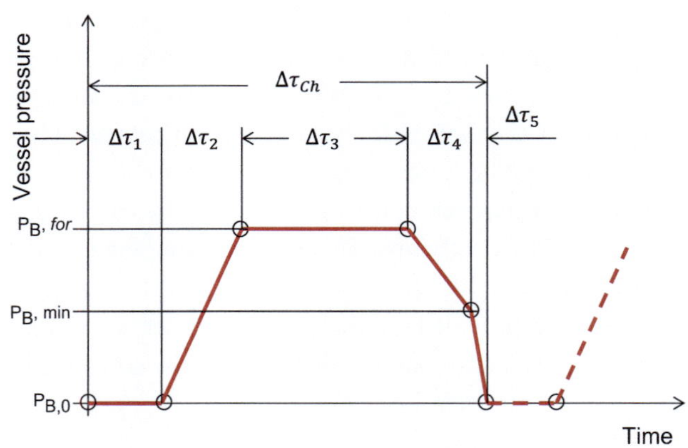

Fig. 7.5 Batch process of a single pressure vessel

and continuous operating compressor as a pressure generator. Deviations will be discussed later.

Step 1: Filling the transmitter with bulk material, duration $\Delta\tau_1$.
The bulk material inlet valves A4 and A5, see Fig. 7.4, as well as the relief valve A3 are open, the outlet valve A6 and the gas valves A2 and A7 are closed, gas valve A1 is open. Thus, the displacement gas flow triggered by the incoming bulk material is discharged from the transmitter via A3, and the gas flow of the compressor bypasses it through the delivery line. The filling time of the pressure vessel with the bulk material accumulated in the surge tank is calculated with the inflow mass flow $\dot{m}_{S,1} = \dot{q}_{S,1} \cdot \varrho_{b,B} \cdot A_{BF}$ to

$$\Delta\tau_1 = \frac{V_{B,\text{Netto}} \cdot \varrho_{b,B}}{\dot{m}_{S,1}} = \frac{V_{B,\text{Netto}}}{\dot{q}_{S,1} \cdot A_{BF}} \tag{7.10}$$

with:

A_{BF} free passage cross-section of the inlet fitting,
$\dot{q}_{S,1}$ specific bulk material volume flow rate during sender filling, density $\varrho_{b,B}$ according to (7.5a) and (7.5b).

$\dot{q}_{S,1}$ -values for fine-grained, aerated bulk materials in the pre-container are generally in the range $\dot{q}_{S,1} \cong (0.8-1.0)\,\text{m}^3/(\text{cm}^2\,\text{h})$, for coarser products this characteristic value decreases to approximately $\dot{q}_{S,1} \cong 0.5\,\text{m}^3/(\text{cm}^2\,\text{h})$. The mentioned values can be used independently of the geometry of the respective inlet cross-section A_{BF}.

Step 2: Pressurizing the transmitter to conveying pressure, duration $\Delta\tau_2$.
For this purpose, valves A5, A4, and A3 are first closed, A6 remains closed. Then opening of the gas valves A2 and A7 (→ quick charging valve). This allows the entire available gas flow $\dot{m}_{F,V}$ to flow into the container filled with bulk material. Its pressure increases to the loading pressure $p_{B,\text{för}}$ set and monitored at pressure sensor D1. D1 triggers the next work step. The time required to pressurize the transmitter is determined as follows:

The transmitter filled with bulk material up to the volume $V_{B,\text{Netto}}$ must be increased from the filling pressure $p_{B,0}$ (→ normally the ambient pressure) to the operating pressure $p_{B,\text{för}}$ must be charged. The volume to be filled with gas for this purpose is $\Delta V_2 = (V_{B,\text{Brutto}} - V_{B,\text{Netto}}) + V_{\text{Poren}} = (V_{B,\text{Brutto}} - V_{B,\text{Netto}}) + \varepsilon_{F,B} \cdot V_{B,\text{Netto}} = V_{B,\text{Brutto}} - V_{B,\text{Netto}} \cdot \varrho_{b,B}/\varrho_S$ and the gas mass that must be introduced into the vessel is $\Delta m_{F,2} = \Delta V_2 \cdot (\varrho_{F,B} - \varrho_{F,0})$. With the respective available gas flow $\dot{m}_{F,V}$ follows from this:

$$\Delta\tau_2 = \frac{\Delta m_{F,2}}{\dot{m}_{F,V}} = \frac{\left(V_{B,\text{Brutto}} - V_{B,\text{Netto}} \cdot \frac{\varrho_{b,B}}{\varrho_S}\right) \cdot (\varrho_{F,B} - \varrho_{F,0})}{\dot{m}_{F,V}} \tag{7.11}$$

$\varrho_{F,B}$ and $\varrho_{F,0}$ are the gas densities at the container pressure $p_{B,\text{för}}$ and at the filling pressure $p_{B,0}$.

Step 3: Emptying the sender, duration $\Delta\tau_3$.
The pressure sensor D1 triggers the closing of the gas valve A7, the opening of A1, and the opening of the bulk material outlet valve A6. The gas flow $\dot{m}_{F,V}$ supplied by the compressor is divided between the beginning of the conveying line and the sender according to the setting of the manual valves A1.1, A2.1, and A7.1. The time to empty the pressure vessel is:

$$\Delta\tau_3 = \frac{V_{B,\text{Netto}} \cdot \varrho_{b,B}}{\dot{m}_S} \quad \rightarrow \quad \text{with: } \dot{m}_S = f_Q \cdot \dot{m}_{S,n} \tag{7.12}$$

Step 4: Pressure reduction through the conveying line, duration $\Delta\tau_4$.
Once the bulk material has completely left the transmitter and is located exclusively in the conveying line, the pressure $p_{B,\text{för}}$ in the vessel drops immediately, see Fig. 7.5 and 7.6. With the compressor gas flow $\dot{m}_{F,V}$ continuing and the valve setting unchanged, the transmitter relaxes to the container pressure $p_{B,\text{min}}$, set at sensor D1, through the conveying line. The subsequent residual relaxation to the filling pressure $p_{B,0}$ is carried out by the separate relief valve A3, see step 5.

The pressure reduction ($p_{B,\text{för}} \rightarrow p_{B,\text{min}}$) is a non-stationary process, whose operating parameters change with time τ or the current container pressure p_B. Since the flow resistance of the delivery line counteracts the movement of the gas/solid mixture and "brakes" it, it influences the respective relaxation time $\Delta\tau_4$. With increasing line pressure loss ($p_{B,\text{för}} - p_E$) the time $\Delta\tau_4$ is extended for a given line. The conveying distance is thereby determined by the gas mass flow

$$\dot{m}_{F,\text{tot}} = \dot{m}_{F,V} + \dot{m}_{F,B} \tag{7.13}$$

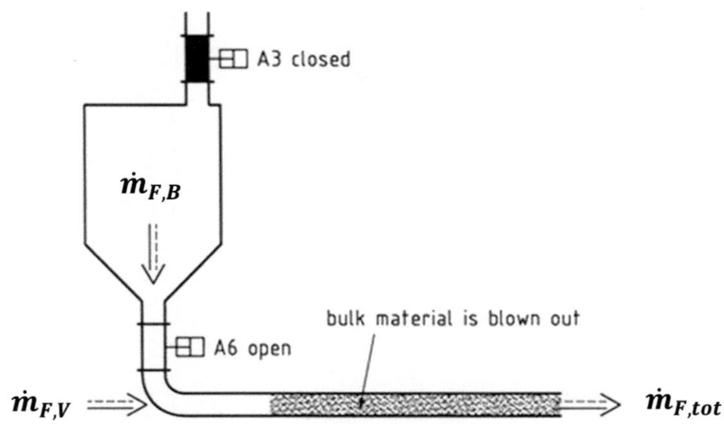

Fig. 7.6 Pressure vessel relaxation through the conveying line

with:

$\dot{m}_{F,B}$ additional gas flow from the expanding pressure vessel

as well as the entrained solid mass flow $\dot{m}_S$ flowing through. The quantities $(\dot{m}_{F,\text{tot}}, \dot{m}_{F,B}, \dot{m}_S)$ change during the expansion. (4.67) provides the relationship between the instantaneous gas mass flow for isothermal state changes $\dot{m}_{F,\text{tot}}$ and the corresponding container pressure p_B for a constant total resistance coefficient $\lambda_{\text{ges}} = (\lambda_F + \mu \cdot \lambda_S)$ along the conveying line. With the solid/gas mixture temperature T_M applies:

$$\dot{m}_{F,\text{tot}} = A_R \cdot \sqrt{(p_B^2 - p_E^2) \cdot \frac{D_R}{\lambda_{\text{ges}} \cdot R_F \cdot T_M \cdot L_R}} \tag{7.14}$$

The application of mass conservation to the gas phase in the conveying system, see Fig. 7.6, results in:

$$\frac{dm_{F,B}}{d\tau} = -(\dot{m}_{F,\text{tot}} - \dot{m}_{F,V}) \tag{7.15}$$

Using the ideal gas law, the change $dm_{F,B}$ of the gas mass in the pressure vessel when the pressure changes by dp_B can be described by

$$dp_B \cdot V_{B,\text{Brutto}} = R_F \cdot T_M \cdot dm_{F,B} \tag{7.16}$$

From the combination of equations (7.14)–(7.16), the differential equation follows:

$$d\tau = -\frac{V_{B,\text{Brutto}}}{R_F \cdot T} \cdot \frac{dp_B}{A_R \cdot \sqrt{(p_B^2 - p_E^2) \cdot \frac{D_R}{\lambda_{\text{ges}} \cdot R_F \cdot T_M \cdot L_R}} - \dot{m}_{F,V}} \tag{7.17}$$

To solve (7.17), the dependency $\lambda_{\text{ges}}(p_B)$ must be given and it must also be known whether the solid is completely discharged from the conveying line before the time $\Delta\tau_4$ has elapsed. All this considering that both the solid filling level of the transport line (→ Loading μ) as well as the flow patterns that occur during relaxation change. This information is generally not available. Even with p_B -independent λ_{ges} can (7.17) only be solved with complex numerical methods. Since $\lambda_{\text{ges}} = (\lambda_F + \mu \cdot \lambda_S)$ with decreasing p_B obviously tends towards the value $\lambda_{\text{ges}} \to \lambda_F$, practice is therefore satisfied with the following approximation equation:

$$\Delta\tau_4 = K_4 \cdot \frac{V_{B,\text{Brutto}}}{R_F \cdot T_M} \cdot \frac{p_{B,\text{för}} - p_{B,\text{min}}}{A_R \cdot \sqrt{(\overline{p}_B{}^2 - p_E^2) \cdot \frac{D_R}{\lambda_F \cdot R_F \cdot T_M \cdot L_R}} - \dot{m}_{F,V}} \tag{7.18}$$

with: $\overline{p}_B = \frac{1}{2} \cdot (p_{B,\text{för}} + p_{B,\text{min}})$.

λ_F is independent of the Reynolds number or gas velocity and thus of p_B in the present case of isothermal flow. The type of averaging for the driving pressure gradient at the conveying line provides only slightly differing values. (7.18) does not contain the additional expansion acceleration of gas and solid. Adjustments to reality are made with the empirical factor $K_4 \geq 1$.

Using the equations (7.15), (7.16) and (7.18), the gas mass flow leaving the conveying pipe and averaged over the relaxation time $\Delta\tau_4$ can now be determined, the so-called final surge. $\overline{\dot{m}}_{F,\text{tot}}$, can be calculated as follows:

$$\overline{\dot{m}}_{F,\text{tot}} = \dot{m}_{F,V} + \frac{V_{B,\text{Brutto}}}{R_F \cdot T_M} \cdot \frac{p_{B,\text{för}} - p_{B,\text{min}}}{\Delta\tau_4} \tag{7.19}$$

Therefore, the following applies: $(\overline{\dot{m}}_{F,\text{tot}}/\dot{m}_{F,V}) = (\overline{\dot{V}}_{F,\text{tot}}/\dot{V}_{F,V}) > 1$, i.e., the filter behind the conveyor system must be designed for the cleaning of a temporarily increased mass flow $\overline{\dot{m}}_{F,\text{tot}} > \dot{m}_{F,V}$ at the end of each pressure vessel cycle. Depending on the operating conditions, values $(\overline{\dot{m}}_{F,\text{tot}}/\dot{m}_{F,V}) \cong 1.5-3.0$ are obtained. The higher values are particularly observed with large pressure differences. $(p_{B,\text{för}} - p_{B,\text{min}})$, large pipe diameters D_R and short conveying distances L_R and possibly exceeded. For the selection and design of suitable separators, see, for example, [2, 3].

Step 5: Pressure reduction through the relief valve, duration $\Delta\tau_5$.
Upon reaching the pressure $p_{B,\text{min}}$, the pressure sensor D1 triggers the closing of the gas valve A2 and the bulk material outlet valve A6, as well as the opening of the relief valve A3. Gas valve A1 remains open, i.e., the compressor gas flow $\dot{m}_{F,V}$ is discharged via the delivery line, see Fig. 7.7. The unsteady pressure reduction $(p_{B,\text{min}} \rightarrow p_{B,0})$ of the transmitter takes place via a separate relief line—diameter D_E, length L_E—into a separate filter, e.g., that of the respective pre-storage tank. Since the vessel no longer contains any bulk material at the beginning of this work step, the process can be calculated as the relaxation of a pure gas. Any entrained wall deposits of fine-grained solids are negligible.

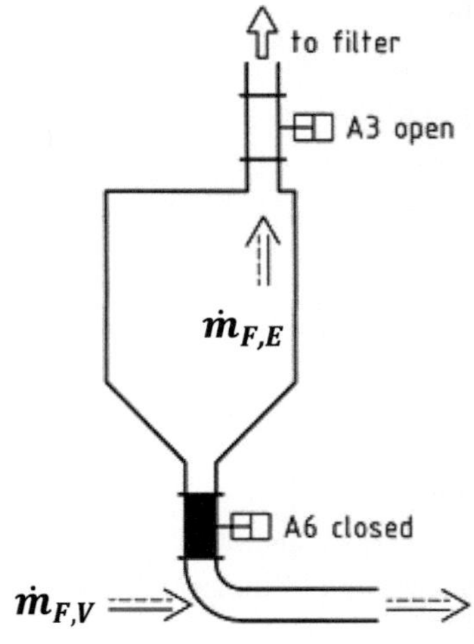

Fig. 7.7 Pressure vessel relaxation through separate relief valve

$p_{B,\min}$ can be set differently. This leads to different effects on individual plant components. If a $p_{B,\min}$ is chosen just below $p_{B,\text{för}}$, almost the entire pressure vessel relaxation takes place with a correspondingly high gas flow $\dot{m}_{F,E}$ via valve A3 and the associated filter, which must be dimensioned accordingly. Relaxation through the conveying line only takes place to a small extent, the resulting final surge is generally negligible and has no influence on the size of the receiving filter. If $p_{B,\min}$ is set close to $p_{B,0}$, then the opposite ratios result. In a corresponding manner, the wear stress on the respective components, which increases with increasing gas/solid velocity, also changes.

The relaxation through A3 and the line (D_E, L_E) is generally an adiabatic process in which no heat exchange with the environment takes place or this is negligibly small. The external work done by the gas is taken from its internal energy. Consequence: When flowing out, the gas temperature T_F decreases, and thus condensation of the moisture contained in the gas becomes possible.

For a given $p_{B,0}$ and large values of $p_{B,\min}$ sound speed can occur at the end of the relaxation line. This happens when the so-called critical pressure ratio $(p_0/p_A)_{\text{crit}}$ is undercut. In the case of frictionless flow through the discharging pipe, $\xi_{\text{ges}} = 0$, this amounts to gases with the adiabatic exponent $\kappa = 1.4$ (→ diatomic gases) $(p_0/p_A)_{\text{crit}} = 0.528$ and decreases according to Fig. 7.8 with increasing resistance coefficient ξ_{ges}. This is defined as $\xi_{\text{ges}} = \lambda_F \cdot (L_E/D_E) + \sum_i \xi_i$, where with ξ_i additional resistances, e.g. deflections, can be taken into account. As can be seen in Fig. 7.8, the pressure drop ($p_{B,\min} \rightarrow p_{B,0}$) of a pressure vessel transmitter through the connected exhaust line runs along one of the curves $\xi_{\text{ges}} = \text{const.}$ from the respective current starting value $(p_0/p_A) = (p_{B,0}/p_{B,\min})$ to $(p_0/p_A) = 1$. If the considered relaxation starts at $(p_{B,0}/p_{B,\min}) < (p_0/p_A)_{\text{crit}}$, then the cal-

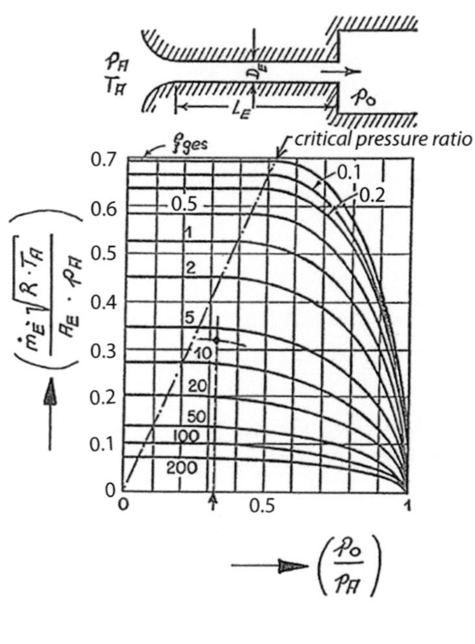

Fig. 7.8 Gas outflow with $\kappa = 1.4$ from a vessel with constant pressure through a connected line, according to [4]

culation of the required relaxation time must be carried out in two steps. Below $(p_0/p_A)_{\text{crit}}$ the constant outflow mass flow $\dot{m}_{F,E,1} = \dot{m}_{F,\text{crit}}$ is established up to the critical pressure ratio. This as well as p_{crit} can be taken from or calculated using Fig. 7.8. The relaxation time up to p_{crit} is then:

$$\Delta\tau_{5.1} = \frac{V_{B,\text{Brutto}}}{R_F \cdot T_A} \cdot \frac{p_{B,\min} - p_{\text{crit}}}{\dot{m}_{F,\text{crit}}} \tag{7.20}$$

with:

T_A ($\cong T_S$), temperature in the vessel.

The further pressure reduction ($p_{\text{crit}} \to p_{B,0}$) occurs with a value corresponding to the current value of $(p_{B,0}/p_B)$ changing gas mass flow $\dot{m}_{F,E}$. The derivation of this adiabatic process, analogous to the approach of step 4, leads to a differential equation that cannot be solved explicitly. However, a closed solution of the corresponding isothermal pressure drop is possible → (7.17) with $\dot{m}_{F,V} = 0$. Comparisons of adiabatic and isothermal relaxation times show that their differences are small and that the isothermal case is on the "safe" side. The duration of an isothermal relaxation starting with the pressure p_{crit} is calculated as:

$$\Delta\tau_{5.2} = \frac{V_{B,\text{Brutto}}}{A_E} \cdot \sqrt{\frac{\xi_{\text{ges}}}{R_F \cdot T_A}} \cdot \ln\left(\frac{p_{\text{crit}} + \sqrt{p_{\text{crit}}^2 - p_E^2}}{p_{B,0} + \sqrt{p_{B,0}^2 - p_E^2}}\right) \tag{7.21}$$

In normal operating conditions, $p_{B,0} = p_E$ (→ in general ≅ ambient pressure p_U). This leads to the usage equation:

$$\Delta\tau_{5.2} = \frac{V_{B,\text{Brutto}}}{A_E} \cdot \sqrt{\frac{\xi_{\text{ges}}}{R_F \cdot T_A}} \cdot \ln\left(\frac{p_{\text{crit}}}{p_E} + \sqrt{\frac{p_{\text{crit}}^2}{p_E^2} - 1}\right) \tag{7.22}$$

The associated, averaged relaxation gas flow over time $\Delta\tau_{5.2}$ is obtained from:

$$\overline{\dot{m}}_{F,E,2} = \frac{V_{B,\text{Brutto}}}{R_F \cdot T_A} \cdot \frac{p_{\text{crit}} - p_{B,0}}{\Delta\tau_{5.2}} \tag{7.23}$$

The total relaxation time in the case of $p_{B,\min} > p_{\text{crit}}$ is thus $\Delta\tau_5 = (\Delta\tau_{5.1} + \Delta\tau_{5.2})$. The downstream dust separator must be designed for $\dot{m}_{F,E,1} = \dot{m}_{F,\text{crit}} > \overline{\dot{m}}_{F,E,2}$. If the operating case $p_{B,\min} \leq p_{\text{crit}}$ is present, the relaxation time is $\Delta\tau_5 = \Delta\tau_{5.2}$ and the average relaxation mass flow $\overline{\dot{m}}_{F,E,2}$. Both are calculated with $p_{B,\min}$ instead of p_{crit} from (7.21) and (7.23). Since these exhaust gas flows only occur briefly at the end of each conveying cycle, they do not represent a continuous load.

The adiabatic temperature reduction of the gas in the expansion line can be determined using the approaches presented in the relevant literature, e.g., [4, 5, 6].

After step 5, the conveying system returns to step 1, i.e., the bulk material inlet valves A4 and A5 open.

Step 0: Switching time of the fittings, duration $\Delta\tau_0$.
The sum of the switching times of the fittings at the transmitter was determined by systematic measurements with

$$\Delta\tau_0 \cong 0.20\,\text{min} \tag{7.24}$$

for a complete pressure vessel cycle. It is distributed proportionally to the individual work steps.

Notes on the vessel cycle/practical experiences:

- The bulk material discharge from the transmitter is adjusted for most transport tasks only via the calculated loading pressure = conveying pressure $p_{B,\text{för}}$, the associated conveying gas flow, and the gas distribution at the vessel. Controls using the material outlet valve or the pressure in the upper vessel chamber are possible but are only used in special cases.
- If the loading pressure at the end of step 2 drops below the required conveying pressure $p_{B,\text{för}}$ after opening valve A6, it must be set correspondingly higher at pressure sensor D1. Pressure reserves should be planned for this purpose.
- If during the conveying in step 3 the pressure $p_{B,\text{för}}$ decreases with time, the partial gas flow into the pressure vessel must be increased, and the one to the beginning of the conveying line must be reduced accordingly. In the case of an $p_{B,\text{för}}$ increase, the procedure is exactly the opposite → one-time adjustment via existing manual valves.
- Single-vessel operation without bulk material outlet valve A6 is possible. In general, this extends the loading time $\Delta\tau_2$ or becomes more difficult to control/calculate. However, a highly stressed valve due to wear is saved.
- For coarse-grained bulk materials, e.g. plastic granules or cylindrical activated carbon with $\varnothing d_S \times L_S \cong (5.0 \times 10)\,\text{mm}$, if the outlet valve A6 is opened at pressures $p_{B,\text{för}} \gtrsim (2.0-3.0)\,\text{bar}$, it often leads to a mechanical wedging of the bulk material in the discharge manifold/at the beginning of the pipeline and thus to a pipeline blockage. To avoid this, the valve A6 must be opened at a lower vessel pressure p_B while the vessel pressurization continues to run.
- Staggered conveying pipes can be represented with sufficient accuracy by their average volume-equivalent diameter $D_R = (\sum_i (D_{R,i}^2 \cdot \Delta L_{R,i})/L_R)^{1/2}$ in (7.14)–(7.22). $(D_{R,i}, \Delta L_{R,i})$ are the individual staggered pipe diameters and lengths.
- To take into account further losses in addition to those caused by friction, instead of $\lambda \cdot (L/D)$ in the preceding equations, $\xi_{\text{ges}} = \lambda_F \cdot (L/D) + \sum_i \xi_i$ can be used. The ξ_i describe additional resistances.
- To cover the indicated calculation uncertainties, a utilization factor of $\xi_{Ch} = 0.95$ is typically used.

e) Part-load behavior:

Pressure vessel systems *must* be designed for the maximum nominal solid mass flow rate $\dot{m}_{S,n}$ to be handled. In this case, one batch is conveyed after another without any waiting time in between, see Fig. 7.5. In part-load operation, $\dot{m}^*_{S,n} < \dot{m}_{S,n}$, the steps 2 to 5 of the conveying cycle remain unchanged in their settings and times. Only the filling time $\Delta\tau^*_1$ of the sender with bulk material is extended. Cause: during the period $\sum_{i=2}^{5}\Delta\tau_i$ (→ material inlet valve of the sender is closed!) less product is accumulated in the surge bin than at full load. The bulk material still required for the complete pressure vessel filling (→ triggering of the level probe U1 required!) passes through the surge bin only with the nominal mass flow $\dot{m}^*_{S,n}$, which is a multiple smaller than the possible withdrawal mass flow from the filled surge bin. This leads to $\Delta\tau^*_1 > \Delta\tau_1$. The transition to partial load operation and back to full load is self-regulating and does not require any external interventions.

The filling time $\Delta\tau^*_1$ at partial load $\dot{m}^*_{S,n} \leq \dot{m}_{S,n}$ can be calculated from that at full load $\Delta\tau_1$. It applies:

$$\Delta\tau^*_1 = \Delta\tau_1 \cdot \frac{\dot{m}^*_{S,n}}{\dot{m}_{S,n}} + V_{B,\text{Netto}} \cdot \varrho_{b,B} \cdot \left(\frac{1}{\dot{m}^*_{S,n}} - \frac{1}{\dot{m}_{S,n}}\right) \tag{7.25}$$

(7.25) results from elementary time and mass balances for $\Delta\tau_1$ and $\Delta\tau^*_1$.

f) Power requirement:

It is still assumed that a directly assigned and continuous compressor is used as a pressure generator and that its current power consumption P_K largely follows the different load changes during a pressure vessel cycle. This is schematically shown in Fig. 7.9, but is not realized by every type of pressure generator.

For the following considerations, it is defined:

$P_{K,\text{för}}$ Compressor coupling power during the delivery phase $\Delta\tau_3$ → installed drive power, $P_{K,\text{för}}(\dot{m}_{F,V}, p_V)$.

$P_{K,\min}$ Coupling power of the compressor during afterblowing through the (empty) conveying line.

$(1/\dot{N}_{Ch})$ $(= \Delta\tau_{\text{zyklus}})$, maximum time duration available for a single conveying cycle. $\dot{N}_{Ch}$ is the batch number per unit of time determined or specified for full load operation. The value of $\Delta\tau_{\text{zyklus}}$ also applies to partial load operation, although the current batch number is then reduced to $\dot{N}^*_{Ch} = N_{Ch} \cdot (\dot{m}^*_{S,n}/\dot{m}_{S,n})$.

As can be seen from Fig. 7.9, the compressor drive must be dimensioned for $P_{K,\text{för}}$, but during a conveying cycle, due to its changing operating conditions, only the average power $\overline{P}_K < P_{K,\text{för}}$ is required. The operator only has to pay for this. For the various load cases, the following applies:

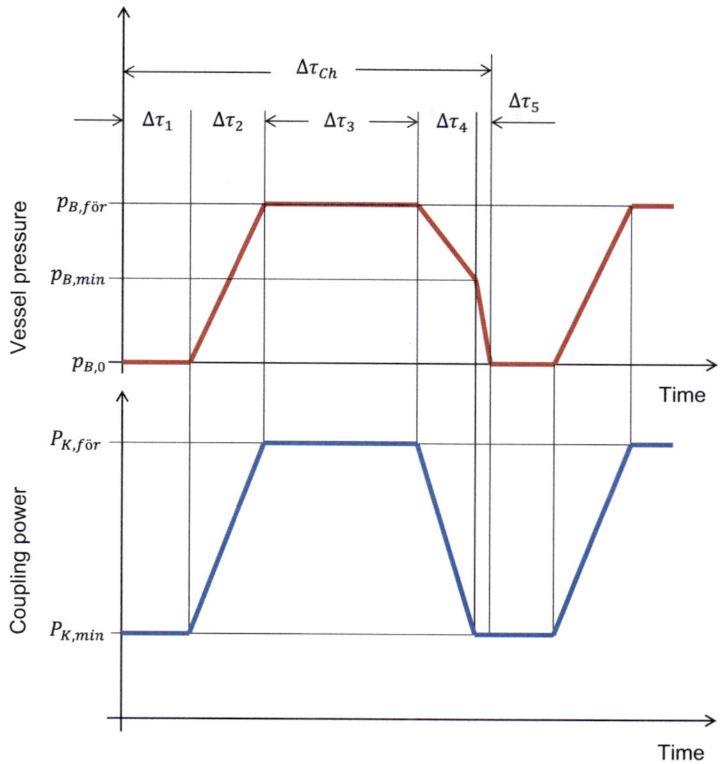

Fig. 7.9 Single pressure vessel cycle with corresponding coupling powers P_K, full load operation

Full load operation: For this operating case, the average drive power of a complete delivery cycle is calculated with some permissible simplifications as:

$$\overline{P}_K = \frac{1}{2} \cdot (P_{K,\text{för}} + P_{K,\text{min}}) \cdot \frac{\Delta\tau_2 + \Delta\tau_4 + \frac{1}{2} \cdot \Delta\tau_0}{\Delta\tau_{\text{zyklus}}} + P_{K,\text{för}} \cdot \frac{\Delta\tau_3 + \frac{1}{2} \cdot \Delta\tau_0}{\Delta\tau_{\text{zyklus}}}$$
$$+ P_{K,\text{min}} \cdot \frac{\Delta\tau_1 + \Delta\tau_5 + (1 - \xi_{Ch}) \cdot \Delta\tau_{\text{zyklus}}}{\Delta\tau_{\text{zyklus}}} \tag{7.26}$$

It is evident that the power demand of the individual work steps is proportional to their time share. $\Delta\tau_i$ is weighted by the total cycle time $\Delta\tau_{\text{zyklus}}$. The term $(1 - \xi_{Ch}) \cdot \Delta\tau_{\text{zyklus}}$ in (7.26) captures the time period during which uncertainties in determining individual $\Delta\tau_i$ should be covered, see (7.6) and section c). Their power requirement is evaluated with $P_{K,\text{min}}$, i.e., it is not used, but is present and extends the blow-off of the conveying line. For the utilization factor ξ_{Ch}, the actually realized value must be used. The total switching time $\Delta\tau_0$ of the fittings on the transmitter is distributed equally to the more power-intensive work steps. Simplification of (7.26) results in:

$$\overline{P}_K = \frac{1}{\Delta\tau_{\text{zyklus}}}\left[\frac{1}{2}\cdot(P_{K,\text{för}} + P_{K,\text{min}})\cdot(\Delta\tau_2 + \Delta\tau_4 + \frac{1}{2}\cdot\Delta\tau_0)\right.$$
$$\left. + P_{K,\text{för}}\cdot(\Delta\tau_3 + \frac{1}{2}\cdot\Delta\tau_0) + P_{K,\text{min}}\cdot(\Delta\tau_1 + \Delta\tau_5)\right] + P_{K,\text{min}}\cdot(1-\xi_{Ch}) \quad (7.27)$$

Partial load operation: With a continuous operating compressor, the filling time $\Delta\tau_1^* > \Delta\tau_1$ of the transmitter increases only in partial load operation (*), all other times remain unchanged. (7.26) and (7.27) can be used to determine the average power requirement $\overline{P}_K^*$ when in these $\Delta\tau_1$ is replaced by $\Delta\tau_1^*$. $\Delta\tau_1^*$ can be calculated using (7.25).

There are further operating modes to reduce the power or energy requirement at partial load. Examples include: speed control, transition to a full load/idle mode, switching off the compressor in extreme underload, etc. The pressure vessel cycle of the latter variant, ($\dot{m}_{S,n}^* \ll \dot{m}_{S,n}$), is schematically shown in Fig. 7.10. After each emptying of the transmitter (→ pressure drop to $p_{B,\text{min}}$) the conveying line is blown up for the adjustable time duration $\Delta\tau_{NB}^*$ in the control. If $\Delta\tau_{NB}^*$ is exceeded and there is no full message from

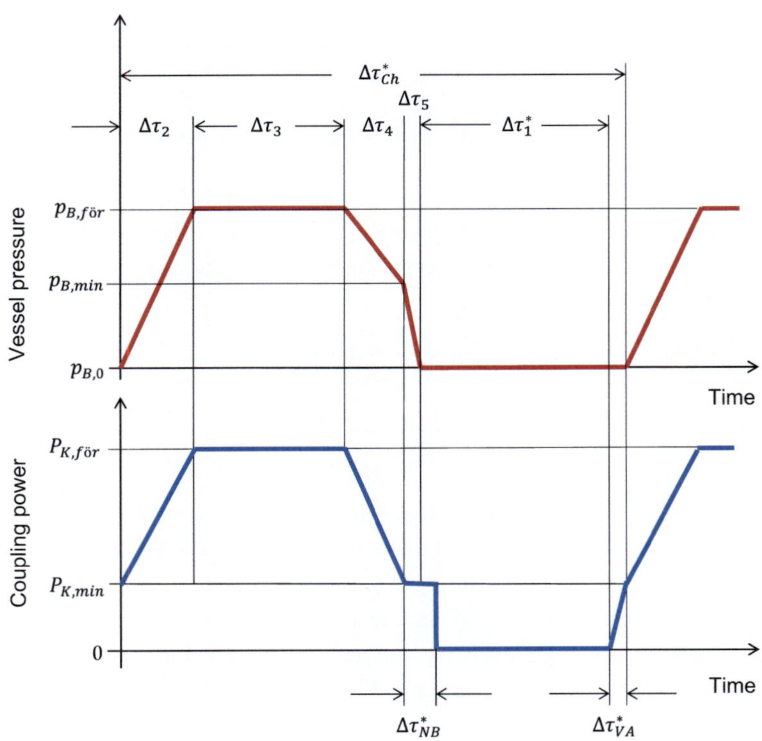

Fig. 7.10 Single pressure vessel cycle during partial load operation with compressor shutdown

the pressure vessel by U1, the compressor is switched off. The next full message triggers the relieved restart of the compressor, which requires the time duration $\Delta\tau^*_{VA}$. During the time $(\Delta\tau^*_1 + \Delta\tau_5 - \Delta\tau^*_{NB} + (1 - \xi_{Ch}) \cdot \Delta\tau_{\text{zyklus}})$ the power requirement of the system is $P_{K,0} = 0$. From this, it follows for the average drive power over a conveying cycle $\Delta\tau_{\text{zyklus}}$:

$$\overline{P}^*_K = \frac{1}{\Delta\tau_{\text{zyklus}}}\left[\frac{1}{2} \cdot (P_{K,\text{för}} + P_{K,\min}) \cdot (\Delta\tau_2 + \Delta\tau_4 + \frac{1}{2} \cdot \Delta\tau_0) + P_{K,\text{för}} \cdot (\Delta\tau_3 + \frac{1}{2} \cdot \Delta\tau_0) + P_{K,\min} \cdot \Delta\tau^*_{NB} + \frac{1}{2} \cdot P_{K,\min} \cdot \Delta\tau^*_{VA}\right] \tag{7.28}$$

It has to be taken into account that pressure generators can only realize a limited number of switch on/off cycles per unit time. In this case a full load/idle circuit system with $P_{K,0} > 0$ is to prefer. Calculation of the average power demand analogously as described above.

7.1.2 Twin Pressure Vessel

Twin conveyors are systems with two equally sized pressure vessels that alternately feed the conveying line. It is therefore a parallel connection of two transmitters, which are usually supplied by only one pressure generator, see Fig. 7.11. The conveying cycles $p_B(\tau)$ that occur during full load operation and gas supply by a single compressor are

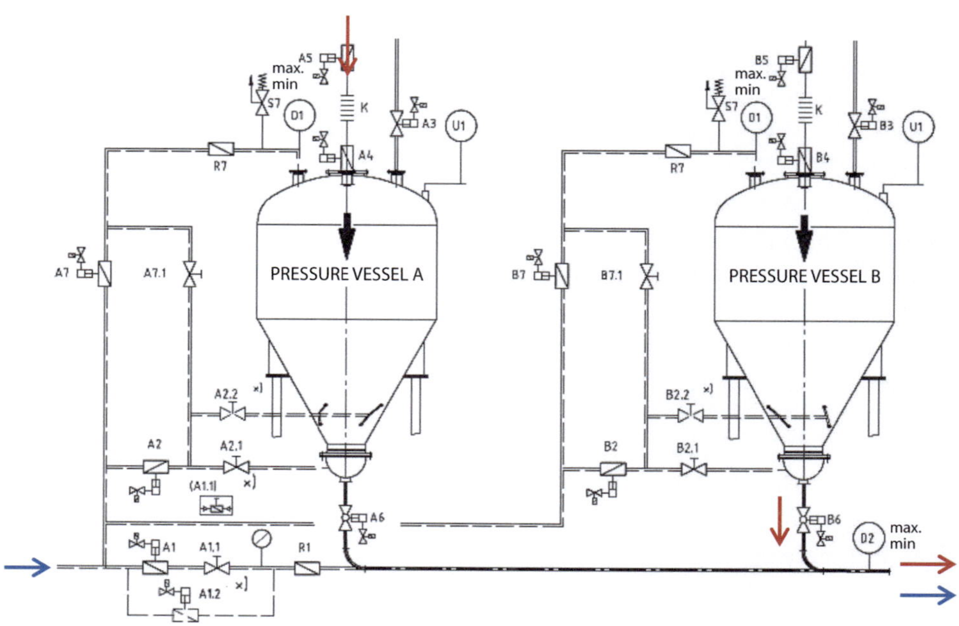

Fig. 7.11 Structure of a twin pressure vessel transmitter; with: *) = optional

shown schematically in Fig. 7.12. As soon as vessel 1 is emptied, relaxed to the pressure $p_{B,\min}$ and its bulk material outlet valve A6 is closed, the compressor is switched to the parallel vessel 2 to pressurize it to the pressure $p_{B,\text{för}}$. The pressure reduction in container 1 and its filling with bulk material, $\Delta\tau_5 + \Delta\tau_1 + 1/2 \cdot \Delta\tau_0$, takes place during the batch time of vessel 2 defined here by $\Delta\tau_{Ch} = \Delta\tau_2 + \Delta\tau_3 + \Delta\tau_4 + 1/2 \cdot \Delta\tau_0$. It is evident that at least during the time $\Delta\tau_5$ bulk material must be accumulated in a pre-container/distributor.

The f_Q-values of the twin pressure vessel with a single compressor are in the range of $f_Q \cong (1.1-1.3)$. By using a second compressor in parallel or by taking gas from a sufficiently large supply network, vessel 2 can already be pressurized to $p_{B,\text{för}}$ during the conveying phase of vessel 1 and start conveying at $p_{B,\min} \cong p_{B,\text{för}}$. This reduces $f_Q \to 1$. In plants that require absolutely continuous operation, e.g., when feeding fine coal into the wind forms of blast furnaces for iron production, the last mentioned mode of operation is realized: The switchover from vessel 1 to vessel 2 takes place while vessel 1 is still partially filled, i.e., at $p_{B,\min} = p_{B,\text{för}}$. The material discharge valves A6 and B6 are regulated in such a way that they are closed or opened so that a constant solid mass flow $\dot{m}_S$ is maintained even during switching.

The arrangement of the bulk material outlet valves (A6, B6) vertically below the vessels, as shown in Fig. 7.11, allows both transmitters to be placed one behind the other on a straight continuous conveyor line. Disadvantage: greater overall height; advantage: largely impact-free routing, thus low wear. If (A6, B6) are installed horizontally, both vessels must be placed side by side and connected to the conveyor line via a V-shaped pipe element. Disadvantage: change of direction/impact point of the solid at the transition to the common conveying section, i.e., higher wear, unused pipeline branch must be kept free of bulk material; advantage: lower overall height.

The net volume of each of the two pressure vessels is calculated analogously (7.4) to:

$$V_{B,\text{Netto}} = \frac{\dot{m}_{S,n}}{\varrho_{b,B} \cdot (2 \cdot \dot{N}_{Ch})} \quad \rightarrow \quad \text{with: } \dot{N}_{Ch} \leq \dot{N}_{Ch,\max} \tag{7.29}$$

It is recommended to choose the vessel size so that even with smaller vessels, the resulting batch time $\Delta\tau_{Ch}$, see Fig. 7.12, does not fall below a value of $\Delta\tau_{Ch} \cong (1.5-2.0)$ min. Shorter times increase the dead time/batch time ratio and thus the surcharge factor f_Q drastically.

The respective calculations of the pressure vessel cycles and the associated average power demand $\overline{P}_K$ can be carried out analogously to the explanations in Sect. 7.1.1.

7.1.3 Double-deck Pressure Vessel

Here, two pressure vessels are arranged one above the other. In this series connection of the vessels, the lower transmitter, vessel 1, continuously conveys, while the upper vessel 2 supplies the bulk material to the lower vessel in batches and on demand. The conveying is therefore carried out with the surcharge factor $f_Q = 1$, i.e. with $\dot{m}_S = \dot{m}_{S,n}$.

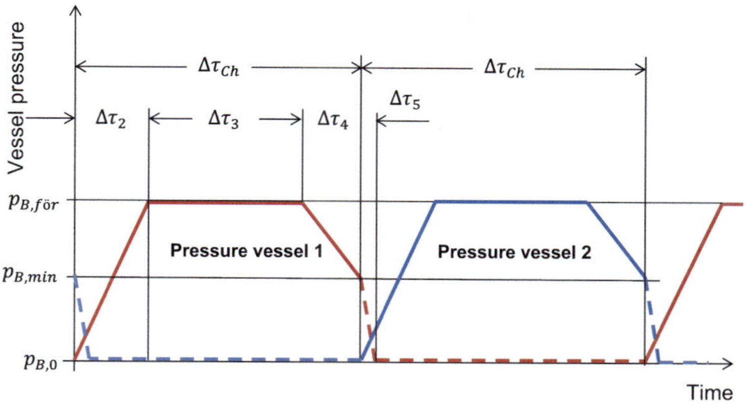

Fig. 7.12 Conveying cycles of a twin pressure vessel with a single compressor, full load operation

The lower vessel, i.e. the actual transmitter, is equipped with a minimum level sensor, which triggers a pressure equalization when it responds and then opens the connection between the two vessels. In special cases, it is necessary to set a slightly higher pressure in the lock vessel 2 than in the transmitter 1. Even before it is completely emptied, the transmitter is refilled with solid material and thus continues to convey continuously. To decouple from an upstream continuously operated system, a buffer container is required above the vessel 2.

The net volume $V_{B,\text{netto}/2}$ of the lock vessel 2 is calculated with (7.4). For a gently stressed intermediate valve and a max-fill level control not required bulk material inlet into the sender 1, this should be designed with a volume larger than the lock vessel 2: $V_{B,\text{netto}/1} > V_{B,\text{netto}/2}$. While the sender 1 is conveying, the vessel 2 must be depressurized from $p_{B,\text{för}}$ to $p_{B,0}$ ($\rightarrow$ duration $\Delta\tau_{5/2}$), and filled with bulk material from the storage container ($\rightarrow$ time duration $\Delta\tau_{1/2}$) and again by means of a separate pressure generator or from a compressed gas network from $p_{B,0}$ to $p_{B,\text{för}}$ be pressurized ($\rightarrow$ time duration $\Delta\tau_{2/2}$). For this, the condition

$$(\Delta\tau_{1/2} + \Delta\tau_{2/2} + \Delta\tau_{5/2} + \frac{1}{2} \cdot \Delta\tau_{0/2}) \leq \Delta\tau_{3/1} = \frac{V_{B,\text{netto}/2} \cdot \varrho_{b,B}}{\dot{m}_{S,n}} \tag{7.30}$$

must be fulfilled. For the entire batch time of vessel 2, it applies:

$$\Delta\tau_{Ch/2} = \Delta\tau_{1/2} + \Delta\tau_{2/2} + \Delta\tau_{3/2} + \Delta\tau_{5/2} + \Delta\tau_{0/2} = \frac{1}{N_{Ch/2}} \tag{7.31}$$

$\Delta\tau_{3/2}$ is the emptying/overflow time of the bulk material from vessel 2 into the transmitter 1 ($\rightarrow$ the conveying time of vessel 2). Fig. 7.13 shows the relationships schematically.

For security reasons, i.e., to avoid emptying the transmitter 1, an additional waiting time $\Delta\tau_{\text{Warte}/2}$ beyond the degree of utilization ξ_{Ch} is often introduced for the filled and pressurized vessel 2. Using (7.30) and (7.31) and applying the explanations of Sect. 7.1.1 analogously, a double-deck system can be dimensioned and, for example, the required

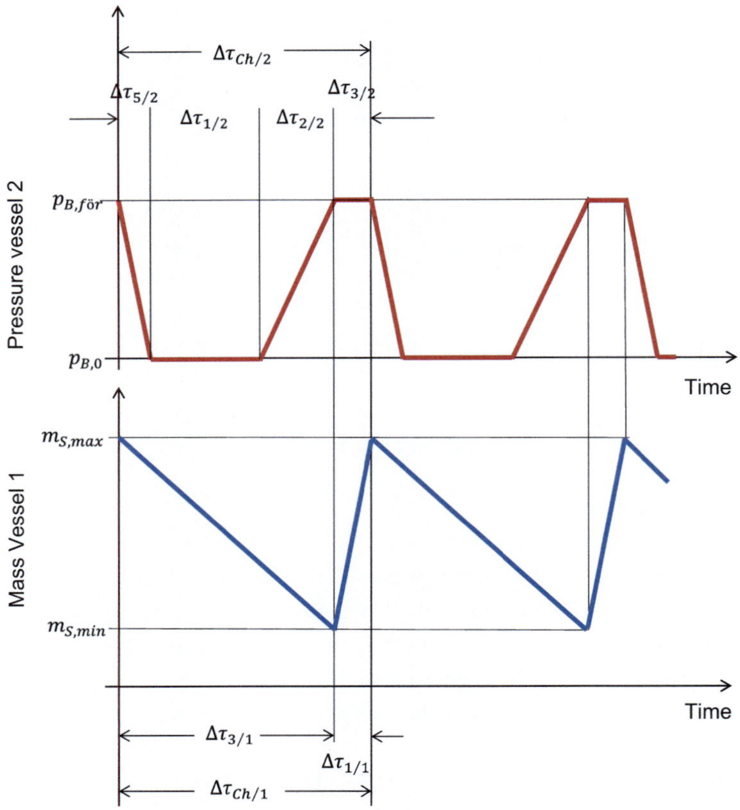

Fig. 7.13 Conveying cycles of a double-deck pressure vessel

gas flow of the separate compressor/pressure gas network for pressurizing the lock vessel 2 can be calculated.

The power consumption of such a conveying system consists of the components

$$\overline{P}_K = P_{K,\text{för}/1} + \overline{P}_{K/2} \tag{7.32}$$

with:

$P_{K,\text{för}/1}$ (constant) drive power of the pressure generator of sender 1,

$\overline{P}_{K/2}$ average drive power of the pressure generator of the lock vessel 2 over the lock cycles

$\overline{P}_{K/2}$ is calculated analogously to the approaches described in Sect. 7.1.1.

The realization of a partial load operation ($\dot{m}_S^* < \dot{m}_S = \dot{m}_{S,n}$) is difficult in double-deck systems. The easiest way is to adjust the conveying pressure $p_{B,\text{för}}$ of the system according to the reduced solid throughput. To avoid the potentially problematic starting processes of a pressurized and material outlet valve A6 opening single-vessel transmitter in cases of very long conveying distances, $L_R \gtrsim 1000\,\mathrm{m}$, and/or high conveying pressure differences, $\Delta p_R \gtrsim 5\,\mathrm{bar}$, double-deck or alternatively twin pressure vessels should be used for these applications.

7.1.4 Multi-pressure Vessel

Multi-pressure vessels are used, among other things, in coal-fired power plants for the removal of fly ash separated in the flue gas cleaning process to the storage silos. Dust separation is generally carried out by electrostatic precipitators, which will be considered here as an example. Electrostatic precipitators are divided into so-called fields along the flow direction of the flue gas, each of which contains several ash collection hoppers. This results in a considerable number of accumulation points from which the fly ash must be removed, see Fig. 7.14. The ash removal can take place parallel or transverse to the flue gas direction.

Fig. 7.15 shows an example of a modern pneumatic conveying system that removes fly ash in rows parallel to the gas flow direction. To reduce investment costs, the number of required plant components, and to simplify the operating mode, all vessel rows feed one common conveying line successively, possibly also on demand. The vessels in a row,

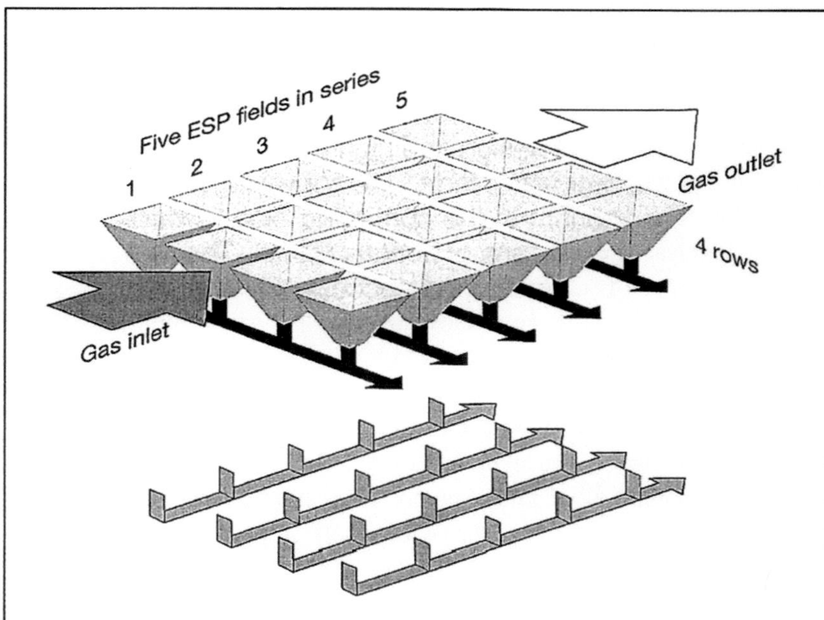

Fig. 7.14 Possibilities of ash removal under an electrostatic precipitator, according to [7]

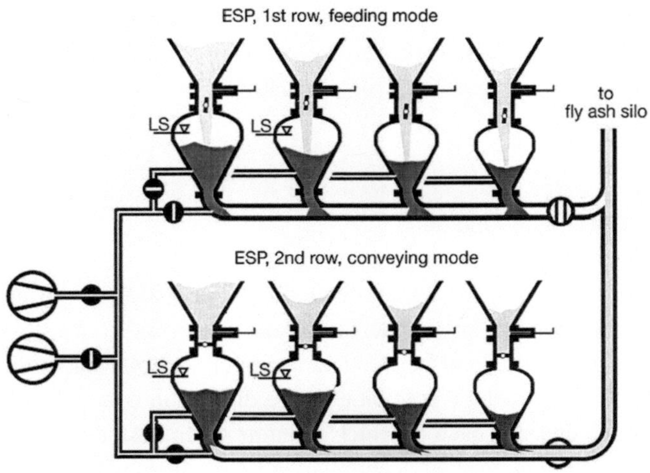

Fig. 7.15 Ash removal in flue gas direction with multi-pressure vessels, according to [7]

in Fig. 7.14 four, have only one single common bulk material outlet valve for the entire group, are filled simultaneously, and are also emptied into the conveying line simultaneously. The whole system essentially represents a large sender divided into individual containers and can be designed analogously. The vessel group (= multi-pressure vessel) is depressurized through the conveying line.

Only in the case of very large conveying distances (→ relatively high pressure loss of the conveying gas) and/or for depressurizing the system in case of malfunctions, the vessel below the first field is equipped with a relief valve. In the application considered here, the fly ash accumulation per field decreases steeply in the flue gas direction. Thus, only the vessel under the first field and, for backup against its failure, the vessel of the second field need to be equipped with a level sensor. For the same reason, the vessels of the first two fields are designed to be the same size, while smaller vessels can be used for the rear fields.

The fly ash removal transverse to the exhaust gas flow is also carried out with multi-pressure vessels, whereby modified requirements must be taken into account here. Examples: approximately the same ash accumulation per unit of time in the individual vessels of a group; each multi-vessel conveys a relatively homogeneous bulk material, that of the first field is significantly coarser than that of the last field. In the ash removal described above parallel to the flue gas direction, a heterogeneous mixture of coarse and fine-grained material must be transported. The advantage is that the generally very cohesive, difficult to pneumatically convey and possibly extremely fine-grained fly ash from the rear fields is placed on the top of the coarser material from the first two fields in the conveying line.

Multi-pressure vessels are, of course, also used for other tasks and bulk materials. In all cases, a stable and constant gas distribution to the individual vessels must be ensured during operation. A prematurely emptied vessel, see Fig. 7.15, must not have an effect like a bypass that leads larger portions of the available gas flow past the other vessels. Applications are known in which up to six individual vessels were connected to form a multi-sender.

7.1.5 Tank vehicle

Fig. 7.16 shows design variants of a road and a rail vehicle/wagon for the transport of pneumatically conveyable bulk materials. These systems, which are manufactured in various designs (→ e.g., as tipping tanks, banana vehicles, various wagon designs), are essentially single pressure vessels that are filled with bulk material at a producer using suitable loading devices and whose current task is completed by the pneumatic emptying of the container, i.e., of a single charge, at the recipient.

There are country-specific regulations worldwide for such road or rail-bound systems, including for maximum axle loads and permissible container pressures. In Europe, for example, the permitted total weight for silo road vehicles with a 2-axle tractor and 3-axle trailer is a maximum of approx. 400 kN (40 t). This results in a payload of approx. 28 t, corresponding to $V_{B,\text{netto}} \cong 30\,\text{m}^3$. The maximum permissible operating pressure is limited to $p_{B,\text{för}} = 2.0\,\text{bar(g)}$. In the USA, for example, they work with $p_{B,\text{för}} = 15\,\text{psi(g)} \cong 1.0\,\text{bar(g)}$. European rail vehicles are generally designed for maximum operating pressures of $p_{B,\text{för}} = 2.5\,\text{bar(g)}$ with vessel volumes in the range of $V_{B,\text{netto}} \cong (80-100)\,\text{m}^3$.

Road vehicles are usually equipped with their own on-board compressor. Its intake volume flow is generally $\dot{V}_{\text{F,V}} \cong 600\,\text{m}^3/\text{h} \cong 720\,\text{kg/h}$ air. This allows pipe diameters up to max. DN100 to be stably charged and a solid throughput of max. $\dot{m}_S \cong 30\,\text{t/h}$ at distances $L_R \lesssim 50\,\text{m}$ can be realized. This results in unloading times in the range of one hour and possibly more. To reduce these, larger diameter pipelines D_R can be used with external gas supply, i.e., with a correspondingly larger stationary pressure generator at the unloading location higher discharge mass flow rates $\dot{m}_S$ can be achieved. In rail vehicles, discharging with external compressors is the standard.

Fig. 7.17 shows an example of the possible structure of a discharge station for road or rail vehicles with an external compressor: The bulk material outlet of the vehicle is connected to the stationary transport line with a flexible conveyor hose and the bulk

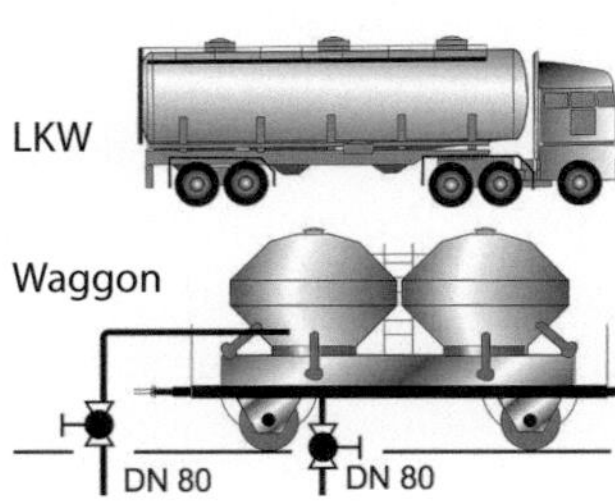

Fig. 7.16 Examples of road and rail vehicles for bulk material transport

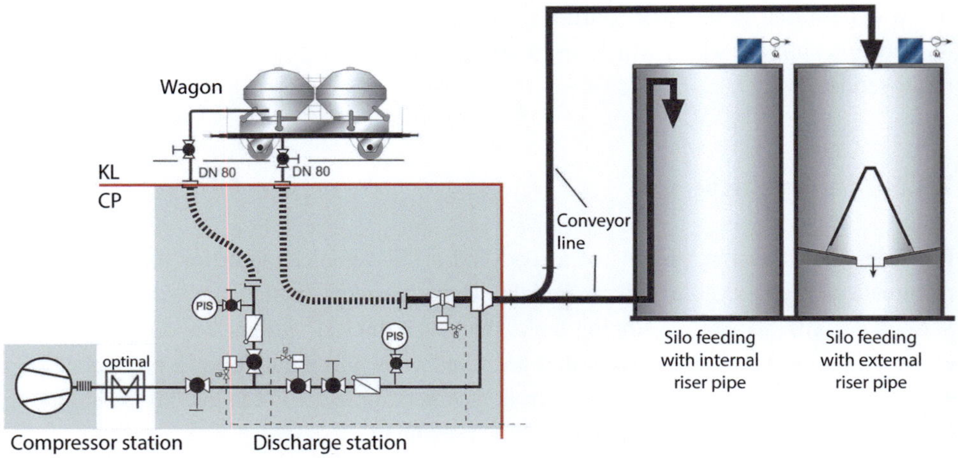

Fig. 7.17 Flow diagram of a pneumatic discharge station for tank vehicles/wagons

material container is connected to the pressure generator via an air hose. In modern systems, DN100 conveyor hoses are preferably used. These allow discharge mass flows up to $\dot{m}_S \cong 80\,\text{t/h}$ for road vehicles and $\dot{m}_S \cong 100\,\text{t/h}$ for rail vehicles and can still be handled by the driver or the unloading personnel. Larger hose diameters require auxiliary constructions for moving and coupling. As with the individual pressure vessel, see Sect. 7.1.1, the gas flow $\dot{m}_{F,V}$ of the compressor is divided solid-specifically to the beginning of the stationary transport line and the pressure vessel. The gas to the vehicle is used entirely or partially for, in general, aeration of the discharge areas/discharge cones of the vehicles, i.e., to support the bulk material outflow. After opening all manually operated valves on the vehicle/wagon, the discharge can be initiated via a local control panel. In modern systems, this process is largely automatic without the intervention of the operating personnel. For road vehicles with multiple discharge cones, for example, the manual closing of an emptied and the opening of the next still full discharge cone may be required by the driver.

The entire emptying process consists of the steps: pressurizing the tank to the pressure $p_{B,\text{för}}$, duration $\Delta\tau_2$, emptying the tank = actual conveying, duration $\Delta\tau_3$, and complete relaxation of the container from $p_{B,\text{för}}$ to $p_{B,\min}$ = ambient pressure p_U through the conveying line, duration $\Delta\tau_4$. The times of the individual steps can be determined using the approaches presented in Sect. 7.1.1. The pressure reduction of the vessel is often calculated as pure gas flow. This leads to a slightly larger final surge volume flow $\overline{\dot{m}}_{F,\text{tot}}$, the consideration of which, however, ensures a safe design of the filter on the receiving silo. A practical design in this regard is illustrated in [8] using examples. Due to the large container volumes of the vehicles and the often short conveying distances, very high surge gas flows occur at the end of the tank vehicle unloading, the effects of which on the receiving silo are often underestimated.

The average power requirement $\overline{P}_K$ of the system can also be determined using the approaches derived in Sect. 7.1.1. In the case of guarantees for the unloading mass flow, a clear distinction must be made between the total batch time $\Delta\tau_{Ch} = (\Delta\tau_2 + \Delta\tau_3 + \Delta\tau_4)$ related solid flow $\dot{m}_{S,n}$ and the one $\dot{m}_S$ during the actual conveying phase $\Delta\tau_3$. Usually, the batch time $\Delta\tau_{Ch}$ is considered as the time difference between switching the compressor on and off. If an average throughput $\dot{m}_{S,n}$ is to be realized, this leads to the design mass flow $\dot{m}_S = (\Delta\tau_{Ch}/\Delta\tau_3) \cdot \dot{m}_{S,n}$ during the actual conveying phase and vice versa.

The pressure loss of the flexible conveying hose between the vehicle and the stationary pipeline, which is included in the conveying line calculation, is approximately two to three times that of a straight steel pipe of the same diameter and length: $\Delta p_{\text{Sch}} \cong (2-3) \cdot \Delta p_{\text{Stahl}}$. Conveying hoses should be kept as short as possible and stronger movement/whipping should be prevented. The container is pressurized with the entire compressor gas flow $\dot{m}_{F,V}$. During the conveying phase = container emptying, at least the minimum gas velocity $v_{F,Sch,\min}$ at the beginning of the conveying hose, see (4.13), must be maintained to ensure an (adequate) safety distance $\Delta v_{F,\text{Sch}}$ exceeding gas flow and additionally the top gas flow $\dot{m}_{F,O}$, cf. (7.8), must be supplied, i.e.:

$$\dot{m}_{F,B} = \varrho_{F,B} \cdot \left(A_{\text{Sch}} \cdot (v_{F,\text{Sch},\min} + \Delta v_{F,\text{Sch}}) + \frac{\dot{m}_S}{\varrho_S} \right) \tag{7.33}$$

Examples of executed discharge systems [9]:

- Cement from rail vehicles; conveying distance: 200 m; unloading capacity per wagon: 150 t/h; conveying hose: DN150; parallel unloading of five wagons via assigned individual lines.
- Lignite coal dust from rail vehicles; conveying distance: 400 m; three wagons are simultaneously connected and simultaneously emptied into a common conveying line; unloading capacity: 150 t/h (→ system analogous to the multi-pressure vessel, see Sect. 7.1.4).
- Cement from rail vehicles; conveying distance: 230 m; six wagons are simultaneously connected and simultaneously emptied into a common FLUIDCON conveying line, see Sect. 6.6; unloading capacity: 230 t/h; there are two such systems that are unloaded one after the other through the same conveying line.

7.2 Screw Feeders

These continuously operating locks, which are very popular in the cement and mineral industries, are high-speed rotating, $n_W \cong (500-1500)\,\text{min}^{-1}$, compression screw systems, whose screw channel is completely filled with the bulk material to be conveyed and in which the sealing of the conveying pressure difference Δp_R is achieved by the

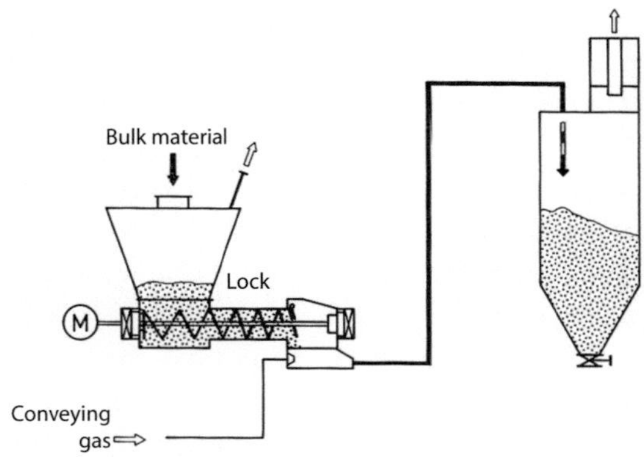

Fig. 7.18 Structure of a conveying system with screw lock

bulk material itself. Fig. 7.18 shows the basic structure of a conveying system with a screw lock.

The advantage of these feeders lies not only in their suitability for the harshest operating conditions and continuous conveying, but also in the fact that, due to their compact design, even with very large solid throughputs, only a small overall height is required, e.g., compared to pressure vessels, which allows for easy positioning below silos without having to raise them. The main disadvantage is the relatively large drive power of the screw compared to competing systems, especially pressure vessels and rotary valve locks.

7.2.1 Designs and Operating Principle

Fig. 7.19 shows designs of screw locks. These essentially differ in the type of screw bearing and the sealing principle during start-up and shutdown. The variants in Fig. 7.19a, c are locks with cantilevered/single-sided mounted screw, while the design in Fig. 7.19b has a double-sided screw bearing. The latter results, among other things, in greater and load-independent smoothness of operation and impact-free operation, but requires special sealing measures for the shaft bearing at the exit of the screw shaft from the dust-filled and pressurized outlet housing.

The bulk material is captured in the inlet area of the feeder by the high-speed rotating compression screw, compressed along the screw channel, introduced into the pressurized outlet housing at the discharge end of the screw channel, and captured there by the conveying gas stream. The start-up and shut-down sealing, i.e., the prevention of gas backflow when the screw channel is not filled with bulk material, is achieved in the designs Fig. 7.19a, b by means of pendulum-mounted and weight, spring, or shock absorber-loaded non-return flaps, while the sealing in design Fig. 7.19c is ensured by a special arrangement of conveying gas nozzle and screw ($\rightarrow$ injector principle).

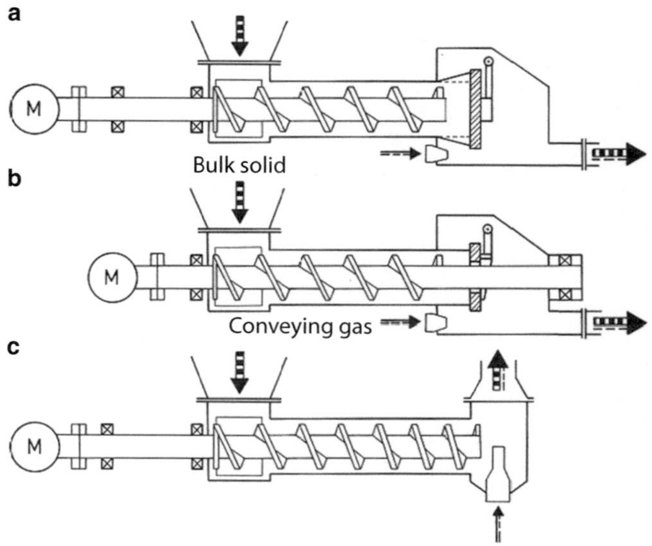

Fig. 7.19 Design variants of screw feeders. **a** floating/single-sided bearing with non-return flap, **b** double-sided bearing with non-return flap, **c** floating (single-sided) bearing without non-return flap

To compress the bulk material, i.e., to reduce the relative void volume ε_F and thus improve the gas sealing, the screws used are generally designed as compression screws. Their inlet pitches $H_{W,1}$ are larger than the pitches $H_{W,2}$ at the screw outlet. The ratio $K_W = H_{W,1}/H_{W,2}$ is referred to as compression: $K_W = 1 \rightarrow$ continuously constant screw pitch, $K_W = 2 \rightarrow$ the pitch decreases along the screw channel to half the initial value. The respective K_W to be used depends on the type and grain size distribution of the current bulk material.

Beyond the sealing effect of the bulk material filling of the screw, the sealing of the conveying pressure is further improved in the designs Fig. 7.19a, b by a material plug building up behind the discharge-side screw end and the compressive effect of the return force of the non-return flap. In the variant Fig. 7.19c, these features are missing. Consequence: while with designs Fig. 7.19a, b, conveying line pressures up to $p_R \cong 2.5\,\text{bar(g)}$, in special cases even higher, can be continuously fed in, with designs Fig. 7.19c only values up to $p_R \cong 1.5\,\text{bar(g)}$ (approx.) are possible.

In order to achieve the mentioned conveying pressures, the bulk materials to be fed in must have a sufficiently high flow through resistance. This is generally ensured for fine-grained products, such as cement or fly ash, and products with a wide particle size distribution. Permissible are, for example, bulk materials with proportions up to approx. 10 mm particle size, if at the same time a sufficient amount of fine material is present that completely fills the void volume between the coarse material. For determining the permissible conveying pressure in individual cases, knowledge of the particle size distribution of the respective bulk material is required.

Further execution variants of screw locks are known [10]. The relationships and measurement results presented below refer to the two-sided bearing design shown in more detail in Fig. 7.20 from Fig. 7.19b [11]. Additional details on the structural design can be found in [10, 12] and Sect. 7.2.2. The lock type considered, known as the X-Pump, covers a standard series with screw diameters of $D_Z = (115-350)$ mm and solid mass flow rates up to approximately $\dot{m}_S \cong 450\,\mathrm{t/h}$. The series is geometrically similar in construction, e.g., the ratio of screw length to screw diameter is $L_W/D_Z \cong 4.5$, that of the inlet pitch to the screw diameter is $H_{W,1}/D_Z = 1.0$, and allows for systematic changes in the plug lengths.

Subsequently, calculation approaches are only presented as far as they are necessary for understanding the functioning of the screw feeder. The approaches used are based on models for describing the processes in the intake zone of single-screw extruders [13, 14, 15, 16] and have been modified accordingly [11, 17].

7.2.2 Feeder Size

First, some notes to clarify the processes/gas sealing in the screw channel: The increasing compression of the screw filling in the transport direction results in an increase in bulk density $\varrho_{b,x}$ and thus a reduction of the volume available to the gap gas. This leads to an increase in local gas pressure $p_{F,x}$ in the gap volume and a backflow of the interstitial gas towards the bulk material inlet, counteracting the solid movement. The actual leakage gas flow is superimposed on this flow. Fig. 7.21 shows measured and with the conveying pressure p_R normalized gas pressure profiles along the screw channel of an X-pump with the screw diameter $D_Z = 150$ mm. With otherwise identical boundary conditions, the bulk material and the plug length L_P were varied. The coordinate $L_{W,x}$ in Fig. 7.21 runs, starting at the plug end $L_{W,x} = 0$, against the transport direction to the bulk material inlet. It can be seen that the pressure gradient responsible for the leakage gas flow at the plug decreases with its increasing length L_P and with a bulk material with a larger relative gap volume $\varepsilon_{F,1}$ at the screw inlet, $\varepsilon_{F,\mathrm{plaster}} > \varepsilon_{F,\mathrm{cement}}$, it also is reduced. The whole thing presents itself as a complex interplay of various operating and bulk material parameters.

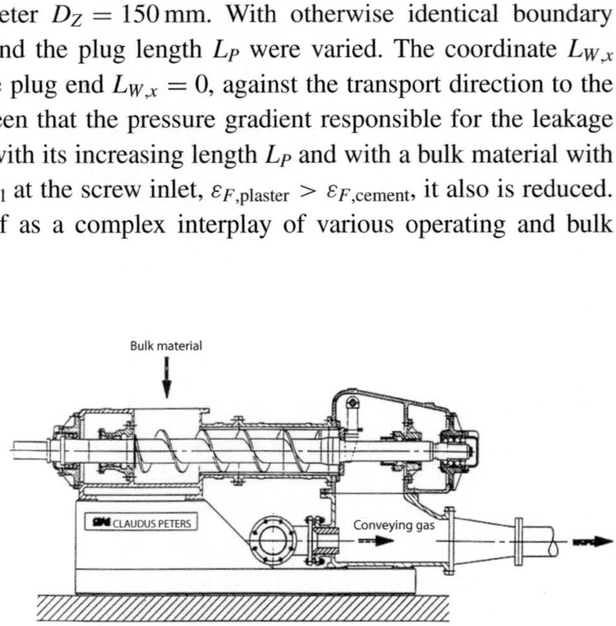

Fig. 7.20 Structure of a screw feeder, type X-pump

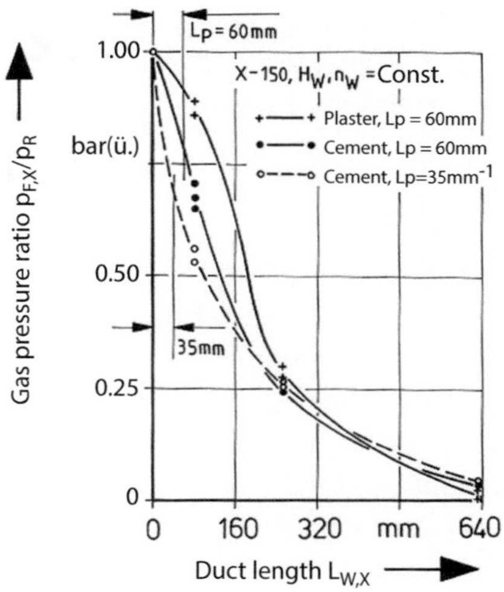

Fig. 7.21 Measured gas pressures along the screw channel of an X-pump, screw diameter $D_Z = 150\,\text{mm}$

Screw locks are volumetric conveyors. The maximum volume flow of bulk material that can be locked or conveyed by them $\dot{V}_{S,1,\max}$ corresponds to the product of the filling volume $\Delta V_{W,1}$ of the first screw spiral below the lock inlet multiplied by the screw speed n_W. With the geometry data shown in Fig. 7.22, the following results:

$$\dot{V}_{S,1,\max} = \dot{V}_{S,1}(\omega_1 = 90^\circ) = \Delta V_{W,1} \cdot n_W = \pi(D_Z - G_W) \cdot G_W \cdot (H_{W,1} - e) \cdot n_W \quad (7.34)$$

$\omega_1 = 90^\circ$ states that the so-called conveying angle ω_x, under which the solid moves with the speed v_ω $(= u_{S,\omega})$ spirally relativ to the stationary screw cylinder, assuming the value 90° in the inlet area, i.e., is aligned parallel to the screw axis. ω_x is, as shown in Fig. 7.22, measured from the circumferential direction and can take values between $\omega_x = 0^\circ$ (→ circulation of the bulk material) and a maximum of $\omega_x = (90^\circ - \varphi_a)$, φ_a = outer helix angle of the screw, for smooth screw cylinders, or $\omega_x = 90^\circ$ for longitudinally grooved cylinders. Regardless of the design of the screw cylinder, $\omega_1 = 90^\circ$, (7.34), is used. ω_x changes along the screw channel. A decrease in the transport direction is normal. The local velocity v_ω results vectorially as the resultant of the circumferential velocity $v_U = D_Z \cdot \pi \cdot n_W$ and the solid velocity v_S in the channel direction [11].

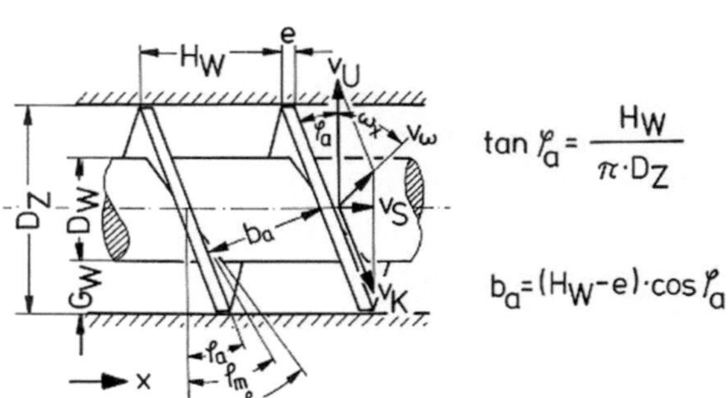

Fig. 7.22 Geometry and operating data of a screw

In practical operation, conveying angles $\omega_1 \ll 90°$ and thus realizable volume flows $\dot{V}_{S,1} \ll \dot{V}_{S,1,\max}$ occur. To take this into account, either the respective conveying angle ω_1 or a volumetric conveying efficiency η_V must be known. It applies:

$$\eta_V = \frac{\dot{V}_{S,1}}{\dot{V}_{S,1,\max}} = \frac{\tan\omega_1}{\frac{H_{W,1}}{\pi \cdot D_Z} + \tan\omega_1} \tag{7.35}$$

In practice, the volumetric efficiency η_V is generally used, which can be calculated using (7.34) and (7.35) from systematically varied measurements of the volume flow rate $\dot{V}_{S,1}$.

Usually, the task is a specified mass flow rate to be introduced = mass flow $\dot{m}_S$ to be conveyed. $\dot{m}_S$ is, as the preceding explanations illustrate, significantly determined by the conditions at the lock inlet. This is therefore the appropriate reference state for the conversion of $\dot{V}_{S,1}$ into $\dot{m}_S$. This results in:

$$\dot{m}_S = \dot{V}_{S,1} \cdot \varrho_{b,1} \quad \rightarrow \quad \text{with: } \varrho_{b,1} = \varrho_{b,\text{fluid}} \tag{7.36}$$

$\varrho_{b,\text{fluid}}$ is in this case the fludized density measured with the standardized fluidized bed apparatus described in Sect. 3.3.2 and illustrated in Fig. 3.12. It takes into account the generally strong loosening of the supplied bulk material. With this measuring procedure, it is possible to capture the loosening behavior of the various conveyed materials in a representative manner. Measures to improve the inflow behavior will be discussed later.

The volumetric screw efficiency η_V is essentially dependent on

a) the characteristics of the bulk material to be conveyed,
b) the pressure difference to be sealed,
c) the screw rotation speed,
d) the geometric design of the screw.

Regarding a): Traditionally, the influence of bulk material properties on η_V is represented as a function of the mass-based specific Blaine surface S_{Blaine}, e.g., [cm^2/g], of the bulk material: η_V decreases with increasing S_{Blaine}, i.e., becoming finer solid material. Fig. 7.23 illustrates that not all bulk materials are covered by the dependency $\eta_V \propto 1/S^n_{\text{Blaine}}$, e.g., the lignite dust in Fig. 7.23 shows a deviating behavior. Further investigations reveal that η_V can be more comprehensively correlated by an Archimedes number Ar, when it is formed with the volume- and surface-equivalent particle diameter $d_{\text{Blaine}} = 6/(S_{\text{Blaine}} \cdot \varrho_P)$. Screw feeders for bulk materials with $S_{\text{Blaine}} \gtrsim 12.500\,\text{cm}^2/\text{g}$ should be tested for their suitability for such a product.

Regarding b) and c): As can be seen from Fig. 7.23 and 7.24, an increasing pressure difference to be sealed off by the lock Δp_R reduces the volumetric efficiency η_V. Furthermore, it can be seen that with decreasing rotation speed n_W also η_V becomes lower. This applies in general. At screw speeds lower than $n_W \cong 450\,\text{min}^{-1}$ a stable and gas backflow-free bulk material acceptance is no longer guaranteed. Preferred speeds are

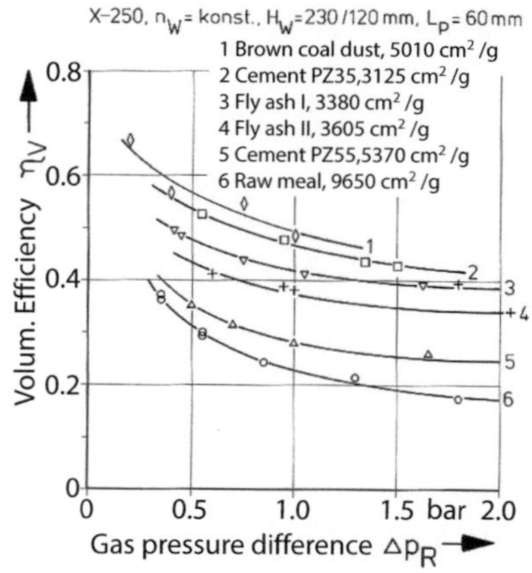

Fig. 7.23 Dependency $\eta_V = f(\Delta p_R, \text{Schüttgut})$, $D_Z = 250\,\text{mm}$

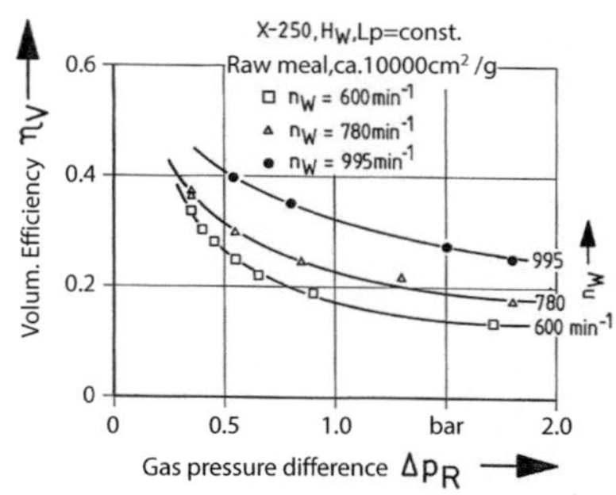

Fig. 7.24 Dependency $\eta_V = f(\Delta p_R, n_W)$, $D_Z = 250\,\text{mm}$

$n_W \cong 750/1000/1500\,\text{min}^{-1}$, as these allow directly coupled drive motors with the appropriate number of poles.

Fig. 7.25 indicates a scale effect when transferring the efficiency measured at smaller locks η_V to larger locks and vice versa. With screw diameters $D_Z \leq 150\,\text{mm}$ η_V increases with increasing screw speed n_W, has a maximum dependent on the current pressure difference Δp_R and then decreases again. For screws $D_Z \geq 200\,\text{mm}$ in the investigated measuring range, a linear increase of the volumetric efficiency η_V with the rotational speed n_W was determined. A decreasing η_V does not necessarily mean a decrease in the solid mass flow $\dot{m}_S$, as at the same time n_W increases.

For d): The influence of the screw geometry on volumetric efficiency η_V and throughput $\dot{m}_S$ is shown in Fig. 7.26. Four screws with the diameter $D_Z = 150\,\text{mm}$, but with different inlet pitches $H_{W,1}$ and compressions $K_W = H_{W,1}/H_{W,2}$ (→ lower abscissa values) are compared. Screws with compression have higher efficiency η_V than screws with constant pitch, while η_V is (in accordance with theory) for screws with $H_{W,1}/D_Z = 0.71$ adjusts to larger values than with $H_{W,1}/D_Z = 1$. The solids mass flow of interest to the user, $\dot{m}_S$, reaches a maximum value for the compression screw with the inlet pitch $H_{W,1}/D_Z = 1$.

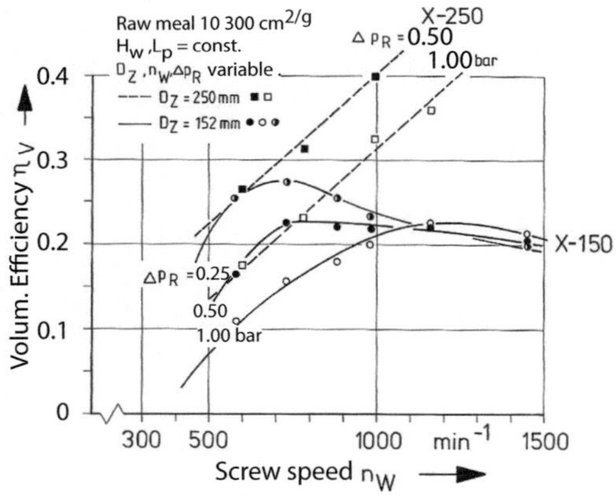

Fig. 7.25 Dependency $\eta_V = f(n_W, D_Z, \Delta p_R)$, $D_Z = 250/152\,\text{mm}$

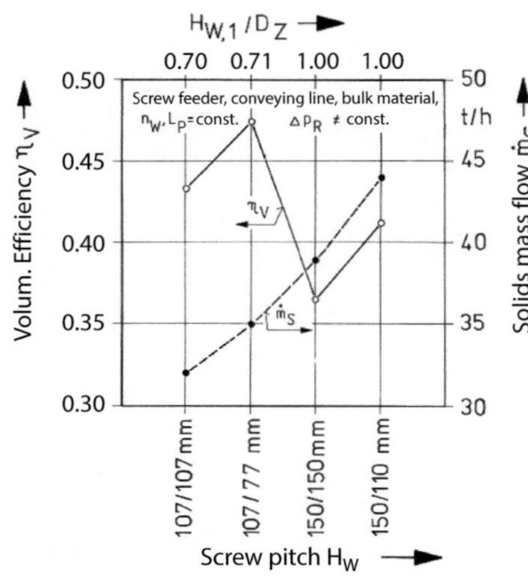

Fig. 7.26 Dependencies $\eta_V, \dot{m}_S = f(\text{screw pitch, screw compression})$

From the above considerations, it follows that, for example, when selecting a screw feeder for a combination of a given bulk material mass flow $\dot{m}_S$ and chosen conveying pipe diameter D_R, i.e., a fixed pressure difference Δp_R at the conveying line ($\rightarrow$ same pressure at the end of the conveying distance and before the feeder), there are several possible implementations: On the one hand, a screw with a larger diameter D_Z and smaller rotational speed n_W, on the other hand, one with a smaller D_Z and correspondingly higher n_W can be chosen. Both variants have advantages and disadvantages: lower wear is opposed by a larger screw lock and vice versa. In principle, the conveying pipe diameter D_R, i.e. the pressure loss Δp_R within the limits controllable by a screw lock, can be considered as a further degree of freedom for system optimization. Of course, it must be defined what the goal of such optimization should be: minimal energy consumption, minimal investment, operating costs, etc.?

Since the bulk material residence time in a screw lock is in the range of seconds, short-term pulsations of the incoming bulk material flow can lead to throughput problems with a lock design for $\dot{V}_{S,1} = \eta_V \cdot \dot{V}_{S,1,\max}$. In practice, a utilization rate of

$$\xi_Z = \frac{\dot{V}_{S,1}}{\dot{V}_{\text{design}}} \lesssim 0.80 \tag{7.37}$$

is assumed. This covers throughput pulsations, inlet densities $\varrho_{b,1} < \varrho_{b,\text{fluid}}$ etc., but also reasonably requires a suitable throughput control, at least, however, the limitation of the maximum inlet mass flow. If the screw lock is designed for $\dot{V}_{\text{design}}$, it is, as desired, capable of temporarily accommodating larger bulk material flows, but will then generate a correspondingly higher line pressure loss in the subsequent conveying section for a short time compared to the actual design. This must be appropriately considered in the system planning.

Partial load behavior of a screw feeder is possible with stable plug formation at the screw end down to the range of $(\dot{m}_{S,\text{Teillast}}/\dot{m}_S) < 0.2$ and possibly even lower. However, screw wear increases disproportionately with decreasing load due to the greater gas backflow.

The bulk materials fed to a screw feeder reach it, depending on the feeding device, e.g., fluidized bed channel, transport screw, drop distance, in a more or less loosened state. To set the inflow density $\varrho_{b,1} = \varrho_{b,\text{fluid}}$ assumed in the design, a so-called venting container is mounted on the feeder inlet, in which the solid is to be separated as far as

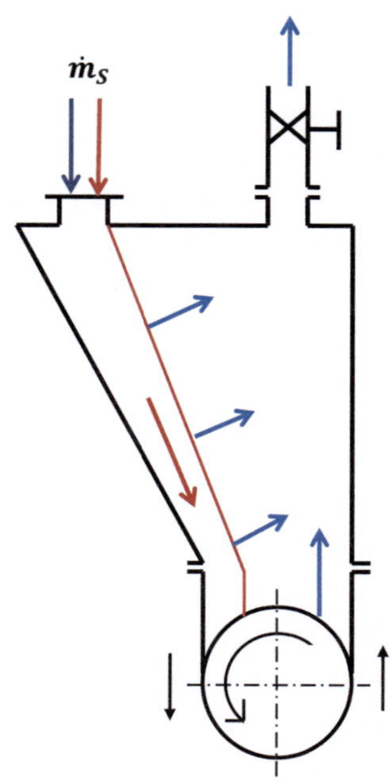

Fig. 7.27 Gas/solid separation in the venting container of a screw lock; *red* = solid, *blue* = gas/air

possible from the entrained gas and the gas emerging from the screw. Fig. 7.27 shows a design variant of the venting container: This is connected to an external, possibly also to an integrated filter in the container, and adjusted to a negative pressure of $p \leq 1.0\,\text{mbar}$ via throttle valves. The bulk material is guided onto an inclined wall surface in such a way that a thin layer of material is formed, which can degas into the adjacent free space. Furthermore, the venting container is arranged so that the bulk material is fed to the screw on the downward turning side. On the rising side, the gas can escape from the screw channel. Complete venting is generally not possible due to the relatively short residence times. Fig. 7.27 illustrates the underlying principles. Other design variants are possible.

If the bulk material is fed to the lock via a very long drop distance, e.g., a drop pipe, it should be equipped with decelerating cascades. For a safe plant design, it is absolutely necessary to know the possible bulk material condition in the lock inlet.

In normal conveying operation, no bulk material must accumulate in the venting container; it must be empty. Accumulating bulk material in the venting container means that the lock cannot convey the supplied bulk material flow: The lock is too small, or the inlet mass flow is larger than planned.

7.2.3 Drive Power

Deviating from the solid throughput, the drive power of a screw lock is influenced by the operating conditions along the entire screw channel. An integral consideration is required here, in which the conveying angle ω_x is included as an essential influencing variable in addition to the wall friction coefficients and the bulk material pressure. The introduced drive power is essentially converted into frictional power between the bulk material and the adjacent channel and screw surfaces, to a lesser extent into internal bulk material friction.

From theoretical considerations [13, 14, 15, 16, 17] it follows that under comparable operating conditions, the drive power requirement P_W of a screw feeder with an increasing (mean) conveying angle $\overline{\omega}_x$, or increasing (mean) efficiency $\overline{\eta}_V$, decreases. This means that the power requirement P_W is reduced by stronger compression, but is increased by longer plugs L_P. The reason for the former is the shorter contact of the bulk material with the touching walls along the screw channel due to the larger $\overline{\omega}_x$ ($\rightarrow$ larger slope of the spiral bulk material circulation). This is confirmed by measurement results.

A summary representation of the required screw drive performances for the introduction of two different bulk materials is shown in Fig. 7.28 as an example. The normalized net power requirement P_{Netto} (= coupling power P_W − idle power P_0) is plotted against the "useful power" ($\dot{V}_{S,1} \cdot \Delta p_R$) (= task). By normalizing, the influences of the geometrically similar lock geometry can be taken into account. $\rightarrow \propto D_Z^3$, the rotational speed $\rightarrow \propto n_W$, as well as back pressure and plug length $\rightarrow \propto p_{S,\max}$ can be recorded. $p_{S,\max}$ is the bulk material pressure at the end of the screw [11]. Each conveyed material is assigned its own performance curve. These differ significantly from one another due to the differing wall friction and flow behavior of the individual bulk

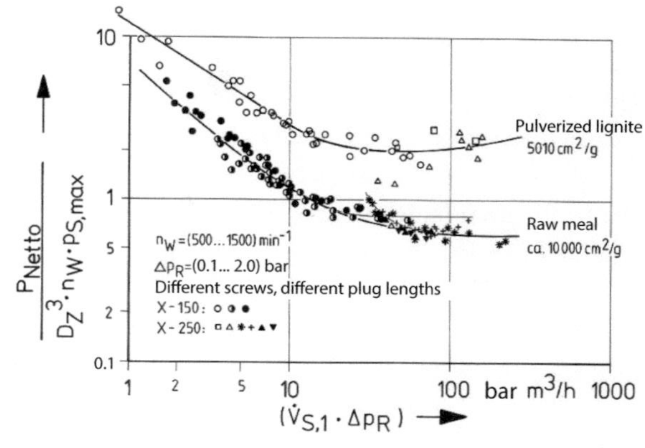

Fig. 7.28 Specific drive power requirement of screw feeders; operating data: $n_W = (600-1500)\ \text{min}^{-1}$, $\Delta p_R = (0.1-2.0)$ bar

materials. A correlation of the power requirements of various conveyed materials is satisfactorily achieved with an Archimedes number Ar formed with the particle diameter $d_{\text{Blaine}} = 6/(S_{\text{Blaine}} \cdot \varrho_P)$. The drive power requirement P_{Netto} of a screw feeder increases with decreasing Blaine surface area S_{Blaine}, i.e., with coarser bulk material.

Particularly advantageous injection conditions in terms of throughput $\dot{m}_S$ and power requirement P_W are achieved when the wall friction coefficient μ_Z between bulk material and cylinder is significantly greater than that μ_W between bulk material and screw surfaces [13, 14, 15, 16, 17]. In extrusion technology, this effect is utilized, among other things, by suitable longitudinal grooving of the extruder channel. Fig. 7.29 shows the results of power measurements that were determined in a comparative study [18] on a screw lock with smooth/ungrooved standard cylinder and alternatively a longitudinally grooved bushing. In the case shown, the longitudinal grooving led to power savings of

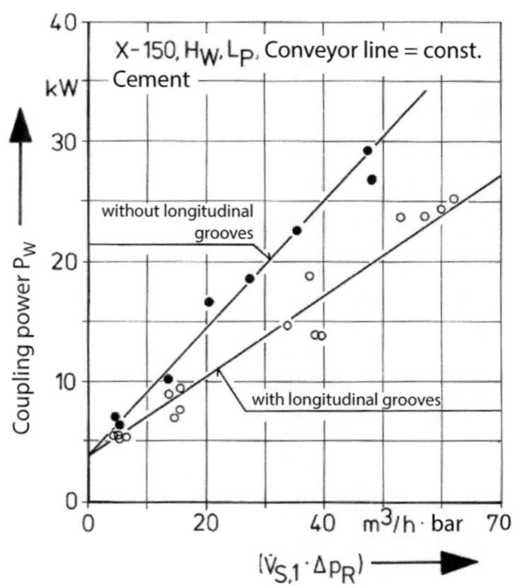

Fig. 7.29 Comparison of drive powers of grooved and ungrooved screw cylinders

up to 30% based on the power requirement of the ungrooved cylinder. An increase in throughput was not observed. Further investigations show that the effect of longitudinal grooving of the screw cylinder is dependent on the solid material. Since screw locks are usually used to transport highly abrasive products through a correspondingly wear-protected screw channel, protected by extremely hard, difficult-to-machine wear bushings or shells, its grooving can only be realized with great effort. Due to this, this possibility of reducing the drive power requirement of screw locks has not yet been widely adopted in practice.

Recent studies [19] with extremely highly compressing screws show that a compression $H_{W,1}/H_{W,2} > 2.0$ no longer provides any energetic/operational advantages.

Note: From the above, it is clear that the functioning of a screw lock is extremely complex. Bulk material properties and especially their inflow state, i.e., the effects of the feeding system in front of it, must be analyzed in detail. Likewise, the expected range of variation of the properties of the bulk materials to be locked in must be known. It is reminded that η_V decreases with finer solid/increasing Blaine surface S_{Blaine}, while the required drive power demand P_W increases in the opposite direction, i.e., with decreasing S_{Blaine}.

7.3 Rotary Valve Locks

These widely used locking systems in practice usually consist of a horizontally arranged rotor divided into chambers, which rotates in a cylindrical stator. Bulk material is fed into the system from the top and removed from the bottom after a 180° rotation. In the pneumatic conveying systems considered here, the solid material falls directly into the transport line or the conveying gas stream via a so-called blow shoe. Fig. 7.30 shows the basic structure of such a system. At the rotary valve, the pressure difference $\Delta p_{ZR} = (\Delta p_R - \Delta p_0)$, Δp_0 = pressure above the valve, dependent on the current conveying pressure Δp_R is applied. The sealing is achieved by the radial and axial gaps between the rotor and stator. Since these, among other things, due to mechanical stress = deflection/shaft lift caused by the applied pressure difference Δp_{ZR} and the different thermal expansion of the stator and rotor when conveying hot bulk materials, cannot be made arbitrarily small, this leads to an increasing gas flow through these gaps in the direction of the rotary valve inlet, i.e., to the gap losses $\dot{m}_{F,\text{Spalt}}$. They overlap with the so-called scooping losses $\dot{m}_{F,\text{Schöpf}}$, which occur because the rotor chambers emptying the bulk material into the conveying line are simultaneously filled with conveying gas at pressure p_R, which is thus also transported in the direction of the valve inlet, pressure $p_0 < p_R$. The sum of the gap and scooping losses is subsequently referred to as leakage gas flow $\dot{m}_{F,\text{Leck}} = (\dot{m}_{F,\text{Spalt}} + \dot{m}_{F,\text{Schöpf}})$. It is obvious that this leakage gas flow will obstruct the bulk material inlet to the rotary valve if it is not properly discharged beforehand. Example: Direct removal of bulk material from a storage container filled with fine-grained

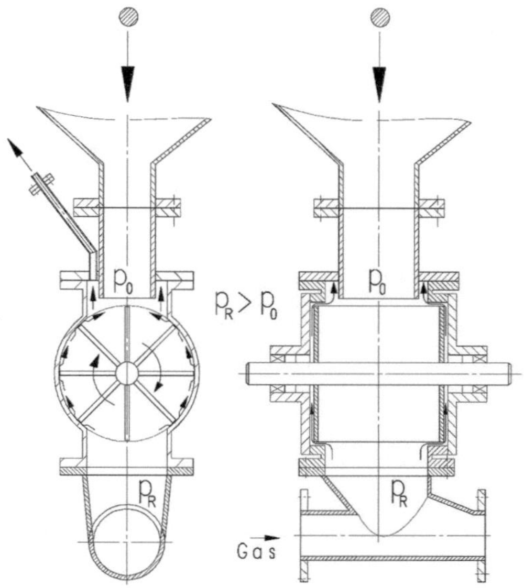

Fig. 7.30 Basic structure of a rotary valve

solid → due to the high flow resistance of the solid bulk material, only a portion of the leakage gas flow can be discharged with the applied pressure difference Δp_{ZR}, the remaining part accumulates below the bulk material inlet to the rotary valve and can support the formation of a bulk material bridge, which then interrupts the material flow.

For rotary valves with a horizontal rotor axis, three basic design variants can be distinguished, see Fig. 7.31:

- Valves with closed cell wheel /rotor, Fig. 7.31a,
- Valves with open cell wheel, Fig. 7.31b,
- Blow-through valves, Fig. 7.31c.

In the blow-through rotary valve shown in Fig. 7.31c, the conveying gas flow guided axially through the solid-discharging rotor chambers supports the bulk material discharge. Designs with radial gas supply into the discharging chambers are also known. For dense phase conveying, rotary valves with a closed rotor on both sides are generally used, as this design variant offers more possibilities for reducing the undesired gap losses $\dot{m}_{F,\mathrm{Spalt}}$.

The rotor chambers are usually designed with radial blades and, in relation to the rotor axis, central bulk material inlet and outlet. The possibility of achieving higher filling degrees of the feeders by using curved blades and/or offset inlets and outlets is generally omitted for cost reasons. In addition to rotary valve feeders with horizontal rotor axis, systems with vertical axis are also offered, see Fig. 7.32, but they have not yet become established. Fundamental investigations on the behavior of rotary valve feeders include, among others, the works [20, 21].

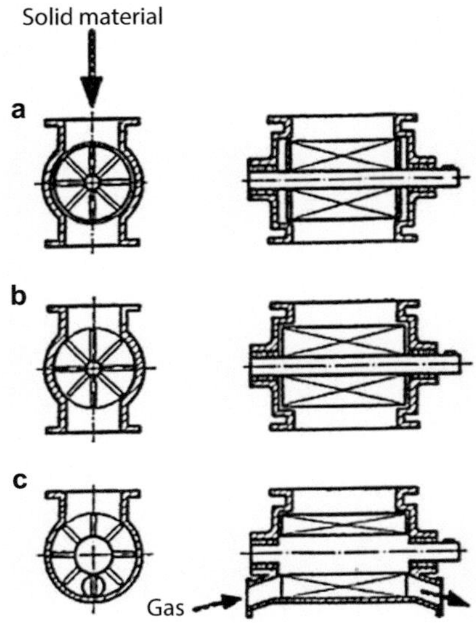

Fig. 7.31 Design variants of rotary valve feeders. **a** feeder with closed rotor, **b** feeder with open rotor, **c** blow-through feeder

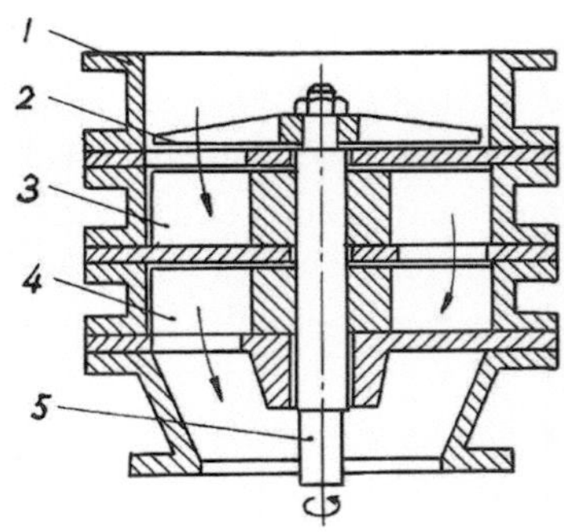

Fig. 7.32 Example of a rotary valve feeder with vertical rotor: 1 = housing, 2 = clearing arm, 3 and 4 = rotor, 5 = common drive shaft, according to [20]

7.3.1 Feeder size, conveying characteristic

Rotary valve feeders are volumetric conveyors. Their bulk material throughput is calculated using (7.38):

$$\dot{V}_{SS} = \frac{\dot{m}_S}{\varrho_{SS}} = \Delta V_{ZR} \cdot \xi_{ZR} \cdot n_{ZR} = \frac{\pi}{4} \cdot D_{ZR}^2 \cdot \varphi_{ZR} \cdot B_{ZR} \cdot \xi_{ZR} \cdot n_{ZR} \tag{7.38}$$

with:

ΔV_{ZR} ($= \pi/4 \cdot D_{ZR}^2 \cdot B_{ZR} \cdot \varphi_{ZR}$), net volume of the cell wheel usable for bulk material,
D_{ZR} rotor diameter,
φ_{ZR} cross-sectional portion of the rotor usable for bulk material,

B_{ZR} width of the rotor,
n_{ZR} design speed of the cell wheel.

The filling degree ξ_{ZR} of the rotary valve is defined by

$$\xi_{ZR} = \frac{\Delta V_{SS}}{\Delta V_{ZR}} = \frac{\Delta V_{SS} \cdot n_{ZR}}{\Delta V_{ZR} \cdot n_{ZR}} = \frac{\dot{V}_{SS}}{\dot{V}_{ZR}} \tag{7.39}$$

ΔV_{SS} is the current bulk material volume in the rotor. For a given design throughput $\Delta \dot{V}_{SS,A}$ and a with a chosen rotational speed $n_{ZR,A}$, the required net volume ΔV_{ZR} of the rotary valve can be calculated and adjusted to the dimensions of a standard series using (7.38). Usually, a filling degree of $\xi_{ZR,A} \cong (0.7-0.8)$ is assumed.

In Box 7.1 a highly simplified model for calculating the dependencies of the degree of filling $\xi_{ZR}(n_{ZR})$ and the $\dot{V}_{SS}(n_{ZR})$-characteristic curve of a given rotary valve is shown. The main relationships are to be demonstrated with this.

Box 7.1: Model for calculating the characteristic curve of a rotary valve

a) Rotary valve speed $n_{ZR} \leq$ critical speed n_{crit}:

Starting from $n_{ZR} = 0$ the speed of a defined rotary valve with the net volume ΔV_{ZR} is continuously increased under otherwise constant boundary conditions. Up to a critical speed n_{crit}, the degree of filling $\xi_{ZR} = 1$ is established. Cause: The filling times/rotor revolution $\tau_{\text{füll}} = (1/n_{ZR})$ are sufficient for a complete rotor filling with bulk material. Its throughput is calculated as follows:

$$\dot{V}_{SS} = \dot{V}_{ZR} = \Delta V_{ZR} \cdot n_{ZR} = \left(\frac{D_{ZR} \cdot \pi \cdot B_{ZR}}{n_{ZR}} \cdot \dot{q}_{\text{füll}} \right) \cdot n_{ZR} \tag{7.40}$$

$\dot{q}_{\text{füll}}$ describes the specific filling volume flow or the filling speed of the bulk material. This is not only dependent on the feed conditions before the lock, but also on the movement conditions in the rotor chambers.

b) Rotary valve speed $n_{ZR} >$ critical speed n_{crit}, $\dot{q}_{\text{füll}} =$ konst.:

Speeds $n_{ZR} > n_{\text{crit}}$ lead to filling degrees $\xi_{ZR} < 1$, since the necessary rotor filling times $\tau_{\text{füll}}$ are greater than the available filling times $(1/n_{ZR})$ are. It applies:

$$\xi_{ZR} = \frac{\Delta V_{SS}}{\Delta V_{ZR}} = \left(\frac{D_{ZR} \cdot \pi \cdot B_{ZR}}{n_{ZR}} \cdot \dot{q}_{\text{füll}} \right) \Big/ \left(\frac{D_{ZR} \cdot \pi \cdot B_{ZR}}{n_{\text{crit}}} \cdot \dot{q}_{\text{füll}} \right) = \frac{n_{\text{crit}}}{n_{ZR}} \tag{7.41}$$

and accordingly (7.38):

$$\dot{V}_{SS} = \Delta V_{ZR} \cdot \xi_{ZR} \cdot n_{ZR} = \Delta V_{ZR} \cdot \left(\frac{n_{\text{crit}}}{n_{ZR}} \right) \cdot n_{ZR} = \Delta V_{ZR} \cdot n_{\text{crit}} \tag{7.42}$$

The bulk material throughput $\dot{V}_{SS}$ remains constant at rotational speeds $n_{ZR} > n_{\text{crit}}$ and is identical to that at the critical rotor speed n_{crit} see Fig. 7.33a, b, curves $r = 0$. The prerequisite for this is the condition $\dot{q}_{\text{füll}} = \text{konst.} \neq f(n_{ZR})$. This is approximately fulfilled when

- the bulk material is fed to the lock in a pre-dosed manner,
- a drop distance that accelerates the solid material is connected upstream of the lock ($\rightarrow$ large $\dot{q}_{\text{füll},0}$),
- a coarse-grained solid material with a narrow particle size distribution ($\rightarrow$ Monograin, low fine content) and high particle density is fed into the lock,
- the leakage gas flow $\dot{m}_{F,\text{Leck}}$ has already been discharged before the lock inlet,
- the cell wheel is already rotating at a speed $n_{ZR} > n_{\text{crit}}$ before the bulk material inlet is opened.

The last-mentioned point avoids the accumulation of solids in the drop distance and thus their removal from a bulk material assembly that reduces the $\dot{q}_{\text{füll}}$ flow rate.

c) Rotary valve speed $n_{ZR} >$ critical speed n_{crit}, $\dot{q}_{\text{füll}} \neq$ konst.:
Practical experience shows that $\dot{q}_{\text{füll}}$ when exceeding n_{crit} also depends on the speed n_{ZR}. For example, if the change in $\dot{q}_{\text{füll}}$ is described by the power approach

$$\dot{q}_{\text{fill}} = \dot{q}_{\text{fill},0} \cdot \left(\frac{n_{\text{crit}}}{n_{ZR}}\right)^{r} \tag{7.43}$$

it follows:

$$\begin{aligned}\xi_{ZR} &= \frac{\Delta V_{SS}}{\Delta V_{ZR}} = \left(\frac{D_{ZR} \cdot \pi \cdot B_{ZR}}{n_{ZR}} \cdot \dot{q}_{\text{fill}}\right) \Big/ \left(\frac{D_{ZR} \cdot \pi \cdot B_{ZR}}{n_{\text{crit}}} \cdot \dot{q}_{\text{fill},0}\right) \\ &= \frac{n_{\text{crit}}}{n_{ZR}} \cdot \left(\frac{n_{\text{crit}}}{n_{ZR}}\right)^{r}\end{aligned} \tag{7.44}$$

and thus:

$$\xi_{ZR} = \left(\frac{n_{\text{crit}}}{n_{ZR}}\right)^{r+1} \tag{7.45}$$

The bulk material throughput can be determined with (7.38). Fig. 7.33a, b shows, as an example, the degree of filling and throughput characteristics of a given rotary valve for different exponents "r".

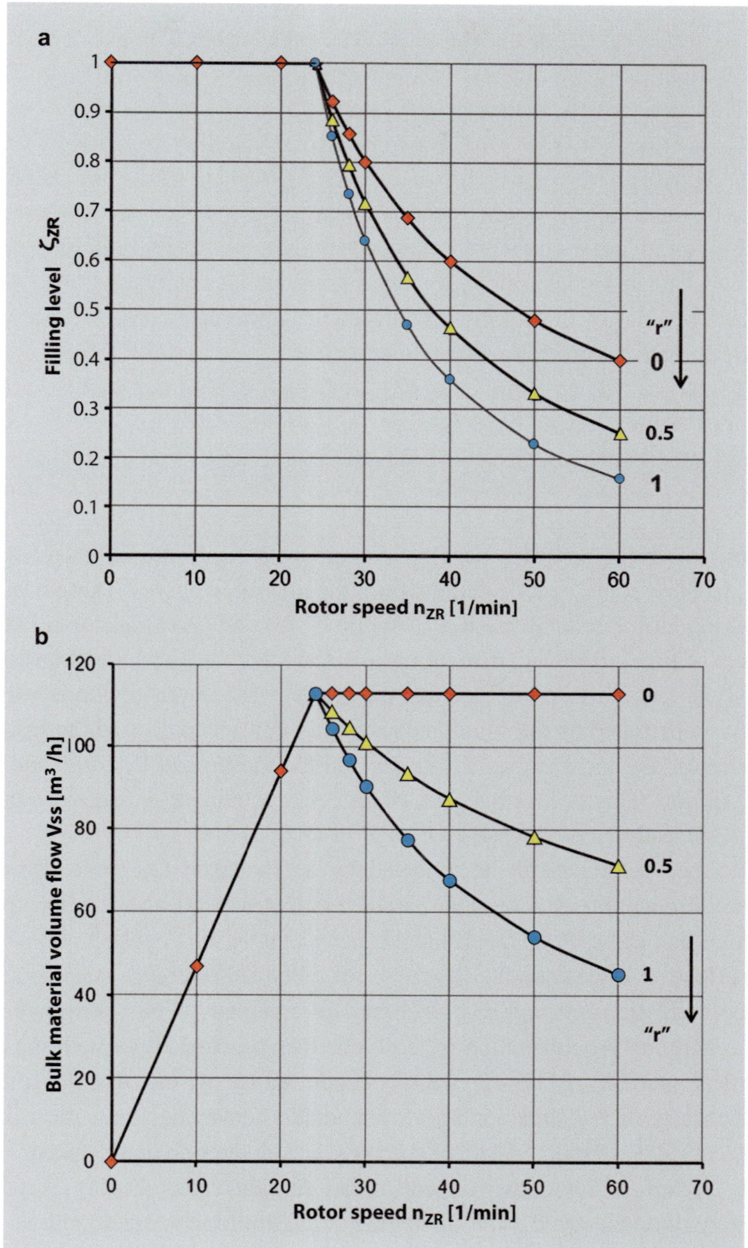

Fig. 7.33 Degrees of filling calculated with the above model ξ_{ZR} (**a**) and bulk material flow rates $\dot{V}_{SS}$ (**b**); $\Delta V_{ZR} = 0.0785\,\mathrm{m}^3, n_{\text{crit}} = 24\,\text{min}^{-1}$

Causes for the dependency $\dot{q}_{\text{füll}}(n_{ZR})$: The bulk material is accelerated in the circumferential direction and, due to its inertia, pressed against the rear chamber walls. This generates a frictional force that increases with n_{ZR} and slows down the bulk material flow. The solid matter is also affected by the radially outward-directed centrifugal force, which has the same tendency. The solid volume flow supplied to the rotor chambers displaces an equal volume of gas flowing against the bulk material, even with a pre-installed leakage gas discharge, and so on.

n_{crit} and r must be determined for the respective operating conditions (→ variable characteristics of the solid, current pressure difference Δp_{ZR}, with or without leakage gas removal, etc.) from experiments.

Measured characteristic curves $\xi_{ZR}(n_{ZR})$ or $\dot{V}_{SS}(n_{ZR})$ illustrate that the processes in the area of the critical speed n_{crit} can only be described inaccurately with the above model. An analysis is carried out in the text section.

Comparisons of the approaches developed there with measured characteristic curves, Fig. 7.34a, b, show significant deviations from the measured curves. These occur particularly in the area of the critical speed n_{crit}. While (7.43) and (7.45) lead to an abrupt drop in the degree of filling from n_{crit} to values $\xi_{ZR} = 1$ to values $\xi_{ZR} < 1$, this transition occurs continuously or smoothly. In the case that the ξ_{ZR} decrease in this region is only slight, it can be overcompensated by the simultaneous increase in speed n_{ZR} and the associated $\dot{V}_{SS}$ -curve increases, as shown in Fig. 7.34b, beyond the throughput $\dot{V}_{SS}(n_{\text{crit}})$ and continues up to a maximum. This always happens when corresponding ξ_{ZR}- and n_{ZR}-values result, which in (7.38) lead to a product $(\xi_{ZR} \cdot n_{ZR}) > (1 \cdot n_{\text{crit}})$.

The described behavior can be modeled by an S-shaped ξ_{ZR}-curve. However, this requires the introduction of additional parameters that depend on the respective operating conditions and can only be determined experimentally. It is considerably simpler and more cost-effective to measure the ξ_{ZR}-curve for a defined reference state and its dimensionless representation $\xi_{ZR} = f(n_{\text{crit}}/n_{ZR})$ can be assumed to be transferable to other boundary conditions. An adaptation to these can then be carried out with the respective current critical speed n_{crit}. This approach is supported by the fact that the rotary valves used for the entry of bulk materials into pneumatic conveying lines are usually operated at speeds $n_{ZR} \lesssim 40\,\text{min}^{-1}$. Critical speeds n_{crit}, i.e., the maximum speeds with a filling degree $\xi_{ZR} = 1$, of rotary valves with radial straight vanes feeding coarse-grained, approximately monodisperse bulk materials, e.g., millimeter-sized plastic granules, are in the range of $n_{\text{crit}} \cong 25\,\text{min}^{-1}$, those feeding dust-fine products, such as cement, at $n_{\text{crit}} \cong 10\,\text{min}^{-1}$. For fine-grained bulk materials, generally significantly smaller filling levels ξ_{ZR} occur than with coarser solids. Increasing pressure differences Δp_{ZR} and/or insufficient leakage gas removal reduce ξ_{ZR}.

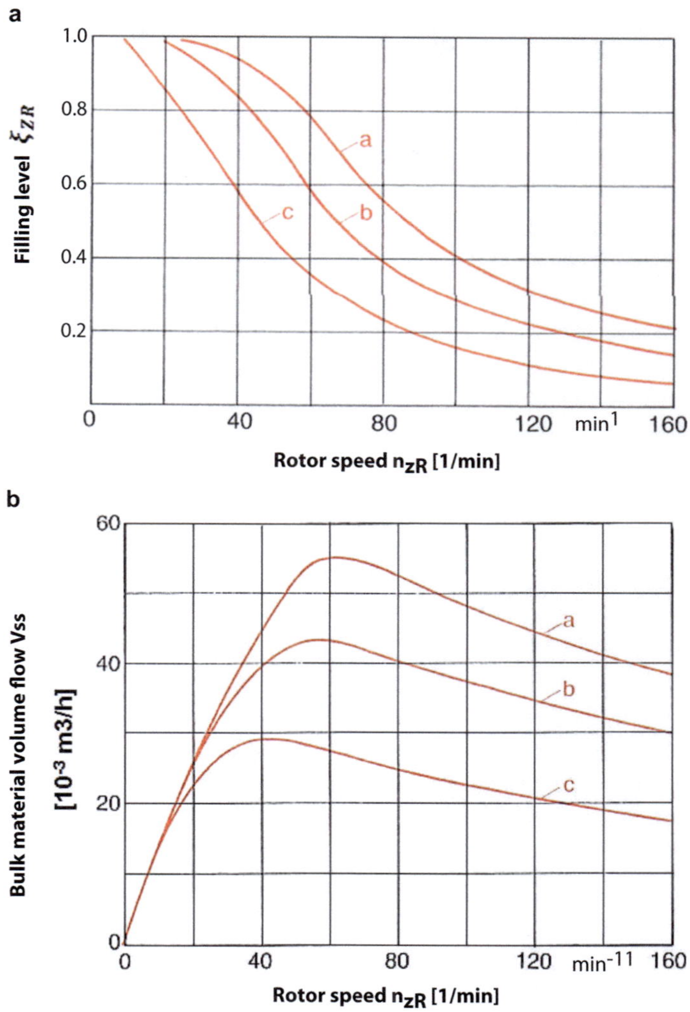

Fig. 7.34 Measured characteristics of a rotary valve $D_{ZR} = 0.20\,\text{m}$, $\Delta p_{ZR} = 0$. **a** Degree of filling, **b** Bulk material throughput. Individual curves: *a* Rapeseed, $d_S = 2.0\,\text{mm}$, $\varrho_{SS} = 675\text{kg/m}^3$, *b* Potash, $d_S = 0.25\,\text{mm}$, $\varrho_{SS} = 1081\,\text{kg/m}^3$, *c* feed meal, $d_S = 0.2\,\text{mm}$, $\varrho_{SS} = 338\,\text{kg/m}^3$, according to [20, 22]

7.3.2 Leakage Gas Flow and Removal

Size and removal of leakage gas flow $\dot{m}_{F,\text{Leck}} = (\dot{m}_{F,\text{Spalt}} + \dot{m}_{F,\text{Schöpf}})$ of a rotary valve have significant effects on its operating behavior:

- The pressure generator of the considered conveying system must be designed for the provision of the conveying gas flow $\dot{m}_F$ plus the leakage gas flow $\dot{m}_{F,\text{Leck}}$, i.e., for $\dot{m}_{F,V} = \dot{m}_F + \dot{m}_{F,\text{leck}}$.
- Since the minimal realizable radial and axial gaps between the rotor and stator, depending among other things on the operating conditions and the size of the cell wheel, are in the range of $\Delta s_{ZR} \cong (0.1-0.2)$ mm, solid particles with particle diameters $d_S < \Delta s_{ZR}$ can be transported by the leakage gas flow through the gaps. Due to their inertia, they cause considerable wear on the rotor and/or stator upon impact, which increases the gap dimensions and thus the leakage gas flow with increasing operating time. The same applies to particle sizes $d_S \cong \Delta s_{ZR}$, which can be trapped in the gap and possibly rotate with the rotor. In the case of abrasive solids, the increase in leakage gas flow caused by wear must be taken into account during plant planning. Plant designs with a ratio $\dot{m}_{F,\text{Leck}}/\dot{m}_F \gtrsim 0.3$ in the new condition should be avoided for the reasons mentioned.
- The leakage gas flow $\dot{m}_{F,\text{Leck}}$ increases with increasing pressure difference Δp_{ZR} at the rotary valve. Since the solid volume flow leaving the rotor chambers must be replaced by an equal gas volume flow, a leakage gas flow also occurs at $\Delta p_{ZR} = 0$ that corresponds to the scooping loss.
- The filling degree ξ_{ZR} of a rotary valve is significantly influenced by the type of leakage gas flow removal. Fig. 7.35 compares, for example, the ξ_{ZR} curves of a valve operating with (a) and without (b) leakage gas discharge at an applied pressure difference of $\Delta p_{ZR} = 0.267$ bar. Curve (c) shows the behavior at $\Delta p_{ZR} = 0$.
- With a large Δp_{ZR} and direct bulk solid removal from a container filled with bulk material, as already indicated above, during operation without leakage gas removal, the high flow resistance of the bulk material can lead to an overpressure above the rotary valve that supports bridge formation and interrupts the flow of bulk material.

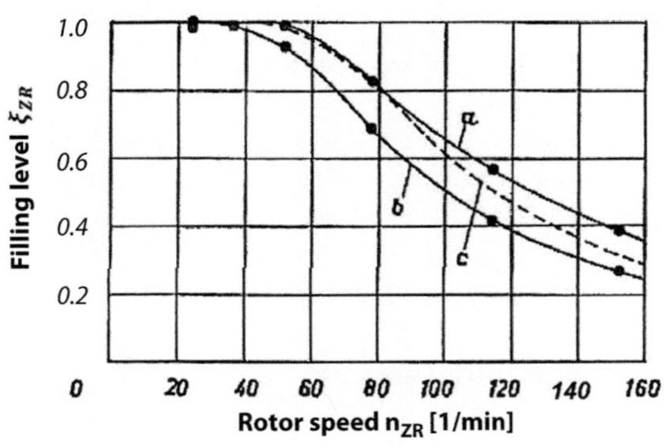

Fig. 7.35 Influence of leakage gas discharge and rotor speed of the rotary valve on its filling degree, $D_{ZR} = 0.20$ m, rapeseed, $d_S = 2.0$ mm, $\varrho_{SS} = 675 \text{kg/m}^3$, **(a)** $\Delta p_{ZR} = 0.267$ bar with leakage gas removal, **(b)** $\Delta p_{ZR} = 0.267$ bar without leakage gas removal, **(c)** $\Delta p_{ZR} = 0$ bar without leakage gas removal, according to [20]

On the other hand, the leakage gas flow, if the bed height is not too large, is also capable of converting fine-grained solids into a liquid-like and thus flowable state.
- The type of bulk material dosing significantly determines the bulk material condition above the rotary valve and thus also the required method of leakage gas removal. A pre-dosed feed, with the correct design of the rotor size, does not cause any bulk material accumulation in front of the valve, while both dosing via variable/adjustable inlet cross-sections and bulk material intake according to the characteristic curve of the valve, (7.38) or Fig. 7.34, cause accumulation above the rotary valve.

a) Scooping losses

From a gas balance around the discharge area of a rotary valve (= difference of the gas masses of an emptied and on p_R pressurized rotor chamber and the previously with bulk material filled chamber at pressure p_0), the mass flow of the scooping loss follows:

$$\dot{m}_{F,\text{Schöpf}} = n_{ZR} \cdot \Delta V_{ZR} \cdot \varrho_{F,R} - \left(n_{ZR} \cdot \Delta V_{ZR} - \frac{\dot{m}_S}{\varrho_P} \right) \cdot \varrho_{F,0} \qquad (7.46)$$

The term $(\dot{m}_S/\varrho_P) = \dot{V}_{SS} \cdot \varrho_{SS}/\varrho_P$ in (7.46) corresponds to the solid volume flow in the downward rotating rotor chambers, which must be additionally filled with gas in the emptied ascending chambers. With $\dot{V}_{SS} = \Delta V_{ZR} \cdot \xi_{ZR} \cdot n_{ZR}$ results from this:

$$\dot{m}_{F,\text{Schöpf}} = n_{ZR} \cdot \Delta V_{ZR} \cdot \left((\varrho_{F,R} - \varrho_{F,0}) + \xi_{ZR} \cdot \frac{\varrho_{SS}}{\varrho_P} \cdot \varrho_{F,0} \right) \qquad (7.47)$$

The conversion to the state p_0, i.e., the gas flow rate $\dot{V}_{F,\text{Schöpf},0}$, at the lock inlet or at the entrance to the leakage discharge line, can be carried out using the ideal gas law, (2.3). If the filling degree $\xi_{ZR} = 0$ is set in (7.47), it provides the extraction losses for pure gas conveying.

b) Gap losses

These consist of radial and axial losses. Calculating the generally significantly larger radial losses with acceptable accuracy is only possible with relatively high effort: The sequence of rotor blades and chambers can be treated analogously to that of a labyrinth seal. The gas pressure loss of a single gap between the rotor and stator results from the sum of the inflow losses into the gap, the friction losses during its flow-through, and the losses due to gas expansion in the free space behind the gap. In this space, the bulk material to be introduced is located on the downward-rotating rotor side, while the upward-rotating rotor half is empty. Depending on the lock design, two to four sealing gaps on each rotor side are simultaneously engaged. Due to the centrifugal acceleration, the bulk material in the downward-rotating rotor chambers lies against the stator wall and must be flowed through by the radial gap gas. Coarse-grained solids with $d_S > \Delta s_{ZR}$ accumulate in front of the gaps, while fine-grained solids with $d_S < \Delta s_{ZR}$ can pass through them. Therefore, a two-phase gas/solid flow of fine-grained solids through the gaps occurs, with a generally significantly higher pressure loss than that of the gas alone and, given Δp_{ZR}, thus a smaller leakage gas flow, see Fig. 7.36. All of this leads to the leakage gas flows through the radial gaps differing on the downward and upward-rotating rotor sides, the difference being greater the finer the respective solid is.

The calculation is further complicated by the fact that the intermediate pressures p_i in the individual rotor chambers, which determine the pressure differences $(p_{i+1} - p_i)$ at the intermediate blades can only be determined iteratively. The prerequisite for this is the condition $\dot{m}_{F,\text{Spalt}} = \text{konst.}$ along the flow path. With the usual or targeted pressure differences $\Delta p_R \cong \Delta p_{ZR} \gg 0.5\,\text{bar}$ the gas must be treated as compressible. Furthermore, when the local critical pressure ratio $(p_{i+1}/p_i)_{\text{crit}}$ is undercut, the speed of sound is reached in the gap. This changes the calculation approaches. In the case of frictionless gap flow and diatomic gases, the $(p_{i+1}/p_i)_{\text{crit}} \cong 0.53$ ($\rightarrow$ without considering possibly transported solids). Values for gas flows with friction can be found in Fig. 7.8. The highest gas velocity always occurs at the outlet of the last gap before the leakage gas removal or, if this is missing, before the bulk material inlet. Details can be found in [20, 23].

To model the axial gap losses, factors such as rotor type, type of axial seal, and other details of the rotary valve design, e.g., a possibly existing bearing seal by barrier gas, must be considered. This is generally only possible with significant idealizations. Overall, the sum of the gap losses can only be calculated theoretically with sufficient accuracy using sophisticated calculation programs. Empirical models, such as those presented in [24, 25, 26], are often rotary valve-specific and relatively inaccurate. It is considerably simpler, safer, and more cost-effective to rely on measured leakage gas curves $\dot{m}_{F,\text{Leck}}(\Delta p_{ZR})$, which are generally also provided by the rotary valve manufacturers. These are always pure gas measurements. It should be questioned how the measurements were taken and whether company-specific safety factors are included in the results, see the following section c) for comparison.

c) Leakage gas measurement

Leakage gas measurements without bulk material can be carried out with a stationary or rotating rotor. With a stationary rotor, the two extreme rotor positions with the minimum and maximum number of sealing blades, i.e., the blades that are minimally/maximally

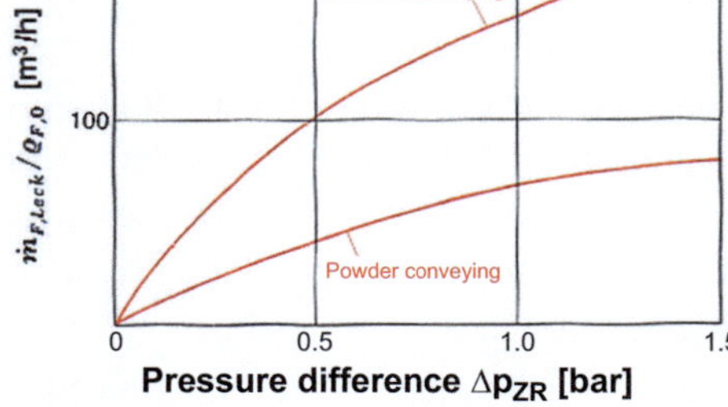

Fig. 7.36 Size of leakage gas flows of a rotary valve during granulate and powder conveying, $D_{ZR} = 250\,\text{mm}$, $\Delta V_{ZR} = 8.0 \cdot 10^{-3}\,\text{m}^3$, $n_{ZR} = 30\,\text{min}^{-1}$, according to [22]

outside the inlet and outlet cross-sections of the stator in engagement, must be examined at different pressure differences Δp_{ZR}. Scooping losses do not occur in this case, only gap losses are determined.

The results of measurements of a rotor rotating at different speeds n_{ZR} are between the min./max. limit values of the non-rotating rotary valve or should be within these limits. The leakage gas flow of the clean gas (RG) measured at various combinations of $(n_{ZR}, \Delta p_{ZR})$, $\dot{m}^{(RG)}_{F,\text{Leck}} = (\dot{m}^{(RG)}_{F,\text{Spalt}} + \dot{m}^{(RG)}_{F,\text{Schöpf}})$, must still be converted to the real operating case with bulk material. In [21, 25, 26] this is done using the approach:

$$\dot{m}_{F,\text{Leck}} = B_S \cdot \dot{m}^{(RG)}_{F,\text{Leck}} \quad \rightarrow \quad \text{when: } \dot{m}_{F,\text{Schöpf}} \ll \dot{m}_{F,\text{Spalt}} \tag{7.48}$$

B_S is a "blocking factor" dependent on the particle size of the solid to be introduced, which can be taken from Fig. 7.37. The determination of B_S was carried out on a rotary valve, in which the leakage gas flow was discharged through the bulk material inlet by an approximately 1.0 m high inflowing bulk material column [26]. Since the gas sealing is very high in this case, it obviously leads to smaller blocking factors B_S than with a separate leakage gas discharge: In this case, larger B_S values are to be expected. Fig. 7.37 confirms the statement of Fig. 7.36 that the leakage gas flow $\dot{m}_{F,\text{Leck}}$ decreases with the introduction of finer-grained products. (7.48) applies to the case where $\dot{m}_{F,\text{Schöpf}} \ll \dot{m}_{F,\text{Spalt}}$ is. Due to the indicated and also visible from Fig. 7.37 uncertainty of the B_S-value and considering the fact that the measured characteristics $\dot{m}^{(RG)}_{F,\text{Leck}}(n_{ZR}, \Delta p_{ZR})$ of a supplier always only describe the new condition of a lock, sufficient safety margins for the leakage gas flow and thus also for the pressure generator design should be planned. This is especially true for conveyances close to the minimum speed $v_{F,A,\text{min}}$. $B_S = 1$ generally provides safe designs.

When using so-called high-pressure rotary valve locks, see Sect. 7.3.4, the boundary condition specified for (7.48) is no longer applicable. Here, B_S should be applied solely to the gap losses. For the conversion of the clean gas losses measured with a rotating rotor to the operating condition with bulk material, the following applies:

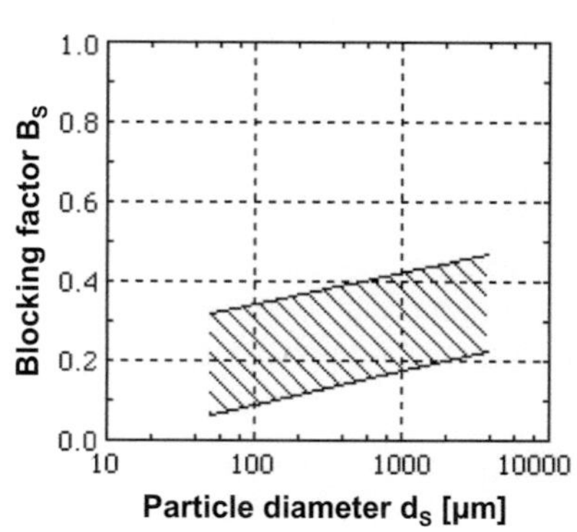

Fig. 7.37 Blocking factor $B_S = \dot{m}_{F,\text{Leck}}/\dot{m}^{(RG)}_{F,\text{Leck}}$ for bulk material injection, $p_0 = 1.0$ bar, $\Delta p_R \leq 1.0$ bar, according to [25, 26]

$$\begin{aligned}\dot{m}_{F,\text{Leck}} &= B_S \cdot \dot{m}^{(RG)}_{F,\text{Spalt}} + \dot{m}_{F,\text{Schöpf}} = B_S \cdot \left(\dot{m}^{(RG)}_{F,\text{Leck}} - \dot{m}^{(RG)}_{F,\text{Schöpf}}\right) + \dot{m}_{F,\text{Schöpf}} \\ &= B_S \cdot \dot{m}^{(RG)}_{F,\text{Leck}} + (1 - B_S) \cdot n_{ZR} \cdot \Delta V_{ZR} \cdot (\varrho_{F,R} - \varrho_{F,0}) + \frac{\dot{m}_S}{\varrho_P} \cdot \varrho_{F,0}\end{aligned} \tag{7.49}$$

$\dot{m}^{(RG)}_{F,\text{Schöpf}}$ is calculated with $\xi_{ZR} = 1$ and $\dot{m}_{F,\text{Schöpf}}$ with the current ξ_{ZR} is calculated from (7.47). Since the rotor speed n_{ZR} influences the bulk material behavior in the rotor chambers through the generated centrifugal force, n_{ZR} should actually be included as an additional parameter in the $B_S(d_S)$-diagram.

Fig. 7.38 shows an experimental setup for determining the clean gas losses $\dot{m}^{(RG)}_{F,\text{Leck}}(n_{ZR}, \Delta p_{ZR})$ of rotary valves. Explanations are provided in the figure caption. Measurements with bulk material, i.e., under operating conditions, are much more complex and require, for example, knowledge of the gas flows supplied to the system by the pressure generator and those leaving the conveying line. Their difference corresponds to the leakage gas flow $\dot{m}_{F,\text{Leck}}$. A direct $\dot{m}_{F,\text{Leck}}$-measurement is more difficult.

In Fig. 7.39 and 7.40, exemplary measurement results are shown, which demonstrate the influence of the radial gap width Δs_{ZR} and the number of rotor vanes on the clean gas leakage flow of a rotary valve $D_{ZR} = 200\,\text{mm}$.

d) Leakage gas removal

When designing a system for leakage gas removal before the bulk material inlet, some bulk material and plant-specific details must be taken into account. These are discussed below.

The leakage gas extraction shown schematically in Fig. 7.30 draws the gas flow around the entire lock inlet. In principle, this is correct, as the gas flows in from all sides

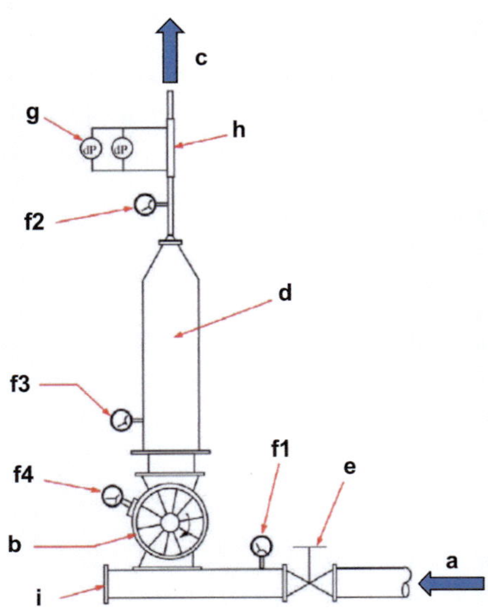

Fig. 7.38 Experimental setup for measuring the clean gas leakage losses of a rotary valve, according to [25]. a—Gas supply, b—Rotary valve, c—Leakage gas outlet, d—Chamber for damping pressure fluctuations, e—Valve for pressure adjustment, f1, f2, f3, f4—Pressure sensors, g—Differential pressure measurement, h—Gas flow rate measurement, i—Blind flange before conveying line

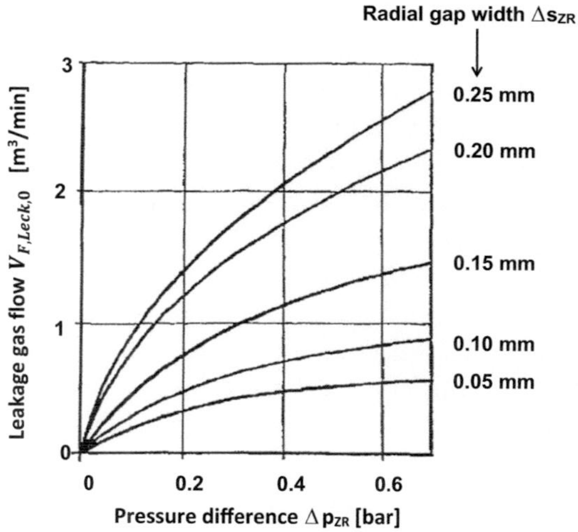

Fig. 7.39 Influence of the radial gap width Δs_{ZR} on the leakage gas flow, $p_0 = 1.0\,\text{bar}$, $T_0 = 15\,°\text{C}$, $D_{ZR} = 200\,\text{mm}$, no special radial and axial seals, according to [26]

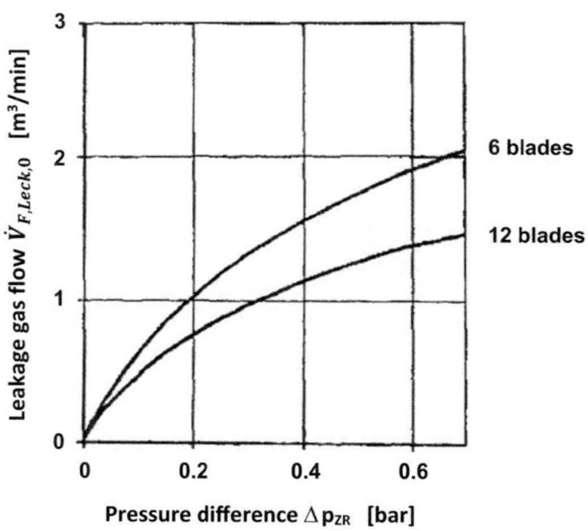

Fig. 7.40 Influence of the number of rotor blades on the leakage gas flow, $p_0 = 1.0\,\text{bar}$, $T_0 = 15\,°\text{C}$, $D_{ZR} = 200\,\text{mm}$, no special radial and axial seals, according to [26]

of the rotor. However, in practice, this design variant can only be used without problems for the feeding of coarse-grained, predominantly monodisperse bulk materials, such as plastic granules. With finer-grained products, such as cement, a short circuit between the bulk material feed and the leakage gas discharge leads to a considerable discharge of fine material through the leakage gas line from the lock. This results in a solid circulation, or its fine fraction, which drastically reduces the filling degree ξ_{ZR} of the rotary valve, partly returning through the leakage gas discharge line to the pre-container or the feeding plant area back into the lock. Measurements in [27] show that when conveying blast furnace slag cement, $d_{S,50} \cong 13.3\,\mu\text{m}$, $\varrho_S \cong 3070\,\text{kg/m}^3$, approximately 30 wt.-% of the solid actually transported through the conveying line was returned to the lock inlet via the leakage gas line.

For fine-grained solids, an arrangement of the leakage gas extraction, as shown in Fig. 7.41, is recommended. Between the trailing edge of the leakage gas discharge in the rotor rotation direction and the leading edge of the bulk material inlet, at least one rotor vane must always seal. This means that the leakage gas extraction does not capture all gas streams flowing from p_R to p_0, but it prevents the more problematic circulation of bulk material.

For pre-dosed bulk material flow and dense phase conveying, a venting container is often connected upstream of the rotary valve, similar to the procedure with screw locks, see Fig. 7.42. Nevertheless, especially for fine-grained products, the leakage gas should also be extracted separately on the rising rotor side, as schematically shown in Fig. 7.41, to prevent its expansion against the incoming bulk material in the container free space. The displacement gas flow induced by the solid feed cannot be prevented in this way. Venting containers can be equipped with their own mounted exhaust gas filter. This is also practiced with screw locks, see Fig. 7.27 and take note of the different rotation directions of screw and rotor.

The leakage gas discharge line should be designed as a dilute phase pneumatic conveying line for the transport of dusty solids/solid fractions, i.e. with gas velocities in the range of $v_{F,\text{Leck}} \cong (10-15)\ \text{m/s}$. The required pipe diameter D_{Leck} can then be calculated with the selected operating speed $v_{F,\text{Leck}}$ from the expected leakage gas volume flow $\dot{m}_{F,\text{Leck}}/\varrho_{F,0}$.

7.3.3 Drive Power

If the rotary valve removes the bulk material from a filled storage container, then the rotor blades must shear off the overlying bulk material column against its internal friction

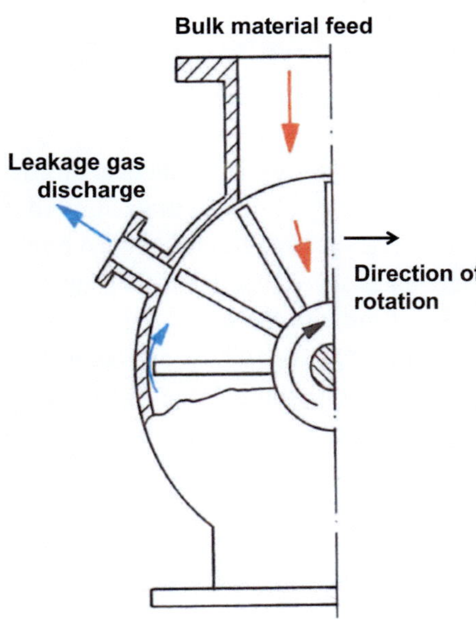

Fig. 7.41 Recommended arrangement of leakage gas extraction for fine-grained bulk materials

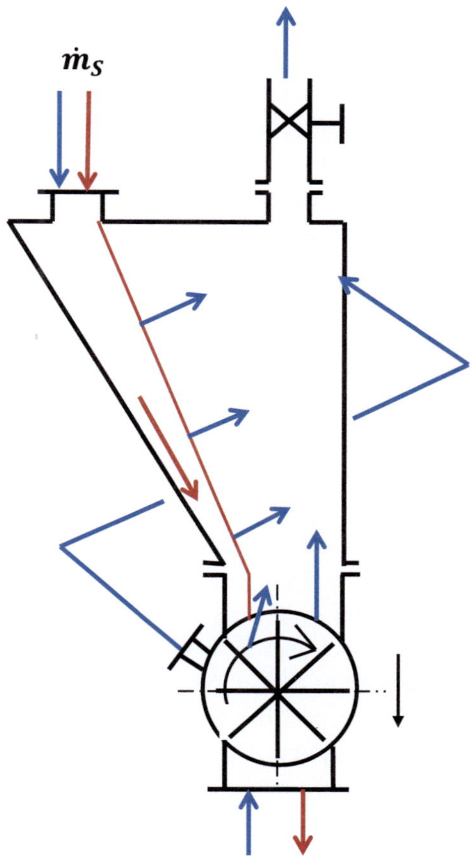

Fig. 7.42 Leakage gas extraction for pre-dosed bulk material flow, *red* = solid, *blue* = gas

coefficient $\mu_{SF} = \tan \varphi_{SF}$, see Sect. 2.2.4, and simultaneously accelerate the bulk material in the circumferential direction to $u_{ZR} = D_{ZR} \cdot \pi \cdot n_{ZR}$. The shearing force can be neglected for pre-dosed, loosened product feed. Due to the centrifugal force, the solid is pressed against the stator wall on the downgoing rotor side. The activated wall friction coefficient $\mu_W = \tan \varphi_W$ must be overcome by the rotary valve drive. The rotor rotates on its downgoing side, starting from the pressure p_0, against the higher pressure p_R, but is supported on the rising side by this pressure difference. Both forces approximately compensate each other. The total drive power P_{ZR} of a rotary valve results from the power components required to overcome the mentioned resistances plus the idle loss of the valve.

Drive powers of rotary valves are larger than those for switching the fittings of a pressure vessel system, but significantly lower than, for example, for driving screw feeders. Example: Conveying of $\dot{m}_S = 50\,\text{t/h}$ cement, conveying pressure difference $\Delta p_R \cong 1.0\,\text{bar}$, ambient pressure $p_0 = 1.0\,\text{bar}$, measured average driving power of a screw feeder: $P_W = 30.2\,\text{kW}$, installed driving power of a rotary valve lock: $P_{ZR} = 2.2\,\text{kW}$. Depending on the lock size, the driving power requirement of rotary valve locks is on the order of $P_{ZR} \cong (1.5{-}7.5)\,\text{kW}$.

7.3.4 High-pressure Locks

For standard locks without special sealing systems, the usable pressure difference Δp_R on the conveying line is limited due to the leakage gas losses increasing with the conveying pressure p_R and/or the also increasing lock wear. Example: Coarse-grained, low-abrasive bulk materials with a narrow particle size distribution, e.g. plastic granules, can economically be conveyed up to $\Delta p_{ZR} \cong 1.0$ bar, fine-grained abrasive products, e.g. fly ashes, can be transported for wear reasons only up to $\Delta p_{ZR} \lesssim 0.3$ bar. In both cases, an increase in the conveying pressure p_R would lead to a reduction of the pipe diameter D_R and in general also of the conveying gas flow $\dot{m}_F$ with all the resulting economic advantages—smaller pressure generator, smaller receiving filter, etc. Corresponding developments have taken place, with their solutions being significantly determined by the particle size distribution and abrasiveness of the currently conveyed bulk material. Thus, two principal development directions can be distinguished:

a) Improvement of sealing systems

For coarse-grained, predominantly monodisperse and low-abrasive bulk materials with particle sizes $d_S > \Delta s_{ZR}$ the axial and radial leakage sealing systems have been improved in such a way that these products can now be conveyed as standard with pressure differences up to $\Delta p_R = \Delta p_{ZR} \cong 3.5$ bar and higher with moderate gas leakages. Fig. 7.43 shows some possible design variants of axial sealing systems taken from the literature or patent literature [23, 28]. Closed rotors are required for these. The radial sealing is usually realized by means of adjustable sealing elements on the individual rotor blades. These elements are wear parts and require maintenance. Some of the axial sealing systems increase the drive power requirement of the locks.

As can be seen in Fig. 7.44, the benefits of such sealing systems can be derived: Solely through the axial sealing, the leakage gas flow $\dot{V}_{F,\mathrm{Leck},0}$ can be reduced to about half, independent of the applied pressure difference Δp_{ZR}. The radial seal brings further improvements, but will probably also disproportionately increase the maintenance effort caused by regular gap inspections and readjustment work (= plant downtime). In Fig. 7.44, clean gas measurements are shown, reference state "0" is the ambient state. Leakage gas measurements on high-pressure locks are reported, for example, in [26].

b) Increase in wear resistance

For fine-grained abrasive bulk materials with grain sizes $d_S < \Delta s_{ZR}$ the sealing systems mentioned above are unsuitable, as the solid particles not only flow through the radial gaps and wear the rotor and stator there, but also penetrate the axial seals due to their fineness and cause damage. Here, the development direction—quite obviously—is towards higher-quality = more wear-resistant stator and rotor materials, often in the form of armor, and correspondingly more complex post-processing to reduce gap widths. Materials such as tungsten carbide for weld overlays, Ni-hard with 500 Brinell hardness,

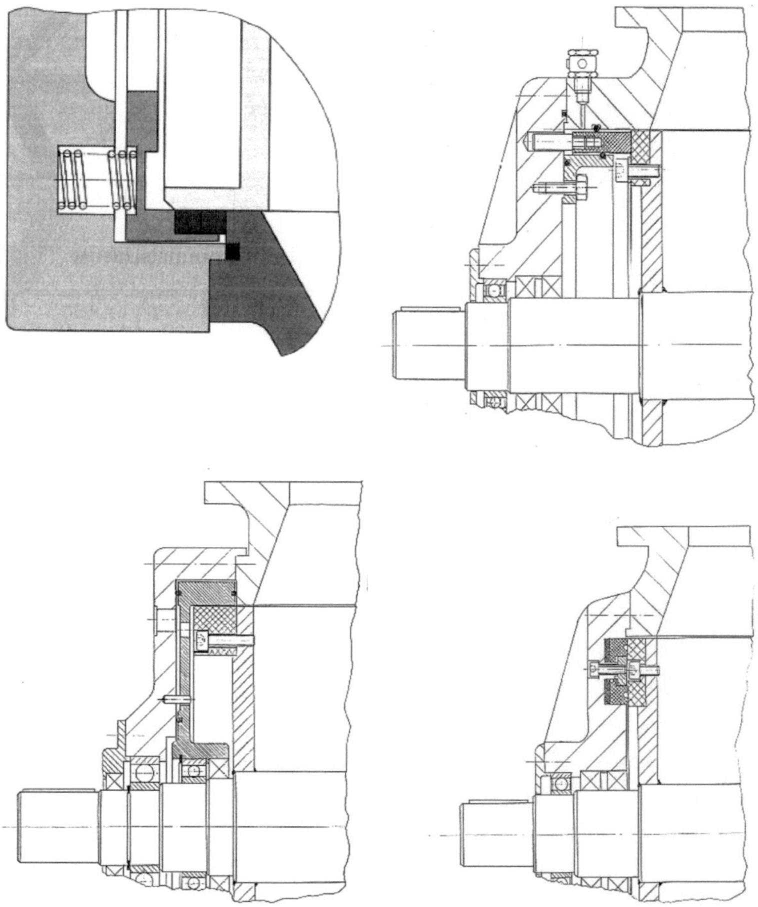

Fig. 7.43 Design variants of axial rotor seals [23, 28]

and in extreme cases, full ceramics are used, graded according to the wear requirements of the respective bulk material and standardized selection/assignment criteria for cost minimization. With such equipped rotary valves, pneumatic conveyance with pressure differences up to $\Delta p_{ZR} \cong 1.5\,\text{bar}$ for fine-grained solids can be economically realized. Fig. 7.45 and 7.46 show some design variants of the wear protection measures mentioned.

Standard for highly abrasive bulk materials are the full linings of the stator with ceramic, as shown in Fig. 7.45, where generally the ridge surfaces of the rotor are also protected with ceramic strips. Fig. 7.46 alternatively shows a special construction with a wear-resistant and externally in its gap width adjustable liner in the upper half of the valve. The lower half of the valve does not seal, but allows for the trouble-free emptying of the rotor chambers, i.e. the removal of the bulk material, see Fig. 7.47. According

Fig. 7.44 Impact of various leakage sealing systems with the same rotary valve and identical operating conditions, according to [22]

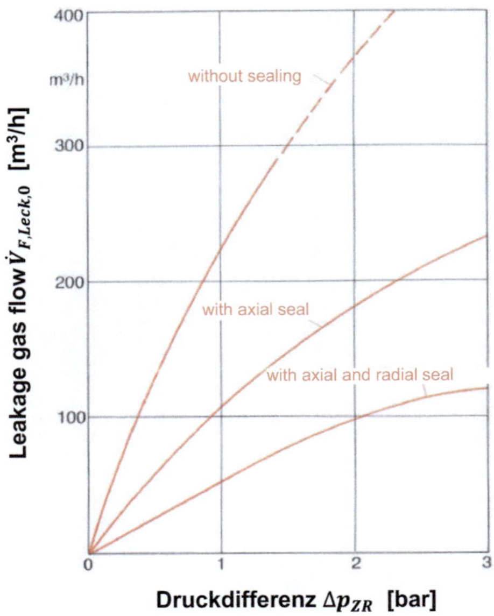

to the author's experience, this design can be economically used up to $\Delta p_{ZR} \cong 1.0$ bar because the number of sealing blades is reduced by the operating principle.

Blow-through rotary valves equipped with the above-mentioned wear protection are increasingly also used for the introduction of fine-grained bulk materials into pneumatic conveyors up to $\Delta p_{ZR} \cong 1.5$ bar, e.g. [29]. Reason: At higher conveying pressures, these valves achieve higher filling degrees than valves with blow-through shoes, a more complete chamber emptying, and a more uniform solid discharge into the transport line.

c) Summary

The rotary valves referred to in the preceding sections as high-pressure locks are complex systems that can continuously introduce bulk materials against relatively high pressures in conveying lines. They therefore compete with pressure vessels and screw feeders: their advantage over pressure vessels is the continuous operation and more compact design, and over screw feeders, the significantly lower energy consumption. A disadvantage is the leakage gas losses increasing with the rotary valve size, i.e., the decreasing efficiency. Pressure vessels achieve, depending on the circuit variant, up to $\dot{m}_S \cong 250$ t/h, screw feeders up to $\dot{m}_S \cong 450$ t/h, high-pressure rotary valves only up to $\dot{m}_S \cong 150$ t/h. Due to the effort to limit gas leakage and wear, rotary valves are highly complicated devices that require permanent control/monitoring/maintenance. All of this is reflected in the price and maintenance costs. Only an economic efficiency analysis, taking into account the current boundary conditions, can decide on the suitability of the most appropriate lock or conveying system for a given task.

Fig. 7.45 Wear protection designs of rotary valves: **a** Housing with Al_2O_3-ceramic lining in monolithic design, wall thickness (10–15) mm, currently producible up to $D_{ZR} \cong 250\,\text{mm}$, **b** Rotor blade surface protected by (10–15) mm tungsten carbide welding, **c** Stator lining with seamlessly glued and post-processed ceramic rods [29]

7.3.5 Practical Aspects

Below are some tips for the effective design of a rotary valve conveying system:

- The optimal number of rotor chambers/blades with regard to gas sealing and resulting chamber geometry is $n_{ZR} = 12$. Larger numbers of blades do indeed improve sealing, but lead to relatively narrow ($\rightarrow$ opening angle $\alpha_{ZR} < 30°$) and deep chambers, in the lower area of which the bulk material can mechanically wedge or settle due to cohesion and rotate with the rotor. This reduces the usable rotary valve volume.

Fig. 7.46 Rotary valve with Ni-hard or ceramic liner and ceramic-protected rotor [30]

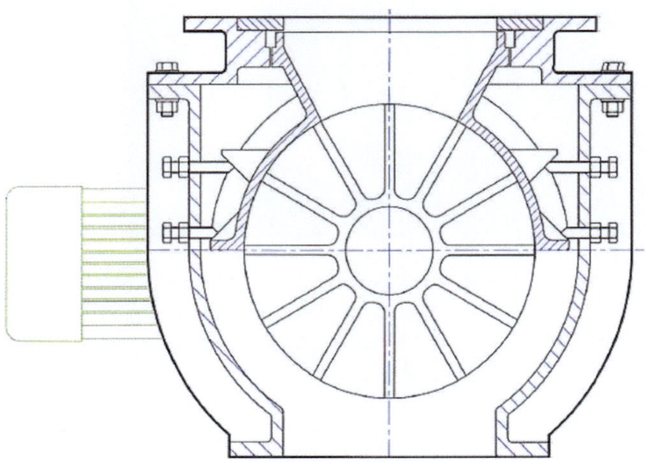

Fig. 7.47 Rotary valve design with adjustable liner according to Fig. 7.46 [30]

- According to investigations in [21], the arrangement of the rotor axis relative to the conveying line axis has no influence on the discharge/operating behavior of locks with blow shoes/feed tees. Also the direction of rotation of the rotor relative to the flow direction of the conveying gas is also said to be without influence. Nevertheless, it seems advisable to impart a velocity component in the flow direction of the conveying gas to the solid when it exits the lock. For this purpose, the rotor axis would have to be arranged transversely to the axis of the conveying line and the rotor itself would have to rotate below its center axis in the direction of gas flow. This process is supported by an appropriate constructive design of the blow shoe, see, for example, Fig. 7.30.

- The blow shoe should have a smooth gas passage without internals, guide plates, etc., and an inlet diameter equal to the subsequent conveying pipe diameter.
- At the rear edge of the bulk material inlet opening in the direction of rotor rotation, the so-called "choping" can occur due to the jamming of coarser particles between the rotor and stator. Depending on the type of solid, this leads to cutting or crushing of the grains and possibly also to blocking of the rotor. Both the resulting change in the bulk material and the local mechanical, often wear-inducing stress on the lock body are undesirable. To avoid choping, covering the rotor with a plate with a V-shaped inlet opening ($\rightarrow$ tip of the V points in the direction of rotation) has proven effective. Alternatively, a plate covering the rear half of the inlet cross-section with a diagonal inlet edge can be used.
- By slowly rotating rotary valves with axially parallel rotor blades, the bulk material is introduced more or less in a thrust-like manner into the conveying section. If this is not permissible, e.g., in the case of reactor charging, a metal sheet with an inclined discharge edge must be installed below the rotor or a rotor with blades inclined relative to the axis must be used.
- When using rotary valves with sealing gas systems for sealing the shaft bearings, the sealing gas pressure and its volume flow should be automatically adjusted according to the current conveying pressure at the beginning of the pipeline as a reference variable. This is particularly necessary in the case of frequent load changes/throughput changes.
- The feeding rotary valves in pneumatic transport of explosive dusts can simultaneously serve as flame barriers for decoupling the different plant areas. A prerequisite is an appropriate certification/approval and the system-related interlocking of these valves.

Note: The filling degree ξ_{ZR} of a rotary valve, as defined by (7.39), corresponds, as the following equation (7.50) shows, to the ratio of the currently introduced bulk material mass flow $\dot{m}_S$ to its maximum possible value $\dot{m}_{S,\max}$:

$$\xi_{ZR} = \frac{\dot{V}_{SS}}{\dot{V}_{ZR}} \cdot \frac{\varrho_{SS}}{\varrho_{SS}} = \frac{\dot{m}_S}{\dot{m}_{S,\max}} = \frac{\dot{V}_{ZR} \cdot \varrho_b}{\dot{V}_{ZR} \cdot \varrho_{SS}} = \frac{\varrho_b}{\varrho_{SS}} \tag{7.50}$$

In the resulting equation $\xi_{ZR} = \varrho_b/\varrho_{SS}$, the bulk density in the rotor chambers $\varrho_b = \Delta m_S/\Delta V_{ZR}$ is, in the first approximation, equal to that of the bulk material mass flow supplied to the rotary valve = inlet density.

7.4 Injector Feeder

Table 7.1 shows that the application range of injector feeders is pneumatic conveying with relatively low solid throughput over short conveying distances. Of the numerous design variants—including driving gas supply via central, ring or multi-nozzles, low to high pressure and suction injectors, see e.g. [22, 31, 32]—injectors with central nozzle and compressed gas supply via rotary lobe blowers, pressure before the driving gas nozzle $p_{\text{vor}} < 1.0\,\text{bar (g)}$, are mainly used in industry. Fig. 7.48 shows the usual design for the injection of fine-grained bulk materials. Their entry is supported by a bottom aeration/fluidization, which is supplied via a manually adjustable gas bypass guided past the driving nozzle. For coarse-grained products, this is not useful due to the required high gas velocities. In the following, only injector feeders with central nozzle for entry into positive pressure conveyances are considered. Basic investigations and measurements on such systems are reported in [31, 33, 34].

7.4.1 Operating Principle

The operating principle of an injector feeder largely corresponds to that of a jet pump: A gas stream $\dot{m}_{F,T}$ compressed to the pre-pressure p_{vor} is fed to a driving nozzle with the outlet diameter D_T and relaxed in this to the pressure p_T which is equal to or slightly smaller than the pressure p_0 in the upstream feed hopper/bulk material feeding system. Since generally only convergent nozzles are used, this, depending on the chosen pressure gradient (p_{vor}/p_T), results in exit velocities in the range of $v_{F,T} \cong (100\,\text{m/s} < \text{sound speed}\ a_{\text{Sch}})$, that is, pressure energy is converted into kinetic energy. The gas jet exiting the nozzle behaves analogously to a free jet [32]. Both the supplied solid and the gas surrounding the jet are accelerated in the conveying direction. The gas acceleration occurs through the shear forces at the jet surface (→ velocity difference), while the solid particles, depending on size/particle mass, penetrate more or less deeply into the jet and are driven there by their flow resistance. The horizontal driving gas jet is increasingly decelerated on its upper side and deflected downwards with increasing loading μ [31]. The opening angle α_T of the jet also shows a dependence on μ: According to the results in [35, 36], in the working range of interest here, however, an average of $\alpha_T \cong 10°$ can be used. For comparison: the opening angle of a free gas jet is $\alpha_{\text{Freistr.}} \cong 18°$.

The highly turbulent gas/solid jet is fed through the so-called catching nozzle, section (A-1) in Fig. 7.48, the subsequent mixing section, section (1–2) in Fig. 7.48, with the diameter D_M and the length L_M. Since the driving nozzle is usually designed to be axially

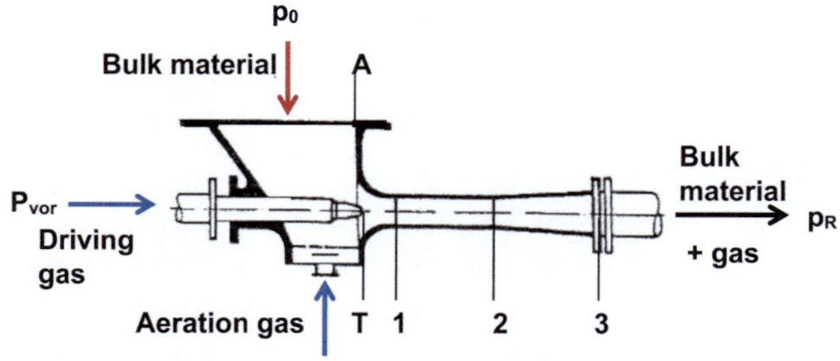

Fig. 7.48 Injector feeder with central nozzle: A—Inlet catching nozzle, T—Outlet driving nozzle, 1—Entry mixing section, 2—Entry diffuser, 3—Entry conveying line

adjustable in practice, it depends on the position of the nozzle outlet where the jet hits the catching nozzle or mixing section wall. This is significant because the gas/solid mixture in the conveying direction behind the impact point causes additional wall friction, which reduces the possible pressure build-up. The driving nozzle position shown in Fig. 7.48 corresponds to the one preferred in practice for conveying fine-grained bulk materials, e.g., cement: driving nozzle outlet $T \cong$ catching nozzle inlet A, distance $(A - T) = a \cong 0$. For coarser products, e.g. plastic granules, the nozzle is generally retracted somewhat further. If this distance is so large that the expanded jet diameter is larger than the catching nozzle or mixing tube diameter, gas is blown out of the feed hopper. As an optimal distance $(A - T)$ is proposed in [33] the value $a_{\text{opt}} \cong 2 \cdot D_T$.

Due to the reduction of the average gas jet velocity due to jet expansion until wall contact, kinetic energy is converted back into pressure energy. The shares for solid and gas acceleration must be deducted from the theoretically maximum possible pressure build-up. In addition, the energy conversion is subject to losses due to the addition of solids and the turbulence caused by the high driving gas velocities in the area of the feed hopper and the mixing section. Behind the wall contact point, pressure losses due to wall friction additionally reduce the pressure build-up. In the mixing tube, the driving jet mixture is converted into a pipe flow. Due to the high turbulence, a relatively uniform solid distribution across the pipe cross-section at the mixing tube end can be assumed. The gas flow through the mixing tube $\dot{m}_F$, i.e. the driving gas flow $\dot{m}_{F,T}\pm$ the gas stream sucked in/blown out through the inlet funnel $\dot{m}_{F,A}$, leaves the mixing section with the velocity $v_{F,2}$ and enters the subsequent diffuser, see Fig. 7.48. In this, it is decelerated to the velocity $v_{F,3}$ with further pressure recovery. Diffusers are usually designed with opening angles of $\alpha_D \cong 7°$. The conversion to pressure takes place here almost without loss.

Under comparable operating conditions, fine-grained solid matter in the mixing section of a given injector is accelerated to higher speeds u_S than coarse product. Cause: The coarse grain needs a longer acceleration path to reach a certain speed, but this is not available. For fine-grained products, this results in a lower pressure build-up in the injector due to the required greater acceleration work than for coarse-grained bulk material. Fig. 7.49 shows the velocity and pressure profiles of gas, fine-grained, and coarse-grained solids along an injector feeder. If at the end of the diffuser the value of $u_{S,3} > v_{F,3}$ is, there is a pressure increase in the conveying pipe behind the diffuser outlet 3, due to the not yet completed deceleration of the solid matter, which is generally negligibly small [37].

From the consulted literature [31, 33, 34, 35, 36, 37], in addition to the already mentioned boundary conditions, the following operating and geometric variables for a first rough design of injector feeders can be derived:

- $L_M/D_M \cong 2.5{-}4.0$.
- Maximum possible load: $\mu_{\text{max.}} \cong 6$, in practical operation: $\mu \leq 4$.
- The maximum pressure increase for the low and medium pressure injectors considered here, $p_{\text{vor}} < 2.0\,\text{bar}$, $p_0 \cong 1.0\,\text{bar}$, is approximately $\Delta p_R = (p_3 - p_0) \cong 0.20\,\text{bar}$ during neutral operation, i.e., without gas suction via the feed hopper. With increasing suction flow $\dot{m}_{F,A}$, Δp_R decreases.

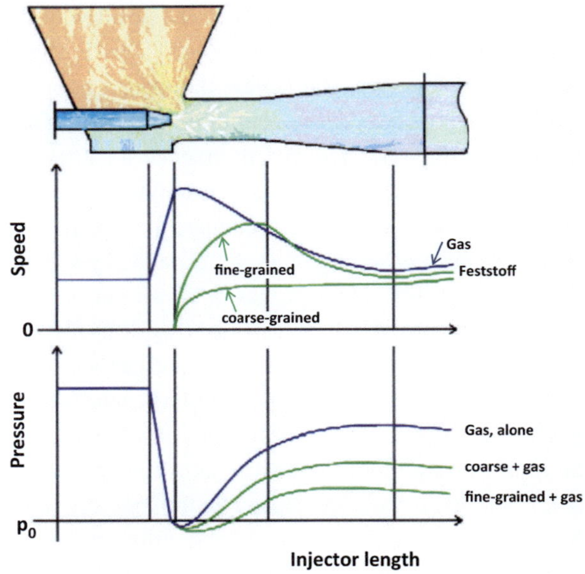

Fig. 7.49 Velocity and pressure profiles of gas as well as fine- and coarse-grained solids along an injector lock

Since all geometry and operating data listed in this section mutually influence each other, there is no one-time optimal design valid for all applications, but it must be found anew for each task. In practice, this encounters certain difficulties because there are injector series with fixed dimensions, in which only the respective driving nozzle can adapted to the specific task.

In Fig. 7.50, some exemplary measurement results from [33] are plotted, which in the lower part of the image, starting at the nozzle outlet, show the build-up of the static gas pressure along the investigated injector and in the upper part the local velocity ratio $(u_S/v_F) < 1$ for different gas flow rates. The conveyed material in this case was a coarse-grained polyethylene granulate. The position designations on the abscissa correspond to those in Fig. 7.48.

7.4.2 Calculation Approaches

Although a bulk material injector does not contain any moving/rotating components, the description of the processes taking place within the device is complex due to gas compressibility, asymmetric bulk material feeding, changing flow cross-sections, and the resulting state changes. Therefore, only the basic principles are presented below. Subsequently, practical designs will be discussed.

For the calculation, the injector is divided, for example, into the areas of driving nozzle, feed hopper, mixing section, diffuser, and possibly the initial area of the conveying line as shown in Fig. 7.48 [33]. The principles of momentum, energy, and mass conservation are then applied to these areas. Simplifying assumptions are required to solve

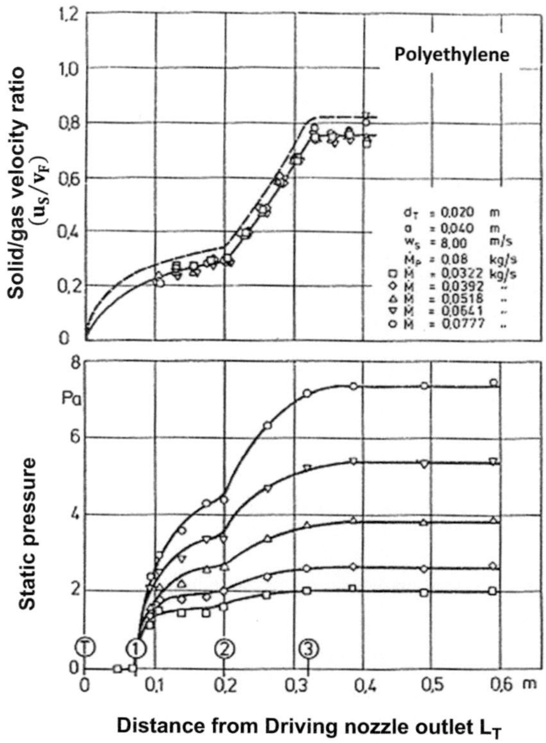

Fig. 7.50 Measured profiles of gas pressure and local solid/gas velocity ratio along an injector feeder during the conveyance of coarse-grained bulk material. The local solid velocity u_S is related to the local gas velocity v_F; bulk material: polyethylene, $d_{S,50} = 3.0\,\text{mm}$, $\varrho_P = 918\text{kg/m}^3$. It means: $d_T = D_T$, $w_S = w_T$, $\dot{M}_P = \dot{m}_S$, $\dot{M} = \dot{m}_F = (\dot{m}_{F,T} + \dot{m}_{F,A})$, according to [33]

these, which can be chosen differently depending on the person working on it, see for example [31, 33, 34]. The operating variables in the individual injector areas can be represented by suitable mean values. With this method, the state profiles along an injector are only determined pointwise at the area transitions. A continuous consideration requires the solution of corresponding differential equations, which is possible for individual grains of the solid, but not for particle clouds, under simplifications.

The driving nozzle can be described as an adiabatic-reversible, i.e., isentropic, gas flow with sufficient rounding/smoothing of the inlet in a good approximation. With the known approaches from the literature, e.g. [4, 5, 6], the mass flow rate is obtained:

$$\dot{m}_{F,T} = \xi_{\text{ges}} \cdot \frac{\pi}{4} \cdot D_T^2 \cdot \psi \cdot \sqrt{2 \cdot p_{\text{vor}} \cdot \varrho_{F,\text{vor}}} \tag{7.51}$$

with:

ξ_{ges}	Discharge coefficient $\xi_{\text{ges}} \leq 1$, here: $\xi_{\text{ges}} \cong 1$,
D_T	Exit diameter of the nozzle,
p_{vor}, $\varrho_{F,\text{vor}}$	Pressure, gas density before the nozzle,
ψ	Discharge function according to (7.52) or Fig. 7.51.

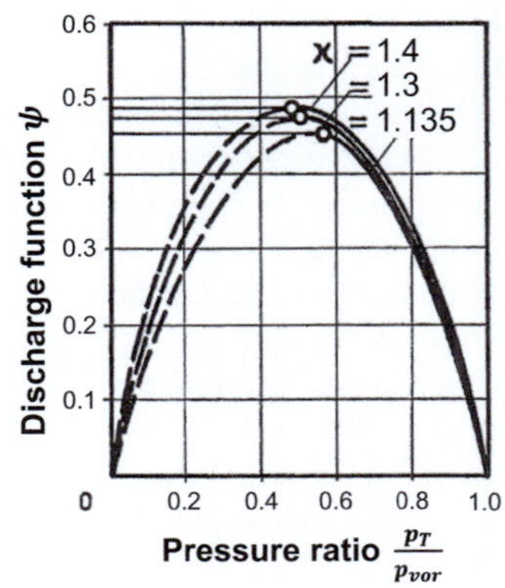

Fig. 7.51 Discharge function $\psi(p_T/p_{\text{vor}}, \kappa)$ of the driving nozzle, p_T = Pressure at the nozzle exit, κ = Adiabatic exponent

$$\psi = \sqrt{\frac{\kappa}{\kappa-1} \cdot \left[\left(\frac{p_T}{p_{\text{vor}}}\right)^{2/\kappa} - \left(\frac{p_T}{p_{\text{vor}}}\right)^{(\kappa+1)/\kappa}\right]} \tag{7.52}$$

In the case of convergent nozzles, the discharge function ψ follows the solid lines in Fig. 7.51. A maximum value $\psi_{\max}$ occurs at the critical pressure ratio $(p_T/p_{\text{vor}})_{\text{crit}}$. $\psi_{\max}$ is valid for convergent nozzles for all values $(p_T/p_{\text{vor}}) \leq (p_T/p_{\text{vor}})_{\text{crit}}$, see also Fig. 7.8 and the explanations in Sect. 7.1.1. Subcritical pressure ratios lead to sonic velocity a_{Sch} in the nozzle exit cross-section. For the gases of interest here, with $\kappa = 1.4$: $(p_T/p_{\text{vor}})_{\text{crit}} = 0.528$, $\psi_{\max} = 0.485$.

In the injector sections following the driving nozzle in the flow direction, kinetic energy of the gas is converted back into static pressure energy. This conversion is associated with losses, which can currently only be determined by measurement and taken into account by introducing appropriately defined efficiency factors η. This is to be demonstrated according to the approaches in [33] using the example of the mixing section, area (1–2) in Fig. 7.52. Its length is L_M, its diameter $D_M = D_2$. The gas jet $\dot{m}_F$ loaded with solid matter $\dot{m}_S$ and still expanding enters with the diameter D_1 through the inlet cross-section 1 and encounters after the distance $(L_m - L_f)$ on the mixing section wall.

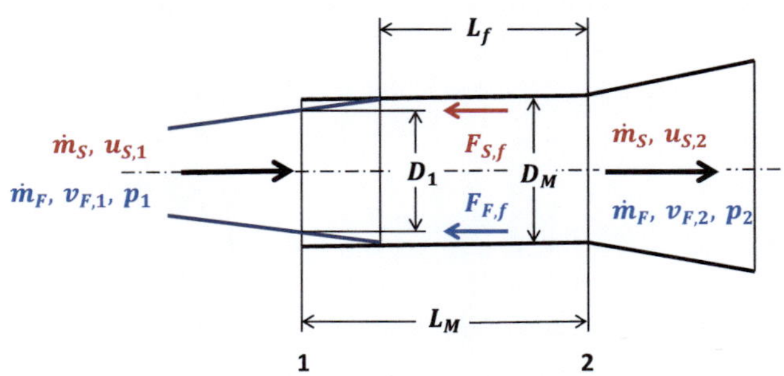

Fig. 7.52 For the calculation of the mixing section (D_M, L_M)

From the energy conservation equation, it follows with the pressure conversion efficiency η_M:

$$p_2 - p_1 = \frac{\varrho_{F,M}}{2}(v_{F,1}^2 - v_{F,2}^2) \cdot \eta_M - \left|(\Delta p_{S,B,M} + \Delta p_{S,R,M} + \Delta p_{F,M})\right| \qquad (7.53)$$

with:

$\varrho_{F,M}$	average gas density in the mixing tube,
$\Delta p_{S,B,M}$, $\Delta p_{S,R,M}$, $\Delta p_{F,M}$	pressure losses due to solid acceleration and wall friction of solid and gas in the mixing section.

The pressure losses ($\Delta p_{S,B,M}$, $\Delta p_{S,R,M}$, $\Delta p_{F,M}$) in the mixing tube can be calculated using the approaches from Sect. 4.7. The friction length is the distance L_f, acceleration length the distance L_M. Taking into account the geometric conditions and introducing the reference speed $v_{F,M} = v_{F,2} = \dot{m}_{F,T}/(\varrho_{F,M} \cdot \pi/4 \cdot D_T^2)$ results after elementary transformations in (7.54):

$$p_2 - p_1 = \frac{\varrho_{F,M}}{2} \cdot v_{F,M}^2 \cdot \left[\left(\frac{D_M}{D_1}\right)^4 - 1\right] \cdot \eta_M - \left|(\Delta p_{S,B,M} + \Delta p_{S,R,M} + \Delta p_{F,M})\right| \qquad (7.54)$$

In the same way, calculation approaches for the expanding jet = feed funnel area with the pressure conversion degree η_T and the diffuser with η_D are introduced. The η-values are highly dependent on operating conditions and also on the type of bulk material. As rough guidelines, the following values can be used: $\eta_T \cong 0.50$, $\eta_M \cong 0.70$, $\eta_D \cong 1.00$. The gas flow $\dot{m}_{F,A}$ sucked in or blown out through the inlet funnel can be roughly estimated according to [33] with (7.55).

$$\dot{m}_{F,A} \cong A_{\text{Inlet}} \cdot \sqrt{2 \cdot \varrho_{F,0} \cdot |p_T - p_0|} \qquad (7.55)$$

The injector design with the above-mentioned model is carried out iteratively, as already apparent from (7.54). It can be used to determine injector characteristics $(p_3 - p_0) = \Delta p_R = f(\dot{m}_S)$ and provides operating points that are set at different load cases by intersections with the characteristic curve of the conveying system. To compare different injectors with each other, it is useful to represent the injector characteristics dimensionless. In [31] this is done via the relationship between the so-called pressure coefficient Π and the loading μ:

$$\Pi = \frac{p_3 - p_T}{\frac{\varrho_{F,T}}{2} \cdot v_{F,T}^2} = f(\mu) \quad \rightarrow \quad \text{with: } p_T \cong p_0 \qquad (7.56)$$

Π is the static pressure difference relative to the dynamic pressure of the jet exiting the nozzle, which is built up in the injector. With increasing μ, Π decreases approximately linearly. One can also refer to Π as the overall efficiency of the injector. Other approaches, e.g. $(p_3 - p_0)/(p_{\text{vor}} - p_3) = f(\mu)$, are possible and commonly used in the industry. A simplified version of the calculation model according to [34] is presented in [37] and explained using example calculations.

If a specialist company offers injector conveying systems, a different approach to the design is necessary due to the generally already existing injector series. The size of these locks is based on the standard pipe diameters available on the market D_R, i.e., injector exit diameter $D_3 = D_R$. Depending on the bulk material class, e.g., coarse or fine-grained solids, different series with fixed geometries may be available. For a given task $(\dot{m}_S, L_R)$, the design process starts with selecting a conveying pressure that can be safely achieved by the injector, $p_R = p_3$, and a suitable conveying gas velocity $v_{F,A} = v_{F,3}$ at the beginning of the pipeline, from this the conveying pipe diameter $D_R = D_3$ and the conveying gas mass flow $\dot{m}_F$ for the maximum expected solid throughput $\dot{m}_{S,\max}$ can be calculated. The loading must meet the condition $\mu \leq 4$. Assuming a neutral driving mode of the injector, with $\dot{m}_{F,T} = \dot{m}_F$ the driving nozzle is designed for $p_T \cong p_0$, see (7.51). The nozzle inlet pressure p_{vor} can be freely chosen, and from its height, different nozzle exit velocities $v_{F,T}$ and nozzle diameters D_T result. Appropriate delivery pressure reserves take into account gas suction via the inlet funnel. Such injector designs are supported by internal design guidelines based on systematic measurements and evaluated operating experience. The geometry of the injectors is only changed in special cases.

7.4.3 Practical Application

Here are some basic guidelines:

- To buffer fluctuations in the inflow of bulk material, a pre-container on the solid inlet of the injector is recommended. The upstream bulk material feeding system, e.g., rotary valve, screw conveyor, fluidized flow channel, feeds into this from above. The buffer container should also be equipped with a check valve in the upper area, which opens when there is negative pressure and closes when there is positive pressure. A justification for this measure follows below, see also [38].
- Injectors should be designed with a slight negative pressure in the injector housing at the design throughput (= maximum solid mass flow). This leads to a permanent low gas suction via the inlet funnel or pre-container and prevents the escape of dust-laden gas. At the same time, this positively supports the solid material inflow, especially of fine-grained products.
- Without the check valve installed in the pre-container, the negative pressure in the injector housing would increase significantly at partial or low load, resulting in increased gas suction via the bulk material feeding system. This can lead to serious problems: For example, fine-grained or specifically lighter solids can be uncontrollably sucked in from the upstream feeding devices. Furthermore, increased wear on the supply devices must be expected with abrasive products. A striking example of this is simple rotary valves not designed for an adjacent pressure difference. Fine-grained

solids or solid fractions are pulled by the gas flow caused by the pressure difference through the radial and axial gaps between the stator and rotor, leading to the wear problems already described in Sect. 7.3.2.

The installed check valve provides the sucked-in gas with a flow path with low resistance, which bypasses the bulk material feeding system. The negative pressure in the injector housing remains approximately at its design value. This avoids the indicated difficulties.

- In case of short-term overload, the bulk material can accumulate in the feed hopper/pre-container. This leads to the blowing out of driving gas into the feeding plant area, accompanied by a pressure increase/positive pressure in the injector housing. The injector continues to convey and adjusts itself to a solid throughput that it can convey without back-blowing. Dust escape is prevented by closing the check valve.
- Due to their simple design and lack of moving parts, injectors can be made from almost any material, e.g., Al_2O_3-ceramics. This allows their use with highly abrasive bulk materials and/or extremely high temperatures.

7.5 Flap Locks

The following discussion focuses exclusively on flap locks suitable for use in pneumatic positive pressure conveying systems. These flap systems separate the area of the plant delivering the bulk material from the higher-pressure conveying system while simultaneously transferring the bulk material from the lower-pressure area to the higher-pressure area. As shown in Figs. 7.1 and 7.53, the operating principle consists of two valves arranged one above the other in a fixed housing, which alternately open and close. Typically, pendulum flaps, rotary flaps, or sliders are used for this purpose. These are gen-

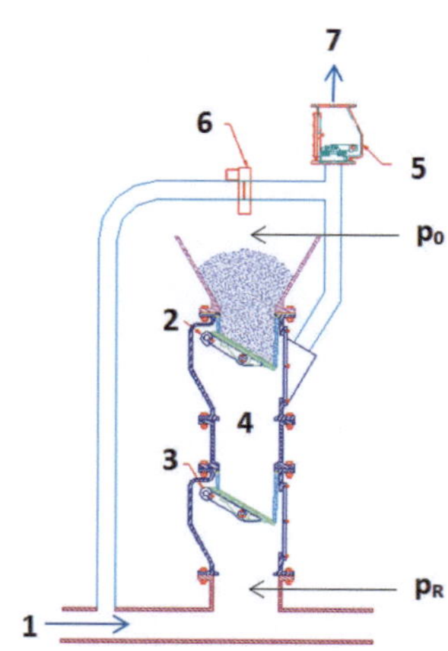

Fig. 7.53 Design variant of a flap lock for feeding bulk material into pneumatic positive pressure conveyors. The meanings are: 1—conveying gas, 2—upper flap, 3—lower flap, 4—filling volume, 5—vent valve, 6—pressure build-up valve, 7—relief/displacement gas to filter, according to [39]

erally pneumatically or mechanically actuated and controlled, e.g., by geared motors. During a lock cycle, one of the two flaps 2 and 3, see Fig. 7.53, is always closed and seals the high-pressure area against the low-pressure area. The following sequence results:

Step 1: Flap 2 opens with flap 3 closed and fills chamber 4 with bulk material.
Step 2: Flap 2 closes, then flap 3 opens and feeds the bulk material to the conveying system. The trailing bulk material accumulates above flap 2.
Step 3: After emptying chamber 4, flap 3 closes, then flap 2 opens again.

And so on.

It is clear that in the case of a simple execution of the flap lock system (i.e., without the additional piping shown in Fig. 7.53), in step 1, the gas flow displaced by the solid flows against it and disturbs its inflow. In step 2, flap 3 opens against the conveying line overpressure, and the space 4 is pressurized to the conveying pressure. In step 3, this gas volume expands against the incoming solid into the upstream plant area. The process is comparable to the intake and discharge of scoop losses in a rotary valve lock. Due to the indicated difficulties and with increasing conveying pressure, the application range of simple locks is limited to pressure differences of $\Delta p_{KS} = (p_R - p_0) \lesssim 0.25$ bar.

The measures required for the safe realization of higher pressure differences/delivery pressures are exemplarily shown in Fig. 7.53: Before flap 2 opens in step 1, the vent valve 5 is opened with the pressure build-up valve 6 closed and remains open during the filling process of the intermediate space 4. This initially leads to a pressure reduction from p_R to p_0 in chamber 4 and then to a discharge of the displacement gas to the exhaust gas filter 7. With the closing of flap 2, the vent valve 5 also closes. After that, the pressure build-up valve 6 opens and chamber 4 is pressurized to the conveying pressure p_R, before flap 3 also opens with a slight delay. During the bulk material discharge from chamber 4, the pressure build-up valve 6 remains open and supports the discharge with the partial gas flow passing through chamber 4. After that, it closes with flap 3. This process is repeated periodically with each introduced product batch.

With the last described lock method, pressure differences up to approx. $\Delta p_{KS} \leq 2.0$ bar (→ some suppliers also specify higher values) can be safely controlled. In general, the bulk material to be sluiced should be dry and flowable. It must be predosed to the flap lock or limited to a maximum inflow rate. Direct withdrawal from the full, e.g., a filled upstream silo, is not possible. The feeding of the bulk material into the conveying line takes place intermittently/batchwise. Cycle numbers/time unit of $n_{KS} \lesssim 6.0\,\text{min}^{-1}$ are usually realized.

7.6 Pneumatic Vertical Conveyors (Airlift)

For the predominantly vertical transport of particularly fine-grained bulk materials, the pneumatic vertical conveyor, also referred to as an Airlift or *gravity feeder,* is used. Fig. 7.54 shows a design variant. A continuously cylindrical container with an aeration

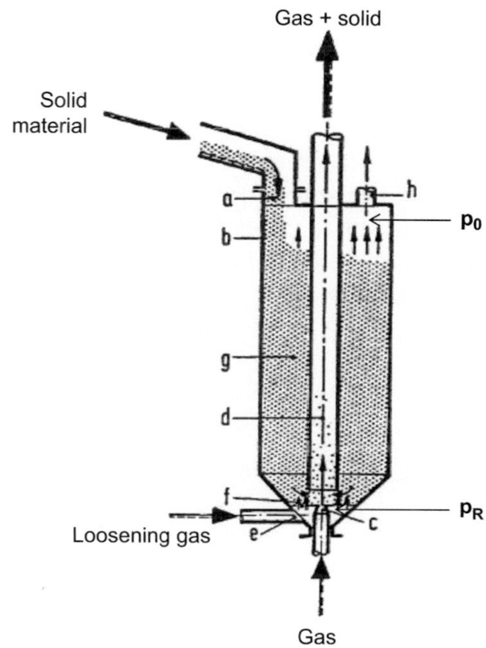

Fig. 7.54 Design variant of a pneumatic vertical conveyor (Airlift). The meanings are: a—bulk material inlet opening, b—lock vessel, c—conveying gas nozzle, d—conveying pipe, e—connection loosening gas, f—aeration floor, g—sealing bulk material layer, h—exhaust gas connection

floor as the lower closure is common. The transport gas is supplied to the conveying pipe via a central nozzle, which ends at a certain operation-specific distance below the conveying pipe inlet. Before the nozzle, an adjustable partial flow for floor aeration is branched off from this gas flow. Due to the aeration and loosening as well as the weight of the overlying bulk material, it flows towards the nozzle, is captured by the nozzle gas flow, and is introduced into the conveying line. The bulk material column in the container seals the pressure difference applied to the lock.

$\Delta p_{SF} = (p_R - p_0)$ is derived. Its height is calculated analogously (7.1) with:

$$H_{SF} = f_{SF} \cdot \frac{\Delta p_{SF}}{g \cdot \varrho_b} \quad \rightarrow \quad \text{with: } f_{SF} \cong 1.2 \tag{7.57}$$

$\varrho_b = (1 - \varepsilon_F) \cdot \varrho_S$ is the density of the moving and gasified/partially fluidized bulk material, f_{SF} takes into account, among other things, possible overloads. (7.57) results in relatively large bulk material and thus container heights. Example: With $\varrho_b = 900\,\text{kg/m}^3$ and $\Delta p_{SF} = \Delta p_R = 50000\,\text{N/m}^2$ follows $H_{SF} = 6.80\,\text{m}$. This limits the practically achievable pressure differences to maximum values of $\Delta p_R \cong (0.5{-}0.6)$ bar. The leakage gas flow $\dot{m}_{F,\text{leck}}$ through the bed can be calculated with the equations from Sects. 3.2.1 and 3.2.2 with knowledge of the flow area $A_{SF} = \pi/4 \cdot (D_{SF}^2 - D_R^2)$, see (7.58), and is generally negligibly small for fine-grained solids due to the high flow resistance.

The minimum width of the annular gap $B_{SF} = (D_{SF} - D_{R,a})/2$ between the container inner diameter D_{SF} and conveying pipe outer diameter $D_{R,a}$ is determined depending on the bulk material properties: For fine-grained cohesive/adhesive products, for example, bridging must be prevented during a time-limited interruption of the conveying without emptying the lock. In addition, practical experience shows that too high downward velocities $u_{S,SF}$ in the container can lead to (periodic) tearing of the solid flow for this bulk material class. In practice, this is avoided by specifying maximum $u_{S,SF}$ values. For fine-grained bulk materials, e.g., raw meal or cement, the following applies: $u_{S,SF} \lesssim 10\,\mathrm{cm/s}$. Cohesionless coarse-grained products, including plastic granules, can be operated at significantly higher velocities $u_{S,SF}$. The diameter of the lock container thus follows from the continuity equation to:

$$D_{SF} \geq \sqrt{\frac{4}{\pi} \cdot \frac{\dot{m}_S}{\varrho_b \cdot u_{S,SF}} + D_{R,a}^2} \tag{7.58}$$

D_{SF} is adapted to the diameter of a generally existing series. This only defines the size of the diameter, while the height of the lock chambers is designed according to the current requirements.

The gas velocities at the nozzle exit in front of the conveying pipe inlets are designed with $v_{F,T} \cong (75-105)\,\mathrm{m/s}$, the associated nozzle pressure loss is then only $\Delta p_T < 0.10\,\mathrm{bar}$, see the calculation approaches in Sect. 7.4.2. In the case of abrasive solids, the inlet area of the conveying pipe can be designed to be replaceable. A suitable check valve must be arranged below the nozzle, which prevents the penetration of bulk material into the clean gas side in the event of shutting down the system without emptying the container. In the vertical conveying pipe D_R a bulk material column of height

$$H_b = \frac{\dot{m}_S \cdot L_v}{\frac{\pi}{4} \cdot D_R^2 \cdot \varrho_{SS} \cdot \overline{C} \cdot \overline{v}_F} \tag{7.59}$$

with:

$\overline{v}_F$ average gas velocity in the conveying pipe,
$\overline{C}$ ($\cong (\overline{u}_S/\overline{v}_F)$), average relative velocity of solid/gas

forms above the non-return valve. The system must overcome its resistance when restarting. In some cases, this can only be achieved with additional measures. Typical aeration velocities for fine-grained bulk materials are in the range of $\dot{q}_{WS} \lesssim 2.0\,\mathrm{m/min}$, while for coarse-grained products, aeration is unnecessary. The conveying gas velocities at the end of the pipeline are usually set with $v_{F,E} \cong (18-22)\,\mathrm{m/s}$. This results in loadings of $\mu \lesssim 35\,\mathrm{kg}\,S/\mathrm{kg}\,\mathrm{F}$. The exhaust gas pipe "h" in Fig. 7.54 must be dimensioned for the sum of leakage and displacement gas flow. So far, airlifts for fine-grained bulk materials with throughputs up to $\dot{m}_S \cong 900\,\mathrm{t/h}$ and lifting heights up to $L_v \cong 125\,\mathrm{m}$ have been built and put into operation. Examples of applications are: filling of cement, raw meal, and alumina silos, feeding of *pre-heater* systems in the cement industry with raw meal. The conveying line can contain pipe switches for the optional feeding of different receivers/silos. Solid/gas separation at the end of the conveying line is generally carried out in two stages: a pre-separator with low pressure loss is followed by a fabric filter. Fig. 7.55 shows an example of an expansion vessel used as a pre-separator [40], in which the solid material is separated from the conveying gas by approximately 80% using a U-shaped deflection profile. The fine material separated in the subsequent filter is returned to the product stream.

At partial or reduced load, i.e., reduced bulk material feed to the lock chamber, the conveying line receives this information with a time delay and initially continues to convey at the original solid mass flow rate. As a result, the bulk material level in the lock chamber decreases until a new equilibrium height with a correspondingly smaller conveying throughput is established according to (7.57). After switching off the bulk material feed, the airlift container can be emptied largely in this way. Accordingly, overload leads to an increase in the filling height in the airlift.

To economically convey coarse-grained bulk materials with high gas permeability using an airlift, it must be equipped with a throttle section to reduce leakage gas losses $\dot{m}_{F,\mathrm{leck}}$: The lower part of the container is drawn in around the conveying line to form a narrow annular gap, while the upper part is designed with a larger diameter for sufficient bulk material buffering. Due to the small cross-sectional area of the throttle section, the leakage gas flow $\dot{m}_{F,\mathrm{leck}}$ remains small. An aeration bottom is not required. Such systems are in operation, for example, as silo or circulation mixers in the plastics industry.

To reduce the height of the lock chamber, it is possible to supply the conveying gas not via a bottom nozzle, but from above via a double jacket arranged around the conveying pipe to the conveying pipe inlet. The gas supply can be provided, for example, by a ring nozzle above the container lid. Information on the design of the gas entry into the conveying line can be found in the following Sect. 7.7. The principles shown there for suction nozzles can be applied analogously.

Pneumatic vertical conveyors are designed with operating temperatures up to 300 °C, possibly even higher (→ in this case: aeration bottom made of metal mesh).

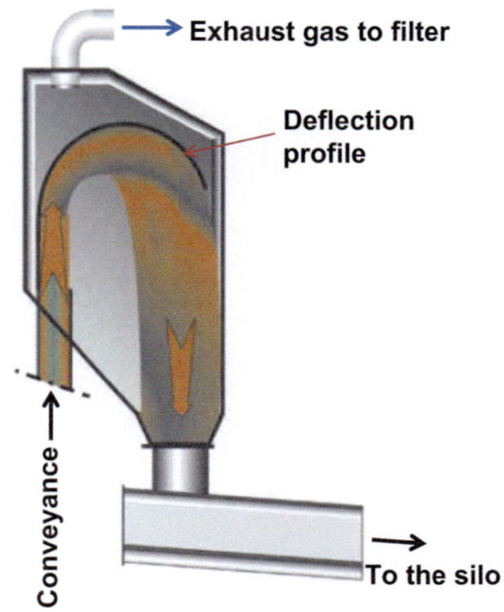

Fig. 7.55 Solid/gas separation by an expansion vessel, according to [40]

7.7 Suction Nozzles

In the case of vacuum- = suction conveying, the bulk material is fed to the conveying line via a suction nozzle. A well-known example of this is pneumatic ship unloaders, which transport solid mass flows up to $\dot{m}_S \cong 600\,\mathrm{t/h}$ through a single conveying pipe. This is done using conveying pipes, which are surrounded by a double jacket supplying the conveying gas flow in the intake area. In Fig. 7.56, an example of such an arrangement, as often shown in textbooks, is depicted [22]. The following questions arise regarding this design: How should the blunt pipe body penetrate the bulk material, possibly compacted by the previous transport? What (extremely) high gas velocities are required to generate a static vacuum $p_{F,\mathrm{in}}$ in the annular gap outlet, which is capable of overcoming the pressure difference $\Delta p_{F,\mathrm{in}} = -(p_{F,\mathrm{in}} - p_0)$, with: p_0 = gas pressure in the bulk material $\cong$ ambient pressure, to lift the solid particles out of the bulk material and into the conveying gas stream? The principle shown is both energetically and wear-wise absolutely unsatisfactory. It is obvious that a gas flow through a bulk material accelerates/drives it more effectively than a gas overflow does.

Fig. 7.57 shows an annular gap nozzle, whose outer tube is positioned differently in the axial direction relative to the conveying tube inlet. In Fig. 7.57a, the conveying tube is retracted relative to the outer tube, i.e., the solid is taken up by gas overflow, while in Fig. 7.57b, due to the conveying tube being advanced relative to the outer tube, the gas must flow through a partial area of the bulk material. The conveying behavior of such nozzles was experimentally investigated, among other things, in [41, 42] with systematically varied positions of the conveying tube – outer tube. The measurement results shown in Fig. 7.58 are taken from [41]. There it means: "$-\Delta x$"—the inlet of the conveyor pipe is located at the amount Δx above the outer pipe end = case (a) in Fig. 7.57, "$+\Delta x$"—the inlet of the conveyor pipe is located at the amount Δx below the outer

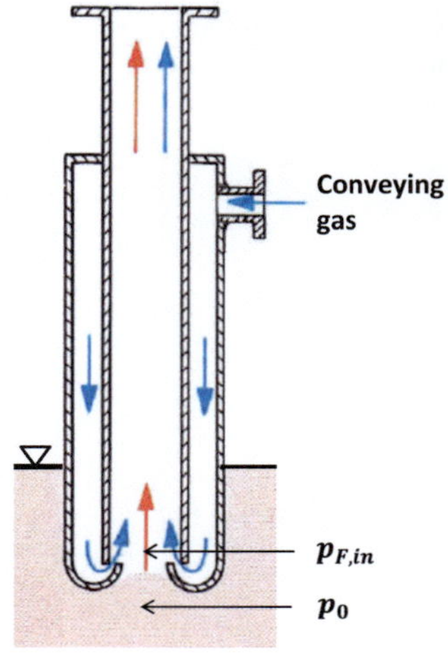

Fig. 7.56 Suction/injector nozzle

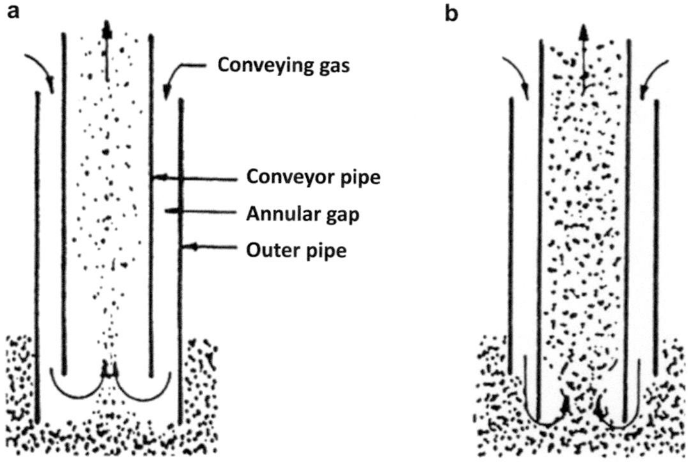

Fig. 7.57 Suction nozzle with different positions of delivery tube—outer tube, according to [41]

pipe outlet = case (b) in Fig. 7.57. The conveyed material was so-called sandy alumina, $d_{S,50} \cong 80\,\mu\text{m}$, which is classified as free-flowing/pourable. Fig. 7.58 illustrates the enormous $\dot{m}_S$-increase with a given gas outlet velocity $v_{F,\text{in}}$ and positive Δx-values, i.e., with the conveying tube advanced relative to the outer tube.

A more detailed analysis of the experimental results shows that for free-flowing/granular bulk material, an automatic continuous flow of solids into the gas stream of the annular channel occurs when the angle $\alpha_{\text{in}} \cong \arctan(2 \cdot \Delta x/(D_M - D_R)) > \alpha_{SS}$, i.e.,

Fig. 7.58 Effect of the adjustment dimension Δx on the solid throughput $\dot{m}_S$ of a given suction nozzle. Conveyed material: sandy clay, $d_{S,50} \cong 80\,\mu\text{m}$

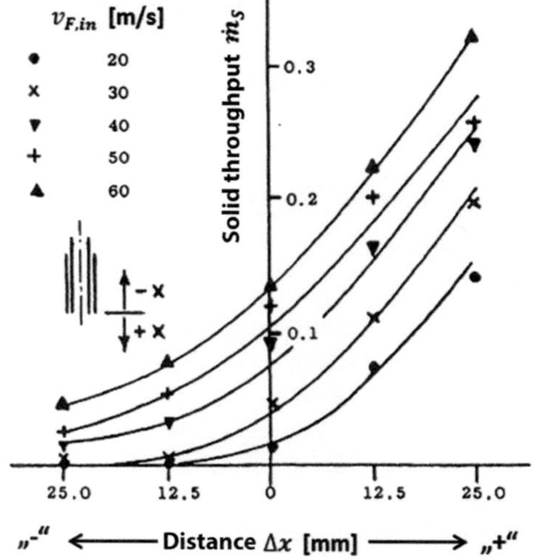

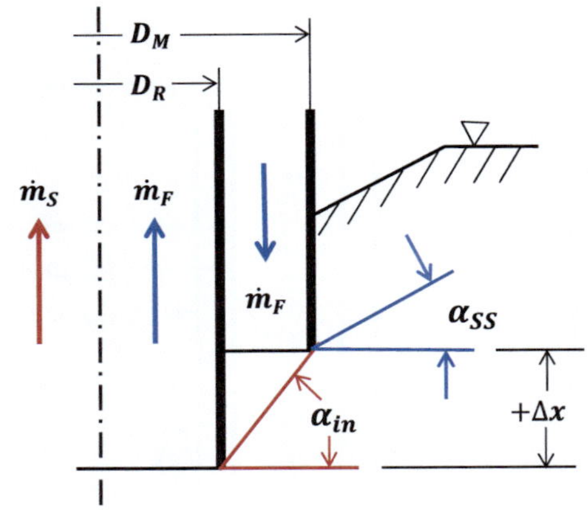

Fig. 7.59 Optimal Δx-setting, see text

greater than the angle of repose α_{SS} of the respective bulk material present is adjusted, see Fig. 7.59 for this. More accurate calculations can take into account the pipe wall thickness. If α_{in} is chosen much larger than the bulk material angle α_{SS}, the then excessive solid feed can cause blockages in the conveying line. With $\dot{m}_S \propto (\Delta x - \Delta x_0)$, $\Delta x_0 =$ height difference at $\alpha_{\text{in}} = \alpha_{SS}$, estimates can be made in this regard.

By throttling the conveying gas flow at the entrance to the annular channel and selecting an optimal Δx-position, the bulk material pick-up mechanisms described above can be combined. Measurement results on this are presented in [41]. Ultimately, the minimum energy requirement for the solid material throughput $\dot{m}_S$ required determines the suitability of the planned conveying system.

The above considerations apply to free-flowing products. In the case of suction conveying of cohesive/sticky/non-flowing bulk materials, it is necessary to loosen them around the suction pipe, as otherwise only holes with vertical walls are created in the product. For this purpose, mechanical discharge aids are usually used, e.g., counter-rotating scraper discs on the outer pipe circumference, which feed the bulk material to the conveying pipe.

The above considerations can be applied analogously to other installation situations, e.g., a horizontal arrangement of the suction conveying line.

References

1. Hilgraf, P.: Optimale Auslegung pneumatischer Dichtstrom-Förderanlagen unter energetischen und wirtschaftlichen Gesichtspunkten. ZKG Int. **39**(8), 439–446 (1986)
2. Löffler, F., Dietrich, H., Flatt, W.: Staubabscheidung mit Schlauchfiltern und Taschenfiltern. Vieweg, Braunschweig (1984)
3. Löffler, F.: Staubabscheiden. Thieme, Stuttgart (1988)
4. Shapiro, A.H.: The dynamics and thermodynamics of compressible fluid flow Bd. 1. The Ronald Press Company, New York (1953)

5. Baehr, H.D.: Thermodynamik: Eine Einführung in die Grundlagen und ihre technische Anwendung. Springer, Berlin (1992)
6. Truckenbrodt, E.: Lehrbuch der angewandten Fluidmechanik. Springer, Berlin (1988)
7. Harder, J.: Dry ash handling systems for advanced coal-fired boilers. Bulk Solids Handl. **17**(1), 13–20 (1997)
8. Schneider, K.: Typische Probleme bei der pneumatischen Silofahrzeug-Entleerung und deren praktische Lösung durch eine optimierte Anlagengestaltung. Vortrag auf dem Schüttgutforum 2012, Würzburg, 23.–24.10.2012. Vortragsmanuskript unter: http://www.enviro-engineering.de/pdf/ManuskriptSchuettgutforum-2012-10.pdf
9. Claudius Peters Projects GmbH: Referenzliste "Fahrzeug-Entleerung". Claudius Peters Projects, Buxtehude (2017)
10. Freimuth, C., Zimmermann, W.: Neue Staubpumpe mit endgelagerter Preßschnecke. ZKG Int. **36**(10), 591–593 (1983)
11. Hilgraf, P., Paepcke, J.: Der Eintrag von Schüttgut in pneumatische Förderstrecken mittels Schneckenschleusen. ZKG Int. **46**(7), 368–375 (1993)
12. Claudius Peters Projects GmbH: Prospekt "CP Pneumatic Conveying Technik". Claudius Peters Projects, Buxtehude (2017). http://www.claudiuspeters.com/en-GB/909/x-pump
13. Schneider, K.: Der Fördervorgang in der Einzugszone eines Extruders. Dissertation, TH Aachen, Aachen (1968)
14. Darnell, W.H., Mol, E.A.J.: Solids conveying in extruders. SPE J., 20–29 (1956)
15. Langecker, G.R.: Untersuchungen zum Stoffverhalten von Kunststoffpulvern in der Einzugszone von Einschneckenmaschinen mit genuteten Buchsen. Dissertation, TH Aachen, Aachen (1977)
16. Rautenbach, R., Pfeiffer, H.: Modellrechnungen zur Auslegung der förderwirksamen genuteten Einzugszone von Einschneckenextrudern. Kunststoffe **72**(3), 137–143 (1982)
17. Claudius Peters AG: Interner Entwicklungsbericht. Buxtehude (unveröffentlicht) (1992)
18. Paepcke, J.: Untersuchung einer hochtourig laufenden Preßschnecke zur Einschleusung von Schüttgütern in Druckräume. Diplomarbeit, Fachhochschule Hamburg, Fachbereich Bio.-Ing.-Wesen, Produktionstechnik und Verfahrenstechnik (unveröffentlicht) (1991)
19. Wäbs, J.-D.: Untersuchung einer Schneckenschleuse mit extremer interner Kompression. Bachelorarbeit, Hochschule für Angewandte Wissenschaften Hamburg (HAW), Studiengang Verfahrenstechnik (unveröffentlicht) (2014)
20. Finkbeiner, Th: Der Mechanismus der Zellenradschleuse für Schüttgut. VDI-Forschungsheft 563. VDI-Verlag, Düsseldorf (1974)
21. Kessel, S.R.: The interaction between rotary valves and pneumatic conveying pipelines. Thames Polytechnic, London (1985). PhD Thesis
22. Siegel, W.: Pneumatische Förderung: Grundlagen, Auslegung, Anlagenbau, Betrieb. Vogel Buchverlag, Würzburg (1991)
23. Schneider, D.: Hochdruck-Zellenradschleuse für feinkörnige und abrasive Schüttgüter—Untersuchungen der Realisierungsmöglichkeiten. Diplomarbeit, TUHH Technische Universität Hamburg-Harburg (unveröffentlicht) (1995)
24. Klinzing, G.E., Rizk, F., Marcus, R., Leung, L.S.: Pneumatic conveying of solids. Springer, Dordrecht (2010)
25. Wypych, P.W., Hastie, D.B.: Theoretical modelling of rotary valve air leakage for pneumatic conveying systems. In: Gostomski, P., Morison, K. (Eds.) Green processing and sustainability, S. 1–13. (2002)
26. Reed, A.R., Kessel, S.R., Pittman, U.K.: Examination of the air leakage of a high pressure rotary valve. Bulk Solids Handl. **8**(6), 725–730 (1988)

27. Schmidt-Jahre, T.: Untersuchung eines pneumatischen Fördersystems mit Zellenradeinschleusung. Bachelorarbeit, Hochschule für Angewandte Wissenschaften Hamburg (HAW), Studiengang Verfahrenstechnik (unveröffentlicht) (2014)
28. Sindermann, M.: Stand der Zellenradschleusentechnik. Waeschle-Schüttgutsymposium, Weingarten, 25.26.1993, S. 79–92 (1993)
29. Fa. Coperion Komponentenpräsentation: Einsatzmöglichkeiten von Coperion-Komponenten für Fa. Claudius Peters Technologies, Buxtehude (2016)
30. Fa. Midland Industrial Designers: Prospekt "Rotary Valves—TWA Extreme Duty/Ceramic lined". Fa. Midland Industrial Designers, Nottingham (2016)
31. Hutt, W.: Untersuchung der Strömungsvorgänge und Ermittlung von Kennlinien an Gutaufgabeinjektoren zur pneumatischen Förderung. Dissertation, Universität Stuttgart, Stuttgart (1983)
32. Ahland, E.O.: Strömungsvorgänge im vertikalen, feststoffbeladenen Förderrohr mit austretendem Freistrahl. Dissertation, Rheinisch-Westfälische Technische Hochschule Aachen, Aachen (1966)
33. Wagenknecht, U.: Untersuchung der Strömungsverhältnisse und des Druckverlaufes in Gas-/Feststoff-Injektoren. TU Braunschweig, Braunschweig (1981). Dissertation
34. Schlag, H.P.: Experimentelle und theoretische Untersuchungen zur Berechnung der Kennlinien von gasbetriebenen Einphaseninjektoren und Gutaufgabeinjektoren. VDI-Fortschrittsberichte, Reihe 3, Nr. 313. VDI, Düsseldorf (1992)
35. Bohnet, M., Teifke, J.: New results on the efficiency on energy transformation in gas-solid-injectors. Proc. Symp. on the Reliable Flow of Particulate Solids, Chr. Michelsen Institute, Bergen, 20.–22. August., S. 1–18 (1985)
36. Kmiec, A., Leschonski, K.: Analysis of two-phase flows in gas-solid injectors. Chem. Eng. J. **45,** 137–147 (1991)
37. VDI-Gesellschaft Verfahrenstechnik und Chemieingenieurwesen: VDI-Wärmeatlas, 10th edn. Abschnitt Lcc1–Lcc17. Springer, Berlin (2006)
38. Schneider, K.: Einsatz des Gas-Feststoff-Injektors als Einschleusorgan bei der pneumatischen Förderung von Schüttgütern. Schüttgut **2**(2), 217–224 (1996)
39. Fa. Plattco Corporation: Animated Plattco DFC-venting. Fa. Plattco Corporation, Plattsburgh (2015)
40. Fa. Claudius Peters Project: Prospekt "CP Pneumatic Conveying Technik". (GB) 03/2017/Issue 2. Fa. Claudius Peters Project, Buxtehude (2017)
41. Reed, A.R., Pittmann, A.N.: Suction nozzle design. An important element in effective operation of pneumatic ship unloaders. Bulk Solids Handl. **12**(3), 409–413 (1992)
42. Levy, A., Jones, M.G., Das, S.: An investigation into the performance of a suction nozzle. Powder Handl. Process. **8**(4), 337–339 (1996)

8 Wear in Conveying Systems

The service life or availability of devices and systems used for the transport or handling of bulk materials is often determined by the wear of the wall surfaces temporarily or permanently contacted by the solid material. However, this does not mean that the bulk material and solid properties alone dominate this process. The specific operating conditions, e.g. relative speed of bulk material/wall, contact pressure, or working temperature, as well as the wall geometry, the structure and properties of the material, and the intermediate space medium also influence the intensity of wear. For a wear analysis, the sole evaluation of the bulk material is not sufficient; rather an analysis of the entire wear system is required in each case [1,2]. It applies: wear is not a material property but a system property.

It is clear that a holistic and systematic analysis of the wear process supports the investigation of causes, allowing the identification of essential influencing factors in the specific case and, based on this, the development of intelligent wear-reducing solutions. These should go beyond simply increasing the wall material quality or the offered wear volume and could, for example, consist of a wall geometry adapted to the current conditions combined with a change in operating conditions: The actual causes of wear must be detected and eliminated or mitigated, with long-term effective solutions being preferable to short-term improvements. The goal is to ensure trouble-free plant operation without wear-related downtimes and minimal wear costs over a predefined, as long as possible, service life.

To perform a wear analysis and develop specific solution approaches, knowledge of the various influencing factors and their effects on the wear process is necessary. In the following, after presenting some fundamental relationships, the essential factors and their dependencies are illustrated. A rough distinction is made between the influence areas of process conditions, bulk material, wall material, and geometry. This attempt at classification particularly highlights the diverse, mutual influences between the individual factors. With regard to the processes in pneumatic conveying systems, the abrasive wear caused by sliding bulk material on a base body and the impact wear triggered by

P. Hilgraf, *Pneumatic Conveying*, https://doi.org/10.1007/978-3-662-67223-5_8

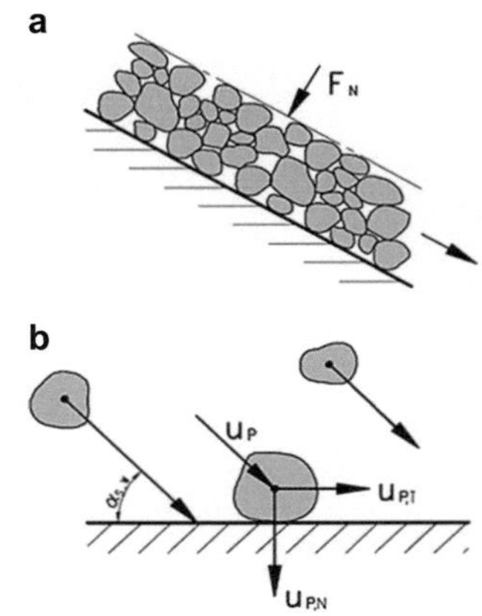

Fig. 8.1 Schematic representation of abrasive (**a**) and impact wear (**b**)

particle bombardment are analyzed in more detail, see Fig. 8.1. A distinction is made between the two limit cases of a base body or wall made of ductile and brittle material.

It should be considered that a pneumatic conveying system does not only consist of the actual conveying line = pipeline and a lock, but at least also includes a feeding system, e.g., a storage container with solid dosage, and a receiving and exhaust gas cleaning area. These are also subject to wear stress.

8.1 Basics of Impact and Sliding Wear

This section presents the basic relationships.

8.1.1 Contact and Fracture Mechanical Influencing Factors

Through the contact of the particles of a bulk material with the surrounding wall, a tool or a counter body, all involved system components are subjected to more or less strong stress. This can lead to both the detachment of wear particles from the base body and the breakage/abrasion of bulk material particles, i.e., wear and grain destruction are competing processes in all manufacturing or processing steps of a bulk material. Sufficiently stable bulk material particles are therefore required to cause wear on the components of a system. The following overview of processes during particle/wall contact assumes such particle stability.

a) Bulk material/wall contact

If a rigid, ideally spherical bulk material particle is pressed against a softer ideally flat elastic base body with different normal forces F_N, neglecting adhesion and friction forces between the contact surfaces, the effects shown in Fig. 8.2 result: The force $F_{N,1}$, see

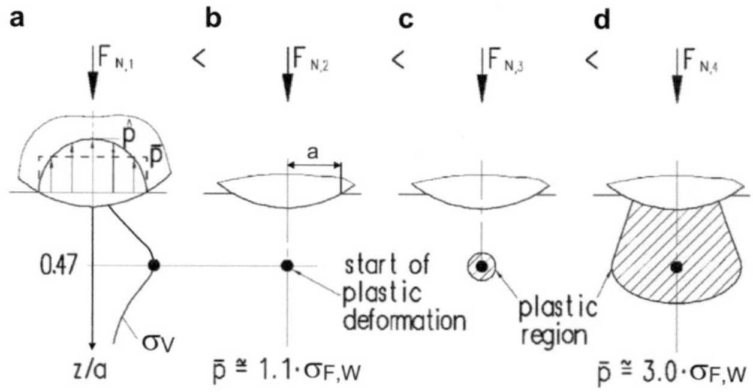

Fig. 8.2 Contact between hard spherical bulk material particle and wall

Fig. 8.2a, leads to a reversible elastic contact, which can be calculated with the Hertz theory. Above the circular contact area with the radius "a", a hemispherical pressure stress distribution with the average stress $\overline{p} = F_N/(\pi \cdot a^2)$ and $\hat{p} = 3/2 \cdot \overline{p}$ as maximum compressive stress appear; at the edge of the contact area $p(a) = 0$. The spatial stress field developing in the base body has a maximum shear stress of $\hat{\tau} \cong 0.31 \cdot \hat{p}$ at a depth of $z = 0.47 \cdot a$ below the contact center, and the largest tensile stresses along the radius a. Uniaxial equivalent stresses calculated with the stress hypotheses according to Tresca or von Mises σ_V provide a maximum value $\hat{\sigma}_V$ at the location of the maximum shear stress $\hat{\tau}$. This means that the highest material stress is not in the contact surface, but within the base body.

Equating the maximum equivalent stress $\hat{\sigma}_V$ with the yield stress of the base material under compression $\sigma_{F,W} = \hat{\sigma}_V$ allows calculating the contact force $F_{N,F}$ or pressure $\overline{p}_F$ at which irreversible plastic deformation occurs in the base body. This happens at an average contact pressure of

$$\overline{p} = \overline{p}_F = 1.1 \cdot \sigma_{F,W} \tag{8.1}$$

and starts at the location of the greatest equivalent- = greatest shear stress, see Fig. 8.2b.

An increase in the normal force above $F_{N,F}$ leads to an enlargement of the plastically deformed material area. At $\overline{p} = \overline{p}_{pl} = 3.0 \cdot \sigma_{F,W}$ the plastic zone reaches the material surface/contact surface, see Fig. 8.2c, d. A further increase of F_N enlarges the flow area further, but leads to no or only a slight increase in the average contact pressure $\overline{p}_{pl}$. This state is comparable to that of a hardness test [3, 4, 5]. It therefore applies:

$$H_W = \overline{p}_{pl} \cong 3.0 \cdot \sigma_{F,W} \tag{8.2}$$

with:

H_W current hardness of the base material formed with the projected contact area (→ in the present text, the Vickers hardness HV is used).

The relationship described by (8.2) is largely independent of the shape of the penetrating body, i.e., in the case of ideal plastic deformation, the normal force required to create a projected contact area A_C or penetration depth z can be calculated using the approach

$F_{pl} = A_C \cdot H_W$. For the combination of ball-plate, for example: $A_C = \pi \cdot a^2 \cong \pi \cdot d_S \cdot z$, with: d_S = particle diameter.

The plastic region is surrounded by elastically deformed material. When the contact is relieved, its stress reduction is hindered, and an elastic residual stress field remains in the base body, i.e., energy is stored locally.

If the spherical bulk material particle is shot at a predetermined speed u_P perpendicular to the flat base body, the processes described above occur in the same way. Using Hertz's theory and (8.1), the impact speed $u_{P,F}$ can be calculated, at which plastic deformation begins in the base body [4, 6]. For this, the following applies:

$$u_{P,F} = 48.65 \cdot k^{5/2} \cdot \left(\frac{R}{d_S}\right)^{3/2} \cdot \frac{\sigma_{F,W}^{5/2}}{\varrho_P^{1/2} \cdot E_\otimes^2} \tag{8.3}$$

with the reduced modulus of elasticity

$$\frac{1}{E_\otimes} = \frac{1}{2} \cdot \left(\frac{1-\nu_W^2}{E_W} + \frac{1-\nu_P^2}{E_P}\right) \tag{8.4}$$

with:

$E_{W/P}$ Elastic modulus of the wall/particle,
$\nu_{W/P}$ Poisson's ratio of the wall/particle,
R Local radius of the contact point,
ϱ_P Particle density,
k Stress coefficient.

A derivation of (8.3) can be found in [6]. Figure 8.3 shows its evaluation with $k = 1.1$ in dimensionless form. In order to estimate the influence of different grain shapes, a dis-

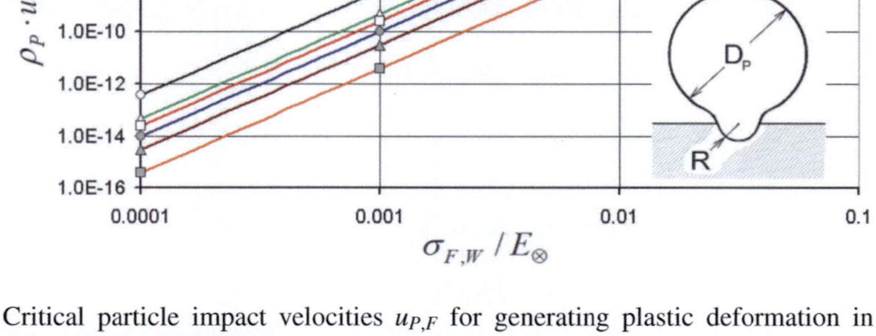

Fig. 8.3 Critical particle impact velocities $u_{P,F}$ for generating plastic deformation in the base material; dimensionless representation

tinction is made between the local particle radius R at the contact point and the diameter d_S of the associated particle, calculated as a volume-equivalent sphere. With sharper bulk material particles, i.e., decreasing ratio R/d_S, the minimum impact speed $u_{P,F}$ required for the initiation of plastic deformation decreases. For the ideally spherical particle, $R/d_S = 1/2$. At constant R/d_S, $u_{P,F}$ is independent of the size of the particle diameter.

Application examples: A steel ball impacts perpendicularly on a flat steel wall, $\sigma_{F,W} = 1000\,\text{MPa}$, $E_W = E_P = 210\,\text{GPa}$, $\nu_W = \nu_P = 0.30$, $d_S = 1.0\,\text{mm}$, $R/d_S = 0.5$, $\varrho_P = 7850\,\text{kg/m}^3 \rightarrow u_{P,F} = 0.146\,\text{m/s}$. If the steel ball is replaced by a quartz ball with $E_P = 140\ GPa$, $\nu_P = 0.3$ and $\varrho_P = 2650\,\text{kg/m}^3$, applies $\rightarrow u_{P,F} = 0.393\,\text{m/s}$. If the quartz particle is flattened at the contact point, $R/d_S = 1.0$, follows $\rightarrow u_{P,F} = 1.113\,\text{m/s}$, if it is sharp-edged, $R/d_S = 0.1$, resulting in $\rightarrow u_{P,F} = 0.035\,\text{m/s}$. Limit velocities for other material combinations are in the same order of magnitude, i.e., collisions between hard particles and a metallic base body almost always lead to plastic deformations in the base body.

If one replaces the factor $k = 1.1$ with the value $k = 3.0$, cf. (8.2), then the impact velocity $u_{P,pl}$ can be estimated, at which the plasticized zone reaches the contact surface. For the above application examples, the results are: steel ball: $u_{P,pl} = 1.79\,\text{m/s}$, quartz sphere: $u_{P,pl} = 4.83\,\text{m/s}$, flattened quartz particle: $u_{P,pl} = 13.67\,\text{m/s}$, sharp-edged quartz particle: $u_{P,pl} = 0.43\,\text{m/s}$. More accurate calculation approaches do not yield significantly different results. The dynamic impact process can be treated as quasi-static, like the penetration of a hardness test body into a surface, for particle velocities $u_P < 100\,\text{m/s}$ [3, 4]. Above $u_P \cong 100\,\text{m/s}$ dynamic effects must be taken into account. This implies that, for example, the wall impact of bulk material particles in a pneumatic conveying line can be calculated quasi-statically.

If a relative movement tangential to the base body surface and thus a friction force $F_T = \mu_W \cdot F_N$ is superimposed on the previously investigated normal-stressed contact ball—base body, the equivalent stress maximum $\hat{\sigma}_V$ within the base body shifts due to the additionally introduced shear stresses with increasing friction coefficient μ_W in the direction of the contact surface and reaches it at a critical value $\mu_{W,\text{crit}}$. The plastic deformation then begins directly in the contact or base body surface. Critical friction coefficients are in the range of $\mu_{W,\text{crit}} \cong 0.30$.

With increasing load F_N, a pronounced work hardening $\sigma_{F,W} \propto F_N^n \propto H_W$ is often observed in plastically deformed materials, especially metals, meaning that the respective material becomes "harder" in the near-surface contact area. Figure 8.4 shows such hardness profiles caused by intensive particle bombardment in a base body made of C-steel [7]. The parameter is the impact angle $\alpha_{S,W}$. The Vickers hardness of the unstressed steel is $H_W \cong 1.60\,\text{kN/mm}^2$. For the previously assumed linear elastic-ideal plastic behavior, $\sigma_{F,W} =$ konst. It is known that the stress speeed can also influence the material behavior.

Wall materials, as shown in Fig. 8.5 [1], have a spatial structure perpendicular to their surface. Following an outer adsorption layer, which forms in a dynamic equilibrium with the adjacent environmental medium, there is an oxide/reaction layer, under which there are different layers resulting from previous processing and shaping. Only after that is

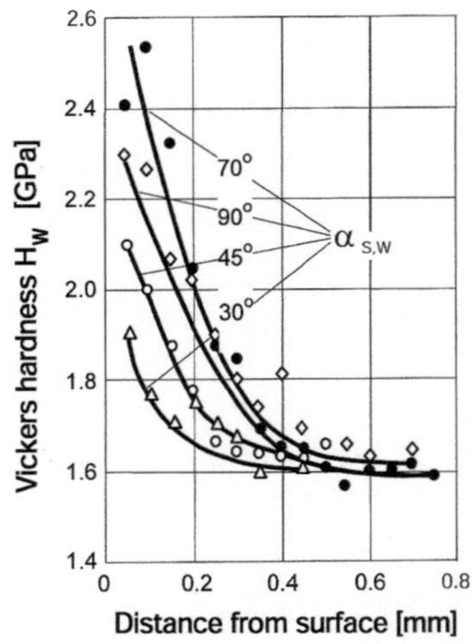

Fig. 8.4 Work hardening of a steel with 0.3 M.-% C by bombardment with cast iron pellets, $d_S = (1.0-1.5)$ mm [7]

the unaffected material structure reached. This results in the material surface showing different behavior/having different material properties than the material interior, i.e., a penetrating particle experiences different material reactions depending on the penetration depth. Example: With few exceptions, the hardness of metal oxides is generally significantly greater than that of metals; for iron, the ratio is $H_{W,\text{Fe-Oxide}}/H_{W,\text{Fe}} \cong 3$ [1]. The problem becomes even more complex when the material consists of several crystal phases, is constructed as a composite material, e.g., matrix with embedded hardening agent, or has undergone surface treatment or coating. It does not need to be emphasized that the bulk material particles are also structured accordingly. Since the base body in the contact area is not ideally flat and bulk material particles, as already discussed above, are neither ideally spherical nor ideally smooth, the contacts between the two are mediated in practice by their surfae roughnesses. This further complicates the description, as, for example, the roughness structure is changed by the contact process itself.

Despite all limitations, it is possible with the approaches mentioned above to estimate the wall material stresses and reactions due to particle/wall contacts, the corresponding penetration depths etc. [3, 4, 5, 8]. Such calculations require idealizing assumptions but still provide practically relevant statements, e.g., regarding possible interactions between the

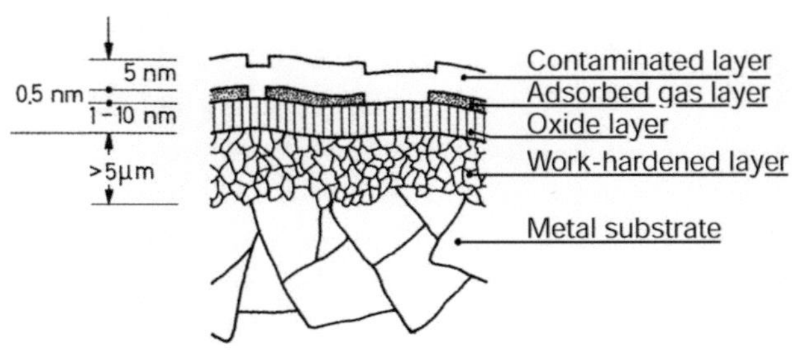

Fig. 8.5 Structure of a metallic material in the surface area

thickness of a planned wear coating, the penetration depth of the bulk material particles, and the resulting stress distribution and location of the stress maximum in the base body.

So far, the case of a "hard" bulk material particle has been considered, which penetrates into a softer base body with the hardness H_W and causes plastic deformations there. Measurements and theoretical analyses show [4, 9, 10] that a particle with a hardness of

$$H_P \geq 1.2 \cdot H_W \tag{8.5}$$

is required. Smaller hardness values H_P lead to plastic deformation of the particle. In the following considerations, unless otherwise noted, hard bulk material particles are assumed.

b) Fracture behavior

If (8.5) is fulfilled, i.e. if the bulk material particles are sufficiently hard and also capable of withstanding the contact loads discussed above without damage, this results in an intensive stress on the base body and high wear on it must be expected. The detachment of a wear particle is equivalent to a local fracture of the base body material. Fractures are caused by tensile stresses (→ Fracture form: Brittle fracture) or shear stresses (→ Fracture form: Toughness fracture). Mixed forms also occur. The precursor of a fracture is the crack, the enlargement of which can lead to fracture. Stable crack growth is distinguished from unstable crack growth. In the former case, the crack only expands with increasing load, while in the latter, it enlarges without further load increase, using the elastic energy stored in the base body.

A brittle fracture separates the material perpendicular to the greatest normal stress and appears, macroscopically, deformation-free. However, on a microscopic level, plastic deformations with the formation of dislocations and dislocation movements occur, which accumulate at obstacles, e.g. inclusions, grain boundaries, hard phases, and promote the formation of microcracks in these areas. Due to the overall low local plastic deformability, the majority of the energy supplied by the external stress is stored elastically and is thus available for initiating a quasi-instantaneous unstable crack growth. A brittle fracture runs along the crystallographic planes (→ transcrystalline fracture), possibly also along the crystal boundaries (→ intercrystalline fracture), if these represent weak points.

In the case of toughness fracture, increasing plastic deformation leads to the exhaustion of the material's capacity for dislocations and subsequently to the formation of micropores around hard inclusions, e.g. carbides, oxides. Further increasing load results in pore enlargement, internal necking, and the merging of cavities. The relatively thin material bridges between the cavities are then sheared off in the direction of the greatest shear stress. Macroscopically, the toughness fracture is characterized by a necking and a fracture direction parallel to the greatest shear stress.

Brittle and toughness fractures can be easily distinguished metallographically. Increasing stress speed and decreasing material temperature shift the fracture type towards brittle fracture.

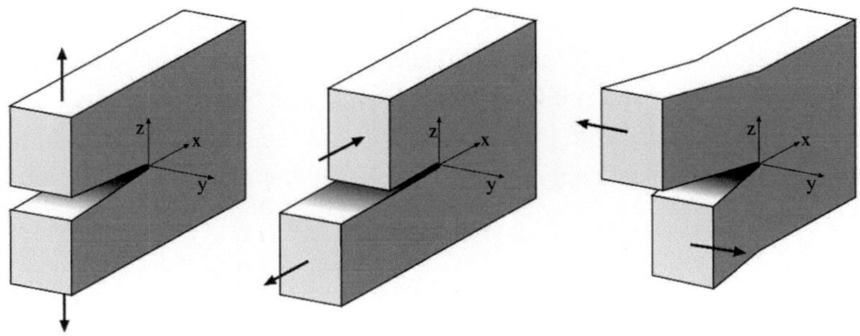

Fig. 8.6 The three fundamental crack opening modes I, II, III

Table 8.1 Fracture toughness of various materials

Material	Fracture toughness K_{Ic} [MPa$\sqrt{mm}$]
High-strength steels	800 - 3000
30CrNiMo8 → (20°C)	3650
30CrNiMo8 → (-20°C)	2000
Structural steels	1000 - 4000
Ti alloys	1200 - 3000
Ti6Al4V	2750
Al alloys	600 - 2000
AlCuMg	900
AlZnMgCu1.5	950
Al_2O_3 ceramic	120 - 300
Marble	40 - 70
Glass	20 - 40
Concrete	5 - 30

The susceptibility to fracture, especially for brittle materials, can be characterized by the so-called fracture toughness K_c. K_c is determined using standardized measurement methods. Since three crack opening modes are possible, the three fracture toughness values K_{Ic}, K_{IIc}, K_{IIIc} are distinguished, see Fig. 8.6. In the following, the important mode I for the present problem is considered, in which the crack is pulled apart perpendicular to the crack surface. Ductile materials have high fracture toughness, while brittle materials have low fracture toughness. Table 8.1 shows examples [11]. K_{Ic} is the critical value of the so-called stress intensity factor K_I, which describes the "strength" of the stress field at the tip of a crack. K_I is given by (8.6)

$$K_I = \sigma_N \cdot y \cdot \sqrt{\pi \cdot c} \qquad \rightarrow \qquad \text{here: } y \cong 1 \tag{8.6}$$

with:

- σ_N externally applied stress,
- c half crack length,
- y geometric shape factor; values for various fracture geometries and stress situations can be found in handbooks; $y \rightarrow 1$ for an infinitely extended base body,

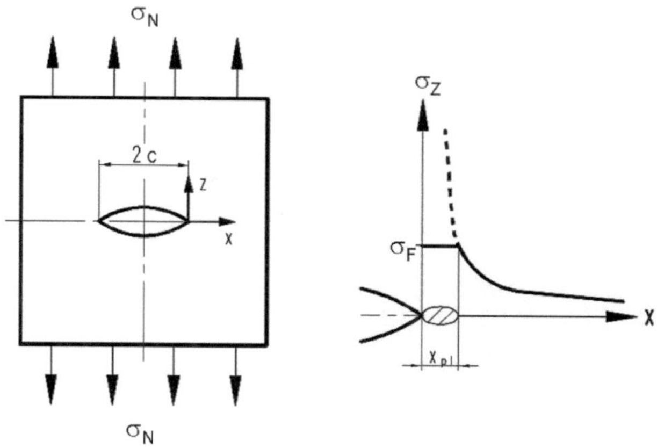

Fig. 8.7 Processes at the crack front

Dimension: K_{I} [Stress] · [Length]$^{1/2}$, e.g. Pa m$^{1/2}$, N/m$^{3/2}$. Figure 8.7 illustrates the processes at a crack front. A crack tip radius $r \to 0$ results in a stress singularity, which is reduced to the flow stress $\sigma_{F,W}$ by plastic deformation. A plastic zone thus forms in front of the crack tip. Its size can be estimated with

$$x_{pl} \cong \frac{1}{\pi}\left(\frac{K_{\mathrm{I}}}{\sigma_{F,W}}\right)^2 \tag{8.7}$$

From an energy balance at a crack of length $2c$ it follows that the specific energy release rate $dW/dc = K_{\mathrm{I}}^2/E_W$ required for a crack enlargement from $2c$ to $(2c + dc)$ ($\to$ creation of a larger crack surface) must exceed a critical value K_{Ic}. It follows: Only when the stress intensity factor K_{I} a material-specific critical size K_{Ic}, the so-called fracture toughness, achieves, crack growth occurs. As (8.7) shows, large $K_{\mathrm{I}c}$ values are associated with large plasticized areas in front of the crack front and vice versa. A ductile material is therefore tougher/less fracture-sensitive than a brittle material because it absorbs more energy irreversibly by building up the plastic zone, which is then no longer available for increasing the crack surface [3, 11].

c) Lateral fracture

A special, particularly important for the wear of brittle wall materials, e.g., ceramic linings, fracture form is the so-called *lateral cracking*. In this case, shallow material chips "burst" from the stressed surface. Lateral fractures occur when a critical contact load is exceeded, which can be estimated with (8.8).

$$F_{P,c} = \alpha \cdot \frac{K_{\mathrm{I},c}^4}{H_W^3} \cdot f\left(\frac{E_W}{H_W}\right) \quad \to \quad \text{with: } \alpha_0 = \alpha \cdot f\left(\frac{E_W}{H_W}\right) \cong 2 \cdot 10^5 \tag{8.8}$$

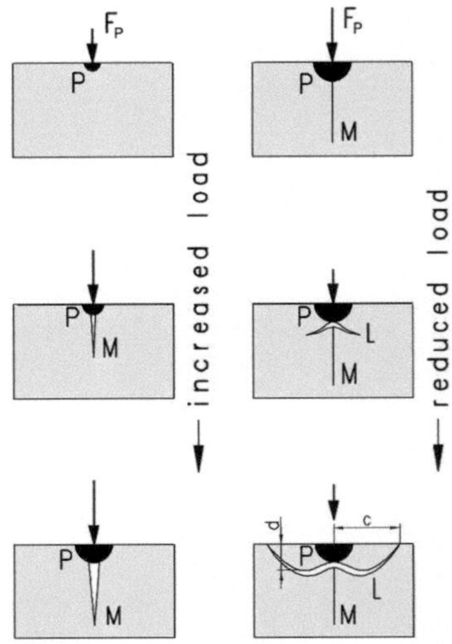

Fig. 8.8 Formation of a lateral fracture

[12] In Fig. 8.8, the processes occurring during fracture are simplified. A sharp-edged particle is pressed with increasing force F_P (→ point load) in the base body, which plastically deforms according to (8.2). The plasticized zone P expands with increasing F_P. When F_P reaches the critical value $F_{P,c}$, (8.8), a vertical relief crack M begins to form below P due to tensile stresses acting there, which expands with increasing F_P. When F_P decreases, M closes again, but at the same time, a transverse crack forms L (→ *lateral crack*). This is triggered by the elastic residual stresses remaining in the base material during the relief of the contact. The ends of the transverse crack bend towards the base material surface when fully relieved and often break through it. This leads to a direct material removal of the size $V \cong \pi \cdot c_l^2 \cdot d_l$, see Fig. 8.8. Another prerequisite for such a fracture formation, in addition to $F_P > F_{P,c}$, is that there are sufficient weak points/defects in the wall material. The vertical relief crack M reduces the material strength of the material, but has no influence on the wear removal [13, 14].

The conditions described above occur, for example, when a bulk material particle is moved across the base body surface under constant load (→ abrasive wear): A locally fixed surface element is initially increasingly loaded by the incoming particle and unloaded again when it runs off. In the case of impact wear, the wall impact of a particle causes a comparable loading and unloading cycle. The processes taking place in the material can be modeled using contact and fracture mechanical approaches, e.g., [6, 14].

8.1.2 Wear Mechanisms

The various types of wear, e.g., abrasive or impact wear, are characterized by their macroscopic properties—kinematics, shape, etc.—while the underlying wear mechanisms define

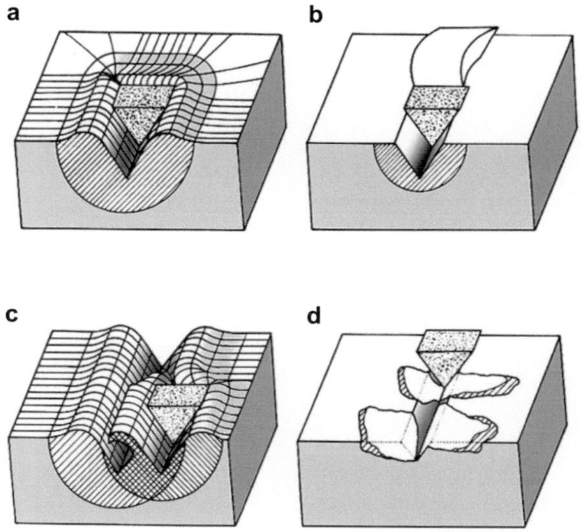

Fig. 8.9 Wear mechanisms [15]. **a** Microplowing, **b** Microcutting, **c** Microfatigue, **d** Microcracking

the relevant interactions in the contact area. Wear mechanisms thus represent a deeper level of description. Different wear mechanisms can be assigned to a single type of wear.

Figure 8.9 shows, for example, the possible, generally overlapping, wear mechanisms [15] in the case of abrasive wear, cf. Fig. 8.1a: microplowing with complete, predominantly lateral material displacement without removal, microcracking in brittle materials, microcutting with complete material removal, and microfatigue due to repeated alternating stress of the displaced, already plastically deformed volume by adjacent particles. The mechanism *wedge formation* [14] represents the transitional form from microplowing to microcutting. In Fig. 8.10, cross-sectional profiles of grooves from four different experiments from [16] are shown for illustration, which were measured with a force electron microscope. The contours are mean values from three repeated experiments each.

Defining, see Fig. 8.10, by means of

$$\phi = \frac{A_v - (A_{d1} + A_{d2})}{A_v} \tag{8.9}$$

the relative removal ϕ = proportion of the detached groove volume, then the following applies: pure microplowing: $\phi = 0$, pure microcutting: $\phi = 1$, *wedge formation*: $0 < \phi < 1$, microcracking: $\phi > 1$. Microcracking includes the aforementioned lateral fracture. In [17], experimentally measured ϕ-values of the size $\phi \cong (0.15-1.00)$ are mentioned. For the grooves in Fig. 8.10, $\phi \cong 0.85$ was determined at the contact load $F_N = 0.5\,\text{N}$ and $\phi \cong 0.90$ at $F_N = 5.0\,\text{N}$ [16].

Microcutting with high wear removal, $\phi \to 1$, occurs at engagement angles above a critical value Θ_c. Here, the engagement angle Θ is defined as the inclination angle of the particle leading edge in the direction of motion against the base body surface. Θ_c is

Fig. 8.10 Measured abrasive wear grooves

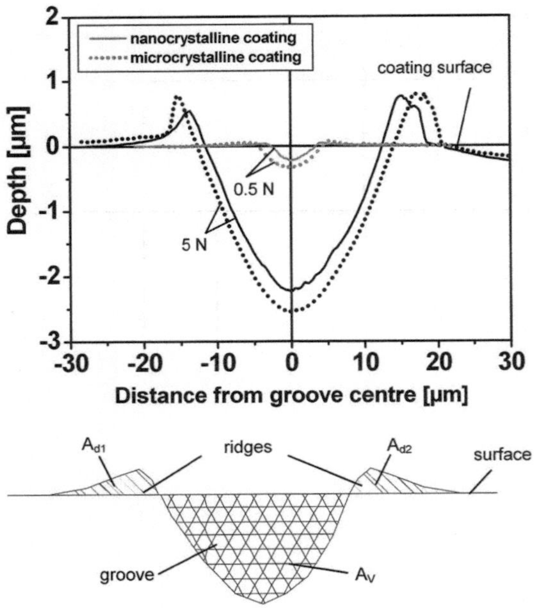

influenced by the shear stress of the material surface, i.e., also the friction coefficient μ_f, and the elastic properties of the base body, described by the ratio (elasticity modulus E_W/ hardness H_W). A lower limit for Θ_c is approximately $\Theta_c \cong 30°$, practically values up to $\Theta_c \cong 90°$ are possible. Large (E_W/H_W)-values result in large Θ_c-values [14, 17].

In the case of jet wear, see Fig. 8.1b, bulk material particles collide with the base material at an angle $\alpha_{S,W}$ measured against the wall surface. The individual particle impact can thus be broken down into a wall-normal and a wall-parallel component. The wear mechanisms possible during the wall-parallel movement largely correspond to those of abrasive wear. One difference is that the contact pressure against the wall is not defined or only low in this phase and changes during contact. The wall-normal component causes the wear mechanisms of micro-fatigue and micro-cracking to become effective. The wall-normal and wall-parallel effects overlap.

When using very heterogeneous base material, e.g. materials composed of a hard coarse-grained phase I distributed in a homogeneous phase II or a particulate hardening agent embedded in a softer matrix, further wear mechanisms can be observed. Figure 8.11 schematically shows the possible extraction or shearing of individual particles of the coarse-grained phase during abrasive wear. Similar effects can also be observed in erosion wear: repeated particle impact leads to a disruption of the base matrix and the bond between matrix and coarse grain in such a way that individual particles of the coarse

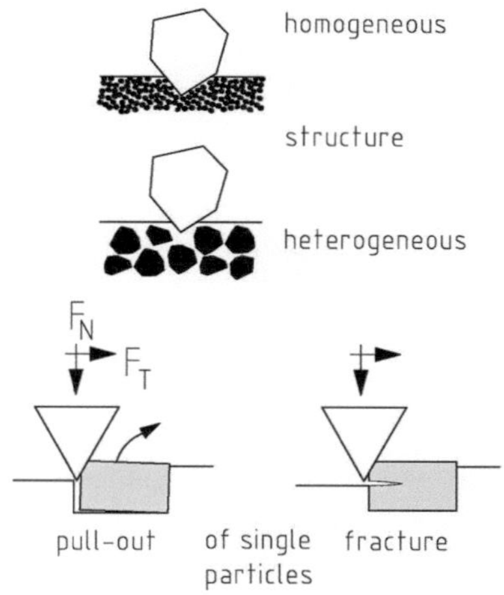

Fig. 8.11 Heterogeneous wall material: stress and wear mechanisms

phase are extracted as a whole. These processes are influenced, among other things, by the ratio of the particle size of the coarse phase to the size of the penetration volume of the bulk material particles, the particle spacing or the volume fraction of the coarse phase in the base material, the bonding of the phases with each other, and the intensity of the stress.

8.1.3 Calculation Approaches/Models

Already the research published in 1995 [18] identifies 182 equations for describing the various types of wear. 28 of these equations specifically deal with wear caused by particle erosion and impact. They are analyzed in more detail in [18]. One of the latest works on the subject of erosion wear is [19]. The basis of all these equations are either theoretical models or description approaches based on measurement results. They are only valid for limited application areas and/or idealized material behavior and must be adapted to the real task by experimentally determined coefficients, constants, etc. Applying them to wear problems far away from the underlying model assumptions generally results in large errors. However, their practical value lies in the fact that they illustrate which of the wall material, bulk material, and operating characteristics dominate the current task in which combination, weighting, and size. This allows for optimizations.

In the following, some of these wear equations are compiled as examples. It is assumed that "hard" bulk material particles are involved. The uniform wear measure used is the easily measurable mass loss W_M, which can be converted into the more practically relevant volume removal $W_V = W_M/\varrho_W$, ϱ_W = density of the wall material. Derivations of the equations and further information can be found in, among others, [6, 10].

a) Abrasive wear, ductile wall material

$$W_{M,l} = \frac{W_M}{l} = X_1 \cdot \varrho_W \cdot \frac{F_N}{H_W} \tag{8.10}$$

with:

- l Sliding path bulk material—wall,
- F_N Total force perpendicular to the bulk material layer with wall contact,
- H_W Hardness of wall material,
- X_1 Wear coefficient.

It is necessary to distinguish between 2- and 3-body abrasive wear in the case of abrasive wear, see Fig. 8.12. In 2-body wear, the bulk material particles are firmly fixed in a counter body, while in 3-body wear they have limited mobility of their own, i.e., the individual particles can rotate and move relative to each other. This results in a wear coefficient X_1, which is about an order of magnitude smaller in 3-body abrasive wear than that of 2-body wear. Wall materials are typically subjected to 3-body abrasive wear when handling bulk materials.

Wear coefficients X_1 are typically between $X_1(\text{2-body}) \cong (6 \cdot 10^{-3} – 6 \cdot 10^{-2})$ for 2-body abrasive wear, and those for 3-body wear are then at about $X_1(\text{3-body}) \cong 0.1 \cdot X_1(\text{2-body})$.

b) Abrasive wear, brittle wall material, lateral fracture

$$W_{M,l} = \frac{W_M}{l} = X_2 \cdot \varrho_W \cdot N_P^{1-k} \cdot \frac{F_N^k}{K_{\mathrm{Ic}}^m \cdot H_W^n} \tag{8.11}$$

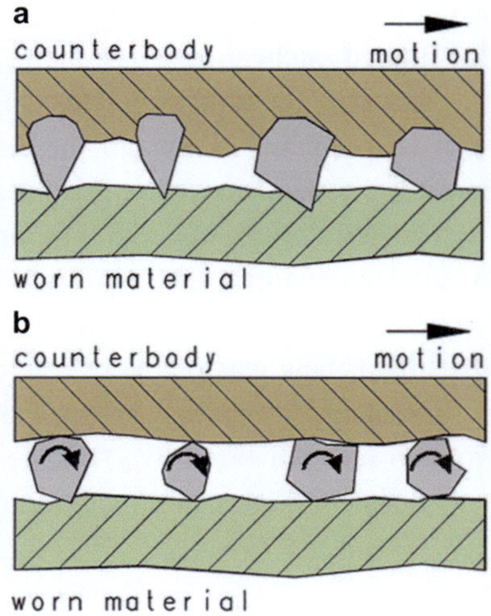

Fig. 8.12 Comparison of 2-body (**a**) and 3-body abrasive wear (**b**)

with: $(k - m - n) = (9/8 - 1/2 - 5/8), (7/6 - 2/3 - 1/2), (5/4 - 3/4 - 1/2) \rightarrow$ Exponents in the models of various authors,

N_P Number of bulk material particles with wall contact,
K_{Ic} Fracture toughness of wall material,

Boundary condition: $F_P = F_N/N_P > F_{P,c}$ according to (8.8).

c) Impact wear, ductile wall material, impact angle $\alpha_{S,W} = 90°$

$$W_{M,M} = \frac{W_M}{m_S} = X_3 \cdot \frac{\varrho_W \cdot u_{P,0}^2}{H_W} \tag{8.12}$$

alternatively:

$$W_{M,M} = \frac{W_M}{m_S} = X_4 \cdot \frac{\varrho_W}{\varepsilon_c^2} \cdot \frac{\varrho_P^{1/2} \cdot u_{P,0}^3}{H_W^{3/2}} \tag{8.13}$$

with:

$u_{P,0}$ Impact velocity of bulk material particles—wall,
ϱ_P Density of bulk material particles,
m_S Bulk material mass impacted during the wear/measurement period,
ε_c critical strain/deformation of the wall material, which leads to the detachment of wear particles.

d) Impact wear, brittle wall material, lateral fracture, impact angle $\alpha_{S,W} = 90°$

$$W_{M,M} = \frac{W_M}{m_S} = X_5 \cdot \varrho_W \cdot \varrho_P^p \cdot d_S^k \cdot \frac{H_W^n}{K_{Ic}^m} \cdot u_{P,0}^q \tag{8.14}$$

with: $p \cong (0.22\text{–}0.60)$, $k = 0.67$, $m = 1.33$, $n \cong ((-0.25)\text{–}0.11)$, $q \cong (2.44\text{–}3.20) \rightarrow$ Exponents in the models of different authors,

d_S Particle diameter of bulk material.

Boundary condition: $F_{P,\max}(u_{P,0}) > F_{P,c}$ according to (8.8).

By means of the wear coefficients X_{1-5} the individual equations can be adapted to measurement results. The approaches show, among other things, that for ductile wall materials, their hardness H_W is obviously relevant, while for brittle wall materials, a combination of wall hardness H_W and fracture toughness K_{Ic} influences the extent of wear.

8.1.4 Dependencies of Wear

The following discusses those influencing factors known to significantly affect the intensity of wear in abrasive and jet stress. The analysis is based on an evaluation of the measurement data available in the relevant literature, including [1, 2, 6, 7, 10, 12, 14–26]. It is subdivided according to operational, bulk material, wall material, and geometry influences. The previous explanations must be taken into account.

I) Operating conditions

a) In general:

- **Duration of stress τ:** Prolonged abrasive and/or impact stress from bulk material particles leads to the compaction/alteration of near-surface wall material layers. This is a running-in process that is completed after a finite period of time τ. Only after the time τ does a constant wear rate $W_{M/\tau} = dW_m/d\tau$ set in. In the case of jet wear, a wear trough that grows in size and depth forms during the running-in phase, see Fig. 8.13. With the jet duration τ, the impact angle β_P increasingly deviates from the jet angle $\alpha_{S,W}$. The running-in process is completed when the impact angle reaches $\beta_P = 90°$.
  Measurements during the running-in phase often yield higher wear values than during later operation, i.e., they are on the "safe" side. In the following, $\beta_P = \alpha_{S,W}$ is set simplistically.
- **Operating temperature T_B**: Increasing working temperature does not necessarily lead to greater wear. However, the expected behavior cannot be determined by simple extrapolation of the temperature dependencies $f(T)$ of the relevant material properties to the operating temperature. Depending on the type of wall material, higher temperatures lead to structural transformations, oxide layer formation, reduction of hardness, etc. In parallel, the corresponding properties of the bulk material particles also change.

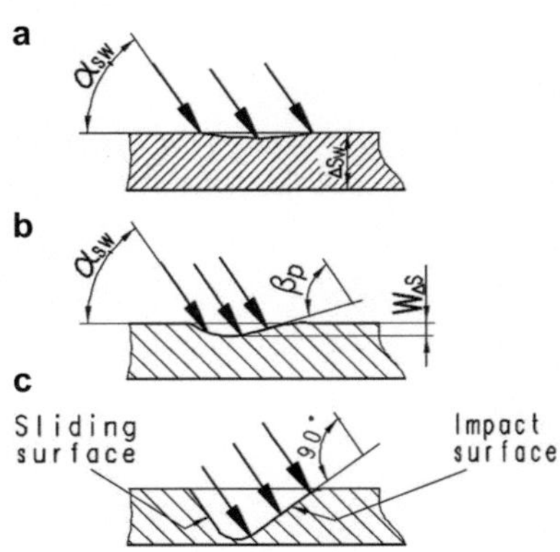

Fig. 8.13 Formation of a wear groove. **a** → **c** increasing operating time

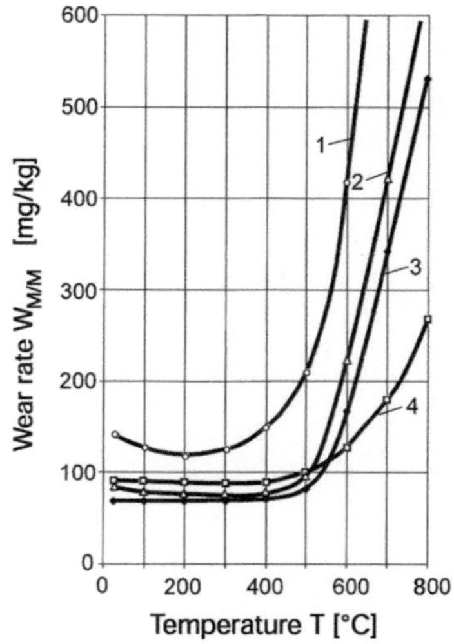

Fig. 8.14 Temperature dependence of impact wear: 1 – steel, 0.2% C, $H_W = 1.30\,\mathrm{kN/mm^2}$, 2 – steel, 0.8% C, $H_W = 1.90\,\mathrm{kN/mm^2}$, 3 – steel, 0.2% C, 4 % Cr, 2.3% V, 9% W, $H_W = 2.40\,\mathrm{kN/mm^2}$, 4 – cast iron, 2% C, 34% Cr, $H_W = 2.90\,\mathrm{kN/mm^2}$; $\alpha_{S,W} = 90°$, $u_{P,0} = 48\,\mathrm{m/s}$, quartz sand $d_S = (0.4–0.6)$ mm, Vickers hardness H_W at room temperature

Fig. 8.14 shows, as an example, the temperature dependence of the wear parameter$W_{M,M} = W_M/\Delta m_S$ of various steel/cast iron materials when shot with quartz sand: particle diameter $d_S = (0.4–0.6)$ mm, angle of impact $\alpha_{S,W} = 90°$, impact speed $u_{P,0} = 48\,\mathrm{m/s}$ [7]. Up to approx. (400–450) °C only insignificant changes can be observed, with normal steel, curve 1, even a slight wear reduction. In the temperature range $T < 450\,°\mathrm{C}$ the forming iron oxide layers, which have a greater hardness than the base material and whose thickness increases with temperature, act as wear protection. At higher temperatures, these layers are oxidized differently over their thickness, and their bond to the base material becomes weaker. They are easily detached by the particle bombardment = wear removal, but they form relatively quickly again, are removed again, etc.

Comparable dependencies, as shown in Fig. 8.15, are also measured in abrasive wear. The wear of the wall material sintered corundum only increases at temperatures $T > 1000\,°\mathrm{C}$ again.

Own practical experiences confirm the above relationships: In the pneumatic transport of iron ore through insulated conveyor pipes made of normal steel at $T_S = 20\,°\mathrm{C}$ and $T_S = 300\,°\mathrm{C}$ it was measured that at 20 °C a greater pipe/bend wear occurred than at 300 °C: $W_M(20\,°\mathrm{C}) > W_M(300\,°\mathrm{C})$.

So far, air has been assumed as the intermediate medium. If the bulk material is treated in another gas atmosphere, completely different temperature dependencies can occur: In an inert atmosphere, for example, no oxide layers are formed.

- **Combination effects:** As indicated in the preceding paragraph, the type and composition of the surrounding medium, among other things, but also the moisture content of the bulk material and/or a possibly parallel corrosion attack influence the amount of wear. Such combination effects often increase the wear removal. Details can be found in [1, 2, 10].

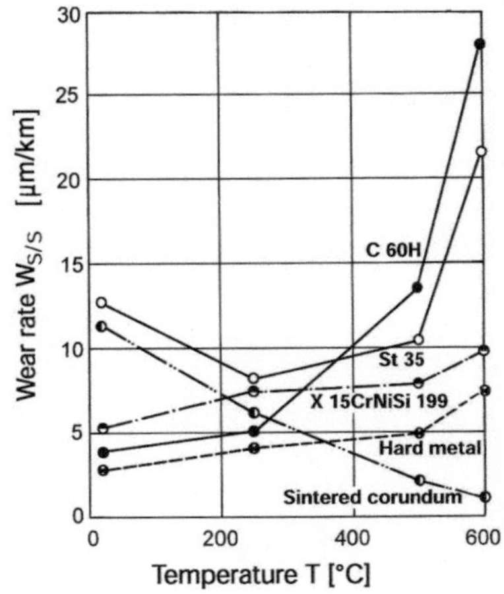

Fig. 8.15 Temperature dependence of sliding wear, measurement in wear pot, sliding speed $u_S = 2.8\,\text{m/s}$, corundum, $d_S = (3–5)$ mm [2]

b) Abrasive wear:

- **Contact force F_N:** The total force F_N perpendicular to the bulk material layer, see Fig. 8.1a, can be converted into a nominal wall normal stress $\sigma_W = F_N/A_W = F_N/(N_P \cdot d_S^2)$. The wear increase with increasing F_N or σ_W, resulting, among other things, from the self-weight of the bulk material, overlying loads, etc., is approached with the assumption

$$W_M \propto \sigma_W^n \quad \rightarrow \quad \text{measured: } 0.7 \leq n \leq 1.2 \tag{8.15}$$

Theory:

$n = 1$, for ductile wall material, (8.10),
$n = (1.125–1.250) > 1$, for brittle wall material, (8.11).

- **Sliding speed u_S** : There are no really clear results available. The friction coefficient μ_f describing the interaction between bulk material and wall is at low sliding speeds, $u_S \lesssim 1\,\text{m/s}$, almost constant and then moderately increasing. A similar wear behavior can be expected. Figure 8.16 shows measurement results from [20]. These lead to $W_{\Delta s} \propto u_S$, i.e. $n \cong 1$. The dependence of wear removal on the sliding path l results here in $W_{\Delta s} \propto l^{0.5}$.

c) Impact wear:

- **Impact speed $u_{P,0}$:** A portion of the kinetic energy $E_P = 1/2 \cdot m_P \cdot u_{P,0}^2$ of a bulk material particle is irreversibly transferred to the wall material upon impact. This local energy input leads to local deformations and/or the disintegration of the wall material, resulting in the detachment of wall components. The influence of $u_{P,0}$ on the wear rate can be described by

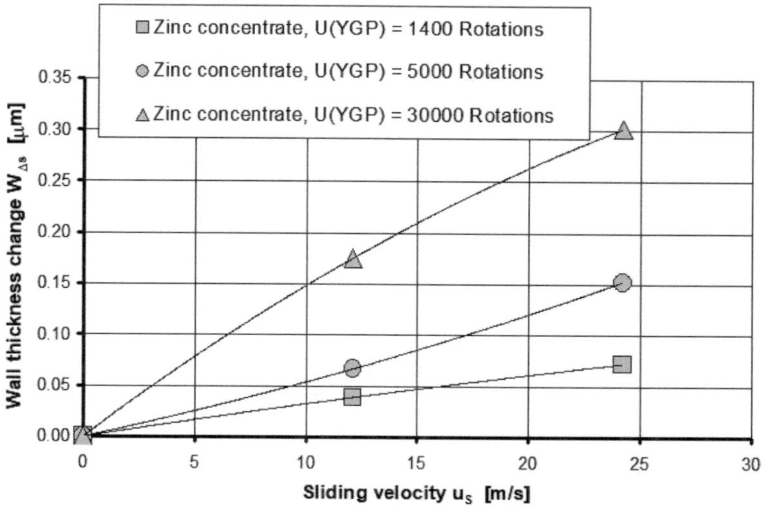

Fig. 8.16 Sliding wear: zinc concentrate against wall material "Creusabro 4000"

$$W_M \propto u_{P,0}^n \quad \rightarrow \quad \text{measured: } 1.8 \leq n \leq 4.0 \tag{8.16}$$

Theory: $W_M \propto E_P \rightarrow n = 2$,

$n = 2\text{–}3$,	for ductile wall material, (8.12) and (8.13),
$n \cong (2.44\text{–}3.20) > 2$,	for brittle wall material, (8.14).

The exponent for common steels is in the range $n \cong (2.2\text{–}2.5)$, for distinctly brittle materials, n approaches $n \cong 4$. Below a critical impact speed $u_{P,\min}$ reduces the wear removal W_M due to the low energy input and falls below the values of (8.16). When exceeding a speed $u_{P,\max}$, particle breakage increases. The energy required for this is missing for wear generation, i.e., measured W_M values are then smaller than calculated with (8.16). Figure 8.17 shows the basic course $W_M(u_{P,0})$. The speeds $u_{P,\min}$ are in the range of fewer (m/s), those of $u_{P,\max}$ are generally significantly larger than approx. 100 m/s.

The value of the exponent n is to a small extent dependent on the angle of impact $\alpha_{S,W}$. For practical purposes, however, this dependency can be neglected. Blowpipe measurements in [22, 27, 28] indicate that n is obviously also dependent on the collision density. $\dot{n}_P = \dot{N}_P/A$, with: $\dot{N}_P$ = particle flow, A = impact area, is influenced (→ see Fig. 8.20). The size of the proportional factor between W_M and $u_{P,0}$, (8.16), depends on the type of wall material, the characteristics of the bulk material, and the angle of impact.

- **Angle of impact** $\alpha_{S,W}$: The dependency $W_M(\alpha_{S,W})$ shows a maximum, the height and position of which are predominantly determined by the properties of the wall material, to a lesser extent by the particle properties, and hardly by the particle impact

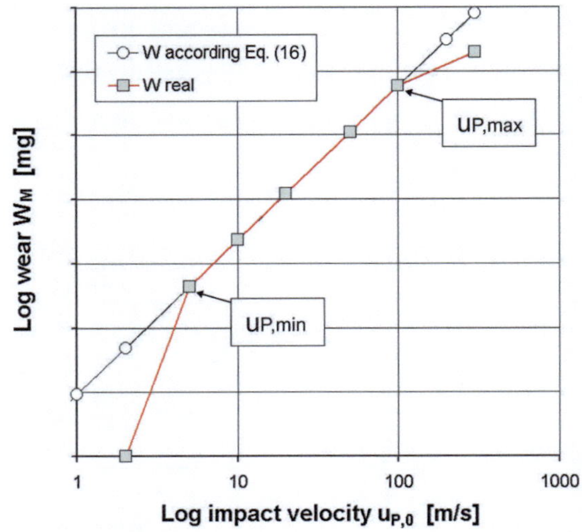

Fig. 8.17 Impact wear: basic course of the dependency $W_M(u_{P,0})$

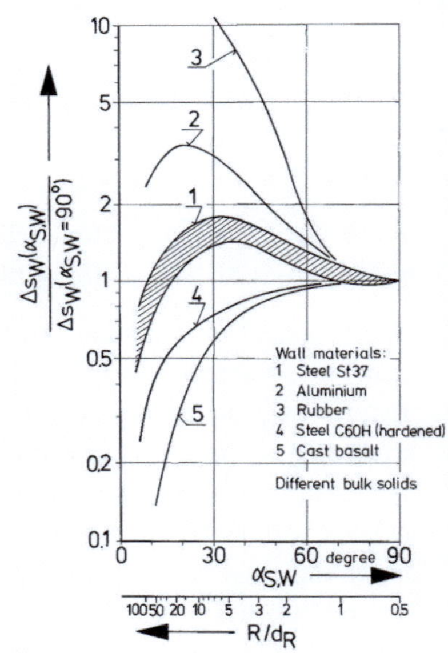

Fig. 8.18 Impact wear: Dependencies of impact angle and wall material

velocity $u_{P,0}$. This is shown in Fig. 8.18. Plotted against the angle of impact $\alpha_{S,W}$ is the wall thickness reduction Δs_W of the base body, normalized with its value at $\alpha_{S,W} = 90°$. $\alpha_{S,W}$-values are assigned to the lower part of Fig. 8.18 for conveying pipe bends with the same impact angle, which are characterized by their ratio of radius to pipe diameter R/D_R, see Sect. 8.2. The different wall materials show wear maxima in various impact angle ranges. It can be roughly divided into the following classes:

– brittle materials, e.g. ceramics, hardened steels, cast basalt:
  $(\alpha_{S,W})_{\max} \to 90°$,

- ductile/brittle materials, e.g. construction and C-steels, cast iron, plastics: $25^\circ \lesssim (\alpha_{S,W})_{\max} \lesssim 55^\circ$,
- ductile and rubber-elastic materials, e.g. aluminum, Al-alloys, rubber: $10^\circ \lesssim (\alpha_{S,W})_{\max} \lesssim 30^\circ$.

Due to the low deformability of brittle materials, particle bombardment at $\alpha_{S,W} = 90^\circ$ in these cases, leads to a disruption of the grain structure in near-surface wall layers and to brittle fracture, while ductile materials are permanently deformed by shallowly impacting particles until the developing lamellae tear off [20–22]. This is shown in Fig. 8.19.

a

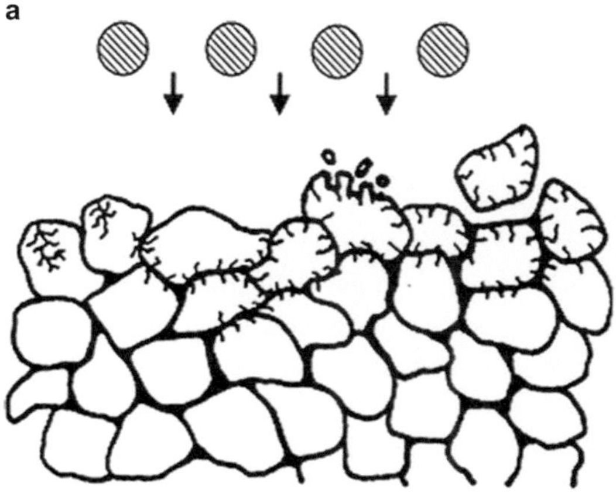

- Induction of microcracks in the grain
- Induction of microcracks at the grain boundaries
- Grain bond fatigue
- Bursting out of grain fragments
- Bursting out of individual grains due to weakening of the grain bond

b

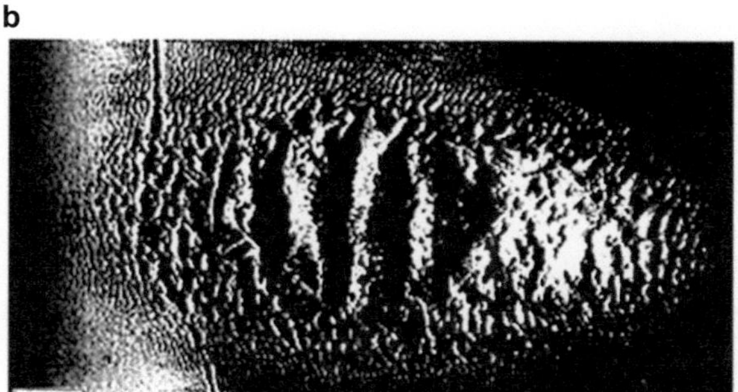

Fig. 8.19 Effects of particle bombardment during impact wear. **a** brittle wall material [21], **b** ductile wall material [22]

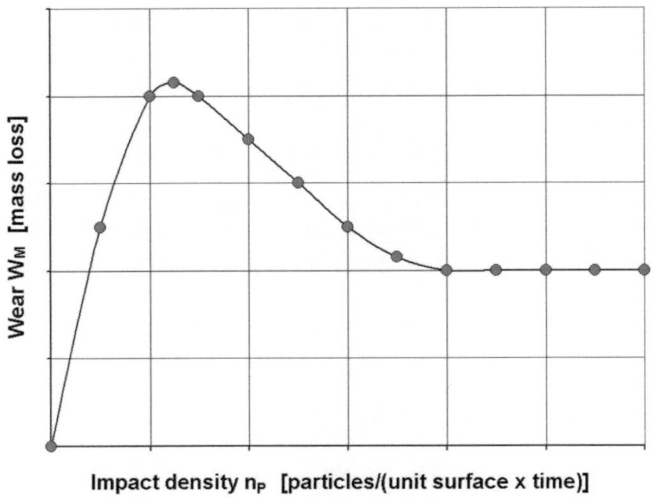

Fig. 8.20 Jet wear: dependence on impact density $\dot{n}_P$

- **Impact density n_P:** This indicates the number of particle impacts per unit of time and area, $\dot{n}_P = \dot{N}_P/A$, and is proportional to the loading μ in a pneumatic conveying system: $\mu = \dot{m}_S/\dot{m}_F = m_P \cdot \dot{N}_P/(A \cdot \varrho_F \cdot v_F) = \dot{n}_P \cdot m_P/(\varrho_F \cdot v_F)$. At low impact densities $\dot{n}_P$, individual bulk material particles hit the base body at greater distances from each other; incoming and rebounding particles do not influence/disturb each other; the wall contacts can be treated as isolated individual contacts. Increasing $\dot{n}_P$ results in increased wear removal W_M due to the larger (impact number/area) and increasing mutual influence of the contacts, e.g., by superimposition of stress fields in the material. With further increased impact density, there are more collisions between incoming and rebounding particles; this reduces the impact velocity of a part of the incoming bulk material flow, i.e., the rebounding particle cloud protects the wall surface; the wear removal W_M decreases again. Figure 8.20 shows the basic course of $W_M(\dot{n}_P)$, which is confirmed by measurements [7].

II) Bulk material properties

a) In general:

- **Particle shape:** Abrasive and impact wear are significantly greater for wall stress caused by compact, angular bulk material particles than for rounded particles. This is due to the additional material removal caused by grooving, i.e., the deeper penetration of the edges into the wall material surface. Elongated or flat particles are less abrasive. They align themselves in preferred directions within the bulk material. It is recommended to distinguish different particle shapes by suitably defined shape factors

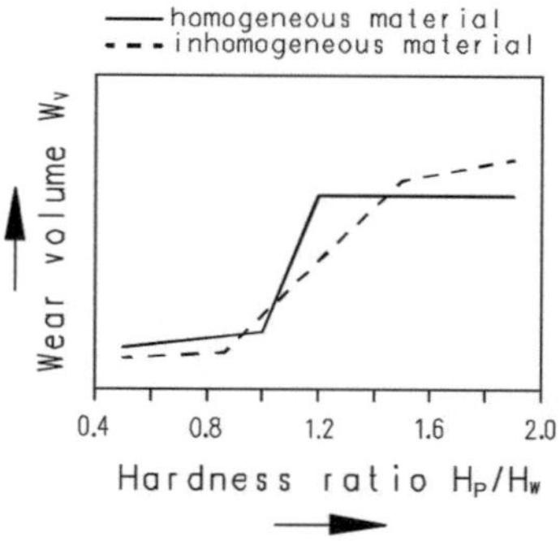

Fig. 8.21 Low level/high level wear characteristic

when evaluating empirical results in order to establish comparability. The current stress situation can also have an influence: In the case of impact wear, for example, a significantly greater grain shape influence is observed at low and medium impingement angles $\alpha_{S,W}$ than at 90°-impact [7].

- **Particle hardness H_P, low level/high level characteristic:** It is obvious that a particle hardness H_P, which is greater than the wall material hardness H_W, causes higher wall wear than in the reverse case $H_W > H_P$. In Fig. 8.21, the wear removal W_V of the stressed wall is plotted against the hardness ratio (H_P/H_W). In a relatively narrow range around $(H_P/H_W) \cong 1$, W_V increases from the so-called low wear level, $(H_P/H_W) < 1$, to the much higher high wear level $(H_P/H_W) > 1$. This is achieved at $H_P \gtrsim 1.2 \cdot H_W$, see (8.5). Along the high and low plateaus, W_V generally changes only slightly with (H_P/H_W). Figure 8.21 indicates that the structure and composition of the wall material also influence the behavior.
- **Particle strength:** Insufficient strength of the bulk material particles can lead to particle breakage, grain abrasion, etc., both in abrasive and impact stress. The energy required for this is missing for wear generation. An increased proportion of particle breakage thus reduces wear removal compared to that of a particle system with high grain strength, i.e., without breakage processes.

b) Abrasive and impact wear:

- **Particle diameter d_S:** Since particle size distributions are generally available, these are described by the mean particle diameter $d_{S,50}$ defined at the sieve residue $R = 50$ M.-%. Measurements with the same solid in different grinding conditions under identical boundary conditions lead to the dependency:

$$W_M \propto d_{S,50}^n \quad \rightarrow \quad \text{measured: } n \leq 1 \tag{8.17}$$

Abrasive wear:
Theory: $n \cong (0.25-0.50) < 1$, for brittle wall material, (8.11) [6], for ductile wall material: d_S does not appear explicitly in the theoretical approaches.

Impact wear:
Theory: $n \cong 0.67 < 1$, for brittle wall material, (8.14), for ductile wall material: d_S does not appear explicitly in the theoretical approaches.
Example of a impact wear measurement [23]: ten different quartz sands with particle diameters ($20\,\mu m \leq d_{S,50} \leq 5000\,\mu m$) against a steel St 52-2, impact velocity $u_{P,0} \cong 11.7\,m/s$, angle of incidence $\alpha_{S,W} = 90°$, measurement under ambient conditions, exponent of (8.17): $n = 0.65$.

III) Wall Material Properties

The range of wear materials used extends from simple unalloyed steels to alloyed steels, corresponding cast materials, hard metals, ceramics, combined composite materials, and polymers and rubber-elastic materials, all of which are either designed as solid base bodies or applied as wear protection layers on them. In the latter case, the quality of the production, e.g., pore-free, bonding with the base material, etc., is essential for the level of wear resistance that is achieved.

Basic structure of a wear resistant material: heterogeneous structure consisting of:

- Base material = Matrix; responsible for strength, stiffness, toughness, resistance to deformation and fracture,
- Hardness carrier = hard phase, hard material component; responsible for resistance to wear. Important: Hardness carrier must be firmly integrated into the matrix.

Example: Increasing C content of a steel and addition of carbide formers, e.g., Cr, Mo, W, promotes the formation of metal carbides = hardness carriers. This leads to an increase in wear resistance while simultaneously reducing toughness.

It is reminded that the surface properties of a wall material differ from those in the material interior after prolonged exposure to a bulk material, see Fig. 8.4, i.e., the characteristics of the unstressed material only approximately describe its wear behavior.

a) In general:

- **Wear resistance:** Under identical stress, measured wear removals W, or wear resistances $1/W$, of various wall materials are exemplarily shown in Fig. 8.22 [20]. On the ordinate, the volume wear normalized with the corresponding value of the steel St 37-2 is displayed, while on the abscissa, the materials characterized by their Vickers hardness $HV = H_W$ are plotted. Measurement conditions: jet wear, bulk material = quartz sand, $d_S = (2–3)\,mm$, $u_{P,0} \cong 11.7\,m/s$, $\alpha_{S,W} = 90°$, cast materials were examined without cast skin. Figure 8.22 illustrates the range of variation with which the different wall materials react to the same stress.
  In the cited example, an impingement angle of $\alpha_{S,W} = 90°$ was measured. Some of the investigated wall materials, as shown above, are unsuitable for this angle. Measurements at a shallower impingement angle $\alpha_{S,W}$ would thus significantly change the relative size of the wear resistances of the individual wall materials among each other.

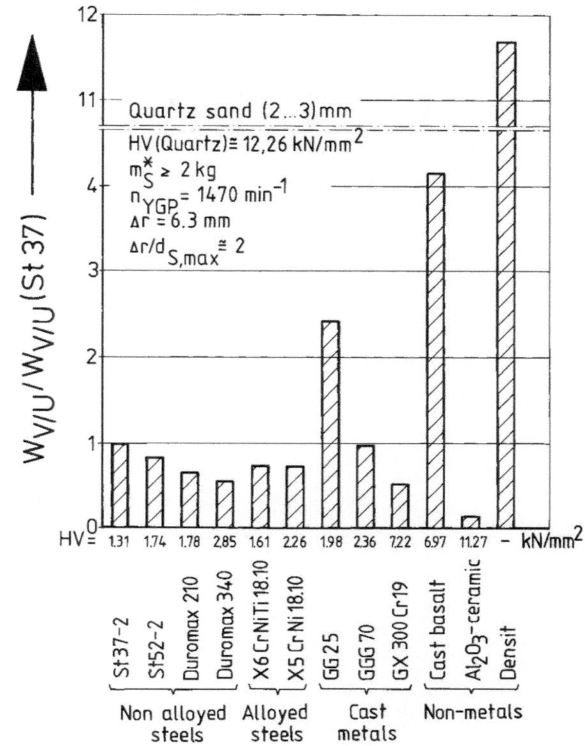

Fig. 8.22 Erosive wear of different wall materials under stress from quartz sand, impingement angle $\alpha_{S,W} = 90°$, $u_{P,0} = \text{const.}$

This applies in general: The ranking of wear resistances $1/W$ of different wall materials depends on the respective prevailing stress conditions. A material assessment must take this into account.

- **Influence of Material Hardness—Fracture Toughness:** The explanations in Sect. 8.1.3 show that the wear rate of ductile wall materials is determined by their hardness H_W, while for brittle wall materials, it is determined by a combination of wall hardness H_W and fracture toughness K_{Ic}. The influence of wall hardness H_W in relation to particle hardness H_P is described by the low level/high level characteristic, see Fig. 8.21. To safely achieve the low level wear area, a hardness ratio $(H_P/H_W) < 0.8$, i.e., a wall hardness $H_W > 1.25 \cdot H_P$, is necessary. In Table 8.2 the Vickers hardness of some bulk materials, H_P, and wall materials, H_W, are compiled. The comparison shows that low wear is difficult to achieve when handling mineral bulk materials with "normal" wall materials, e.g., simple steels.

  If, in addition to the hardness H_W of the base body, its fracture toughness K_{Ic} also influences the amount of wear, see (8.11) and (8.14), then it should be considered that hardness H_W and fracture toughness K_{Ic} are inversely proportional to each other: with increasing H_W, K_{Ic} decreases. The resulting relationships are shown in the measurement-generated Fig. 8.23 [15, 24]. The wear resistance $1/W$ increases with increasing fracture toughness K_{Ic} at first, but after exceeding a maximum value, it decreases

Table 8.2 Vickers hardness of various bulk materials and wall materials

Bulk solid / wall material	Vickers hardness H_P / H_W [kN/mm²]
Limestone	0.90 - 1.40
Hard coal	0.30 - 1.20
Gypsum	≅ 0.60
Anhydrite	≅ 1.60
Dolomite	1.50 - 3.70
Coke	2.50 - 6.00
Felspar	6.00 - 7.50
Blast furnace slag	≅ 6.40
Quartz	9.00 - 13.00
Corundum	≅ 18.00
Construction steel	0.95 - 2.00
Non-alloyed cast steel	1.15 - 2.10
High-alloyed cast steel	2.10 - 6.30
Cast iron	1.05 - 3.15
Cast basalt	≅ 7.40
Alloyed hard metal / white iron	4.50 - 7.65
Sintered hard metal	11.00 - 18.50
Al_2O_3-oxide ceramics	12.00 - 24.00
Steel phase: cementite	≅ 8.60
Steel phase: ferrite	≅ 1.07

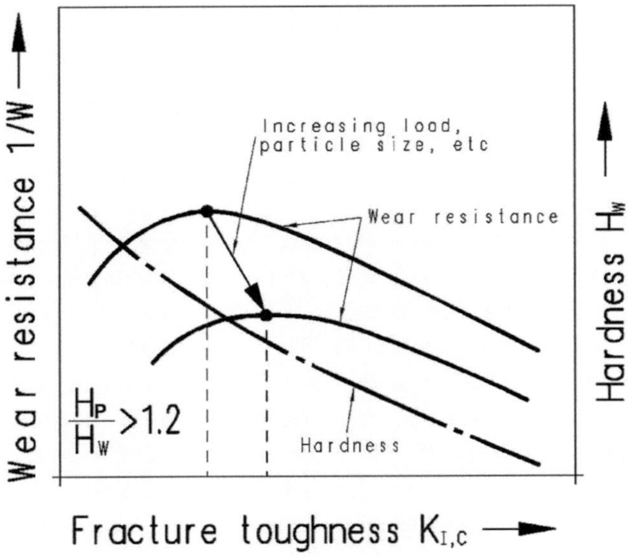

Fig. 8.23 Wear resistance as a function of fracture toughness and material hardness

again with further K_{Ic} increase. The reason for this is the continuously decreasing hardness H_W in the same direction. Thus, there exists an optimal (H_W, K_{Ic})-combination, at which the wear resistance becomes maximal, but which depends on the current stress situation. This fact applies to both abrasive and erosive wear and is confirmed by practical experience.

- **Microstructure:** Its influence on wear behavior is exemplified by a composite material, consisting, for example, of a metallic matrix with embedded ceramic particles as hardening carriers. This combination combines the ductility and high fracture toughness of the metal with the great hardness and high elastic modulus of the ceramic. The resulting wear resistance can be influenced by:
  - Hardness, size, shape, volume fraction, and distribution of the dispersed hardness carrier,
  - Properties of the matrix material,
  - Strength of the bond between the two phases.

  When subjected to bulk material, the wear behavior is, among other things, determined by the ratio of the diameter of the hardness carrier particles to the size of the wall contact caused by the bulk material particles, e.g., depth and width of a groove in abrasive wear. Approximation: contact size $\propto$ bulk material particle size. It applies, see Fig. 8.11:
  - Contact size $\gg$ diameter of hardness carrier particles: The composite material behaves like a hard homogeneous material. The ratio of the hardness of the solid particles H_P to that of the hardness carrier particles $H_{W,HT}$ as well as the fracture toughness of the composite material determine the size of the wear.
  - Contact size $\leq$ diameter of hardness carrier particles: Matrix and hardening carrier react specifically, depending on the stress conditions. If $H_P < H_{W,HT}$, the matrix is preferentially destroyed by wear removal. In the case of $H_P > H_{W,HT}$ and simultaneously high load, the hardness carrier particles are increasingly prone to fracture. As already shown in Sect. 8.1.2 and Fig. 8.11, the strength of the bond between the hardness carrier particles and the matrix also determines the size of the wear removal. The detachment of hardness carrier particles represents a significant material loss.

Material selection for abrasive wear:
For lower wall stresses, $F_P < F_{P,c}$, (8.8), wall materials with high hardness H_W should be used. For higher stress, the material must also have a sufficiently high fracture toughness K_{Ic}.

Material selection for impact/erosive wear:
For shallow to medium impingement angles, $\alpha_{S,W} \lesssim 60°$, wall materials with high hardness H_W are suitable; the dominant wear mechanism is micro-cutting. For impingement angles ($60° \lesssim \alpha_{S,W} \leq 90°$) micro-cracking and material disruption are dominant; they require sufficient fracture toughness K_{Ic} and ductility. In many applications, simple rolled steels have proven to be effective.

IV) Geometric/constructional influences

a) In general:

- **Avoidance of disturbance points:** Fig. 8.24 shows examples of disturbance points in pneumatic conveying lines, which often cause considerable local wear. Triggers are unwanted deflections of particle paths, impact edges due to misalignment of the base body, etc. Comparable effects are also observed in abrasive stress. Examples: Longitudinal grooves parallel to the sliding direction due to a coarser surface treatment of the base body or due to gaps in the installation of wear plates can act as forced guides for wall-close particles. These particles lay their wear-generating edges in the predetermined grooves and deepen them further. Cross-sectional constrictions in the sliding direction result in increased wall pressure, etc.

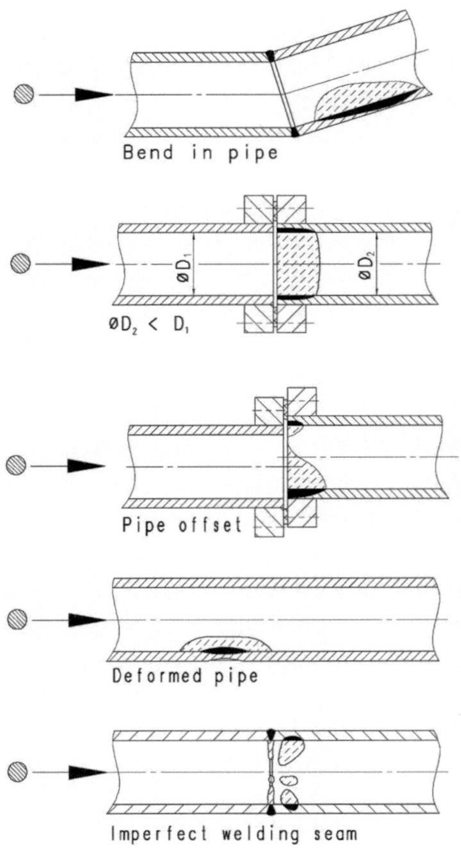

Fig. 8.24 Examples of disturbance points in a pneumatic conveying line

b) Abrasive wear:

- **Autogenous wear protection**: The wear system is designed in such a way that a stationary bulk material layer forms between the wall and the solid material flow sliding over it. Thus, bulk material slides against bulk material, relieving the wall. Figure 8.25a shows the principle, which is also used, for example, in the design of transport grates in clinker cooler construction.
  Fluidisation: In many applications, a drastic reduction in wear is possible through fluidisation of the (→ fine-grained) bulk material. For this purpose, the stressed wall must be designed as a gas-permeable fluidisation floor, e.g., by means of an applied, possibly metallic, mesh. Figure 8.25b illustrates the design. The bulk material "floats" over the wall. The service life of such fluidized floors is in the range of several years.
- **Avoidance of jamming effects:** Gaps between moving/rotating equipment internals and the adjacent equipment wall, e.g., mixer blades or rotary valve blades and the surrounding housing, must be either much smaller or significantly larger than the particle size of the bulk material. The highest wear occurs when gap width Δh and particle diameter d_S are of approximately the same size, see Fig. 8.26. The bulk material is then neither passed through nor rejected by the gap, but jams and moves relative to the wall.

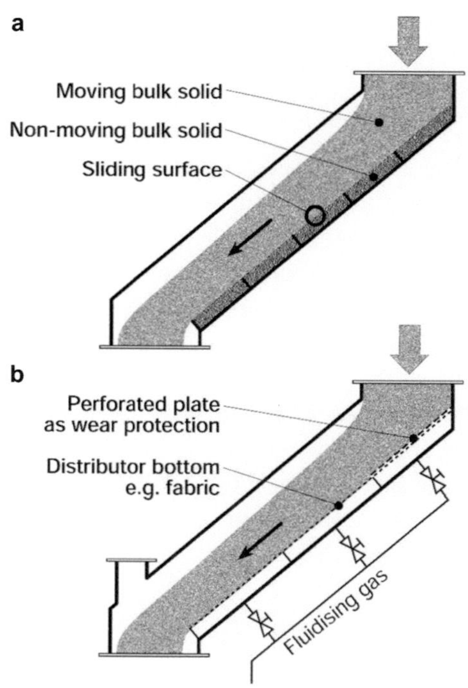

Fig. 8.25 Wear protection under sliding stress. **a** Autogenous protection, **b** Fluidisation

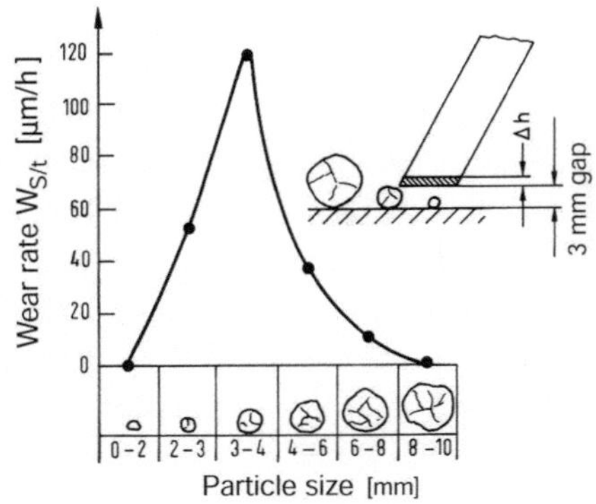

Fig. 8.26 Wear due to jamming effects [25]

c) Impact wear:

- **Autogenous wear protection**: In Fig. 8.27b–e, variants of the principle "bulk material protects against bulk material" are shown using the example of 90° deflections in a pneumatic conveying line. A buffering solid layer is built up between the pipe wall and the impacting bulk material flow. Thus, the bulk material hits bulk material, protecting the pipe wall [20].
- **Adjustment of impact surface orientation** etc.: Fig. 8.27f shows a 90° bend lined with abrasion-resistant polyurethane, whose impact side is designed with an impact/inclination angle $\alpha_{S,W}$ optimally, i.e., wear-minimally, adapted to the wear properties of this wall material. Comparable solutions with other wall materials are also possible in other impact situations.

The above explanations describe the general and essential tendencies of wear behavior for the user. In detail, surprising deviations are possible. Example: Moderate stress on a brittle wall material by a fine-grained bulk material, usually $d_S \lesssim 10\,\mu\text{m}$, can lead to ductile wear behavior of the otherwise brittle material. When bombarding a brittle wall plate with the fine-grained solid, a wear maximum occurs at a shallow bombardment angle $\alpha_{S,W}$, similar to a ductile material, while bombardment with coarse bulk material shows the expected brittle behavior with $(\alpha_{S,W})_{\max} \to 90°$ [10]. The reason for this is that the penetration volume of the fine-grained particles is smaller than the low, but existing, local plastic deformability of the brittle material and can therefore be absorbed by it.

It should be noted that the size of the exponents n in the equations discussed above can also depend on the chosen wear characteristic, e.g., W_M, $W_{M,M}$, $W_{M,\tau}$. Details can be found in [23].

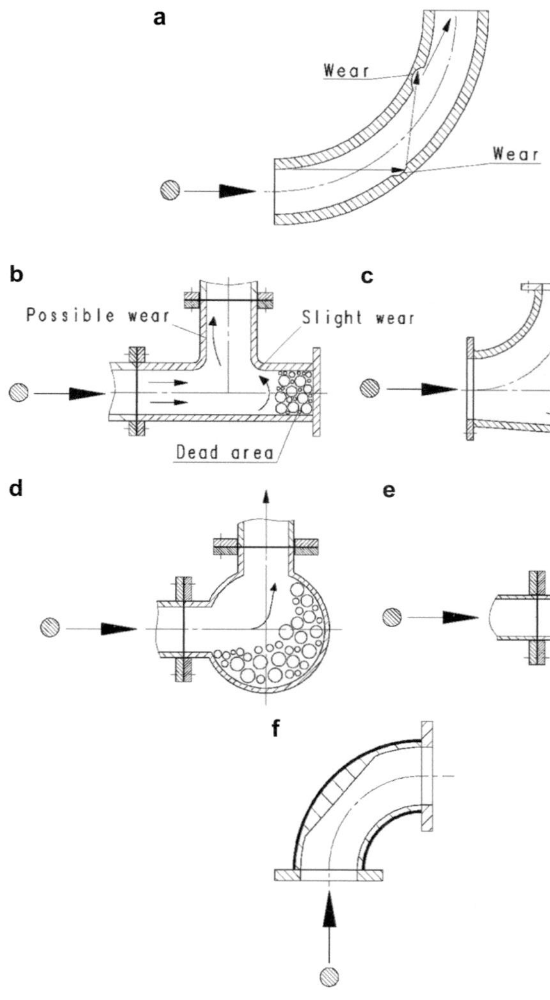

Fig. 8.27 Impact wear: Variants of 90° deflections

8.2 Wear in Pneumatic Conveying Systems

The following considerations serve to clarify/apply the above explanations to various conveying line elements and some selected locks of pneumatic conveying systems.

8.2.1 Conveying Line

Due to the possible different flow conditions in the conveying sections, various stress and wear mechanisms occur in them.

a) Vertical straight pipe

In the dilute phase pneumatic conveying area, the bulk material is relatively evenly distributed across the pipe cross-section. The particles collide with the surrounding pipe wall as a result of lateral movements caused by the generally turbulent gas flow, mutual collisions, or reflections from previous wall impacts. Predominantly, sliding jet stress occurs with a relatively shallow impact angle and low particle impact density. Bulk material strands do not rest on the pipe wall. The axial mechanical forces on a plug during plug conveying and thus the wall normal stresses σ_W are small. Consequence: only minor and evenly distributed surface wear over the pipe circumference.

b) Horizontal straight pipe

Due to the gravity acting perpendicular to the conveying direction, the impact density during pneumatic conveying in the lower third of the pipe is higher than on the rest of the pipe surface. Moving strands also rest on the wall in this pipe sector. Both result in a relatively low surface wear in the lower third of the conveying pipe, see Fig. 8.28a, b.

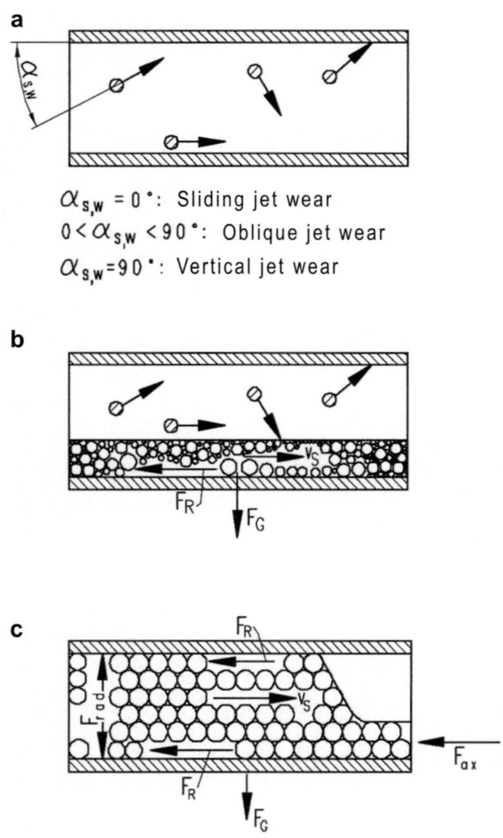

Fig. 8.28 Flow patterns and associated wear mechanisms in horizontal conveying. **a** Dilute pneumatic conveying, **b** Strand conveying, **c** Plug conveying

In the plug conveying of cohesionless bulk materials, a stationary strand is deposited between the individual plugs. Each of the plugs picks up bulk material from this strand at its front side and releases about the same amount at its back side. The result is, among other things, a mechanical force on the plug front, which generates considerable wall normal stresses $\sigma_{S,\mathrm{rad}} \cong \sigma_W$. The stress component acting on the lower pipe area from the plug weight is generally small compared to $\sigma_{S,\mathrm{rad}}$. Consequence: Especially with hard and angular coarse-grained bulk material particles, high and evenly along the pipe circumference distributed groove wear due to microcutting is possible, see Fig. 8.28c.

Practical example of extreme groove wear [29]: Plug conveying of hard, pressed activated carbon particles: cylindrical particles, $\varnothing 2.5\,\mathrm{mm} \times (3-10)\,\mathrm{mm}$, sharp-edged. Plug velocities in the range of a few meters per second. Shortly after commissioning, the original stainless steel conveying line had to be completely replaced by a line protected by a cast basalt lining. Cause of wear: The first longitudinal grooves and furrows in the pipe surface, torn open by the particle edges, acted like forced guides for the wall-close particles of subsequent plugs. These particles placed their wear-generating edges in the predetermined grooves and deepened them further, i.e., the wear was locally concentrated. The use of cast basalt increased the hardness of the pipe wall from $H_W(\text{stainless steel}) \cong 1.6\,\mathrm{kN/mm^2}$ to $H_W(\text{cast basalt}) \cong 7.0\,\mathrm{kN/mm^2}$, i.e., the hardness ratio (H_S/H_W) shifted towards the low level area ($\rightarrow$ for comparison: $H_S(\text{coke}) \cong 5.9\,\mathrm{kN/mm^2}$) and on the other hand, the usable wear volume/wall thickness increased.

c) Deflections

First, 90° pipe bends are considered. In deflections, larger centrifugal forces act on bulk material particles due to their greater mass than on the carrier gas. The result is a largely separation of both phases. In dilute phase pneumatic conveying, a locally impacting bulk material jet forms on the pipe wall, creating a local wear trough. Strands entering a pipe bend, depending on the spatial orientation of the deflection, are pressed more strongly against the outer pipe wall by the centrifugal force ($\rightarrow$ e.g., in a bend from the horizontal to the vertical upwards) or locally impact on it ($\rightarrow$ e.g., in a bend from the vertical upwards to the horizontal), i.e., both sliding and impact wear are possible here. Compact bulk material plugs experience an additional mechanical jamming force on their front side in deflections, resulting in an increase in the wall normal stress σ_W.

In general, the local trough wear caused by jet formation is critical. This is examined in more detail: Solid particles uniformly distributed over the entrance pipe cross-section of the 90° bend shown in Fig. 8.29 hit the outer pipe wall at different impact angles $\beta_P = \alpha_{S,W}$. The most stressed point of the developing wear trough is located approximately on the straight extended trajectory of the particles entering the bend on the inner pipe side, see Fig. 8.29. The associated critical impact angle is:

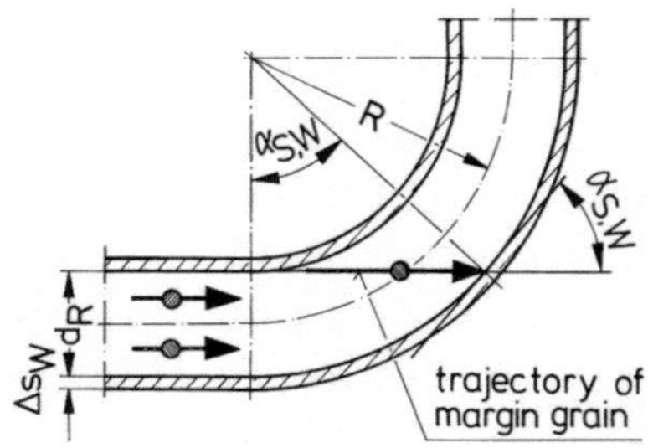

Fig. 8.29 Position of the fracture site or the primary impact point in a 90° bend

$$\cos \alpha_{S,W} = \frac{2 \cdot \frac{R}{D_R} - 1}{2 \cdot \frac{R}{D_R} + 1} \quad \rightarrow \quad \text{with: } \frac{\Delta s_W}{D_R} \leq 0.01 \tag{8.18}$$

(8.18) establishes a relationship between the geometric data of a 90° bend (→ R = bend radius, D_R = pipe inner diameter) and the impact angle. $\alpha_{S,W}$. In Fig. 8.18, the wall thickness changes Δs_W related to the case of $\alpha_{S,W} = 90°$ of such bends against the impact angle $\alpha_{S,W}$ and the corresponding (R/D_R)-ratios are plotted for different wall materials. The critical, i.e., wear-promoting, (R/D_R)-ratios can be directly taken from the diagram for the various wall materials: In case of impact wear and a given pipe diameter D_R narrow bending radii R for brittle wall materials, very large radii R for ductile materials, and for brittle/ductile wall materials, e.g., normal steel, (R/D_R) ratios in the range $(R/D_R) \cong (5-10)$ should be avoided for wear-related reasons.

Fig. 8.30 shows an example of the wear profile along a 180° Plexiglas bend shot with steel sand with $(R/D_R) = 3.33$ [30]. (8.18), applied to its inlet side, predicts the wear trough, i.e., the greatest wear, at a circumferential angle $\alpha_U = \alpha_{S,W} = 42.4°$ (→ see also Fig. 8.29). The measured wear trough lies in a narrow range around $\alpha_U \cong 40°$. Figure 8.30 clearly illustrates that deflections in a pneumatic conveying line represent the wear-critical components.

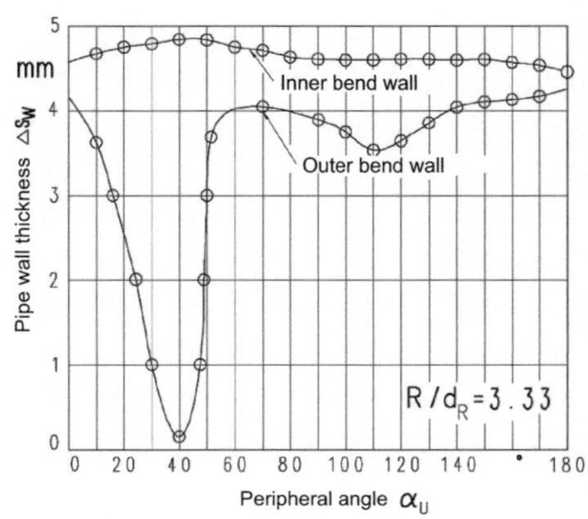

Fig. 8.30 Wear profile along a 180° bend

In Fig. 8.30 at $\alpha_U \cong 110°$ a smaller second wear trough can be seen. Such secondary wear due to particle reflections is also observed in 90° bends, especially with large (R/D_R) ratios, see Fig. 8.27a. Higher loadings μ can lead to wear-reducing coverage of part of the bend surface due to a bulk material strand formed by gas/solid separation. Since gas/solid separations in deflections affect the subsequent conveying pipe sections, it is absolutely necessary to design a pipe area of at least $\Delta L_R \gtrsim (5 \cdot D_R)$ behind bends to be particularly wear-resistant. Practical example: A 90° bend lined with cast basalt is always followed by an approximately 1.0 m long cast basalt pipe section.

An alternative approach to reducing wear in deflections is the so-called autogenous wear protection, already shown in Fig. 8.27b–e, in which the respective deflection is designed in such a way that a buffering bulk material layer forms between the pipe wall and the impacting solid flow. Thus, bulk material hits bulk material, relieving the pipe wall. Own investigations on T-Bends (→ Fig. 8.27b), Vortex elbows (→ Fig. 8.27c) and deflection pots (→ Fig. 8.27d) confirm that a stationary bulk material cushion with little exchange builds up in these, thus effectively forming a bend with a small (R/D_R)-ratio and an impact surface made of bulk material. Such systems can be used for both dilute and dense phase conveying. Their pressure loss is equal to or only slightly higher than that of alternative 90°-bends. When arranging these deflectors at the end of the vertical section of a vertically upward and then horizontally flowed through conveying pipe, pressure and thus throughput fluctuations may occur. Cause: periodic build-up and breakdown of the bulk material cushion in the deflector due to too low solid loading μ and/or too low gas velocity v_F. The incoming gas/solid mixture cannot hold the full weight of the bulk material cushion in the deflector. The resulting pressure pulsations can be significant and may force throughput reductions. Furthermore, it must be considered that such deflection systems are not completely residue-free.

The effect of a concrete-filled U-profile on the outside of a pipe bend is also based on the self-protection discussed above, see Fig. 8.27e: In the concrete, which has only a low particle wear resistance, a trough-shaped washout is formed by the particle bombardment, which is filled with bulk material particles if it is of sufficient size. For this practical solution, a minimum thickness of the concrete layer is required.

The following Table 8.3 compares exemplarily relative service lives of different 90° deflections, which were stressed in the same operating conveyor system by a zircon compound [26]. It shows that autogenous wear protection (→ T-Bend) in this specific application case leads to an enormous extension of service life compared to the various 90°-pipe bends. Further information on deflectors can be found in, among others, [31].

d) Flow obstacles/interference points

Flow obstacles and other interference points in the conveying line generally lead to a change in the trajectory of the bulk material particles and, as a result, to high and locally concentrated wall wear. Figure 8.24 shows typical errors with their effects, and Fig. 8.31 a practical example.

Table 8.3 Relative service lives of 90°-deflections; pipe bends made of normal steel

90° deflection	Relative service life
Pipe bend, $R/D_R = 8$	1
Pipe bend, $R/D_R = 12$	1.8
Pipe bend, $R/D_R = 16$	1.9
Pipe bend, $R/D_R = 24$	3.3
T-Bend	60.9

Fig. 8.31 Pipeline wear in the area of an interference point

Considerations analogous to the above can also be applied to the switch systems commonly used in conveying lines.

8.2.2 Rotary Valve

As already described in Sect. 7.3, the design variants of rotary valves used in pneumatic conveyors seal the pressure difference Δp_{ZR}, which generally corresponds to the conveying line pressure loss Δp_R, by more or less tight radial and axial gaps. Δp_{ZR} causes a gas flow from the valve outlet to the inlet. Fig. 7.30 shows this process schematically. For the gap size/clearance between rotor and stator, values smaller than $\Delta s_{ZR} \cong (0.1-0.2)$ mm are not feasible. Increasing valve pressure difference Δp_{ZR} or pressure ratio (p_R/p_0) results in high gas velocities in the gaps, which reach at approximately $(p_R/p_0) \cong 2$ the speed of sound. This especially in the gap that separates the last rotary valve chamber from the rotary valvel inlet or the leakage gas discharge pipe.

When fine-grained bulk materials with $d_S < \Delta s_{ZR}$, e.g., fly ash, coal dust, cement, are introduced into pneumatic conveying lines, solid material is transported together with the leakage gas flow through the gaps. The flow deflection in the area of the gap inlet and the high particle velocities u_S lead to significant jet wear in the case of abrasive/hard bulk materials. The stator generally wears more than the rotor. Consequently, to achieve an acceptable service life when introducing fine-grained abrasive solids into pneumatic conveying lines using unarmored rotary valve feeders, the applied pressure difference Δp_{ZR} should be as small as possible, preferably $\Delta p_{ZR} \lesssim 0.25$ bar. The realization of higher pressure differences requires special armored locks. When using rotary valve systems with ceramic lining of the stator and a covering of the gap-side of the blade surfaces with ceramic strips/tiles, pressure differences up to $\Delta p_{ZR} \cong 1.5$ bar are possible and offered. Further details on this can be found in Sect. 7.3.4.

For bulk materials with particle dimensions d_S significantly larger than the gap size Δs_{ZR}, the gap wear is subordinate. Rotary valve lockings against pressure differences $\Delta p_{ZR} > 1.0$ bar are common here, e.g., in plug conveying of predominantly monodisperse and low-abrasive plastic granules values up to $\Delta p_{ZR} \cong 3.5$ bar are realized. The bulk material is pushed in front of the rotor blades. If the particle diameter and the blade/stator gap width are of the same order of magnitude, a wear-intensive three-body stress occurs: rotor blade, stator wall, bulk material particles clamped in between. The result is intense furrow wear, see Fig. 8.26. Constructive countermeasures are possible.

The individual rotary valve blades are moved by the rotation of the cell wheel towards the edge of the housing inlet. Especially with coarser bulk materials, particles can become jammed and sheared off or the rotor can come to a stop. This so-called "chopping" must be avoided for process and wear reasons. A number of possibilities, e.g., tangential lock inlet, inlet plates with an discharge edge inclined to the rotor axis, resiliently movable shear edges, etc., are known and tested.

8.2.3 Screw Feeder

The screw feeder shown in Fig. 7.20 is a high-speed rotating compression screw system supported on both sides, with its screw channel completely filled with bulk material, see Sect. 7.2. The sealing of the conveying pressure p_R is achieved by the flow resistance of the moving bulk material filling. Along the screw channel, the bulk material is compressed in the conveying direction, i.e., increasingly pressed against the screw surface and the surrounding channel wall. In doing so, it moves at different speeds relative to the screw surface and relative to the channel wall. Both components are thus subjected to abrasive sliding wear, which increases in the transport direction. On the pressure side of the screw blade, i.e., on the side pushing the bulk material, generally stronger wear is observed than on the corresponding backside. On the screw outer side facing the channel wall, erosion wear caused by gas backflow can occur.

The greatest wear on a screw feeder is always measured at the last screw spiral before the bulk material is discharged into the pressure chamber/into the conveying gas stream. Due to the high stress, this area wears out many times faster than all other vulnerable components. The so-called "end wing" is therefore designed as a easily removable replacement part. In the same way, wear shells arranged on the inside of the screw channel can be quickly replaced without much effort. In general, a large number of wear protection measures, e.g., hardening, coatings, partial use of special materials, etc., are known and tested for screw feeders.

The amount of wear can, of course, also be influenced by the process engineering plant design: lowering the conveying pressure p_R and reducing the screw speed n_W

Fig. 8.32 Wear pattern of the end wing of a screw feeder

reduce the wear rate in some cases disproportionately. The same effect is achieved by assigning the correct screw type to the respective bulk material, e.g., by selecting a suitable compression (→ reduction of the screw pitch in the conveying direction). Note: Increasing screw compression does not necessarily lead to an increase in wear, as intuitively expected. The stronger compression changes the conveying angle ω_x, which describes the spiral path of the solid particles through the screw channel, in such a way that the bulk material passes through the screw on a shorter/more direct path [32]. This reduces the potential wear. Practical experience and measurements on this are available.

A service life of 4000 operating hours for the wear-critical "end wing" in the standard version is usually always achievable. In Fig. 8.32, the heavily worn end wing of a screw conveyor used for conveying lignite dust, $d_{S,50} \cong 58\,\mu\text{m}$, particle sizes up to $d_{S,\max} \cong 1100\,\mu\text{m}$, is shown as an example.

8.2.4 Pressure vessel lock

Pressure vessels, see Fig. 7.4, should be designed as mass flow containers (→ uniform, pulsation-free and dead zone-free bulk material discharge). In this case, the solid slides along the container wall. This must be taken into account by adding wear surcharges to the wall thickness (→ pressure vessel regulation). When supporting the bulk material outflow by punctual gas addition via nozzles in the container wall, local wall wear due to jet deflection, reflection or turbulence must be avoided. A correctly dimensioned and positioned flat aeration pad is preferable here.

Wear-critical components of a pressure vessel are the valves switching in bulk material or in bulk material-laden gas streams. In Fig. 7.4 thus: inlet valve A4 (A5), outlet valve A6, and relief valve A3. These valves must fulfill both the function of shutting off the bulk material/gas flow and sealing the overpressure in the sender. Practice uses a variety of different, on the one hand, market-available standard valves, e.g., rotary flaps, ball valves, pinch valves, and on the other hand, specially adapted special constructions, e.g., dome valves, rotary slide valves. Their selection must be adapted to the specific properties of the bulk material in question, including grain size and hardness, adhesion behavior, etc. Service live times are generally given in the form of achievable switching cycles. To avoid temporarily narrow flow cross-sections, i.e., high flow velocities, valve opening and closing times $\Delta\tau_{auf/zu} \leq 2\,s$ should be aimed for.

A significant extension of the valve service life is possible by dividing the two functions "sealing" and "shutting off" between two different valves. In Fig. 7.4 for example, valve A4 takes over the sealing function and A5 the shutting off of the bulk material flow. To fill the pressure vessel, A4 opens before A5 releases the bulk material flow and closes again only after A5 has been closed first. Valve A5 can be a non-gas-tight, relatively simple valve.

If one of the above-mentioned valves becomes leaky, e.g., due to local seal wear/ breakage, it must be replaced or repaired immediately. The jet wear caused by the leak results in excessive material removal, which destroys the valve's main body in a short time. Detection is possible, for example, by an automatic, regular tightness check of the sender during operation: The filled and pressurized pressure vessel is kept closed for a predetermined time; the pressure drop occurring during this time is a measure of leaks.

A longer valve service life does not necessarily mean lower wear costs if, in addition to the expenses for wear parts, those for removal and installation of the valves, duration of operational interruption, etc., are taken into account. Example: Two differently designed outlet valves A6, see Fig. 7.4, for test reasons installed in a twin pressure vessel conveying system for cement achieved the following service lives: metallic sealing ball valve: 120,000 cycles, mechanical pinch valve: 70,000 cycles. The total costs of the specific pinch valve accumulated over several operating years were still significantly lower than those of the ball valve. While the defective ball valve had to be replaced by a second valve from the operator's warehouse and sent to the manufacturer for repair/overhaul, the (cost-effective and available in stock) rubber diaphragm of the pinch valve could be quickly and easily replaced by the operating personnel.

Figure 8.33 shows the wear on a cone lift valve in the pressure vessel inlet, caused by an extremely abrasive and coarser fly ash, which was fed asymmetrically to the inlet as a compact jet and with a relatively large drop height.

Fig. 8.33 Wear on the inlet cone valve of a pressure vessel

8.3 Wear Measurement and Prediction

Predictions about the wear to be expected in a planned pneumatic conveying system require representative wear measurements. Representative in this context means that the later stress state must be mapped in the investigations, taking into account all operational boundary conditions. This is only possible in principle in the operating system itself. The closest to this are tests in pilot conveying systems with sufficiently practical dimensions. Here, essential parameters of the operating system, such as identical lock, conveying pressure, gas velocity profile along the conveying path, and loading, can be set and examined with the original bulk material. However, the geometric dimensions of the test conveying system will generally not match those of the operating system. This results, among other things, in a different solid velocity profile along the conveying line than in later operation. At the same time, such tests are costly and time-consuming, as on the one hand, the bulk material must be regularly replaced to keep its conveying and wear properties constant, and on the other hand, long test times are necessary to generate measurable wear. Bulk material quantities of several cubic meters are required.

The wear assessment in practice, often based on the weight loss of defined wear bodies introduced into the gas-solid flow, provides only very rough qualitative comparisons. For the reasons indicated, wear measurements are carried out in engineering practice on measuring devices on a laboratory scale, in which certain stress situations, e.g., impact or sliding stress, are mapped. Their measurement results are then converted or transferred to the real operating conditions using empirical corrections. It is obvious that the stress mechanisms of the large-scale plant must be represented as realistically as possible in the respective device in order to arrive at simple and, above all, safe methods of result transfer. Each stress mechanism thus requires its own measuring device. Theoretical calculation models can support the transfer of test results [33].

There is a wide variety of impact and sliding wear measuring devices available on the market, some of which are standardized in various countries in terms of design and measuring procedures, see e.g. [7, 34]. However, in industry, devices specifically designed for certain measuring tasks are often used. The following examples illustrate investigations of wear behavior of deflections in conveying lines using laboratory, pilot plant, and operational measurements, as well as their interpretation.

a) Laboratory measurement

An impact wear measuring device used by the author and the transfer rule developed for pipe deflections are presented here. A detailed description of the impact wear measuring device shown in Fig. 8.34 can be found in [23]. In the container filled with bulk material, a rotor equipped with four impact plates rotates. The impact plates are replaceable and made of the respective wall material to be examined. By changing the rotor speed n_{YGP}

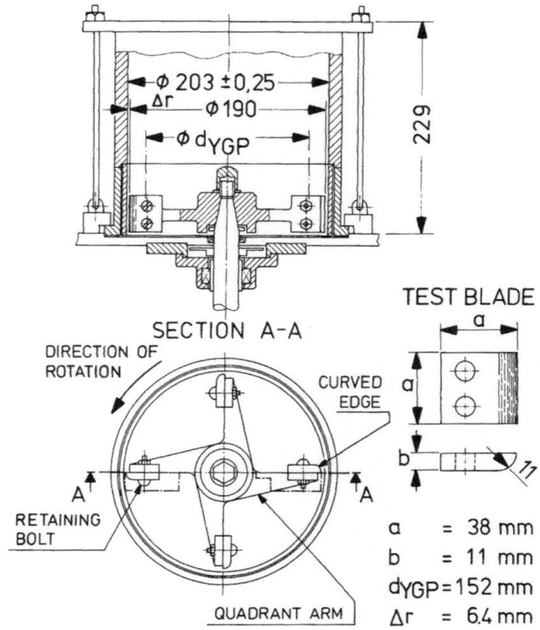

Fig. 8.34 Structure of the modified YGP impact wear tester [23]

the impact speed u_S of the solid can be adjusted over a wide range. 90° impact wear is measured. The measured variable is the weight loss of the impact plates W_M, from which the wear characteristic $W_{M/U} = (W_M/U_{\text{YGP}})$, with: U_{YGP} = number of rotor revolutions, is formed. $W_{M/U}$ can be converted into any other characteristic values. The measuring procedure is designed in such a way that during the measurement there is neither a grain destruction influencing the result nor a rotation of the bulk material changing the impact speed. This is exemplified by Fig. 8.35, which shows that up to a maximum number of $U_{\text{YGP}} \lesssim 750$ rotor revolutions, a constant wear value $W_{M/U}$ independent of U_{YGP} is measured. This specific measured value is used for all considerations/plant calculations. Further details can be found in [23]. The measuring device is a modified version of the YGP tester standardized in some countries [35].

Fig. 8.36 shows the wear rates $W_{M,U} = W_M/U_{\text{YGP}}$ of the wall material St 52-2 under stress by quartz sand of particle size $d_S = (1-3)$ mm with variation of the impact velocity $u_{P,0}$. The dependency shown there $W_{M,U}(u_{P,0})$ is described by the exponent $n_U = 1.35$. Comparable exponents are found in [22, 27, 28]. Converted to the approach according to (8.16) ($\rightarrow W_M = W_{M/\tau}$, i.e., the same stress duration τ per measurement point) results in $n = 2.35$, see Box 8.1 [23]. Figure 8.22 shows further exemplary results obtained with the device.

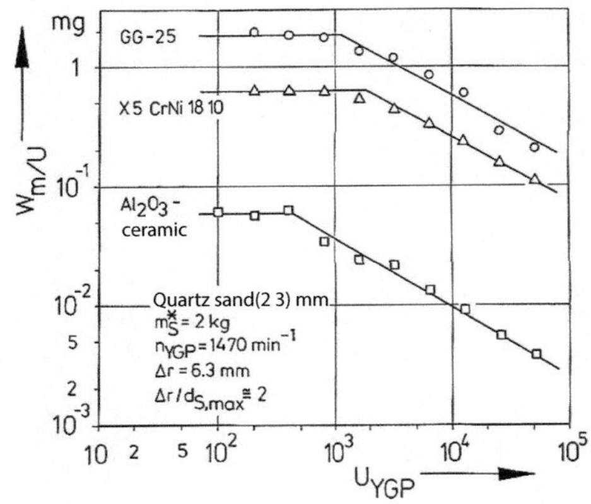

Fig. 8.35 Dependencies $W_{M,U}(U_{\text{YGP}})$ of various plate/wall materials; solid: quartz sand with $d_S = (2-3)$ mm

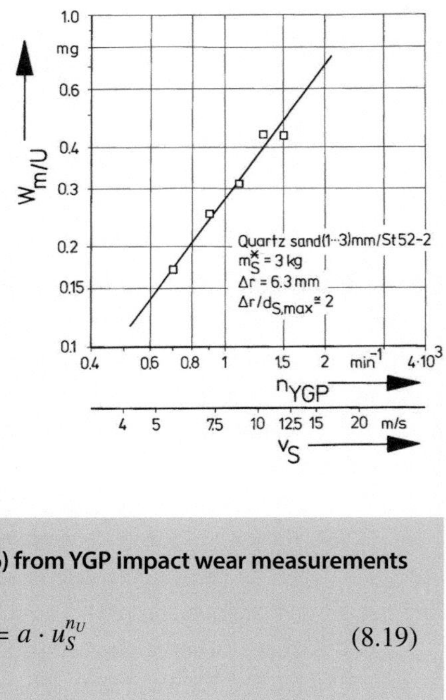

Fig. 8.36 Wear rate $W_{M,U} = f$(Rotor speed n_{YGP} or impact velocity $u_{P,0}$); Solid: quartz sand with $d_S = (1-3)$ mm

Box 8.1: Determination of the exponent n of (8.16) from YGP impact wear measurements

$$W_{M/U} = \frac{W_M}{U_{\text{YGP}}} = a \cdot u_S^{n_U} \tag{8.19}$$

with

$$U_{\text{YGP}} = n_{\text{YGP}} \cdot \Delta\tau, \tag{8.20}$$

$$u_{P0} = D_{\text{YGP}} \cdot \pi \cdot n_{\text{YGP}} \tag{8.21}$$

follows from this

$$W_{M/\tau} = \frac{W_M}{\Delta\tau} = a \cdot u_{P,0}^{n_U} \cdot n_{\text{YGP}} = a \cdot u_{P,0}^{n_U} \cdot \frac{u_{P,0}}{D_{\text{YGP}} \cdot \pi} = a^* \cdot u_{P,0}^{n_U+1} = a^* \cdot u_{P,0}^{n} \tag{8.22}$$

Thus:

$$n = n_U + 1 \tag{8.23}$$

In order to apply the characteristics measured with the impact wear measuring device to defined wear situations in operating systems, appropriate corrections are necessary. When transferring to pipeline bends, the following procedure has proven successful: The with original bulk material, the planned wall material, and the generally largest occurring solid-≅ impact velocity u_S measured wear coefficient $W_{M,U}$ is converted into a wall thickness change per unit mass of impacted bulk material $W_{\Delta s/M}$ and using (8.24)

$$\frac{\Delta s_W}{\Delta m_S} = \frac{\Delta s_W}{\dot{m}_S \cdot \Delta \tau_W} = f(\alpha_{S,W}) \cdot f(\mu) \cdot W_{\Delta s/M} \tag{8.24}$$

with:

Δs_W	Wall thickness of the planned bend,
Δm_S	Bulk material mass pushed through until the bend breaks,
$\Delta \tau_W$	Service life of the bend until it breaks,
$\dot{m}_S$	Solid mass flow rate through the system,
$f(\alpha_{S,W})$	Correction of the impact angle from $\alpha_{S,W} = 90°$ to the existing angle $\alpha_{S,W}$,
$f(\mu)$	Correction of the solid concentration in the measuring device to the loading μ in operation

the required bend wall thickness Δs_W with a given service life $\Delta \tau_W$ or vice versa is calculated. The correction function $f(\alpha_{S,W})$ corresponds to the dependency shown in Fig. 8.18, $f(\mu)$ is the impact density or loading normalized with its value in the measuring device in Fig. 8.20 [23]. Practical experience shows an accuracy of lifetime predictions of approximately ±25 %, which could not be further improved by including additional corrections, e.g., taking into account the spatial deflector arrangement.

b) Pilot plant measurement

To measure the wear of the bend in the area of solid impact on the outer pipe wall, selected pipe bends of the pilot plant conveying lines can be equipped with replaceable wear measuring bodies. The arrangement is shown schematically in Fig. 8.37. Five individual wear bodies, made of St 37, are arranged gap-free one behind the other on the outer pipe center and are adapted to the local pipe geometry. During conveying tests, their weight loss W_M is measured at regular time intervals and assigned to the bulk material mass Δm_S conveyed up to that point.

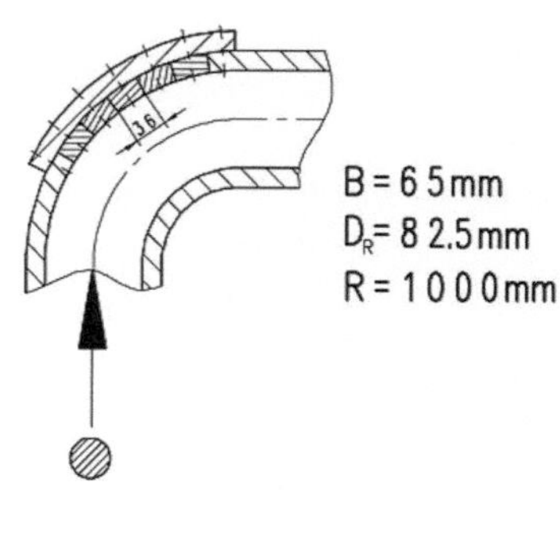

Fig. 8.37 Example of a wear measurement arrangement in conveying lines

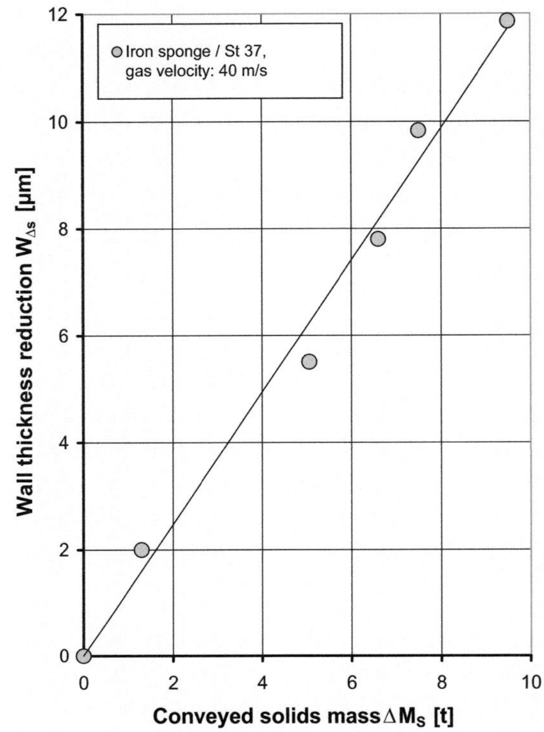

Fig. 8.38 Wear on a conveying pipe bend during the transport of coarse-grained iron sponge

In Fig. 8.38, measurement results are plotted, which were determined on a horizontally arranged and at the end of the conveying line located 90°-bend, $D_R = 82.5\,\text{mm}$, $R = 1.0\,\text{m}$. The solid was pelletized iron sponge, $d_S \cong (5-20)\,\text{mm}$. The gas velocities at the entrance of the bend were in the range $v_F \cong 40\,\text{m/s}$, the corresponding particle velocities were at about half the gas velocity. Since uniform material removal was measured over the entire wear body surface, in Fig. 8.38 on the ordinate, the more illustrative wall thickness decrease $W_{\Delta s}$ is represented. The curve $W_{\Delta s}(\Delta m_S)$ runs linearly in the measuring range. Its gradient is: $W_{\Delta s/M} = (W_{\Delta s}/\Delta m_S) = 1.226\,\mu\text{m/t}$. This is an extremely high wear value, which leads to a pipe bend with a wall thickness of $\Delta s = 10\,\text{mm}$ already worn through after $\Delta m_S \cong 8156\,\text{t}$ of conveyed bulk material or its service life at a mass flow of $\dot{m}_S = 100\,\text{t/h}$ only about $\tau \cong 81.5\,\text{h}$ would be.

c) Operational measurement

Fig. 8.39 shows exemplary results of long-term wear measurements at an operating plant, in which fly ash, $d_S \cong 25\,\mu\text{m}$, is conveyed from the filter of a coal-fired power plant. Plotted as a function of the operating time τ_B is the wall thickness decrease of various 90°-bends in the front areas of the parallel conveying routes there. The gas velocity is

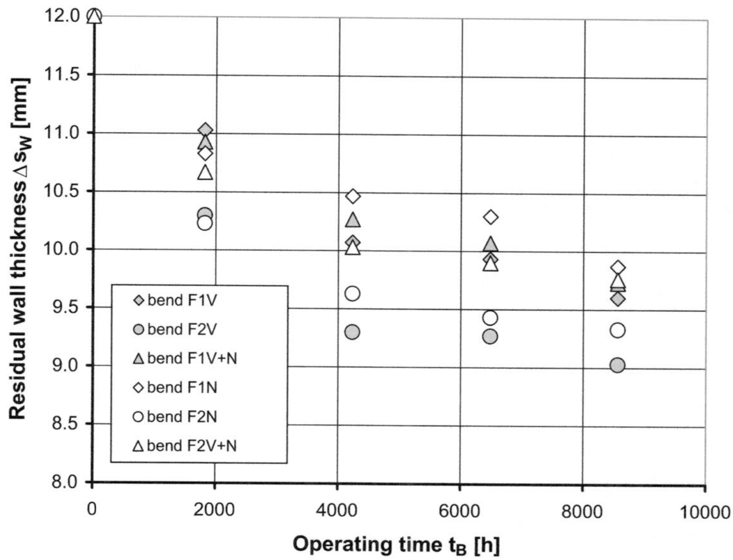

Fig. 8.39 Wear on conveying pipeline bends during the transport of fine-grained fly ash

on average $v_F \cong 10\,\text{m/s}$. It involves conveyance with pressure vessel lock systems. The results show that the wear rate $W_{\Delta s/\tau} = (W_{\Delta s}/\tau_B)$ decreases with increasing runtime τ_B.

The measurements were carried out in cooperation between the operator and the plant supplier according to a jointly developed measurement and evaluation scheme. They also served to determine the time interval between control examinations/maintenance intervals for the further operating time. At each measurement, in order to capture the influences of changes in the bulk material (→ coals of different origins were burned), fly ash samples were taken and their wear-relevant characteristics determined.

d) Wear protection systems

Providers of pneumatic systems or components are forced to equip them with a wear protection concept tailored to the current operating situation, depending on the abrasiveness of the bulk material to be transported. To reduce the number of design variants and facilitate the selection of units, various companies have introduced classification systems that assign specific design variants of wear protection to the wear properties of the bulk materials. Figure 8.40 shows an example of this. However, this approach only leads to satisfactory results if the solid is not only described by its hardness, e.g., the Mohs hardness scale, but also by grain size, shape, and all those properties that influence the service life of the component under consideration, see Sect. 8.1.4. Such an assignment must therefore always be created specifically for the unit.

Wear protection packages for rotary valves

Mineral powders	Chalk*	Coal dust*	Ashes*	Cement*	Alumina*	Protection package	Housing	ROTOR
Non abrasive						Standard	GG/Steel	Steel
Low abrasive						DuroProtect 1	DuroChrom	Steel
Easy abrasive						DuroProtect 2	DuroChrom	Steel
Medium abrasive						DuroProtect 3	DuroChrom	DuroDox
Very abrasive						DuroProtect 4	DuroChrom	DuroKarb
High abrasive						DuroProtect 5	DuroKera	DuroKarb

Fig. 8.40 Wear protection packages for rotary valves, according to [36, 37]. Application area: mineral powders; meaning: suffix -Chrom = Hard chrome coating, -Karb = Tungsten carbide coating, -Kera = Ceramic lining

8.4 Calculation Example 12: Wear Analysis of a 90° Deflector

The wear-related design of the 90°-deflections in a pneumatic conveying line are to be analyzed. The conveying system is designed as a positive pressure conveying system. The pressure at the end of the line is equal to the ambient pressure $p_E = 1.0\,\text{bar}$. The following data is available:

Conveying gas: Ambient air, temperature before conveying system $T_F \cong 110\,°\text{C}$.

Bulk material: Relatively coarse fly ash, quartz content $x_{QS} \cong 70$ M.-%, $T_S \cong 80\,°\text{C}$, $d_{S,50} \cong 40\,\mu\text{m}$, $\varrho_P \cong 2400\,\text{kg/m}^3$, $H_P \cong 8.5\,\text{kN/mm}^2$.

Conveyor line: Length $L_R = 100\,\text{m}$ including $\Delta L_v = 20\,\text{m}$ vertical height distributed over 2 height sections, conveying pipe ($\varnothing 168.3 \times 6.3$) mm $\rightarrow$ $D_R = 155.7\,\text{mm}$, number of 90° deflections $n_{U,90°} = 8$. For structural reasons, bending radii R of pipe bends larger than $R_{\max} = 2.0\,\text{m}$ not realizable.

Operating data: Solid mass flow $\dot{m}_S = 25\,\text{t/h}$, pressure loss in the conveying pipe $\Delta p_R \cong 0.8\,\text{bar}$, initial gas velocity $v_{F,A} = 11.0\,\text{m/s}$, final gas velocity $v_{F,E} = 19.8\,\text{m/s}$, loading $\mu = 18.4\,\text{kg}\,S/\text{kg}\,F$, flow pattern: dilute phase pneumatic conveying with beginning strand formation.

The most heavily loaded deflector is due to the increasing conveying gas velocity v_F towards the end of the pipeline obviously the last bend seen in the transport direction. This is considered here. The associated particle impact velocity $u_{P,0}$ can be determined with a local velocity ratio $C = u_S/v_F \cong 0.84$ to $u_{P,0} \cong 16.6\,\text{m/s}$

Execution as a 90° pipe bend: The most stressed point of a pipe bend is located approximately on the straight extended trajectory of the particles entering the bend on the inner side of the pipe, see Fig. 8.29. The associated critical impingement angle can be estimated for various deflector materials using (8.18). (8.18) describes the relationship between the geometric data of a 90° bend and the impingement angle $\alpha_{S,W}$. It is represented in Fig. 8.18 by the two opposing abscissas.

Wall material steel St 37: $H_W \cong 1.31\,\mathrm{kN/mm^2}$, $H_P/H_W \cong 6.49 \rightarrow$ High level wear. Low wear impact angles are in the range ($60° < \alpha_{S,W} < 15°$), see Fig. 8.18. This leads to R/D_R-ratios of ($20 < R/D_R < 2$) and thus bend radii of ($3.115\,\mathrm{m} < R < 0.311\,\mathrm{m}$). Thus, only radii $R < 0.311\,\mathrm{m}$ are feasible.

Wall material cast basalt: $H_W \cong 7.0\,\mathrm{kN/mm^2}$, $H_P/H_W \cong 1.21 \rightarrow$ High level wear near the transition from high level to low level. Low-wear impact angles are in the range $\alpha_{S,W} < 30°$. This leads to values of $R/D_R > 7$. The maximum achievable radius $R = R_{\max} = 2.0\,m$ results in $R/D_R \cong 12.8$. It should be used.

Wall material Al_2O_3*:* $K_{Ic} \cong 5.0\,\mathrm{MPa}\,\sqrt{\mathrm{m}}$, $H_W \cong 15.0\,\mathrm{kN/mm^2}$, $H_P/H_W \cong 0.57 \rightarrow$ distinct low level wear. Impact angles as with ast basalt. Bend radius $R = 2.0\,\mathrm{m}$ is optimal. The critical load for the initiation of a lateral fracture according to (8.8) is calculated to be $F_{P,c} \cong 37\,\mathrm{N}$. Applying the equations/relationships derived in [6] for the maximum force occurring during particle impact $F_{P,\max}$ yields the result $F_{P,\max} \ll F_{P,c}$, i.e., lateral fractures do not occur.

Based on Figs. 8.18 and 8.22, under simplifying assumptions, it can be roughly estimated that steel St 37 and cast basalt have approximately the same service life with *identical* wall thickness. The service life of the Al_2O_3-ceramic is about 40 times longer. For more accurate statements, reference measurements are required.

In the 90° bend, the bulk material in question leads to a separation of gas and solid. The bulk material forms a strand on the outer wall of the bend. Since this only dissolves again behind the deflection, it causes abrasive wall wear in the subsequent pipe. A pipe section of length $\Delta L_R > 5 \cdot D_R$ must therefore also be designed to be wear-resistant. For coarse-grained bulk materials, this measure is also necessary, among other things, due to increased particle rebound.

Execution with autogenous wear protection: Since the given task does not require an absolutely residue-free conveying path, it can also be equipped with the deflection elements shown in Fig. 8.27b–e: $H_P = H_W =$ bulk material. Such systems can be used for both dilute and dense phase conveying. It should be noted that the solid content of deflection pots, which are flowed through vertically upwards into the horizontal, must be kept in balance by the impacting bulk material flow. If this does not happen, e.g., due to insufficient loading and/or gas velocity, strong pressure and thus throughput fluctuations may occur. In the present case, this is not to be expected.

Based on measurements on an operating conveying system, it is reported in [26] that the service life of a T-bend, see Fig. 8.27b and Table 8.3, is approximately 60 times longer than that of a 90° pipe bend with $R/D_R = 8$ made of normal steel. A zircon compound was conveyed.

Which of the discussed systems should be used must be decided by means of a total cost analysis, which takes into account the costs for acquisition and maintenance influenced by the expected service lives as well as possible production interruptions. A safeguard by wear measurements on a scale-up-capable measuring device at laboratory scale would be desirable for a new plant. The increased operating temperature has no influence on the decision in the present example. This also applies to the effects on the energy requirements of the plant: The pressure losses of the various deflectors are different, e.g., those of the deflection pots are about 20% larger than those of the bends, but they cause only a small proportion of the total pressure loss. However, a review in individual cases is useful. A reduction of the particle impact velocity by lowering the gas velocity level is not recommended for the present bulk material in general and also with regard to possibly changing product properties, e.g., the particle size distribution.

References

1. Czichos, H., Habig, K.-H.: Tribologie-Handbuch, 2. Aufl. Vieweg, Wiesbaden (2003)
2. Uetz, H. (Hrsg.): Abrasion und Erosion. Hanser, München (1986)
3. Fischer-Cripps, A.C.: Introduction to contact mechanics. Springer, New York (2000)
4. Johnson, K.L.: Contact mechanics. Cambridge University Press, Cambridge (1985)
5. Hill, R.: The mathematical theory of plasticity. Clarendon Press, Oxford (1950)
6. Hilgraf, P.: Wear in bulk materials handling. Bulk Solids Handl. **27**(7), 464–477 (2007)
7. Kleis, I., Kulu, P.: Solid particle erosion. Springer, London (2008)
8. Popov, V.L.: Kontaktmechanik und Reibung. Springer, Berlin (2009)
9. Chaudhri, M.M., Hutchings, I.M., Makin, P.L.: Plastic compression of spheres. Philos. Mag. A **49**(4), 493–503 (1984)
10. Hutchings, I.M.: Tribology: friction and wear of engineering materials. Edward Arnold, London (1992)
11. Gross, D., Selig, Th : Bruchmechanik, 4. Aufl. Springer, Berlin (2007)
12. Wear mechanisms in ceramics. In: Evans, A.G., Marshall, D.B., Rigney, D.A. (Hrsg.) Fundamentals of friction and wear of materials ASM. S. 439–452. (1981)
13. Wiederhorn, S.M., Lawn, B.R.: Strength degradation of glass impacted with sharp particles: I, annealed surfaces. J. Am. Ceram. Soc. **62**, 66–70 (1979)
14. Hutchings, I.M.: Mechanisms of wear in powder technology: a review. Powder Technol. **76**, 3–13 (1993)
15. Zum Gahr, K.-H.: Microstructure and wear of materials. Elsevier, Amsterdam (1987)
16. Xiumei, Q.: Wear mechanisms of nano- and microcrystalline TiC-Ni based thermal spray coatings. GKSS-Forschungszentrum Geesthacht GmbH, Geesthacht (2005)
17. Zum Gahr, K.-H.: Wear by hard particles. Tribol. Int. **31**(10), 587–596 (1998)

18. Meng, H.C., Ludema, K.C.: Wear models and predictive equations: their form and content. Wear **181–183**, 443–457 (1995)
19. Huang, C., Chiovelli, S., Minev, P., Luo, J., Nandakumar, K.: A comprehensive phenomenological model for erosion of materials in jet flow. Powder Technol. **187**, 273–279 (2008)
20. Hilgraf, P.: Wear in pneumatic conveying systems. Powder Handl. Process. **17**(5), 272–284 (2005)
21. Hoppert, H.: Betriebserfahrungen mit verschleißfester Aluminiumoxid-Keramik in der Grundstoffindustrie. ZEMENT-KALK-GIPS **42**(11), 567–573 (1989)
22. Glatzel, W.-D., Brauer, H.: Prallverschleiß. Chem.-Ing.-Techn. **50**(7), 487–497 (1978)
23. Hilgraf, P., Cohrs, H.: The practicability of the YGP erosion tester for the determination of erosion in pneumatic conveying systems. Powder Handl. Process. **4**(2), 189–197 (1992)
24. Zum Gahr, K.-H.: Zusammenhang zwischen abrasivem Verschleiß und der Bruchzähigkeit von metallischen Werkstoffen. Z. Met. **69**(10), 643–650 (1978)
25. Wahl, H.: Verschleiß metallischer Gleitflächenpaarungen unter Wirkung festkörniger Zwischenstoffe (Metall-Korn-Gleitverschleiß). Aufbereitungs Tech. **10**(6), 305–322 (1969)
26. Hilbert, J.D.: Alternatives in pneumatik conveying pipeline bends. Bulk Solids Handl. **4**(3), 657–660 (1984)
27. Arundal, P.A., Tayler, I.A., Dean, W., Mason, J.S., Doran, T.E.: The rapid erosion of various pipewall materials by a stream of abrasive alumina particles. Pneumotransport 2, BHRA Conference, Guildford, 09.1973. Paper E1, England, 1973
28. Wellinger, K., Uetz, H.: Gleitverschleiß, Spülverschleiß, Strahlverschleiß unter der Wirkung von körnigen Stoffen. VDI-Forschungsheft 449. VDI-Verlag, Düsseldorf (1955)
29. NN.: Rauchgasreinigungsanlage des EVO-Kraftwerk Arzberg wieder in Betrieb. VGB Kraftwerkstechnik **68**(2), 209 (1988)
30. Kriegel, E.: Druckverlust und Verschleiß in Rohrkrümmern beim pneumatischen Transport. Verfahrenstechnik **4**(8), 333–339 (1970)
31. Dhodapkar, S., Solt, P., Klinzing, G.: Understanding bends in pneumatic conveying systems. Chem. Eng., 53–60 (2009). www.che.com
32. Paepcke, J.: Untersuchung einer hochtourig laufenden Preßschnecke zur Einschleusung von Schüttgütern in Druckräume. Diplomarbeit, Fachhochschule Hamburg, Fachbereich Bio.-Ing.-Wesen, Produktionstechnik und Verfahrenstechnik 1991, unveröffentlicht
33. Hanson, R., Allsopp, D., Deng, T., Smith, D., Bradley, M.S.A., Hutchings, I.M., Patel, M.K.: A model to predict the life of pneumatic conveyor bends. Proc. Instn. Mech. Eng. **216**, 143–149 (2002). Part E: J. Process Mechanical Engineering
34. Hilgraf, P.: Verschleiß durch Schüttgüter. Lehrgangshandbuch zum gleichnamigen Seminar. Technische Akademie Wuppertal, Wuppertal (2018)
35. BS 1016, Part 111: Determination of abrasion index of coal. British Standards Institution, 1990
36. Fa. Coperion: Komponentenpräsentation: Einsatzmöglichkeiten von Coperion-Komponenten für Fa. Claudius Peters Technologies. Fa. Coperion, Buxtehude (2016)
37. Fa. Coperion GmbH: Prospekt „Komponenten für Mineralstoffe. Fa. Coperion GmbH, Weingarten (2016)

Design of a Conveying System 9

The procedure is demonstrated using a specific example.

9.1 Task Description

Quartz powder is to be pneumatically transported over the conveying route shown in Fig. 9.1. The bulk material continuously falls from the separator of an upstream production plant. The bulk material throughput of this plant is a maximum of $\dot{m}_{S,P} = 22.5\,\mathrm{t/h}$, the solid temperature $T_S = 60\,°\mathrm{C}$. The end point of the conveying is an existing storage silo with a dedusting filter. The transport should be a low-wear one, product-friendly, and with the lowest possible energy consumption. Conveying gas: ambient air. The supplier is provided with a 10 l material sample.

The conveying gas should be taken from an existing compressed air network. Max./min. network pressures: $p_{N,\max} = 4.50\,\mathrm{bar}$, $p_{N,\min} \cong 4.00\,\mathrm{bar}$; ambient pressure $p_0 = 1.0\,\mathrm{bar}$; ambient temperature $T_0 = 20\,°\mathrm{C}$.

9.2 Clarification of the Task

- Upstream production plant: grinding plant with external classifier circuit; product separation by means of pressure pulse-cleaned bag filter and discharge via collecting screw, maximum solid throughput of this screw $\dot{m}_{S,\mathrm{Sch}} \cong 30\,\mathrm{t/h}$; free available construction height below screw outlet $\Delta H_{\mathrm{Bau}} \cong 10\,\mathrm{m}$.
- Existing receiving silo: net filling volume $V_{\mathrm{Silo}} = 300\,\mathrm{m}^3$, cone opening angle $\alpha_{\mathrm{Silo}} = 60°$, cylinder diameter $D_{\mathrm{Silo}} = 4.0\,\mathrm{m}$; bag filter with row-by-row pressure pulse cleaning as top-mounted filter, usable filter area $A_{\mathrm{EFilter}} = 10\,\mathrm{m}^2$; planned clear-

P. Hilgraf, *Pneumatic Conveying*, https://doi.org/10.1007/978-3-662-67223-5_9

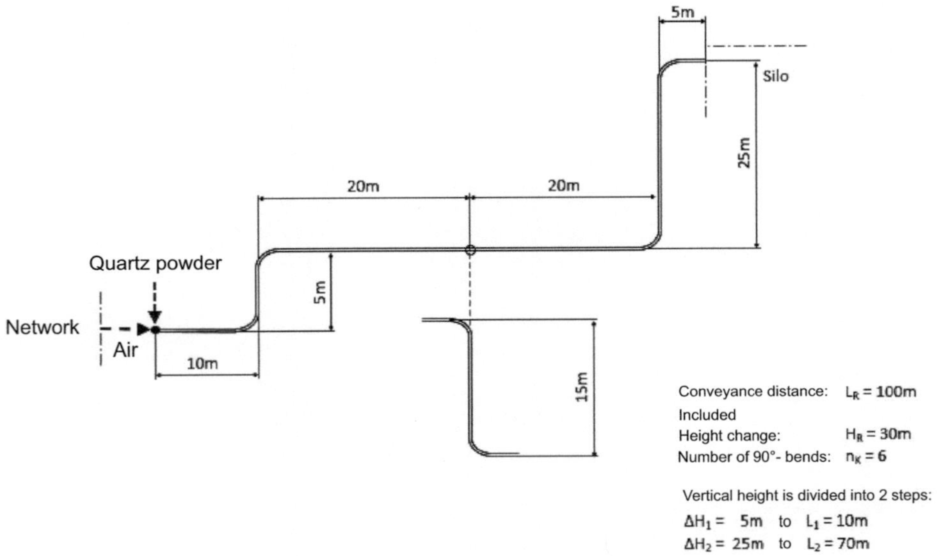

Fig. 9.1 Given conveying route

ance above and below the conveying line connection should be > 0.5 m each ($\rightarrow$ distance to the silo ceiling and to the max. level sensor).

- No alternative conveying routes possible, as the use of an existing pipe/belt bridge is required.
- Planned future expansions/modifications:
  - Possible installation of additional product silos: conveying path to the intended location is shorter than in the present task $\rightarrow$ no consequences.
  - Production of finer quartz powder qualities: leads to a lower solid throughput of the grinding plant $\rightarrow$ possibly required adjustment of the conveying gas flow is possible through network extraction.

Consequences for problem-solving:

- The operating behavior of grinding/classifier circuits allows fluctuations of the product discharge $\dot{m}_{S,P}$ of $\pm 10\,\%$ to be expected. Since $\dot{m}_{S,\mathrm{Sch}} > \dot{m}_{S,P}$, these propagate to the conveying system. Design throughput of the conveying system thus:

$$\dot{m}_{S,n} = 1.1 \cdot \dot{m}_{S,P} = 1.1 \cdot 22.5\,\mathrm{t/h} = 24.75\,\mathrm{t/h}$$

  Selected: $\dot{m}_{S,n} = 25.0\,\mathrm{t/h}$
- The filter area of the existing filter on the receiving silo must be checked for sufficient size according to the design of the conveyor system.

- The conveying line opens directly into the headspace of the receiving silo: arrangement, positioning, and height of the required clearance must be determined/checked for the expected operating conditions.
- Future planned plant expansions do not need to be considered in the design of the conveyor system.

9.3 Evaluation of the Conveyed Material

- Tab. 9.1 summarizes the characteristics determined from the provided 10 l bulk material sample. Fig. 9.2 shows the classification of the conveyed material in the extended Geldart diagram, see Sect. 4.5. It is assumed that the material sample was a *representative* sample.

Consequences for problem-solving:

- The present quartz powder is ideally suited for conventional dense phase/slow conveying. Conventional means that dense phase conveying can be achieved through a simple conveying pipe without the support of auxiliary devices, e.g., gas addition from a bypass pipe arranged parallel to the conveying line.
  The position in the left hatched area of the extended Geldart diagram identifies the quartz powder as a bulk material with high gas holding capacity, see Tab. 9.1: The venting time of 2.0 kg product in a ∅100 mm-fluidized bed is $\Delta\tau_E = 95$ s. In dense

Tab. 9.1 Characteristics of the conveyed material

Solid material		Quartz flour
Bulk density[a]	ρ_{SS}	1210 kg m^{-3}
Vibration density	ρ_{SR}	1620 kg m^{-3}
Particle density[b]	ρ_S	2610 kg m^{-3}
Angle of repose	α_{SS}	47 degrees
Average particle ø[c]	$d_{S,50}$	39 µm
Inclination of the RRS straight line	α_{RRS}	51 degrees
Spec. surface[d]	A_S	1928 cm^3 g^{-l}
Venting time[e]	τ_E	95 s
Fluidized channel slope[f] at 1 m^3 m^{-2} min^{-1}	α_R	3.4 degrees
Particle shape		predominantly elongated (~2:1), sharp-edged

[a]according to DIN 1060
[b]with Beckman air pygnometer
[c]from sieve analysis at 50% residue
[d]according Blaine
[e]Venting time of a fluidized bed
[f]Slope of a fluidized bed chute

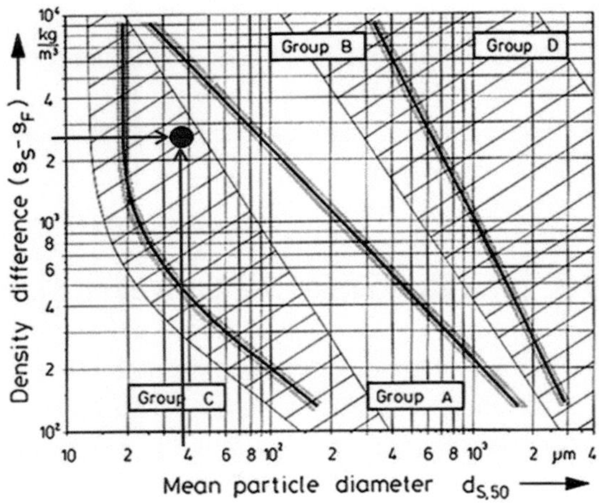

Fig. 9.2 Classification of the quartz powder in the extended Geldart diagram

phase conveying, a highly loaded strand conveying will therefore develop along the conveying route.

- The Vickers hardness of pure quartz is $H_P \cong (9-13)\,\text{kN/mm}^2$, those of some wall materials: $H_W(\text{St }37\text{-}2) \cong 1.3\,\text{kN/mm}^2$, $H_W(\text{Cast basalt}) \cong 7.4\,\text{kN/mm}^2$, $H_W(Al_2O_3\text{-Ceramic}) \cong (12-24)\,\text{kN/mm}^2$. It is, Al_2O_3-ceramics excluded, thus so-called high-level wear ($\rightarrow$ hardness of conveyed material $H_P >$ hardness of wall material H_W) to be expected, see also Sects. 8.1.4 and 8.4. In order to achieve economical service life of critical plant components, such as deflections, in addition to a suitable material selection, the most extensive reduction of conveying speed is an effective measure.

9.4 Selection of the Solid Lock

- The conveyed material is fine-grained, extremely abrasive and should be conveyed with the lowest possible gas velocities in dense phase. Solid locks suitable for dense phase conveying in principle are: rotary valve, screw feeder and pressure vessel lock.
- Evaluation:
  - Rotary valve lock: fine-grained product, high wear attack due to gap/leakage gas flow; ceramic wear protection is mandatory; even then, limited to a maximum $\Delta p_{ZR} \cong 1.0\,\text{bar}$ pressure difference at the lock = conveying line pressure difference. Lock type is only suitable to a limited extent.
  - Screw lock: increased wear due to the required high screw speed; proven wear methods for screws, very easily replaceable wear parts; limited to a maximum $\Delta p_{ZR} \cong 1.0\,\text{bar}$ pressure difference at the lock = conveying line pressure difference. Lock type is only suitable to a limited extent.

  - Pressure vessel lock: low sliding speed of the bulk material along the vessel walls; conveying pressure differences $\Delta p_R > 1.0$ bar are relatively unproblematic; problematic are the pressure vessel fittings closing and opening in the bulk material, whose stress, however, can be managed with the opening/closing times and easy replaceability mentioned in Sect. 8.2.4.
- Selection: Use of a pressure vessel lock: hardly any restrictions regarding a dense phase design; lowest expected operating costs in terms of wear and energy ($\rightarrow$ no lock drive).
- Type of pressure vessel circuit: Possible design variants are:
  - Single vessel, i.e., discontinuous conveying operation,
  - Parallel connection of two vessels, i.e., quasi-continuous operation; vessels are side by side; one conveys while the other fills,
  - Series connection of two vessels, i.e., continuous operation; vessels are stacked, the lower vessel conveys continuously, while the upper one sluices bulk material into the lower one on demand.

Decision

Single vessel with buffer container $\rightarrow$ Product is produced continuously and accumulation in the separator is not possible, but with the interposition of a buffer- = pre-silo, continuous removal is not required $\rightarrow$ simplest, most cost-effective solution; construction height must be checked.

9.5 Design of the Conveying Route

- Design solid throughput: The conveying cycle of a single pressure vessel contains dead times $\Delta\tau_{tot}$, during which no conveying takes place; i.e., to achieve the required, average solid throughput over the operating time $\dot{m}_{S,n}$, during the actual conveying time $\Delta\tau_{för}$ a solid throughput increased by the factor

$$f_Q = \frac{\Delta\tau_{Ch}}{\Delta\tau_{för}} = \frac{\Delta\tau_{för} + \Delta\tau_{tot}}{\Delta\tau_{för}} = 1 + \frac{\Delta\tau_{tot}}{\Delta\tau_{för}} \tag{9.1}$$

$\dot{m}_S$ must be conveyed, i.e.:

$$\dot{m}_S = f_Q \cdot \dot{m}_{S,n} \tag{9.2}$$

Compare this to Sect. 7.1.1. f_Q cannot be calculated in advance and must therefore be initially estimated based on experience values and later checked, and if necessary, corrected. For the present case, it is estimated:

$$f_Q = 1.30$$

From this, the solid throughput for the conveying line design follows:

$$\dot{m}_S = 1.30 \cdot 25.0\,\text{t/h} = 32.5\,\text{t/h} = 9.028\,\text{kg/s}$$

- Usable conveying line pressure difference: available network pressure: $p_{N,\min} \cong 4.00\,\text{bar}$; gas extraction from the network via flow control station, pressure loss: $\Delta p_{\text{Regel}} \cong 0.50\,\text{bar}$; distribution losses at the pressure vessel: $\Delta p_B \cong 0.30\,\text{bar}$; supply losses to the transmitter: $\Delta p_V \cong 0.10\,\text{bar}$; pressure in the receiving silo: $p_E \cong p_0 = 1.0\,\text{bar}$. The usable pressure difference at the conveying line is thus:

$$\begin{aligned}\Delta p_R &\leq p_{N,\min} - (\Delta p_{\text{Regel}} + \Delta p_B + \Delta p_V) - p_0 \\ &= 4.00\,\text{bar} - (0.50 + 0.30 + 0.10)\,\text{bar} - 1.0\,\text{bar} = 2.10\,\text{bar}\end{aligned}$$

For the design chosen:

$$|\Delta p_R| = 2.0\,\text{bar}$$

- Design temperature T_R: For fine-grained solids, due to the large heat exchange surface, see specific quartz powder surface according to Blaine in Table 9.1, already after a few meters of conveying distance/almost abruptly, the mixing temperature T_M of bulk material and conveying gas is reached. A significant cooling along the conveying distance does normally not occur, since the pipe outer surface and the external heat transfer coefficient pipe outer wall → ambient air are too low in relation to the heat capacity of the mixture. In the present case, it can be calculated sufficiently accurately with $T_R = T_M$. The mixing temperature T_M follows from (2.14). In this, the loading μ is still unknown. Since a dense phase conveying is to be designed, values $\mu \gtrsim 30\,\text{kg S/kg F}$ are to be expected, at the same time the ratio $(c_{p,S}/c_{p,F})$ is on the order of 1. Regardless of the conveying gas temperature T_F it follows that $T_M \cong T_S$ is valid. Thus, it holds:

$$T_R = 60\,^\circ\text{C} = 333\,\text{K}$$

- Friction coefficient λ_S of the bulk material: It is assumed that the supplier does not have its own measurement results with the bulk material quartz powder and relies on a suitable approach from the literature. The approach according to Stegmeyer [1], (4.47), is used:

$$\lambda_S = 2.1 \cdot \frac{Fr_{S,T}^{0.25}}{Fr_R} \cdot \frac{\left(\frac{D_R}{d_{S,50}}\right)^{0.1}}{\mu^{0.3}} \quad \rightarrow \quad \text{with: } Fr_R = \frac{v_F^2}{g \cdot D_R} \text{ and } Fr_{S,T} = \frac{w_T^2}{g \cdot d_{S,50}}$$

The properties of the conveyed material are contained in $Fr_{S,T}$ and $d_{S,50}$. Since the bulk material is available with its characteristic values, these can be incorporated in the λ_S -equation. For the terminal velocity w_T, a value in the (laminar) Stokes range is expected. For this, (3.14) and $Re_S = w_T \cdot d_{S,50}/\nu_F \lesssim 1$. With the known solid and gas data at $p = p_0 = 1.0\,\text{bar}$

$$d_{S,50} = 39\,\mu\text{m} = 39 \cdot 10^{-6}\,\text{m} \quad \rightarrow \quad \text{compare Tab. 9.1},$$
$$\varrho_S = 2610\,\text{kg/m}^3 \quad \rightarrow \quad \text{compare Tab. 9.1},$$
$$\varrho_F = \varrho_F(60\,°\text{C}) = \varrho_{\text{F},0} \cdot \text{T}_0/\text{T}_\text{R} \quad \rightarrow \quad \varrho_{\text{F},0} = 1.20\,\text{kg/m}^3,$$
$$\varrho_F = 1.20\,\text{kg/m}^3 \cdot 293\,\text{K}/333\,\text{K} = 1.056\,\text{kg/m}^3 \quad \rightarrow \quad \text{Gas-/air density},$$
$$\nu_F = \nu_F(60\,°\text{C}) = 18.88 \cdot 10^{-6}\,\text{m}^2/\text{s}$$
$$\rightarrow \quad \text{kinematic viscosity air, } \nu_F = \eta_F/\varrho_F,$$

follows for the particle terminal velocity:

$$w_T = \frac{1}{18} \cdot \frac{g \cdot d_S^2}{\nu_F} \cdot \frac{\varrho_P - \varrho_F}{\varrho_F}$$
$$= \frac{1}{18} \cdot \frac{9.81\,\frac{\text{m}}{\text{s}^2} \cdot (39 \cdot 10^{-6}\,\text{m})^2}{18.88 \cdot 10^{-6}\,\text{m}^2/\text{s}} \cdot \frac{(2610 - 1.056)\,\frac{\text{kg}}{\text{m}^3}}{1.056\,\frac{\text{kg}}{\text{m}^3}} = 0.1085\,\frac{\text{m}}{\text{s}}$$

Verification of the validity range:

$$Re_S = \frac{w_T \cdot d_{S,50}}{\nu_F} = \frac{0.1085\,\frac{\text{m}}{\text{s}} \cdot 39 \cdot 10^{-6}\,\text{m}}{18.88 \cdot 10^{-6}\,\frac{\text{m}}{\text{s}^2}} = 0.224 < 1$$
$$\rightarrow \quad \text{Boundary condition fulfilled}$$

Thus:

$$Fr_{S,T} = \frac{w_T^2}{g \cdot d_{S,50}} = \frac{\left(0.1085\,\frac{\text{m}}{\text{s}}\right)^2}{9.81\,\frac{\text{m}}{\text{s}^2} \cdot 39 \cdot 10^{-6}\,\text{m}} = 30.77$$

Inserted into the λ_S-equation follows:

$$\lambda_S = 2.1 \cdot \frac{Fr_{S,T}^{0.25}}{Fr_R} \cdot \frac{\left(\frac{D_R}{d_{S,50}}\right)^{0.1}}{\mu^{0.3}} = 2.1 \cdot \frac{30.77^{0.25}}{Fr_R} \cdot \frac{\left(\frac{D_R}{39 \cdot 10^{-6}\,\text{m}}\right)^{0.1}}{\mu^{0.3}}$$
$$\lambda_S = 13.65\,\text{m}^{-0.1} \cdot \frac{D_R^{0.1}}{Fr_R \cdot \mu^{0.3}} \tag{9.3}$$

Further resolution is not yet possible.

- Estimation of the conveying line diameter D_R: The pressure loss Δp_R of the conveying section results accordingly (4.39) as the sum of the individual pressure losses Δp_i:

$$\Delta p_R = \sum_{i=1}^{n} \Delta p_i$$

With the Δp_i -equations from Sect. 4.7 it follows:

$$-\Delta p_R = \left(\lambda_S \cdot \frac{\Delta L_h}{D_R} + \frac{1}{2} \cdot \lambda_S \cdot \frac{H_R}{D_R} + \frac{2}{C} \cdot \frac{g \cdot H_R}{v_F^2} + 2 \cdot n_U \cdot K_U \cdot C + 2 \cdot C\right) \cdot \mu \cdot \frac{\varrho_F}{2} \cdot v_F^2 + |\Delta p_F| \tag{9.4}$$

Neglecting the frictional pressure loss of the gas, this can be written as

$$-\Delta p_R = W \cdot \mu \cdot \frac{\varrho_F}{2} \cdot v_F^2$$

The resistance value W corresponds to the bracket expression in (9.4). If the loading $\mu = \dot{m}_S/\dot{m}_F = \dot{m}_S/(\pi/4 \cdot D_R^2 \cdot \varrho_F \cdot v_F)$ is set, the conveying pipe diameter D_R can be solved for. It applies:

$$D_R = \sqrt{\frac{2}{\pi} \cdot W \cdot \frac{\dot{m}_S \cdot v_F}{|\Delta p_R|}} \tag{9.5}$$

To solve (9.5), values for W and v_F must be determined or estimated. The conveying gas velocity at the end of the pipeline is set to

$$v_{F,E} \leq 18.0\,\frac{\mathrm{m}}{\mathrm{s}}$$

for wear reasons. With $|\Delta p_R| = 2.0\,\mathrm{bar}$ this results in an initial gas velocity of:

$$v_{F,A} = v_{F,E} \cdot \frac{p_E}{p_A} = 18.0\,\frac{\mathrm{m}}{\mathrm{s}} \cdot \frac{1.0\,\mathrm{bar}}{2.0\,\mathrm{bar} + 1.0\,\mathrm{bar}} = 6.0\,\frac{\mathrm{m}}{\mathrm{s}}$$

This should be achievable, but still needs to be verified. As a reference speed v_F when estimating D_R the geometric mean of the initial and final speeds is used:

$$v_F = \sqrt{v_{F,A} \cdot v_{F,E}} = \sqrt{6.0\,\frac{\mathrm{m}}{\mathrm{s}} \cdot 18.0\,\frac{\mathrm{m}}{\mathrm{s}}} = 10.4\,\frac{\mathrm{m}}{\mathrm{s}}$$

To determine W, a conveying pipe diameter must first be estimated: $D_R = 0.10\,\mathrm{m} \rightarrow$ Estimate.
Furthermore, the following applies:

$\Delta L_h = L_R - H_R = (100\text{–}30)\,\mathrm{m} = 70\,\mathrm{m} \quad \rightarrow \quad$ horizontal conveying distance,

$H_R = 30\,\mathrm{m} \quad \rightarrow \quad$ vertical conveying distance,

$C = u_S/v_F \cong 0.70 \quad \rightarrow \quad$ average velocity ratio, estimated value,

$n_U = 6 \quad \rightarrow \quad$ number of 90°-deflections,

$K_U = \Delta u_S/u_{S,\mathrm{in}} \cong 0.50 \quad \rightarrow \quad$ relative solid deceleration in the bend, (4.56),

Note: For the D_R-estimate, we initially assume 90°-bends as deflection elements.
To calculate the mean resistance coefficient λ_S in W, the pipe Froude number must be known in (9.3). It applies:

$$Fr_R = \frac{v_F^2}{g \cdot D_R} = \frac{\left(10.4\,\frac{\text{m}}{\text{s}}\right)^2}{9.81\,\frac{\text{m}}{\text{s}^2} \cdot 0.10\,\text{m}} = 110.25$$

Furthermore, for the determination of the loading μ in (9.3) the gas density at the average pipeline pressure is required:

$$\varrho_F = \varrho_F(60\,°\text{C}) \cdot \frac{\sqrt{p_A \cdot p_E}}{p_0} = 1.056\,\frac{\text{kg}}{\text{m}^3} \cdot \frac{\sqrt{3.0\,\text{bar} \cdot 1.0\,\text{bar}}}{1.0\,\text{bar}} = 1.829\,\frac{\text{kg}}{\text{m}^3}$$

$$\mu = \frac{\dot{m}_S}{\pi/4 \cdot D_R^2 \cdot \varrho_F \cdot v_F} = \frac{9.028\,\frac{\text{kg}}{\text{s}}}{\frac{\pi}{4} \cdot (0.1\,\text{m})^2 \cdot 1.829\,\frac{\text{kg}}{\text{m}^3} \cdot 10.4\,\frac{\text{m}}{\text{s}}} = 60.5\,\frac{\text{kg}\,S}{\text{kg}\,F}$$

Inserted into (9.3), this results in:

$$\lambda_S = 13.65\,\text{m}^{-0.1} \cdot \frac{D_R^{0.1}}{Fr_R \cdot \mu^{0.3}} = 13.65\,\text{m}^{-0.1} \cdot \frac{(0.10\,\text{m})^{0.1}}{110.25 \cdot 60.5^{0.3}} = 0.0287$$

This results in W as:

$$\begin{aligned} W &= \lambda_S \cdot \frac{\Delta L_h}{D_R} \cdot \left(1 + \frac{1}{2} \cdot \frac{H_R}{\Delta L_h}\right) + \frac{2}{C} \cdot \frac{g \cdot H_R}{v_F^2} + 2 \cdot C \cdot (n_U \cdot K_U + 1) \\ &= 0.0287 \cdot \frac{70\,\text{m}}{0.10\,\text{m}} \cdot \left(1 + \frac{1}{2} \cdot \frac{30\,\text{m}}{70\,\text{m}}\right) + \frac{2}{0.70} \cdot \frac{9.81\,\frac{\text{m}}{\text{s}^2} \cdot 30\,\text{m}}{\left(10.4\,\frac{\text{m}}{\text{s}}\right)^2} \\ &\quad + 2 \cdot 0.70 \cdot (6 \cdot 0.50 + 1) \\ &= 37.77 \end{aligned}$$

and the conveying pipe diameter to:

$$D_R = \sqrt{\frac{2}{\pi} \cdot W \cdot \frac{\dot{m}_S \cdot v_F}{|\Delta p_R|}} = \sqrt{\frac{2}{\pi} \cdot 37.77 \cdot \frac{9.028\,\frac{\text{kg}}{\text{s}} \cdot 10.4\,\frac{\text{m}}{\text{s}}}{200000\,\frac{\text{N}}{\text{m}^2}}} = 0.106\,\text{m}$$

Selected:

$$\text{Pipe } \varnothing 114.3\,\text{mm} \times 5.0\,\text{mm} \quad \rightarrow \quad D_R = 104.3\,\text{mm} = 0.1043\,\text{m}$$

An iteration does not seem necessary, as the D_R-estimate is sufficiently close to the calculated D_R-value.

- Verification of the conveying gas velocity: An appropriate equation from the literature is used. In [2], (4.29), for the calculation of the choking velocity of fine-grained bulk materials, the approach

$$v_{F,st} \cong 0.25 \cdot \frac{\sqrt{g \cdot D_R}}{\sqrt{\varrho_F/\varrho_{b,\mathrm{Str}}}} \quad \rightarrow \quad \text{with: } \mu \cdot \frac{\varrho_F}{\varrho_{b,\mathrm{Str}}} < 0.75$$

is suggested. This is used by the provider. The critical condition = smallest conveying gas velocity occurs at the beginning of the pipeline "A":

$$\varrho_F = \varrho_{F,A} = \varrho_F(60\,°\mathrm{C}, 3.0\,\mathrm{bar}) = \varrho_\mathrm{F}(60\,°\mathrm{C}) \cdot \frac{p_A}{p_0}$$

$$= 1.056\,\frac{\mathrm{kg}}{\mathrm{m}^3} \cdot \frac{3.0\,\mathrm{bar}}{1.0\,\mathrm{bar}} = 3.168\,\frac{\mathrm{kg}}{\mathrm{m}^3}$$

$$\varrho_{b,\mathrm{Str}} \cong \varrho_{SS} = 1210\,\frac{\mathrm{kg}}{\mathrm{m}^3} \quad \rightarrow \quad \text{strand density}$$

Actual gas mass flow rate:

$$\dot{m}_F = \frac{\pi}{4} \cdot D_R^2 \cdot \varrho_{F,E} \cdot v_{F,E} = \frac{\pi}{4} \cdot (0.1043\,\mathrm{m})^2 \cdot 1.056\,\frac{\mathrm{kg}}{\mathrm{m}^3} \cdot 18.0\,\frac{\mathrm{m}}{\mathrm{s}}$$

$$= 0.1623\,\frac{\mathrm{kg}}{\mathrm{s}} = 584.4\,\frac{\mathrm{kg}}{\mathrm{h}}$$

$$\mu = \frac{\dot{m}_S}{\dot{m}_F} = \frac{9.028\,\frac{\mathrm{kg}}{\mathrm{s}}}{0.1623\,\frac{\mathrm{kg}}{\mathrm{s}}} = 55.63\,\frac{\mathrm{kg}\,S}{\mathrm{kg}\,F}$$

Inserted into the $v_{F,st}$ -equation applies:

$$v_{F,st} \cong 0.25 \cdot \frac{\sqrt{g \cdot D_R}}{\sqrt{\varrho_F/\varrho_{b,\mathrm{Str}}}} = 0.25 \cdot \sqrt{9.81\,\frac{\mathrm{m}}{\mathrm{s}^2} \cdot 0.1043\,\mathrm{m} \cdot \frac{1210\,\frac{\mathrm{kg}}{\mathrm{m}^3}}{3.168\,\frac{\mathrm{kg}}{\mathrm{m}^3}}} = 4.94\,\frac{\mathrm{m}}{\mathrm{s}}$$

Verification of the validity range:

$$\mu \cdot \frac{\varrho_F}{\varrho_{b,\mathrm{Str}}} = 55.63 \cdot \frac{3.168\,\frac{\mathrm{kg}}{\mathrm{m}^3}}{1210\,\frac{\mathrm{kg}}{\mathrm{m}^3}} = 0.146 < 0.75 \quad \rightarrow \quad \text{Boundary condition fulfilled}$$

From the above consideration, it follows that $v_{F,A} = 6.0\,\mathrm{m/s} > \mathrm{v_{F,st}} = 4.94\,\mathrm{m/s}$, thus the chosen conveying gas velocities are feasible.

- Determination of the 90°-deflection elements: In principle, autogenously protected deflection pots and T-bends as well as 90° -pipe bends are possible, see Sect. 8.2.1 and Fig. 8.27. With regard to the future planned production of alternative quartz powder qualities, a residue-free operating conveying system is required. This is only possible with pipe bends.
- From Fig. 8.18, it can be seen that for the wall material steel, minimal wear occurs when the condition $20 \lesssim (R/D_R) \lesssim 1.5$ is met. A $(R/D_R) \gtrsim 20$ is not practical, pipe bends $(R/D_R) \cong 1.5$ with $\varnothing 114.3\,\mathrm{mm} \times 6.3\,\mathrm{mm}$, $D_R = 101.7\,\mathrm{mm}$ are available according to DIN 2605, Part 1. A smooth transition must be provided on the inlet side.

Decision:
Use of 90° -pipe bends made of steel with a ratio $(R/D_R) \cong 1.5$ and welded outer wall reinforcement.

- Recalculation of the conveying line pressure loss Δp_R: The pipeline is subsequently recalculated piece by piece incompressibly from the line end (→ State $p_E = 1.0\,\text{bar}$, $v_{F,E} = 18.0\,\text{m/s}$ is known) to the line beginning. The input values in the respective calculation sections are assumed to be constant over the section. The calculated final values are simultaneously the entry values in the next section (→ Replacement of continuous changes by a stair step). An increase in accuracy is possible by increasing the number of calculation sections and/or performing the calculation again in further iteration steps with the mean values from the initial and final states of a section and correcting it. In view of the relatively uncertain data basis (→ λ_S-function from the literature) the calculation is done relatively roughly in the present case.
- Further simplification of the λ_S-function: With the data known so far and

$$Fr_R = \frac{v_F^2}{g \cdot D_R} = \frac{v_F^2}{9.81\,\frac{\text{m}}{\text{s}^2} \cdot 0.1043\,\text{m}} = \frac{v_F^2}{1.0232\,\frac{\text{m}^2}{\text{s}^2}}$$

follows from (9.3):

$$\lambda_S = 13.65\,\text{m}^{-0.1} \cdot \frac{D_R^{0.1}}{Fr_R \cdot \mu^{0.3}}$$

$$= 13.65\,\text{m}^{-0.1} \cdot \frac{(0.1043\,\text{m})^{0.1} \cdot 1.0232\,\frac{\text{m}^2}{\text{s}^2}}{v_F^2 \cdot 55.63^{0.3}} = \frac{3.3368\,\frac{\text{m}^2}{\text{s}^2}}{v_F^2} \tag{9.6}$$

$$\lambda_S = \frac{3.3368\,\frac{\text{m}^2}{\text{s}^2}}{v_F^2}$$

- Determination of the solid/gas velocity ratio C: In the respective vertical and horizontal pipe sections,

$$C_v = 1 - \frac{w_{T,\varepsilon}}{v_F} \quad \rightarrow \quad \text{with: } w_{T,\varepsilon} = \text{settling velocity of a particle cloud}$$

$$C_h = 1 - \frac{v_{F,A,\min}}{v_F} \quad \rightarrow \quad \text{with: } v_{F,A,\min} = \text{minimum conveying gas velocity}$$

is established. The provider has no information about $v_{F,A,\min}$; the above calculated $v_{F,st}$-value is still significantly above the conveying limit. To determine $w_{T,\varepsilon}$, the terminal velocity of a particle cloud, and from this C_v an older equation from [3] is used:

$$w_{T,\varepsilon} = w_T \cdot (1 + 3 \cdot k \cdot \sqrt{\mu}) \quad \rightarrow \quad \text{with: } k = 1 \text{ if } Re_{S,\varepsilon} < 10 \tag{9.7}$$

$$w_{T,\varepsilon} = 0.1085\,\frac{\text{m}}{\text{s}} \cdot (1 + 3 \cdot \sqrt{55.63}) = 2.54\,\frac{\text{m}}{\text{s}}$$

$$Re_{S,\varepsilon} = \frac{w_{T,\varepsilon} \cdot d_{S,50}}{\nu_F} = \frac{2.54\,\frac{\text{m}}{\text{s}} \cdot 39 \cdot 10^{-6}\,\text{m}}{18.88 \cdot 10^{-6}\,\frac{\text{m}}{\text{s}^2}} = 5.24 < 10 \rightarrow \text{Boundary condition fulfilled}$$

$$C_v = 1 - \frac{2.54\,\frac{\text{m}}{\text{s}}}{v_F}$$

Practical experience shows that, in general, with sufficient accuracy

$$C = C_v = C_h$$

can be set. This is also assumed here. Because of $C(v_F)$ also applies $C(L_R)$, i.e., C changes along the conveying path.

Note: $w_{T,\varepsilon}$ can be described more accurately by newer equations, e.g., (4.85), (4.38). The minimum conveying speed $v_{F,A,\min}$ of the present quartz powder is approximately $v_{F,A,\min} \cong 3\,\text{m/s}$, see Fig. 4.16. This confirms the assumption $C_v \cong C_h$.

- Bend loss coefficient K_U: This can be calculated with (4.58). With the deflection angle $\alpha_U = 90°$ and a wall friction coefficient of quartz powder against St 37 of $\beta_R = \tan \varphi_W = \tan 25° = 0.466$ follows:

$$K_U = 1 - \exp\left(-\frac{\alpha_U}{180°} \cdot \pi \cdot \beta_R\right) = 1 - \exp\left(-\frac{90°}{180°} \cdot \pi \cdot 0.466\right) = 0.519$$

  No differentiation is made with regard to the spatial position of the deflections.
- Calculation of pressure losses $\Delta p(Ai)$ of the individual calculation sections i: The inlet values in the respective calculation section receive index 1, the outlet values receive index 2. The direction of the calculation is opposite to the flow direction from the pipe end, index E, to the pipe beginning, index A. The losses of the conveying gas itself are determined separately after the overall calculation is completed.
- 1. Calculation section: horizontal pipe section from the silo inlet to the 90°-bend, see Fig. 9.1 → horizontal solid friction according to (4.46). With (9.6) and the constant values $\mu = 55.63\,\text{kg}\,S/\text{kg}\,F$, $D_R = 0.1043\,\text{m}$, the following results from this:

$$\begin{aligned} -\Delta p_{S,R,h} &= \mu \cdot \lambda_S \cdot \frac{\Delta L_h}{D_R} \cdot \frac{\rho_F}{2} \cdot v_F^2 \\ &= 55.63 \cdot \frac{3.3368\,\frac{\text{m}^2}{\text{s}^2}}{v_F^2} \cdot \frac{\Delta L_h}{0.1043\,\text{m}} \cdot \frac{\rho_F}{2} \cdot v_F^2 = 889.7\,\frac{\text{m}}{\text{s}^2} \cdot \Delta L_h \cdot \rho_F \end{aligned} \tag{9.8}$$

(9.8) shows no influence of the conveying speed on the horizontal frictional pressure loss of the bulk material. This is consistent with the theoretical expectations for the dense phase conveying range, see Sect. 4.7.1: $\Delta p_{S,R,h}$ is established there proportional to the speed-independent weight force of the bulk material. For the pressure loss of the 1st calculation section, with $\Delta L_h(A1) = 5.0\,\text{m}$ and $\varrho_{F,1}(A1) = \varrho_{F,E} = \varrho_F(60\,°\text{C}) = 1.056\,\text{kg/m}^3$:

$$-\Delta p(A1) = 889.7\,\frac{\text{m}}{\text{s}^2} \cdot \Delta L_h(A1) \cdot \varrho_{F,1}(A1)$$
$$= 889.7\,\frac{\text{m}}{\text{s}^2} \cdot 5.0\,\text{m} \cdot 1.056\,\frac{\text{kg}}{\text{m}^3} = 4698.5\,\frac{\text{N}}{\text{m}^2}$$

Pressure at the end of the section, $p_1(A1) = p_E = 100000\,\text{N/m}^2$:

$$p_2(A1) = p_1(A1) - \Delta p(A1) = (100000 + 4698.5)\,\frac{\text{N}}{\text{m}^2} = 104698.5\,\frac{\text{N}}{\text{m}^2}$$

Gas velocity at the end of the section, $v_{F,1} = v_{F,E} = 18.0\,\text{m/s}$:

$$v_{F,2}(A1) = v_{F,E} \cdot \frac{p_E}{p_2(A1)} = 18.0\,\frac{\text{m}}{\text{s}} \cdot \frac{100000\,\frac{\text{N}}{\text{m}^2}}{104698.5\,\frac{\text{N}}{\text{m}^2}} = 17.19\,\frac{\text{m}}{\text{s}}$$

Gas density at the end of the section:

$$\varrho_{F,2}(A1) = \varrho_{F,E} \cdot \frac{p_2(A1)}{p_E} = 1.056\,\frac{\text{kg}}{\text{m}^3} \cdot \frac{104698.5\,\frac{\text{N}}{\text{m}^2}}{100000\,\frac{\text{N}}{\text{m}^2}} = 1.1056\,\frac{\text{kg}}{\text{m}^3}$$

- 2. Calculation section: 90°-pipe bend, vertically upwards into the horizontal flow → Pressure loss due to re-acceleration of the decelerated solid matter. With the constant values along the conveying route $\mu = 55.63\,\text{kg}\,S/\text{kg}\,F$, $K_U = 0.519$, follows from (4.57):

$$-\Delta p_{S,U} = K_U \cdot C_U \cdot \mu \cdot \varrho_{F,U} \cdot v_{F,U}^2 = 0.519 \cdot C \cdot 55.63 \cdot \varrho_F \cdot v_F^2$$
$$= 28.87 \cdot C \cdot \varrho_F \cdot v_F^2$$

For the pressure loss of the 2nd calculation section, with $v_{F,1}(A2) = v_{F,2}(A1) = 17.19\,\text{m/s}$, $\varrho_{F,1}(A2) = \varrho_{F,2}(A1) = 1.1056\,\text{kg/m}^3$ as well as $C_1(A2) = 1 - (2.54\,\text{m/s}/v_{F,1}(A2)) = 1 - (2.54\,\text{m/s}/17.19\,\text{m/s}) = 0.852$ follows:

$$-\Delta p(A2) = 28.87 \cdot C_1(A2) \cdot \varrho_{F,1}(A2) \cdot (v_{F,1}(A2))^2$$
$$= 28.87 \cdot 0.852 \cdot 1.1056\,\frac{\text{kg}}{\text{m}^3} \cdot \left(17.19\,\frac{\text{m}}{\text{s}}\right)^2 = 8035.9\,\frac{\text{N}}{\text{m}^2}$$

Operating parameters at the end of the section:

$$p_2(A2) = p_2(A1) - \Delta p(A2) = (104698.5 + 8035.9)\,\frac{\text{N}}{\text{m}^2} = 112734.4\,\frac{\text{N}}{\text{m}^2}$$

$$v_{F,2}(A2) = v_{F,E} \cdot \frac{p_E}{p_2(A2)} = 18.0\,\frac{\text{m}}{\text{s}} \cdot \frac{100000\,\frac{\text{N}}{\text{m}^2}}{112734.4\,\frac{\text{N}}{\text{m}^2}} = 15.97\,\frac{\text{m}}{\text{s}}$$

$$\varrho_{F,2}(A2) = \varrho_{F,E} \cdot \frac{p_2(A2)}{p_E} = 1.056\,\frac{\text{kg}}{\text{m}^3} \cdot \frac{112734.4\,\frac{\text{N}}{\text{m}^2}}{100000\,\frac{\text{N}}{\text{m}^2}} = 1.1905\,\frac{\text{kg}}{\text{m}^3}$$

- 3. Calculation section: $1/2 \times$ height of the vertical conveyor section, from the upper bend downwards $\rightarrow$ vertical solid friction + solid lift. For the vertical solid friction, with (4.52) and (9.8):

$$-\Delta p_{S,R,v} = \frac{1}{2} \cdot \Delta p_{S,R,h} = 444.9\,\frac{\text{m}}{\text{s}^2} \cdot \Delta L_v \cdot \varrho_F$$

For the lift pressure loss of the solid, with (4.53) and the known constants along the conveyor section:

$$-\Delta p_{S,H} = \frac{g}{\overline{C}_v} \cdot \mu \cdot \overline{\varrho}_F \cdot \Delta L_v = \frac{9.81\,\frac{\text{m}}{\text{s}^2}}{C} \cdot 55.63 \cdot \varrho_F \cdot \Delta L_v = \frac{545.73\,\frac{\text{m}}{\text{s}^2}}{C} \cdot \varrho_{F,1} \cdot \Delta L_v$$

For the pressure loss of the 3rd calculation section, the following applies with $\Delta L_v(A3) = 1/2 \cdot 25\,\text{m} = 12.5\,\text{m}$, $\varrho_{F,1}(A3) = \varrho_{F,2}(A2) = 1.1905\,\text{kg/m}^3$, $C_1(A3) = 1 - (2.54\,\text{m/s}/v_{F,2}(A2)) = 1 - (2.54\,\text{m/s}/15.97\,\text{m/s}) = 0.841$:

$$\begin{aligned}-\Delta p(A3) &= \left(444.9\,\frac{\text{m}}{\text{s}^2} + \frac{545.73\,\frac{\text{m}}{\text{s}^2}}{C_1(A3)}\right) \cdot \Delta L_v(A3) \cdot \varrho_{F,1}(A3)\\ &= \left(444.9\,\frac{\text{m}}{\text{s}^2} + \frac{545.73\,\frac{\text{m}}{\text{s}^2}}{0.841}\right) \cdot 12.5\,\text{m} \cdot 1.1905\,\frac{\text{kg}}{\text{m}^3} = 16277.7\,\frac{\text{N}}{\text{m}^2}\end{aligned}$$

Operating parameters at the end of the section:

$$p_2(A3) = p_2(A2) - \Delta p(A3) = (112734.4 + 16277.7)\,\frac{\text{N}}{\text{m}^2} = 129012.1\,\frac{\text{N}}{\text{m}^2}$$

$$v_{F,2}(A3) = v_{F,E} \cdot \frac{p_E}{p_2(A3)} = 18.0\,\frac{\text{m}}{\text{s}} \cdot \frac{100000\,\frac{\text{N}}{\text{m}^2}}{129012.1\,\frac{\text{N}}{\text{m}^2}} = 13.95\,\frac{\text{m}}{\text{s}}$$

$$\varrho_{F,2}(A3) = \varrho_{F,E} \cdot \frac{p_2(A3)}{p_E} = 1.056 \cdot \frac{129012.1\,\frac{\text{N}}{\text{m}^2}}{100000\,\frac{\text{N}}{\text{m}^2}} = 1.3624\,\frac{\text{kg}}{\text{m}^3}$$

- 4. Calculation section: $1/2 \times$ height of the vertical conveyor section, from the center to the lower bend $\rightarrow$ vertical solid friction + solid lift. With $\Delta L_v(A4) = 1/2 \cdot 25\,\text{m} = 12.5\,\text{m}$, $C_1(A4) = 1 - (2.54\,\text{m/s}/v_{F,2}(A3)) = 1 - (2.54\,\text{m/s}/13.95\,\text{m/s}) = 0.818$, $\varrho_{F,1}(A4) = \varrho_{F,2}(A3) = 1.3624\,\text{kg/m}^3$ the pressure loss of this pipe section is calculated as:

$$-\Delta p(A4) = \left(444.9\,\frac{\mathrm{m}}{\mathrm{s}^2} + \frac{545.73\,\frac{\mathrm{m}}{\mathrm{s}^2}}{C_1(A4)}\right) \cdot \Delta L_v(A4) \cdot \varrho_{F,1}(A4)$$
$$= \left(444.9\,\frac{\mathrm{m}}{\mathrm{s}^2} + \frac{545.73\,\frac{\mathrm{m}}{\mathrm{s}^2}}{0.818}\right) \cdot 12.5\,\mathrm{m} \cdot 1.3624\,\frac{\mathrm{kg}}{\mathrm{m}^3} = 18938.8\,\frac{\mathrm{N}}{\mathrm{m}^2}$$

Operating parameters at the end of the section:

$$p_2(A4) = p_2(A3) - \Delta p(A4) = (129012.1 + 18938.8)\,\frac{\mathrm{N}}{\mathrm{m}^2} = 147950.9\,\frac{\mathrm{N}}{\mathrm{m}^2}$$
$$v_{F,2}(A4) = v_{F,E} \cdot \frac{p_E}{p_2(A4)} = 18.0\,\frac{\mathrm{m}}{\mathrm{s}} \cdot \frac{100000\,\frac{\mathrm{N}}{\mathrm{m}^2}}{147950.9\,\frac{\mathrm{N}}{\mathrm{m}^2}} = 12.17\,\frac{\mathrm{m}}{\mathrm{s}}$$
$$\varrho_{F,2}(A4) = \varrho_{F,E} \cdot \frac{p_2(A4)}{p_E} = 1.056\,\frac{\mathrm{kg}}{\mathrm{m}^3} \cdot \frac{147950.9\,\frac{\mathrm{N}}{\mathrm{m}^2}}{100000\,\frac{\mathrm{N}}{\mathrm{m}^2}} = 1.5624\,\frac{\mathrm{kg}}{\mathrm{m}^3}$$

- 5. Calculation section: 90°-pipe bend, horizontal and then upwards into the vertical flow → Pressure loss due to re-acceleration of the decelerated solid. With $v_{F,1}(A5) = v_{F,2}(A4) = 12.17\,\mathrm{m/s}$, $\varrho_{F,1}(A5) = \varrho_{F,2}(A4) = 1.5624\,\mathrm{kg/m^3}$ as well as $C_1(A5) = 1 - (2.54\,\mathrm{m/s}/v_{F,1}(A5)) = 1 - (2.54\,\mathrm{m/s}/12.17\,\mathrm{m/s}) = 0.791$ applies:

$$-\Delta p(A5) = 28.87 \cdot C_1(A5) \cdot \varrho_{F,1}(A5) \cdot (v_{F,1}(A5))^2$$
$$= 28.87 \cdot 0.791 \cdot 1.5624\,\frac{\mathrm{kg}}{\mathrm{m}^3} \cdot \left(12.17\,\frac{\mathrm{m}}{\mathrm{s}}\right)^2 = 5284.4\,\frac{\mathrm{N}}{\mathrm{m}^2}$$

Operating parameters at the end of the section:

$$p_2(A5) = p_2(A4) - \Delta p(A5) = (147950.9 + 5284.4)\,\frac{\mathrm{N}}{\mathrm{m}^2} = 153235.3\,\frac{\mathrm{N}}{\mathrm{m}^2}$$
$$v_{F,2}(A5) = v_{F,E} \cdot \frac{p_E}{p_2(A5)} = 18.0\,\frac{\mathrm{m}}{\mathrm{s}} \cdot \frac{100000\,\frac{\mathrm{N}}{\mathrm{m}^2}}{153235.3\,\frac{\mathrm{N}}{\mathrm{m}^2}} = 11.75\,\frac{\mathrm{m}}{\mathrm{s}}$$
$$\rho_{F,2}(A5) = \rho_{F,E} \cdot \frac{p_2(A5)}{p_E} = 1.056\,\frac{\mathrm{kg}}{\mathrm{m}^3} \cdot \frac{153235.3\,\frac{\mathrm{N}}{\mathrm{m}^2}}{100000\,\frac{\mathrm{N}}{\mathrm{m}^2}} = 1.6182\,\frac{\mathrm{kg}}{\mathrm{m}^3}$$

- 6. Calculation section: horizontal pipe section up to the next 90°-bend in the horizontal → horizontal solid friction. With $\Delta L_h(A6) = 20.0\,\mathrm{m}$ and $\varrho_{F,1}(A6) = \varrho_{F,2}(A5) = 1.6182\,\mathrm{kg/m^3}$ follows:

$$-\Delta p(A6) = 889.7\,\frac{\mathrm{m}}{\mathrm{s}^2} \cdot \Delta L_h(A6) \cdot \varrho_{F,1}(A6)$$
$$= 889.7\,\frac{\mathrm{m}}{\mathrm{s}^2} \cdot 20.0\,\mathrm{m} \cdot 1.6182\,\frac{\mathrm{kg}}{\mathrm{m}^3} = 28799.8\,\frac{\mathrm{N}}{\mathrm{m}^2}$$

Operating parameters at the end of the section:

$$p_2(A6) = p_2(A5) - \Delta p(A6) = (153235.3 + 28799.8)\,\frac{\mathrm{N}}{\mathrm{m}^2} = 182035.1\,\frac{\mathrm{N}}{\mathrm{m}^2}$$

$$v_{F,2}(A6) = v_{F,E} \cdot \frac{p_E}{p_2(A6)} = 18.0\,\frac{\mathrm{m}}{\mathrm{s}} \cdot \frac{100000\,\frac{\mathrm{N}}{\mathrm{m}^2}}{182035.1\,\frac{\mathrm{N}}{\mathrm{m}^2}} = 9.89\frac{\mathrm{m}}{\mathrm{s}}$$

$$\varrho_{F,2}(A6) = \varrho_{F,E} \cdot \frac{p_2(A6)}{p_E} = 1.056\,\frac{\mathrm{kg}}{\mathrm{m}^3} \cdot \frac{182035.1\,\frac{\mathrm{N}}{\mathrm{m}^2}}{100000\,\frac{\mathrm{N}}{\mathrm{m}^2}} = 1.9223\frac{\mathrm{kg}}{\mathrm{m}^3}$$

- 7. Calculation section: 90°-pipe bend, flowed through horizontally → pressure loss due to the re-acceleration of the decelerated solid. With the data $v_{F,1}(A7) = v_{F,2}(A6) = 9.89\,\mathrm{m/s}$, $\varrho_{F,1}(A7) = \varrho_{F,2}(A6) = 1.9223\,\mathrm{kg/m^3}$ and $C_1(A7) = 1 - (2.54\,\mathrm{m/s}/v_{F,1}(A7)) = 1 - (2.54\,\mathrm{m/s}/9.89\,\mathrm{m/s}) = 0.743$ the pressure loss amounts to:

$$\begin{aligned} -\Delta p(A7) &= 28.87 \cdot C_1(A7) \cdot \varrho_{F,1}(A7) \cdot (v_{F,1}(A7))^2 \\ &= 28.87 \cdot 0.743 \cdot 1.9223\,\frac{\mathrm{kg}}{\mathrm{m}^3} \cdot \left(9.89\,\frac{\mathrm{m}}{\mathrm{s}}\right)^2 = 4033.2\,\frac{\mathrm{N}}{\mathrm{m}^2} \end{aligned}$$

Operating parameters at the end of the section:

$$p_2(A7) = p_2(A6) - \Delta p(A7) = (182035.1 + 4033.2)\,\frac{\mathrm{N}}{\mathrm{m}^2} = 186068.3\,\frac{\mathrm{N}}{\mathrm{m}^2}$$

$$v_{F,2}(A7) = v_{F,E} \cdot \frac{p_E}{p_2(A7)} = 18.0\,\frac{\mathrm{m}}{\mathrm{s}} \cdot \frac{100000\,\frac{\mathrm{N}}{\mathrm{m}^2}}{186068.3\,\frac{\mathrm{N}}{\mathrm{m}^2}} = 9.67\,\frac{\mathrm{m}}{\mathrm{s}}$$

$$\varrho_{F,2}(A7) = \varrho_{F,E} \cdot \frac{p_2(A7)}{p_E} = 1.056\,\frac{\mathrm{kg}}{\mathrm{m}^3} \cdot \frac{186068.3\,\frac{\mathrm{N}}{\mathrm{m}^2}}{100000\,\frac{\mathrm{N}}{\mathrm{m}^2}} = 1.9649\,\frac{\mathrm{kg}}{\mathrm{m}^3}$$

- 8. Calculation section: horizontal pipe section between the two horizontally flowed through 90°-bends → horizontal solid friction. With $\Delta L_h(A8) = 15.0\,\mathrm{m}$ and $\varrho_{F,1}(A8) = \varrho_{F,2}(A7) = 1.9649\,\mathrm{kg/m^3}$ follows:

$$\begin{aligned} -\Delta p(A8) &= 889.7\,\frac{\mathrm{m}}{\mathrm{s}^2} \cdot \Delta L_h(A8) \cdot \varrho_{F,1}(A8) \\ &= 889.7\,\frac{\mathrm{m}}{\mathrm{s}^2} \cdot 15.0\,\mathrm{m} \cdot 1.9649\,\frac{\mathrm{kg}}{\mathrm{m}^3} = 26227.6\,\frac{\mathrm{N}}{\mathrm{m}^2} \end{aligned}$$

Operating parameters at the end of the section:

$$p_2(A8) = p_2(A7) - \Delta p(A8) = (186068.3 + 26227.6)\,\frac{\mathrm{N}}{\mathrm{m}^2} = 212295.9\,\frac{\mathrm{N}}{\mathrm{m}^2}$$

$$v_{F,2}(A8) = v_{F,E} \cdot \frac{p_E}{p_2(A8)} = 18.0\,\frac{\mathrm{m}}{\mathrm{s}} \cdot \frac{100000\,\frac{\mathrm{N}}{\mathrm{m}^2}}{212295.9\,\frac{\mathrm{N}}{\mathrm{m}^2}} = 8.48\,\frac{\mathrm{m}}{\mathrm{s}}$$

$$\rho_{F,2}(A8) = \rho_{F,E} \cdot \frac{p_2(A8)}{p_E} = 1.056\,\frac{\mathrm{kg}}{\mathrm{m}^3} \cdot \frac{212295.9\,\frac{\mathrm{N}}{\mathrm{m}^2}}{100000\,\frac{\mathrm{N}}{\mathrm{m}^2}} = 2.2418\,\frac{\mathrm{kg}}{\mathrm{m}^3}$$

- 9. Calculation section: 90°-pipe bend, flowed through horizontally → Pressure loss due to the re-acceleration of the decelerated solid. With the data $v_{F,1}(A9) = v_{F,2}(A8) = 8.48\,\text{m/s}$, $\varrho_{F,1}(A9) = \varrho_{F,2}(A8) = 2.2418\,\text{kg/m}^3$ and $C_1(A9) = 1 - (2.54\,\text{m/s}/v_{F,1}(A9)) = 1 - (2.54\,\text{m/s}/8.48\,\text{m/s}) = 0.701$ applies:

$$-\Delta p(A9) = 28.87 \cdot C_1(A9) \cdot \varrho_{F,1}(A9) \cdot (v_{F,1}(A9))^2$$
$$= 28.87 \cdot 0.701 \cdot 2.2418\,\frac{\text{kg}}{\text{m}^3} \cdot \left(8.48\,\frac{\text{m}}{\text{s}}\right)^2 = 3262.5\,\frac{\text{N}}{\text{m}^2}$$

Operating parameters at the end of the section:

$$p_2(A9) = p_2(A8) - \Delta p(A9) = (212295.9 + 3262.5)\,\frac{\text{N}}{\text{m}^2} = 215558.4\,\frac{\text{N}}{\text{m}^2}$$
$$v_{F,2}(A9) = v_{F,E} \cdot \frac{p_E}{p_2(A9)} = 18.0\,\frac{\text{m}}{\text{s}} \cdot \frac{100000\,\frac{\text{N}}{\text{m}^2}}{215558.4\,\frac{\text{N}}{\text{m}^2}} = 8.35\,\frac{\text{m}}{\text{s}}$$
$$\varrho_{F,2}(A9) = \varrho_{F,E} \cdot \frac{p_2(A9)}{p_E} = 1.056\,\frac{\text{kg}}{\text{m}^3} \cdot \frac{215558.4\,\frac{\text{N}}{\text{m}^2}}{100000\,\frac{\text{N}}{\text{m}^2}} = 2.2763\,\frac{\text{kg}}{\text{m}^3}$$

- 10. Calculation section: horizontal pipe section up to the next 90°-bend vertically downward → horizontal solid friction. With $\Delta L_h(A10) = 20.0\,\text{m}$ and $\varrho_{F,1}(A10) = \varrho_{F,2}(A9) = 2.2763\,\text{kg/m}^3$ follows:

$$-\Delta p(A10) = 889.7\,\frac{\text{m}}{\text{s}^2} \cdot \Delta L_h(A10) \cdot \varrho_{F,1}(A10)$$
$$= 889.7\,\frac{\text{m}}{\text{s}^2} \cdot 20.0\,\text{m} \cdot 2.2763\,\frac{\text{kg}}{\text{m}^3} = 40512.2\,\frac{\text{N}}{\text{m}^2}$$

Operating parameters at the end of the section:

$$p_2(A10) = p_2(A9) - \Delta p(A10) = (215558.4 + 40512.2)\,\frac{\text{N}}{\text{m}^2} = 256070.6\,\frac{\text{N}}{\text{m}^2}$$
$$v_{F,2}(A10) = v_{F,E} \cdot \frac{p_E}{p_2(A10)} = 18.0\,\frac{\text{m}}{\text{s}} \cdot \frac{100000\,\frac{\text{N}}{\text{m}^2}}{256070.6\,\frac{\text{N}}{\text{m}^2}} = 7.03\,\frac{\text{m}}{\text{s}}$$
$$\varrho_{F,2}(A10) = \varrho_{F,E} \cdot \frac{p_2(A10)}{p_E} = 1.056\,\frac{\text{kg}}{\text{m}^3} \cdot \frac{256070.6\,\frac{\text{N}}{\text{m}^2}}{100000\,\frac{\text{N}}{\text{m}^2}} = 2.7041\,\frac{\text{kg}}{\text{m}^3}$$

- 11. Calculation section: 90° -pipe bend, vertically upwards into the horizontal flow → Pressure loss due to the re-acceleration of the decelerated solid. With $v_{F,1}(A11) = v_{F,2}(A10) = 7.03\,\text{m/s}$, $\varrho_{F,1}(A11) = \varrho_{F,2}(A10) = 2.7041\,\text{kg/m}^3$ as well as $C_1(A11) = 1 - (2.54\,\text{m/s}/v_{F,1}(A11)) = 1 - (2.54\,\text{m/s}/7.03\,\text{m/s}) = 0.639$ applies:

$$-\Delta p(A11) = 28.87 \cdot C_1(A11) \cdot \varrho_{F,1}(A11) \cdot (v_{F,1}(A11))^2$$
$$= 28.87 \cdot 0.639 \cdot 2.7041\,\frac{\text{kg}}{\text{m}^3} \cdot \left(7.03\,\frac{\text{m}}{\text{s}}\right)^2 = 2465.4\,\frac{\text{N}}{\text{m}^2}$$

Operating parameters at the end of the section:

$$p_2(A11) = p_2(A10) - \Delta p(A11) = (256070.6 + 2465.4)\,\frac{\mathrm{N}}{\mathrm{m}^2} = 258536.0\,\frac{\mathrm{N}}{\mathrm{m}^2}$$

$$v_{F,2}(A11) = v_{F,E} \cdot \frac{p_E}{p_2(A11)} = 18.0\,\frac{\mathrm{m}}{\mathrm{s}} \cdot \frac{100000\,\frac{\mathrm{N}}{\mathrm{m}^2}}{258536.0\,\frac{\mathrm{N}}{\mathrm{m}^2}} = 6.96\,\frac{\mathrm{m}}{\mathrm{s}}$$

$$\varrho_{F,2}(A11) = \varrho_{F,E} \cdot \frac{p_2(A11)}{p_E} = 1.056\,\frac{\mathrm{kg}}{\mathrm{m}^3} \cdot \frac{258536.0\,\frac{\mathrm{N}}{\mathrm{m}^2}}{100000\,\frac{\mathrm{N}}{\mathrm{m}^2}} = 2.7301\,\frac{\mathrm{kg}}{\mathrm{m}^3}$$

- 12. Calculation section: vertical pipe section between two 90°-bends → vertical solid friction + solid lift. With $\Delta L_v(A12) = 5.0\,\mathrm{m}$, $\varrho_{F,1}(A12) = \varrho_{F,2}(A11) = 2.7301\,\mathrm{kg/m^3}$ as well as $C_1(A12) = 1 - (2.54\,\mathrm{m/s}/v_{F,1}(A11)) = 1 - (2.54\,\mathrm{m/s}/6.96\,\mathrm{m/s}) = 0.635$ follows:

$$-\Delta p(A12) = \left(444.9\,\frac{\mathrm{m}}{\mathrm{s}^2} + \frac{545.73\,\frac{\mathrm{m}}{\mathrm{s}^2}}{C_1(A12)}\right) \cdot \Delta L_v(A12) \cdot \varrho_{F,1}(A12)$$
$$= \left(444.9\,\frac{\mathrm{m}}{\mathrm{s}^2} + \frac{545.73\,\frac{\mathrm{m}}{\mathrm{s}^2}}{0.635}\right) \cdot 5.0\,\mathrm{m} \cdot 2.7301\,\frac{\mathrm{kg}}{\mathrm{m}^3} = 17805.1\,\frac{\mathrm{N}}{\mathrm{m}^2}$$

Operating parameters at the end of the section:

$$p_2(A12) = p_2(A11) - \Delta p(A12) = (258536.0 + 17805.1)\,\frac{\mathrm{N}}{\mathrm{m}^2} = 276341.1\,\frac{\mathrm{N}}{\mathrm{m}^2}$$

$$v_{F,2}(A12 = v_{F,E} \cdot \frac{p_E}{p_2(A12)} = 18.0\,\frac{\mathrm{m}}{\mathrm{s}} \cdot \frac{100000\,\frac{\mathrm{N}}{\mathrm{m}^2}}{276341.1\,\frac{\mathrm{N}}{\mathrm{m}^2}} = 6.51\,\frac{\mathrm{m}}{\mathrm{s}}$$

$$\varrho_{F,2}(A12) = \varrho_{F,E} \cdot \frac{p_2(A12)}{p_E} = 1.056\,\frac{\mathrm{kg}}{\mathrm{m}^3} \cdot \frac{276341.1\,\frac{\mathrm{N}}{\mathrm{m}^2}}{100000\,\frac{\mathrm{N}}{\mathrm{m}^2}} = 2.9182\,\frac{\mathrm{kg}}{\mathrm{m}^3}$$

- 13. Calculation section: 90°-pipe bend, horizontally flowing into the vertical → Pressure loss due to re-acceleration of the decelerated solid. With the values $v_{F,1}(A13) = v_{F,2}(A12) = 6.51\,\mathrm{m/s}$, $\varrho_{F,1}(A13) = \varrho_{F,2}(A12) = 2.9182\,\mathrm{kg/m^3}$, $C_1(A13) = 1 - (2.54\,\mathrm{m/s}/v_{F,1}(A13)) = 1 - (2.54\,\mathrm{m/s}/6.51\,\mathrm{m/s}) = 0.610$ applies:

$$-\Delta p(A13) = 28.87 \cdot C_1(A13) \cdot \varrho_{F,1}(A13) \cdot (v_{F,1}(A13))^2$$
$$= 28.87 \cdot 0.610 \cdot 2.9182\,\frac{\mathrm{kg}}{\mathrm{m}^3} \cdot \left(6.51\,\frac{\mathrm{m}}{\mathrm{s}}\right)^2 = 2178.0\,\frac{\mathrm{N}}{\mathrm{m}^2}$$

Operating parameters at the end of the section:

$$p_2(A13) = p_2(A12) - \Delta p(A13) = (276341.1 + 2178.0)\,\frac{\text{N}}{\text{m}^2} = 278519.1\,\frac{\text{N}}{\text{m}^2}$$

$$v_{F,2}(A13) = v_{F,E} \cdot \frac{p_E}{p_2(A13)} = 18.0\,\frac{\text{m}}{\text{s}} \cdot \frac{100000\,\frac{\text{N}}{\text{m}^2}}{278519.1\,\frac{\text{N}}{\text{m}^2}} = 6.46\,\frac{\text{m}}{\text{s}}$$

$$\rho_{F,2}(A13) = \rho_{F,E} \cdot \frac{p_2(A13)}{p_E} = 1.056\,\frac{\text{kg}}{\text{m}^3} \cdot \frac{278519.1\,\frac{\text{N}}{\text{m}^2}}{100000\,\frac{\text{N}}{\text{m}^2}} = 2.9412\,\frac{\text{kg}}{\text{m}^3}$$

- 14. Calculation section: horizontal pipe section from the vertical to the beginning of the line → horizontal solid friction. With $\Delta L_v(A14) = 10.0\,\text{m}$ and $\varrho_{F,1}(A14) = \varrho_{F,2}(A13) = 2.9412\text{kg/m}^3$ follows:

$$-\Delta p(A14) = 889.7\,\frac{\text{m}}{\text{s}^2} \cdot \Delta L_h(A14) \cdot \varrho_{F,1}(A14)$$
$$= 889.7\,\frac{\text{m}}{\text{s}^2} \cdot 10.0\,\text{m} \cdot 2.9412\,\frac{\text{kg}}{\text{m}^3} = 26172.9\,\frac{\text{N}}{\text{m}^2}$$

Operating parameters at the end of the section:

$$p_2(A14) = p_2(A13) - \Delta p(A14) = (278519.1 + 26172.9)\,\frac{\text{N}}{\text{m}^2} = 304692.0\,\frac{\text{N}}{\text{m}^2}$$

$$v_{F,2}(A14) = v_{F,E} \cdot \frac{p_E}{p_2(A14)} = 18.0\,\frac{\text{m}}{\text{s}} \cdot \frac{100000\,\frac{\text{N}}{\text{m}^2}}{304692.0\,\frac{\text{N}}{\text{m}^2}} = 5.91\,\frac{\text{m}}{\text{s}}$$

$$\varrho_{F,2}(A14) = \varrho_{F,E} \cdot \frac{p_2(A14)}{p_E} = 1.056\,\frac{\text{kg}}{\text{m}^3} \cdot \frac{304692.0\,\frac{\text{N}}{\text{m}^2}}{100000\,\frac{\text{N}}{\text{m}^2}} = 3.2175\,\frac{\text{kg}}{\text{m}^3}$$

- 15. Calculation section: Initial acceleration of the solid from the pressure vessel into the 14th calculation section → Pressure loss due to solid acceleration. With $v_{F,1}(A15) = v_{F,2}(A14) = 5.91\,\text{m/s}$, $\varrho_{F,1}(A15) = \varrho_{F,2}(A14) = 3.2175\text{kg/m}^3$ as well as $C_1(A15) = 1 - (2.54\,\text{m/s}/v_{F,1}(A15)) = 1 - (2.54\,\text{m/s}/5.91\,\text{m/s}) = 0.570$ follows from (4.55):

$$-\Delta p_{S,B} = C_A \cdot \mu \cdot \varrho_{F,A} \cdot v_{F,A}^2 = C_1(A15) \cdot \mu \cdot \varrho_{F,1}(A15) \cdot (v_{F,1}(A15))^2$$
$$-\Delta p(A15) = 0.570 \cdot 55.63 \cdot 3.2175\,\frac{\text{kg}}{\text{m}^3} \cdot \left(5.91\,\frac{\text{m}}{\text{s}}\right)^2 = 3563.5\,\frac{\text{N}}{\text{m}^2}$$

Operating parameters at the end of the section:

$$p_2(A15) = p_2(A14) - \Delta p(A15) = (304692.0 + 3563.5)\,\frac{\text{N}}{\text{m}^2} = 308255.5\,\frac{\text{N}}{\text{m}^2}$$

$$v_{F,2}(A15) = v_{F,E} \cdot \frac{p_E}{p_2(A15)} = 18.0\,\frac{\text{m}}{\text{s}} \cdot \frac{100000\,\frac{\text{N}}{\text{m}^2}}{308255.5\,\frac{\text{N}}{\text{m}^2}} = 5.84\,\frac{\text{m}}{\text{s}}$$

$$\varrho_{F,2}(A15) = \varrho_{F,E} \cdot \frac{p_2(A15)}{p_E} = 1.056\,\frac{\text{kg}}{\text{m}^3} \cdot \frac{308255.5\,\frac{\text{N}}{\text{m}^2}}{100000\,\frac{\text{N}}{\text{m}^2}} = 3.2552\,\frac{\text{kg}}{\text{m}^3}$$

- Pressure loss of the conveying gas. From (4.65) it follows for the present case:

$$-\Delta p_F = \left(\lambda_F \cdot \frac{\Delta L}{D_R} + n_U \cdot \xi_U\right) \cdot \frac{\varrho_F}{2} \cdot v_F^2 + \varrho_{F,\mathrm{Hub}} \cdot g \cdot H_R$$

Simplifying here, $v_F = v_{F,E} = 18.0\,\mathrm{m/s}$, $\varrho_F = \varrho_{F,\mathrm{Hub}} = \varrho_{F,E} = 1.056\,\mathrm{kg/m^3}$ is set. With $\Delta L = L_R = 100\,\mathrm{m}$, $H_R = 30\,\mathrm{m}$, the resistance coefficient of pipe friction $\lambda_F = 0.02$, the resistance coefficient of a 90°-gas deflection $\xi_U = 0.17$ follows for the gas pressure loss:

$$\begin{aligned}-\Delta p_F &= \left(0.02 \cdot \frac{100\,\mathrm{m}}{0.1043\,\mathrm{m}} + 6 \cdot 0.17\right) \cdot \frac{1.056\,\frac{\mathrm{kg}}{\mathrm{m^3}}}{2} \cdot \left(18.0\,\frac{\mathrm{m}}{\mathrm{s}}\right)^2 \\ &\quad + 1.056\,\frac{\mathrm{kg}}{\mathrm{m^3}} \cdot 9.81\,\frac{\mathrm{m}}{\mathrm{s^2}} \cdot 30\,\mathrm{m} \\ &= 3780.9\,\frac{\mathrm{N}}{\mathrm{m^2}}\end{aligned}$$

- Assessment of the conveying line calculation: The pressure at the beginning of the conveying pipe $\cong$ Pressure in the transmitter is

$$p_A = p_2(A15) - \Delta p_F = (308255.5 + 3780.9)\frac{\mathrm{N}}{\mathrm{m^2}} = 312036.4\,\frac{\mathrm{N}}{\mathrm{m^2}} = 3.12\,\mathrm{bar},$$

the pressure difference of the conveyance is thus:

$$\Delta p_R = (312036.4 - 100000)\frac{\mathrm{N}}{\mathrm{m^2}} = 212036.4\,\frac{\mathrm{N}}{\mathrm{m^2}} = 2.12\,\mathrm{bar}$$

Lowest conveying gas velocity in the system:

$$v_{F,A} = v_{F,E} \cdot \frac{p_E}{p_A} = 18.0\,\frac{\mathrm{m}}{\mathrm{s}} \cdot \frac{1.0\,\mathrm{bar}}{3.12\,\mathrm{bar}} = 5.77\,\frac{\mathrm{m}}{\mathrm{s}} > v_{F,st} = 4.94\,\frac{\mathrm{m}}{\mathrm{s}}$$

Result: The conveying pressure difference is with $\Delta p_R = 2.12\,\mathrm{bar}$ slightly higher than the pressure difference of $\Delta p = 2.0\,\mathrm{bar}$ chosen for the design, but is still within the usable limit pressure difference of $\Delta p = 2.1\,\mathrm{bar}$ which is identical with the smallest stable network pressure $p_{N,\mathrm{min}}$. The gas velocity $v_{F,A}$ is within the permissible range. A recalculation with a larger conveying line diameter is not considered necessary. In Fig. 9.3, the curves of the conveying gas and solid velocities as well as the conveying gas pressure along the conveying route are shown schematically. The solid velocities u_S were calculated using the corresponding C_1- and K_U values.

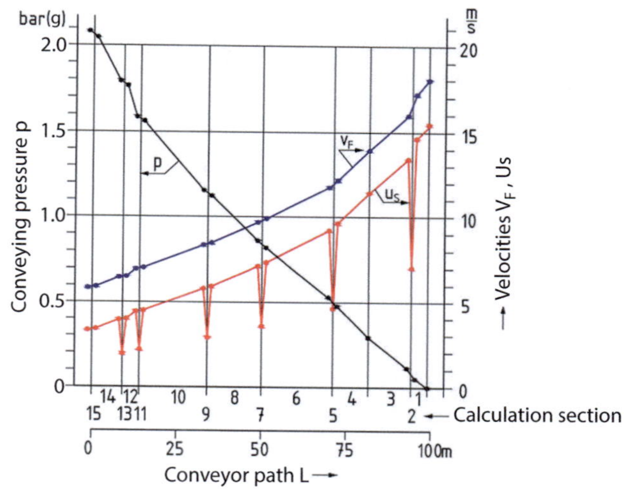

Fig. 9.3 Pressure and velocity curves along the conveying route

9.6 Design of the Pressure Vessel Lock

Displacement gas flow: In addition to the conveying gas flow through the conveying line, here: $\dot{m}_F = 0.1623\,\text{kg/s} = 584.4\,\text{kg/h}$, the so-called top gas or displacement gas flow $\dot{m}_{F,O}$ must be supplied to the pressure vessel. This replaces the solid volume flow $(\dot{m}_S/\varrho_S)$ removed from the pressure vessel during conveying and thereby maintains the container pressure p_B constant. With the gas density in the sender

$$\varrho_{F,B} = \varrho_{F,E} \cdot \frac{p_B}{p_E} = 1.056\,\frac{\text{kg}}{\text{m}^3} \cdot \frac{3.12\,\text{bar}}{1.0\,\text{bar}} = 3.2947\,\frac{\text{kg}}{\text{m}^3}$$

follows from (7.8):

$$\dot{m}_{F,O} = \left(\frac{\dot{m}_S}{\varrho_S}\right) \cdot \varrho_{F,B} = \left(\frac{32.5\,\frac{\text{t}}{\text{h}}}{2.610\,\frac{\text{t}}{\text{m}^3}} \cdot 3.2947\,\frac{\text{kg}}{\text{m}^3}\right) = 41.0\,\frac{\text{kg}}{\text{h}}$$

$\dot{m}_{F,O}$ does not participate in the conveying and remains in the pressure vessel until the pressure reduction. The total gas flow to be supplied to the conveyor system or to be withdrawn from the network is thus obtained as follows:

$$\dot{m}_{F,V} = \dot{m}_F + \dot{m}_{F,O} = (584.4 + 41.0)\,\frac{\text{kg}}{\text{h}} = 625.4\,\frac{\text{kg}}{\text{h}}$$

- Selection of the vessel design: Pressure vessels with solid discharge downwards (→ *bottom discharge*) or upwards (→ *top discharge*) are possible, see Sect. 7.1. Vessels with head discharge are not residue-free (→ planned product changes) and on the part of the conveying pipe lying in the container, wear occurs on both sides. Also at risk of

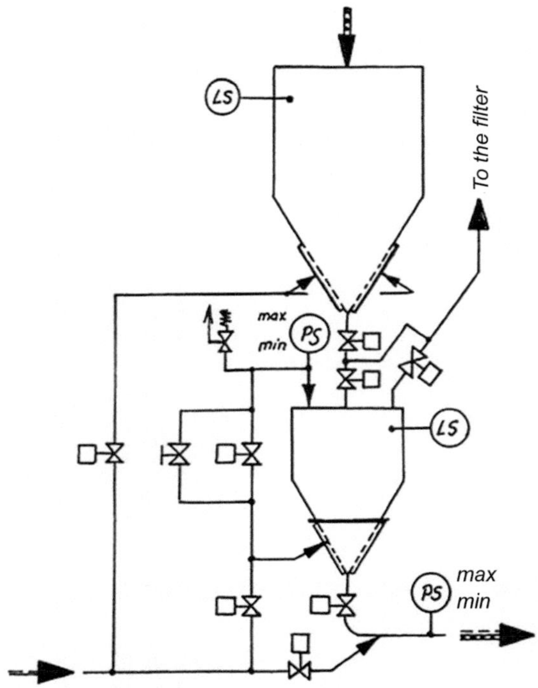

Fig. 9.4 Structure of the pressure vessel system; hand valves, compensators, etc. not shown

wear is the conveying line inlet (→ these are all not visible or easily accessible components). A transmitter with bottom-side discharge according to Fig. 9.4 is chosen.

- Valves: Valves that switch in bulk material or in bulk material-laden gas streams are subject to wear, while valves through which clean gas flows are not critical.
  - Bulk material inlet: The product inlet valve must open and close in the bulk material flow. To increase operational reliability/lifetime in the case of the extremely abrasive quartz powder, the functions "shutting off the bulk material flow" and "sealing the pressure vessel" normally combined in one valve are divided into two individual valves. The upper valve blocks the bulk material flow, the lower one seals. To fill the transmitter, first the lower sealing valve is opened, followed by the upper shut-off valve. When closing the vessel filled with bulk material, the order is reversed.
  - Bulk material outlet: The product outlet valve opens in the bulk material flow and closes without bulk material (→ pressure vessel is empty). A high-quality single valve, e.g., a double-sided mounted ball valve with a ceramic ball and seats or a steel ball and ceramic seats, is used.
  - Relief valve: This is periodically flowed through by a gas stream loaded with more or less fine particles of the product at high velocities (→ sandblasting effect). Depending on the pressure level, sonic speed may occur briefly. The finest product particles can adhere to the container walls. Use of a high-quality single valve similar to the bulk material outlet valve.
  - For the above-mentioned valves, opening and closing times $\Delta\tau_{\mathrm{auf/zu}} \leq 2\,\mathrm{s}$ should be aimed for, see Sect. 8.2.4.

Note: In principle, a single pressure vessel can also be operated without a bulk material outlet valve. Advantage: What is not there cannot wear out or fail. Disadvantage: less controllable/adjustable pressure vessel cycle.

- Determination of vessel size: From (7.4) follows with $\dot{m}_{S,n} = 25.0\,\text{t/h}$, $\varrho_{b,B} \cong 0.9 \cdot \varrho_{SS} = 0.9 \cdot 1210\,\text{kg/m}^3 = 1089\,\text{kg/m}^3$ (→ bulk material is aerated), $\dot{N}_{Ch} = 10\,\text{h}^{-1} < \dot{N}_{Ch,\max} \cong 12\,\text{h}^{-1}$ (→ bulk material-specific empirical value):

$$V_{B,\text{Netto}} \geq \frac{\dot{m}_{S,n}}{\varrho_{b,B} \cdot \dot{N}_{Ch}} = \frac{25 \cdot 10^3\,\frac{\text{kg}}{\text{h}}}{1089\,\frac{\text{kg}}{\text{m}^3} \cdot 10\,\text{h}^{-1}} = 2.30\,\text{m}^3$$

Selected:
Standard vessel: $V_{B,\text{Netto}} = 2.50\,\text{m}^3$
with:

$V_{B,\text{Netto}} = 2.50\,\text{m}^3$	usable filling volume = net volume,
$V_{B,\text{Brutto}} = 2.84\,\text{m}^3$	total volume = gross volume,
$H_{B,\text{ges}} \cong 3.5\,\text{m}$	total construction height from installation level to the upper edge of the upper product inlet fitting.

The actual hourly batch number is therefore:

$$\dot{N}^*_{Ch} = \frac{\dot{m}_{S,n}}{\varrho_{b,B} \cdot V_{B,\text{Netto}}} = \frac{25 \cdot 10^3\,\frac{\text{kg}}{\text{h}}}{1089\,\frac{\text{kg}}{\text{m}^3} \cdot 2.50\,\text{m}^3} = 9.183\,\text{h}^{-1}$$

- Size of the pre-container. The pre-container must be able to hold at least the volume of one pressure vessel charge $V_{B,\text{Netto}}$. Selected:

$$V_{\text{Vor}} = 5.0\,\text{m}^3$$

The pre-container fits together with the pressure vessel in the available construction height of $\Delta H_{Bau} \cong 10\,\text{m}$. The cone of the pre-container is to be equipped with a discharge aid, preferably a flat aeration unit. This is only activated when the transmitter is to be filled (→ Start-up support of the bulk material). Interlocking with the pressure vessel circuit: When the upper product inlet valve opens, the discharge aid is activated and remains in operation until the upper inlet valve closes at the latest. Fig. 9.4 shows the integration of the pre-container aeration unit into the system diagram.
- Calculation of the pressure vessel cycle: This has been described in Sect. 7.1.1 and is shown in Fig. 7.5. The sum of the individual times $\Delta\tau_{Ch}$ must be less than or equal to the available batch time $\Delta\tau^*_{Ch}$. It applies:

$$\sum_i \Delta\tau_i = \Delta\tau_{Ch} \leq \Delta\tau^*_{Ch} = \frac{60\,\frac{\text{min}}{\text{h}}}{\dot{N}^*_{Ch}} = \frac{60\,\frac{\text{min}}{\text{h}}}{9.183\,\text{h}^{-1}} = 6.534\,\text{min}$$

- Filling time $\Delta\tau_1$ of the sender with bulk material: With $\dot{q}_{S,1} = 0.90\,\text{m}^3/(\text{cm}^2\,\text{h})$, $A_{B,\text{zu}} = 706.5\,\text{cm}^2$ ($\rightarrow$ inlet diameter DN 300) follows from (7.10):

$$\Delta\tau_1 = \frac{V_{B,\text{Netto}}}{\dot{q}_{S,1} \cdot A_{BF}} = \frac{2.50\,\text{m}^3}{0.90\,\frac{\text{m}^3}{\text{cm}^2\,\text{h}} \cdot 706.5\,\text{cm}^2} \cdot 60\,\frac{\text{min}}{\text{h}} = 0.236\,\text{min}$$

- Pressurizing time $\Delta\tau_2$ of the transmitter to conveying pressure: (7.11) provides with $\varrho_{F,0} = \varrho_{F,E}$ and the other known data:

$$\Delta\tau_2 = \frac{(V_{B,\text{Brutto}} - V_{B,\text{Netto}} \cdot \frac{\varrho_{b,B}}{\varrho_S}) \cdot (\varrho_{F,B} - \varrho_{F,0})}{\dot{m}_{F,V}}$$

$$= \frac{\left(2.84\,\text{m}^3 - 2.50\,\text{m}^3 \cdot \frac{1089\,\frac{\text{kg}}{\text{m}^3}}{2610\,\frac{\text{kg}}{\text{m}^3}}\right) \cdot (3.2947 - 1.056)\,\frac{\text{kg}}{\text{m}^3}}{625.4\,\frac{\text{kg}}{\text{h}}} \cdot 60\,\frac{\text{min}}{\text{h}}$$

$$= 0.382\,\text{min}$$

- Emptying time $\Delta\tau_3$ of the sender: This follows from (7.12) to:

$$\Delta\tau_3 = \frac{V_{B,\text{Netto}} \cdot \varrho_{b,B}}{\dot{m}_S} = \frac{2.50\,\text{m}^3 \cdot 1089\,\frac{\text{kg}}{\text{m}^3}}{32.5 \cdot 10^3\,\frac{\text{kg}}{\text{h}}} \cdot 60\,\frac{\text{min}}{\text{h}} = 5.026\,\text{min}$$

Pressure reduction time $\Delta\tau_4$ through the conveying line: Beginning pressure drop in the sender indicates that the vessel is completely empty/bulk material-free. In the present case, the pressure set during the conveying in the sender $p_{B,\text{för}} \cong p_A = 3.12\,\text{bar}$, should initially up to a value adapted to the respective operating conditions $p_{B,\text{min}}$ be decreased via the conveying line. It is set to $p_{B,\text{min}} = 2.00\,\text{bar}$. With the already known and the following data $p_{B,\text{för}} \cong p_A = 3.12\,\text{bar}$, $R_F = 287.1\,\text{J}/(\text{kg K})$, $A_R = \pi/4 \cdot D_R^2 = \pi/4 \cdot (0.1043\,\text{m})^2 = 0.008544\,\text{m}^2$, $\overline{p}_B = 1/2 \cdot (p_{B,\text{för}} + p_{B,\text{min}}) = 1/2 \cdot (3.12 + 2.00)\,\text{bar} = 2.56\,\text{bar}$, $\dot{m}_{F,V} = 625.4\,\text{kg/h} = 0.1737\,\text{kg/s}$, $\lambda_F = 0.02$, $T_R = 60\,°\text{C} = 333\,\text{K}$ and the empirical value $K_4 \cong 4.0$ follows from (7.18):

$$\Delta\tau_4 = K_4 \cdot \frac{V_{B,\text{Brutto}}}{R_F \cdot T_R} \cdot \frac{p_{B,\text{för}} - p_{B,\text{min}}}{A_R \cdot \sqrt{(\overline{p}_B{}^2 - p_E^2) \cdot \frac{D_R}{\lambda_F \cdot R_F \cdot T_R \cdot L_R}} - \dot{m}_{F,V}}$$

$$= 4.0 \cdot \frac{2.84\,\text{m}^3}{287.1\,\frac{\text{J}}{\text{kg K}} \cdot 333\,\text{K}}$$

$$\cdot \frac{(3.12 \cdot 10^5 - 2.00 \cdot 10^5)\,\frac{\text{N}}{\text{m}^2}}{0.008544\,\text{m}^2 \cdot \sqrt{((2.56 \cdot 10^5)^2 - (1.00 \cdot 10^5)^2)\,\frac{\text{N}^2}{\text{m}^4} \cdot \frac{0.1043\,\text{m}}{0.02 \cdot 287.1\,\frac{\text{J}}{\text{kg K}} \cdot 333\,\text{K} \cdot 100\,\text{m}}} - 0.1737\,\frac{\text{kg}}{\text{s}}}$$

$$\cdot \frac{1\,\text{min}}{60\,\text{s}} = 0.169\,\text{min}$$

Using (7.19), the increased gas mass flow rate $\Delta\tau_4$ averaged over the relaxation time, leaving the conveying pipe, can be calculated as $\overline{\dot{m}}_{F,\text{tot}}$, the so-called final surge. It holds:

$$\overline{\dot{m}}_{F,\text{tot}} = \dot{m}_{F,V} + \frac{V_{B,\text{Brutto}}}{R_F \cdot T_R} \cdot \frac{p_{B,\text{för}} - p_{B,\min}}{\Delta\tau_4}$$

$$= 625.4\,\frac{\text{kg}}{\text{h}} + \frac{2.84\,\text{m}^3}{287.1\,\frac{\text{J}}{\text{kg K}} \cdot 333\,\text{K}} \cdot \frac{(3.12 \cdot 10^5 - 2.00 \cdot 10^5)\,\frac{\text{N}}{\text{m}^2}}{0.169\,\text{min}} \cdot \frac{60\,\text{min}}{\text{h}}$$

$$= 1806.6\,\frac{\text{kg}}{\text{h}}$$

During the limited time period $\Delta\tau_4$ the gas flow burdening the receiving filter increases on average by a factor of $(\overline{\dot{m}}_{F,\text{tot}}/\dot{m}_{F,V}) = (1806.6\,\text{kg/h}/625.4\,\text{kg/h}) = 2.89$ compared to that of the steady state conveying.

- Pressure reduction time $\Delta\tau_5$ through the expansion/vent valve: The transmitter must reduce the pressure from $p_{B,\min} = 2.00\,\text{bar}$ through the expansion valve and the downstream expansion line to the filling pressure $p_{B,0} \cong p_E = 1.00\,\text{bar} \rightarrow$ Dimensions of the expansion line: inner diameter $D_E = 0.080\,\text{m}$, length $L_E = 15\,\text{m}$. It must first be checked whether over- or under-critical expansion is present, see Sect. 7.1.1. Here, with:

$$\frac{p_E}{p_{B,\min}} = \frac{1.00\,\text{bar}}{2.00\,\text{bar}} = 0.50$$

$$\xi_{\text{ges}} = \lambda_F \cdot \frac{L_E}{D_E} + \sum_i \xi_i = 0.02 \cdot \frac{15\,\text{m}}{0.080\,\text{m}} + 5 = 8.75$$

$$\rightarrow \quad \text{with: } \sum_i \xi_i \cong 5, \quad \text{practical value}$$

From Fig. 7.8 or 9.5, it can be seen that there is a over-critical relaxation and for the calculation of the relaxation time $\Delta\tau_5$ (7.22) must be used. With $T_A \cong T_R = 333\,\text{K}$ applies:

$$\Delta\tau_5 = \frac{V_{B,\text{Brutto}}}{\frac{\pi}{4} \cdot D_E^2} \cdot \sqrt{\frac{\xi_{\text{ges}}}{R_F \cdot T_A}} \cdot \ln\left(\frac{p_{B,\min}}{p_E} + \sqrt{\frac{p_{B,\min}^2}{p_E^2} - 1}\right)$$

$$= \frac{2.84\,\text{m}^3}{\frac{\pi}{4} \cdot (0.080\,\text{m})^2} \cdot \sqrt{\frac{8.75}{287.1\,\frac{\text{J}}{\text{kg K}} \cdot 333\,\text{K}}}$$

$$\cdot \ln\left(\frac{2.00\,\text{bar}}{1.00\,\text{bar}} + \sqrt{\left(\frac{2.00\,\text{bar}}{1.00\,\text{bar}}\right)^2 - 1}\right) \cdot \frac{1\,\text{min}}{60\,\text{s}}$$

$$= 0.119\,\text{min}$$

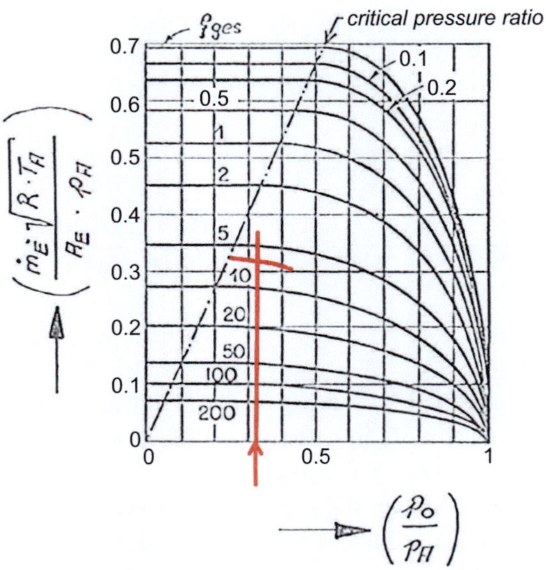

Fig. 9.5 Gas outflow with $\kappa = 1.4$ from a container

The corresponding, over time $\Delta\tau_5$ averaged expansion gas flow results from (7.23):

$$\bar{\dot{m}}_{F,E} = \frac{V_{B,\text{Brutto}}}{R_F \cdot T_A} \cdot \frac{p_{B,\min} - p_E}{\Delta\tau_5}$$
$$= \frac{2.84\,\text{m}^3}{287.1\,\frac{\text{J}}{\text{kg K}} \cdot 333\,\text{K}} \cdot \frac{(2.00 - 1.00) \cdot 10^5\,\frac{\text{N}}{\text{m}^2}}{0.119\,\text{min}} \cdot \frac{60\,\text{min}}{\text{h}} = 1497.8\,\frac{\text{kg}}{\text{h}}$$

- Sum of switching times $\Delta\tau_0$ of the fittings at the transmitter: This was determined according to (7.24) to:

$$\Delta\tau_0 = 0.20\,\text{min}$$

- Actual batch time $\Delta\tau_{Ch}$ and throughput reserves. It applies:

$$\Delta\tau_{Ch} = \sum_i \Delta\tau_i = (0.236 + 0.382 + 5.026 + 0.169 + 0.119 + 0.200)\,\text{min}$$
$$= 6.132\,\text{min} < \tau^*_{\text{Ch}} = 6.534\,\text{min}$$

The temporal utilization of the conveyor system/pressure vessel is therefore:

$$\xi_{Ch} = \frac{\Delta\tau_{Ch}}{\Delta\tau^*_{Ch}} = \frac{6.132\,\text{min}}{6.534\,\text{min}} = 0.938 = 93.8\,\%$$

i.e., there are approximately 6.2% time reserves available. When utilizing these reserves ($\rightarrow \xi_{Ch} = 100\,\%$) the maximum possible hourly batch number is:

$$\dot{N}_{Ch,\max} = \frac{60\,\frac{\text{min}}{\text{h}}}{\Delta\tau_{Ch}} = \frac{60\,\frac{\text{min}}{\text{h}}}{6.132\,\text{min}} = 9.785\,\text{h}^{-1}$$

and the maximum possible nominal solid throughput:

$$\begin{aligned}\dot{m}_{S,n,\max} &= \dot{N}_{Ch,\max} \cdot V_{B,\text{Netto}} \cdot \varrho_{b,B} \\ &= 9.785\,\text{h}^{-1} \cdot 2.50\,\text{m}^3 \cdot 1089\,\frac{\text{kg}}{\text{m}^3} \cdot \frac{1\,\text{t}}{1000\,\text{kg}} = 26.64\,\frac{\text{t}}{\text{h}}\end{aligned}$$

Result: The system is designed with sufficient safety.

9.7 Filter Verification

- Pressure expansion gas filter: The average pressure relief gas flow $\overline{\dot{m}}_{F,E} = 1497.8\,\text{kg/h}$ falls intermittently through the separate relief valve for $\Delta\tau_5 = 0.119\,\text{min}$ at intervals of $\Delta\tau_{Ch} = 6.132\,\text{min}$. In the present case, $\overline{\dot{m}}_{F,E}$ is mixed with the exhaust gas flow/into the mill filter of the upstream grinding system. Since $\overline{\dot{m}}_{F,E} \ll \dot{m}_{F,\text{Mill}}$ is, a review of the filter area is not necessary.
- Conveying gas filter on receiving silo: During the actual conveying, a gas flow of $\dot{m}_F = 584.4\,\text{kg/h}$ ($\rightarrow \dot{m}_{F,0}$ remains in the vessel), at the end of each batch, however, for $\Delta\tau_4 = 0.169\,\text{min}$ a gas flow of $\overline{\dot{m}}_{F,\text{tot}} = 1806.6\,\text{kg/h}$ must be dedusted. The basic design is carried out for the gas flow $\dot{m}_F$ ($\rightarrow$ continuous load), then $\overline{\dot{m}}_{F,\text{tot}}$ ($\rightarrow$ short-term periodic peak load) is checked.
  The filter design is carried out using the method described, among others, in [4, 5]. The permissible specific filter area load under continuous stress

$$\dot{q}_F = \frac{\dot{V}_F(\text{Operating condition})}{A_{\text{Filter}}}, \tag{9.9}$$

  in which optimal operating conditions regarding residual dust content, pressure loss, cleaning, etc., can be determined by the following approach:

$$\dot{q}_F = \dot{q}_{F,0} \cdot A_i \cdot B \cdot C \cdot D \cdot E \cdot F \cdot G \cdot H \tag{9.10}$$

  The individual factors ($A_i - H$) as well as the basic value $\dot{q}_{F,0}$ can be taken from tables/diagrams in [4, 5]. Based on the values proposed in [5], the following applies to the present operating case:
  - Basic value $\dot{q}_{F,0}$ of the specific filter area load: For quartz powder, it is suggested: $\dot{q}_{F,0} = (2.1-2.8)\,\text{m}^3/(\text{m}^2 \cdot \text{min})$, chosen: $\dot{q}_{F,0} = 2.40\,\text{m}^3/(\text{m}^2 \cdot \text{min})$.
  - Factor A_i for the type of filter system: Jet bag filter with Venturi tubes, $A_i = A_1 = 1.00$.
  - Factor B for the application area: Product recovery: $B = 0.90$.

- Factor C for the particle size distribution of the bulk material to be separated: From Table 9.1 and the measured particle size distribution, it follows that significantly more than 50 wt.-% of the quartz powder has a particle size between $d_S = (10-50)\,\mu\text{m}$. For this, the following applies: $C = 1.00$.
- Factor D for the raw gas loading with bulk material in $[g_S/m_F^3]$: It is estimated that approximately 99% of the pneumatically conveyed quartz powder is deposited in the free space of the receiving silo. Conveying gas volume flow at the filter inlet:

$$\dot{V}_{F,E} = \frac{\dot{m}_F}{\varrho_{F,E}} = \frac{584.4\,\frac{\text{kg}}{\text{h}}}{1.056\,\frac{\text{kg}}{\text{m}^3}} = 553.4\,\frac{\text{m}^3}{\text{h}}$$

Dust loading of the gas stream:

$$\mu_{\text{Filter}} = \frac{0.01 \cdot \dot{m}_S}{\dot{V}_{F,E}} = \frac{0.01 \cdot 32.5 \cdot 10^3\,\frac{\text{kg}}{\text{h}}}{553.4\,\frac{\text{m}^3}{\text{h}}} \cdot 1000\,\frac{\text{g}}{\text{kg}} = 587.3\,\frac{\text{g}}{\text{kg}}$$

From this, it follows: $D = 0.80$

- Factor E for the gas temperature: For $t_F = 60\,°\text{C}$ amounts to $E = 0.95$.
- Factor F for the bulk density: $\varrho_{SS} = 1210\,\text{kg/m}^3 > 600\,\text{kg/m}^3 \rightarrow F = 1.00$.
- Factor G for the filter inflow: bunker top filter with raw gas inflow from bottom to top at $\dot{q}_{F,0} = 2.4\,\text{m}^3/(\text{m}^2 \cdot \text{min}) \rightarrow G = 0.875$.
- Factor H for the climatic conditions: installation of the system in Central Europe $\rightarrow$ $H = 1.00$.
- Permissible specific filter surface load $\dot{q}_F$: With the above values, the following applies:

$$\begin{aligned}\dot{q}_F &= 2.40\,\frac{\text{m}^3}{\text{m}^2\,\text{min}} \cdot 1.00 \cdot 0.90 \cdot 1.00 \cdot 0.80 \cdot 0.95 \cdot 1.00 \cdot 0.875 \cdot 1.0\\ &= 1.44\,\frac{\text{m}^3}{\text{m}^2\,\text{min}}\end{aligned}$$

- Required filter area A^*_{Filter}: For this, the following applies with

$$A^*_{\text{Filter}} = \frac{\dot{V}_{F,E}}{\dot{q}_F} = \frac{553.4\,\frac{\text{m}^3}{\text{h}}}{1.44\,\frac{\text{m}^3}{\text{m}^2\,\text{min}}} \cdot \frac{1\,\text{h}}{60\,\text{min}} = 6.405\,\text{m}^2$$

$$A^*_{\text{Filter}} = 6.405\,\text{m}^2 < A_{\text{Filter}} = 10\,\text{m}^2 \quad \rightarrow \quad \text{safe operation with basic design.}$$

- Specific area load during short-term peak gas flow $\overline{\dot{m}}_{F,\text{tot}}$:

$$\dot{q}_{F,\text{tot}} = \frac{\overline{\dot{m}}_{F,\text{tot}}}{\varrho_{F,E} \cdot A_{\text{Filter}}} = \frac{1806.6\,\frac{\text{kg}}{\text{h}}}{1.056\,\frac{\text{kg}}{\text{m}^3} \cdot 10\,\text{m}^2} \cdot \frac{1\,\text{h}}{60\,\text{min}} = 2.85\,\frac{\text{m}^3}{\text{m}^2 \cdot \text{min}}$$

$\dot{q}_{F,\text{tot}}$ is at the upper limit of the basic value $\dot{q}_{F,0} = 2.80\,\text{m}^3/(\text{m}^2\,\text{min})$. Consultation with the filter supplier provides the statement that this stress is safely manageable.

Result: The existing filter system of the receiving silo is designed to be sufficiently safe for conveying. Relaxation shocks are also controlled.

Note 1: The determination of the pipe pressure loss in the above example is based on a correlation equation taken from the literature [1]. Accordingly, more or less large inaccuracies must be expected. Comparisons with the quartz powder measurement results shown in Fig. 4.16 as well as in the resulting print out of a calculation program based on the scale-up model described in Sect. 5.3, see Fig. 9.6, show that the design carried out above provides too high pressure losses and is therefore on the safe side, but not energetically/economically optimal.

Note 2: The above calculation is based on a constant operating temperature along the conveying path $T_B \cong$ solid temperature T_S. This is sufficiently accurate in many cases, as the cooling of the solid/gas mixture is only slight for short pipe lengths and moderate bulk material entry temperatures into the transport system. In cases where the reduction in gas velocity caused by a temperature drop compared to the isothermal case can influence the conveying behavior, the mixture cooling along the pipeline must be taken into account. From an energy balance on an infinitesimal pipe element and its integration over the pipe length L_x the following calculation equation results:

$$T(L_x) = T_U + (T_M - T_U) \cdot \exp\left(- \frac{k \cdot D_R \cdot \pi}{\dot{m}_S \cdot c_{p,S} + \dot{m}_F \cdot c_{p,F}} \cdot L_x \right) \tag{9.11}$$

with:

L_x	considered pipe length,
T_U, T_M	ambient temperature, mixture inlet temperature $\cong$ mixing temperature,
$c_{p,F}, c_{p,S}$	specific isobaric heat capacities of gas and solid,
k	$(= f(\alpha_{\text{inside}}, \alpha_{\text{outside}}, \lambda_W/\Delta s_W))$, heat transfer coefficient related to the inner pipe diameter D_R (or another reference diameter).

If the pipe is insulated, instead of $(\lambda_W/\Delta s_W)$ the value $\sum(\lambda_W/\Delta s_W + \lambda_{\text{Iso}}/\Delta s_{\text{Iso}})$ must be used. The determination of the resulting heat transfer coefficients $\alpha_{\text{inside}}, \alpha_{\text{outside}}$ etc. can be done, for example, with the approaches in [6].

Depending on whether the conveyor pipe runs partially inside or outside of buildings, k must be adjusted accordingly. The calculation of the conveying data in the individual calculation sections is then carried out with the current local temperatures.

```
CPP CLAUDIUS PETERS                   BUXTEHUDE,  23.04.2018
PNEUMATIC CONVEYING                   OFFER-/ORDER-NO. Quartz powder
V9.17                                 PERSON RESPONSIBLE.. P. Hilgraf
*******************************************************************************

NEW CALCULATION OF A CONVEYING SYSTEM
TYPE OF FEEDER.. PRESSURE VESSEL
METHOD.. (1/KR) CALCULATION
SOLID...      QUARTZ POWDER
              RESIDUE AT 0.09 MM                    R90      =  15.00 WT.%
              PARTICLE DIAMETER AT 50 WT.% RESIDUE  D(R50) =  0.039 MM
              PARTICLE DIAMETER AT 0.1 WT.% RESIDUE D(R0.1)=  0.300 MM
              BULK DENSITY ACCORDING TO DIN 1060     RHO(S) = 1210.0 KG/M3
              PARTICLE DENSITY                      RHO(P) = 2610.0 KG/M3
              WATER CONTENT OF BULK SOLID           W(S)   <   0.50 WT.%
              ABRASION CLASSIFICATION NO.           STG    =      1
CONVEYING
GAS..         AIR
              STANDARD DENSITY          RHO(F) =  1.293 KG/M3
              GAS CONSTANT              R(GAS) =  287.0 J/(KG K)

CONVEYING
DATA..        SOLID THROUGHPUT          MS(GES)=  25.00 T/H
              TRUE CONVEYING DISTANCE   L(GES) =  100.0 M    (INCL. HEIGHT)
              CONVEYING HEIGHT          H(GES) =   30.0 M
              CONVEYING HEIGHT DECOMPOSED  NH  =      2 STEPS

              HEIGHT of  1.STEP         H( 1)  =    5.0 M
               1.STEP STARTS AFTER      L( 1)  =   10.0 M

              HEIGHT of  2.STEP         H( 2)  =   25.0 M
               2.STEP STARTS AFTER      L( 2)  =   70.0 M

              NUMBER OF 90-DEGREE-BENDS NK     =    6.0

              DESIGN TEMPERATURE        T(X)   =   60.0 DEG C

              ALTITUDE                  H(ORT) =  LOWER THAN 500 M
              AMBIENT PRESSURE          P(UM)  =  1.013 BAR(ABS)
              COUNTER-PRESS. OF CONVEY. P(GEG) =  1.013 BAR(ABS)

BULK DENSITY IN VESSEL RHO2 = 1089.0 KG/M3
TYPE OF PRESSURE VESSEL... STANDARD-TYPE

|VESSEL-NO. |   VESSEL-SIZE    |  NO.OF CYCLES/H   | TIME/CYCLES    | LOADSHARE |
|           |   VN(I)  (M3)    |   NCH(I)  (1/H)   | TCH(I)  (MIN)  | ME(I)  (%) |
|-----------|------------------|-------------------|----------------|-----------|
|  1        |    2.0           |  11.48            |   5.04         |  100.0    |

CONVEYING PIPE..       114.3 MM *  5.0 MM , DR(I) =  104.3 MM    ( GIVEN )

TOTAL GAS FLOW              VF(GES) =     525.7 M3/H ( AT 20 DGR C, 1.0 BAR(ABS))
CONSIST. OF.. CONVEY. GAS   VF(F)  =     496.9 M3/H ( AT 20 DGR C, 1.0 BAR(ABS))
              DISPLAC. GAS  VF(O)  =      28.8 M3/H ( AT 20 DGR C, 1.0 BAR(ABS))

PRESSURE DROP IN CONVEYING PIPE     DP(GES) =  1.58 BAR
CONSISTING OF: SOLID FRICTION LOSS  DP(R)   =  1.02 BAR
               LOSS DUE TO LIFT     DP(HUB) =  0.30 BAR
               LOSS DUE TO BENDS    DP(K)   =  0.26 BAR
PRESSURE DROP/100M DISTANCE    DP(GES)/100M =  1.578 BAR/100 M

PRESSURE AT START OF PIPE      P(R)   =  2.59 BAR(ABS)
PRESSURE IN VESSEL             P(B)   =  2.79 BAR(ABS)
PRESSURE BEFORE VESSEL         P(VOR) =  3.19 BAR(ABS)  =   2.18 BAR(G)

AVAILABLE CONVEYING TIME   T(AVAILABLE) = 60.00 MIN
NEEDED CONVEYING TIME      TCH(GES)     = 57.88 MIN
UTILIZATION RATIO          ETA(NUTZ)    =  96.5 %
```

Fig. 9.6 Calculation of quartz meal conveying using a calculation program based on the scale-up model described in Sect. 5.3 → result printout

```
INITIAL CONVEYING GAS VELOCITY            WF(A)  =   7.09 M/S
FINAL CONVEYING GAS VELOCITY              WF(E)  =  18.12 M/S
INITIAL VELOCITY AT 20 DGR C              WF(0)  =   6.24 M/S
CRITICAL DISTANCE TO CONV. BOUNDARY       DWF(A) =   2.00 M/S
REFERENCE VELOCITY                        WF(B)  =   5.21 M/S

SOLID/AIR RATIO          MUE = 51.2 KG(S)/KG(F)
             OR          MUE = 61.6 KG(S)/(M3(F) AT 20 DGR C,1 BAR(ABS))

CYCLE FACTOR                              FQ     =   1.225
SOLID THROUGHPUT DURING CONVEYING PHASE  MS(GES) =   30.62 T/H

F-FAKTOR (FOR CONTROL, RECALCULATED FROM AEQUIVALENT LENGTH) = 2096.6
GAS SUPPLY WITHOUT RECEIVER..

ADDITIONAL INFORMATION
======================

BREAKDOWN OF PRESSURE VESSEL-CYCLE..
CYCLE TIME...    TIME/CHARGE          TCH   =  5.04 MIN
                 PRESSURIZATION TIME  TL    =  0.24 MIN
                 CONVEYING TIME       TB    =  4.27 MIN
                 PRESSURE-STRIP-TIME  TBMIN =  0.07 MIN,PB(MIN)= 2.0 BAR(ABS)
                 VENTING TIME         TE    =  0.08 MIN
                 FILLING TIME         TBF   =  0.19 MIN,QBF= 0.90 M3/(CM2*H)
                 SWITCH TIME          TS    =  0.20 MIN
                 EXTRA WAITING TIME   TW    =  0.00 MIN

VENTING TIME TE CALCULATED FOR..
D(E) =  80. MM, L(E) =  15. M

FILLING TIME CALCULATED FOR AN UNRESTRICTED FLOW AREA..
A(BF) =  707. CM2

MATERIAL INLET IS NO STANDARD VALUE
STANDAD DIAMETER : A(STA) =  410. CM2
SELECTED         : A(BF)  =  707. CM2

PRESSURE DIFFERENCE VESSEL-CONVEYING PIPE..
MINIMUM/STANDARD         DP(G-R) =  0.20 BAR
THEORETIC VALUE          DP(B)   =  0.10 BAR

***END OF OUT PRINT***
```

Fig. 9.6 (continued)

References

1. Stegmaier, W.: Zur Berechnung der horizontalen pneumatischen Förderung feinkörniger Stoffe. F+h Fördern Heb. **28**(5/6), 363–366 (1978)
2. Muschelknautz, E., Wojahn, H.: Auslegung pneumatischer Förderanlagen. Chem.-Ing.-Tech **46**(6), 223–235 (1974)
3. VDI-Gesellschaft Verfahrenstechnik und Chemieingenieurwesen (Eds.): VDI-Wärmeatlas, 6th edn. Abschnitt Lh1-Lh13. VDI, Düsseldorf (1991)
4. Löffler, F., Dietrich, H., Flatt, W.: Staubabscheidung mit Schlauchfiltern und Taschenfiltern. Vieweg, Braunschweig (1984)
5. Löffler, F.: Staubabscheiden. Thieme, Stuttgart (1988)
6. VDI-Gesellschaft Verfahrenstechnik und Chemieingenieurwesen (Ed.): VDI-Wärmeatlas, 10th edn. Springer, Berlin (2006)

MIX
Papier aus verantwortungsvollen Quellen
Paper from responsible sources
FSC® C105338

If you have any concerns about our products,
you can contact us on
ProductSafety@springernature.com

In case Publisher is established outside the EU,
the EU authorized representative is:
Springer Nature Customer Service Center GmbH
Europaplatz 3, 69115 Heidelberg, Germany

Printed by Libri Plureos GmbH
in Hamburg, Germany